Modern Actuarial Risk Th...

Rob Kaas • Marc Goovaerts
Jan Dhaene • Michel Denuit

Modern Actuarial Risk Theory

Using **R**

Second Edition

Professor Rob Kaas
UvA / KE
Roetersstraat 11
1018 WB Amsterdam
The Netherlands
r.kaas@uva.nl

Professor Jan Dhaene
AFI (Accountancy, Finance, Insurance)
Research Group
Faculteit Economie en Bedrijfswetenschappen
K.U. Leuven
Naamsestraat 69
3000 Leuven
Belgium
jan.dhaene@econ.kuleuven.be

Professor Marc Goovaerts
AFI (Accountancy, Finance, Insurance)
Research Group
Faculteit Economie en
Bedrijfswetenschappen
K.U. Leuven
Naamsestraat 69
3000 Leuven
Belgium
UvA / KE
Roetersstraat 11
1018 WB Amsterdam
The Netherlands
marc.goovaerts@econ.kuleuven.be

Professor Michel Denuit
Institut de Statistique
Université Catholique de Louvain
Voie du Roman Pays 20
1348 Louvain-la-Neuve
Belgium
michel.denuit@uclouvain.be

2nd ed. 2008. Corr. 2nd Printing

ISBN: 978-3-642-03407-7   e-ISBN: 978-3-540-70998-5
DOI 10.1007/978-3-540-70998-5
Springer Heidelberg Dordrecht London New York

Library of Congress Control Number: 2009933607

Cover design: WMXDesign GmbH, Heidelberg

Printed on acid-free paper

Springer is part of Springer Science+Business Media (www.springer.com)

Foreword to the First Edition

Study the past if you would define the future —
Confucius, 551 BC - 479 BC

Risk Theory has been identified and recognized as an important part of actuarial education; this is for example documented by the Syllabus of the *Society of Actuaries* and by the recommendations of the *Groupe Consultatif*. Hence it is desirable to have a diversity of textbooks in this area.

This text in risk theory is original in several respects. In the language of figure skating or gymnastics, the text has two parts, the compulsory part and the free-style part. The compulsory part includes Chapters 1–4, which are compatible with official material of the Society of Actuaries. This feature makes the text also useful to students who prepare themselves for the actuarial exams. Other chapters are more of a free-style nature, for example the chapter on *Ordering of Risks*, a speciality of the authors. And I would like to mention the chapters on *Generalized Linear Models* in particular. To my knowledge, this is the first text in risk theory with an introduction to these models.

Special pedagogical efforts have been made throughout the book. The clear language and the numerous exercises are an example for this. Thus the book can be highly recommended as a textbook.

I congratulate the authors to their text, and I would like to thank them also in the name of students and teachers that they undertook the effort to translate their text into English. I am sure that the text will be successfully used in many classrooms.

Lausanne, 2001   *Hans Gerber*

Preface to the Second Edition

When I took office, only high energy physicists had ever heard of what is called the Worldwide Web ... Now even my cat has its own page — Bill Clinton, 1996

This book gives a comprehensive survey of non-life insurance mathematics. Originally written for use with the actuarial science programs at the Universities of Amsterdam and Leuven, it is now in use at many other universities, as well as for the non-academic actuarial education program organized by the Dutch Actuarial Society. It provides a link to the further theoretical study of actuarial science. The methods presented can not only be used in non-life insurance, but are also effective in other branches of actuarial science, as well as, of course, in actuarial practice.

Apart from the standard theory, this text contains methods directly relevant for actuarial practice, for example the rating of automobile insurance policies, premium principles and risk measures, and IBNR models. Also, the important actuarial statistical tool of the Generalized Linear Models is studied. These models provide extra possibilities beyond ordinary linear models and regression, the statistical tools of choice for econometricians. Furthermore, a short introduction is given to credibility theory. Another topic that always has enjoyed the attention of risk theoreticians is the study of ordering of risks. The book reflects the state of the art in actuarial risk theory; many results presented were published in the actuarial literature only recently.

In this second edition of the book, we have aimed to make the theory even more directly applicable by using the software R. It provides an implementation of the language S, not unlike S-Plus. It is not just a set of statistical routines but a full-fledged object oriented programming language. Other software may provide similar capabilities, but the great advantage of R is that it is open source, hence available to everyone free of charge. This is why we feel justified in imposing it on the users of this book as a de facto standard. On the internet, a lot of documentation about R can be found. In an Appendix, we give some examples of use of R. After a general introduction, explaining how it works, we study a problem from risk management, trying to forecast the future behavior of stock prices with a simple model, based on stock prices of three recent years. Next, we show how to use R to generate pseudo-random datasets that resemble what might be encountered in actuarial practice.

Models and paradigms studied

The time aspect is essential in many models of life insurance. Between paying premiums and collecting the resulting pension, decades may elapse. This time element is less prominent in non-life insurance. Here, however, the statistical models are generally more involved. The topics in the first five chapters of this textbook are basic for non-life actuarial science. The remaining chapters contain short introductions to other topics traditionally regarded as non-life actuarial science.

1. *The expected utility model*

The very existence of insurers can be explained by the expected utility model. In this model, an insured is a risk averse and rational decision maker, who by virtue of Jensen's inequality is ready to pay more than the expected value of his claims just to be in a secure financial position. The mechanism through which decisions are taken under uncertainty is not by direct comparison of the expected payoffs of decisions, but rather of the expected utilities associated with these payoffs.

2. *The individual risk model*

In the individual risk model, as well as in the collective risk model below, the total claims on a portfolio of insurance contracts is the random variable of interest. We want to compute, for example, the probability that a certain capital will be sufficient to pay these claims, or the value-at-risk at level 99.5% associated with the portfolio, being the 99.5% quantile of its cumulative distribution function (cdf). The total claims is modeled as the sum of all claims on the policies, which are assumed independent. Such claims cannot always be modeled as purely discrete random variables, nor as purely continuous ones, and we use a notation, involving Stieltjes integrals and differentials, encompassing both these as special cases.

The individual model, though the most realistic possible, is not always very convenient, because the available dataset is not in any way condensed. The obvious technique to use in this model is convolution, but it is generally quite awkward. Using transforms like the moment generating function sometimes helps. The Fast Fourier Transform (FFT) technique gives a fast way to compute a distribution from its characteristic function. It can easily be implemented in R.

We also present approximations based on fitting moments of the distribution. The Central Limit Theorem, fitting two moments, is not sufficiently accurate in the important right-hand tail of the distribution. So we also look at some methods using three moments: the translated gamma and the normal power approximation.

3. *Collective risk models*

A model that is often used to approximate the individual model is the collective risk model. In this model, an insurance portfolio is regarded as a process that produces claims over time. The sizes of these claims are taken to be independent, identically distributed random variables, independent also of the number of claims generated. This makes the total claims the sum of a random number of iid individual claim amounts. Usually one assumes additionally that the number of claims is a Poisson variate with the right mean, or allows for some overdispersion by taking a negative binomial claim number. For the cdf of the individual claims, one takes an average of the cdfs of the individual policies. This leads to a close fitting and computationally tractable model. Several techniques, including Panjer's recursion formula, to compute the cdf of the total claims modeled this way are presented.

For some purposes it is convenient to replace the observed claim severity distribution by a parametric loss distribution. Families that may be considered are for example the gamma and the lognormal distributions. We present a number of such distributions, and also demonstrate how to estimate the parameters from data. Further, we show how to generate pseudo-random samples from these distributions, beyond the standard facilities offered by R.

4. *The ruin model*
The ruin model describes the stability of an insurer. Starting from capital u at time $t = 0$, his capital is assumed to increase linearly in time by fixed annual premiums, but it decreases with a jump whenever a claim occurs. Ruin occurs when the capital is negative at some point in time. The probability that this ever happens, under the assumption that the annual premium as well as the claim generating process remain unchanged, is a good indication of whether the insurer's assets match his liabilities sufficiently. If not, one may take out more reinsurance, raise the premiums or increase the initial capital.

Analytical methods to compute ruin probabilities exist only for claims distributions that are mixtures and combinations of exponential distributions. Algorithms exist for discrete distributions with not too many mass points. Also, tight upper and lower bounds can be derived. Instead of looking at the ruin probability $\psi(u)$ with initial capital u, often one just considers an upper bound e^{-Ru} for it (Lundberg), where the number R is the so-called adjustment coefficient and depends on the claim size distribution and the safety loading contained in the premium.

Computing a ruin probability assumes the portfolio to be unchanged eternally. Moreover, it considers just the insurance risk, not the financial risk. Therefore not much weight should be attached to its precise value beyond, say, the first relevant decimal. Though some claim that survival probabilities are 'the goal of risk theory', many actuarial practitioners are of the opinion that ruin theory, however topical still in academic circles, is of no significance to them. Nonetheless, we recommend to study at least the first three sections of Chapter 4, which contain the description of the Poisson process as well as some key results. A simple proof is provided for Lundberg's exponential upper bound, as well as a derivation of the ruin probability in case of exponential claim sizes.

5. *Premium principles and risk measures*
Assuming that the cdf of a risk is known, or at least some characteristics of it like mean and variance, a premium principle assigns to the risk a real number used as a financial compensation for the one who takes over this risk. Note that we study only risk premiums, disregarding surcharges for costs incurred by the insurance company. By the law of large numbers, to avoid eventual ruin the total premium should be at least equal to the expected total claims, but additionally, there has to be a loading in

the premium to compensate the insurer for making available his risk carrying capacity. From this loading, the insurer has to build a reservoir to draw upon in adverse times, so as to avoid getting in ruin. We present a number of premium principles, together with the most important properties that characterize premium principles. The choice of a premium principle depends heavily on the importance attached to such properties. There is no premium principle that is uniformly best.

Risk measures also attach a real number to some risky situation. Examples are premiums, infinite ruin probabilities, one-year probabilities of insolvency, the required capital to be able to pay all claims with a prescribed probability, the expected value of the shortfall of claims over available capital, and more.

6. *Bonus-malus systems*

With some types of insurance, notably car insurance, charging a premium based exclusively on factors known a priori is insufficient. To incorporate the effect of risk factors of which the use as rating factors is inappropriate, such as race or quite often sex of the policy holder, and also of non-observable factors, such as state of health, reflexes and accident proneness, many countries apply an experience rating system. Such systems on the one hand use premiums based on a priori factors such as type of coverage and list-price or weight of a car, on the other hand they adjust these premiums by using a bonus-malus system, where one gets more discount after a claim-free year, but pays more after filing one or more claims. In this way, premiums are charged that reflect the exact driving capabilities of the driver better. The situation can be modeled as a Markov chain.

The quality of a bonus-malus system is determined by the degree in which the premium paid is in proportion to the risk. The Loimaranta efficiency equals the elasticity of the mean premium against the expected number of claims. Finding it involves computing eigenvectors of the Markov matrix of transition probabilities. R provides tools to do this.

7. *Ordering of risks*

It is the very essence of the actuary's profession to be able to express preferences between random future gains or losses. Therefore, stochastic ordering is a vital part of his education and of his toolbox. Sometimes it happens that for two losses X and Y, it is known that every sensible decision maker prefers losing X, because Y is in a sense 'larger' than X. It may also happen that only the smaller group of all risk averse decision makers agree about which risk to prefer. In this case, risk Y may be larger than X, or merely more 'spread', which also makes a risk less attractive. When we interpret 'more spread' as having thicker tails of the cumulative distribution function, we get a method of ordering risks that has many appealing properties. For example, the preferred loss also outperforms the other one as regards zero utility premiums, ruin probabilities, and stop-loss premiums for compound distributions with these risks as individual terms. It can be shown that the collective model of Chapter 3 is more spread than the individual model it approximates, hence using the collective model, in most cases, leads to more conservative decisions regarding premiums to be asked, reserves to be held, and values-at-risk. Also, we can prove that the stop-loss insurance, demonstrated to be optimal as regards the variance of the retained risk in Chapter 1, is also preferable, other things being equal, in the eyes of all risk averse decision makers.

Sometimes, stop-loss premiums have to be set under incomplete information. We give a method to compute the maximal possible stop-loss premium assuming that the mean, the variance and an upper bound for a risk are known.

In the individual and the collective model, as well as in ruin models, we assume that the claim sizes are stochastically independent non-negative random variables. Sometimes this assumption is not fulfilled, for example there is an obvious dependence between the mortality risks of a married couple, between the earthquake risks of neighboring houses, and between consecutive payments resulting from a life insurance policy, not only if the payments stop or start in case of death, but also in case of a random force of interest. We give a short introduction to the risk ordering that applies for this case. It turns out that stop-loss premiums for a sum of random variables with an unknown joint distribution but fixed marginals are maximal if these variables are as dependent as the marginal distributions allow, making it impossible that the outcome of one is 'hedged' by another.

In finance, frequently one has to determine the distribution of the sum of dependent lognormal random variables. We apply the theory of ordering of risks and comonotonicity to give bounds for that distribution.

We also give a short introduction in the theory of ordering of multivariate risks. One might say that two randoms variables are more related than another pair with the same marginals if their correlation is higher. But a more robust criterion is to restrict this to the case that their joint cdf is uniformly larger. In that case it can be proved that the sum of these random variables is larger in stop-loss order. There are bounds for joints cdfs dating back to Fréchet in the 1950s and Höffding in the 1940s. For a random pair (X, Y), the copula is the joint cdf of the *ranks* $F_X(X)$ and $F_Y(Y)$. Using the smallest and the largest copula, it is possible to construct random pairs with arbitrary prescribed marginals and (rank) correlations.

8. *Credibility theory*
The claims experience on a policy may vary by two different causes. The first is the quality of the risk, expressed through a risk parameter. This represents the average annual claims in the hypothetical situation that the policy is monitored without change over a very long period of time. The other is the purely random good and bad luck of the policyholder that results in yearly deviations from the risk parameter. Credibility theory assumes that the risk quality is a drawing from a certain structure distribution, and that conditionally given the risk quality, the actual claims experience is a sample from a distribution having the risk quality as its mean value. The predictor for next year's experience that is linear in the claims experience and optimal in the sense of least squares turns out to be a weighted average of the claims experience of the individual contract and the experience for the whole portfolio. The weight factor is the credibility attached to the individual experience, hence it is called the credibility factor, and the resulting premiums are called credibility

premiums. As a special case, we study a bonus-malus system for car insurance based on a Poisson-gamma mixture model.

Credibility theory is actually a Bayesian inference method. Both credibility and generalized linear models (see below) are in fact special cases of so-called Generalized Linear Mixed Models (GLMM), and the R function glmm is able to deal with both the random and the fixed parameters in these models.

9. *Generalized linear models*

Many problems in actuarial statistics are Generalized Linear Models (GLM). Instead of assuming a normally distributed error term, other types of randomness are allowed as well, such as Poisson, gamma and binomial. Also, the expected values of the dependent variables need not be linear in the regressors. They may also be some function of a linear form of the covariates, for example the logarithm leading to the multiplicative models that are appropriate in many insurance situations.

This way, one can for example tackle the problem of estimating the reserve to be kept for IBNR claims, see below. But one can also easily estimate the premiums to be charged for drivers from region i in bonus class j with car weight w.

In credibility models, there are random group effects, but in GLMs the effects are fixed, though unknown. The glmm function in R can handle a multitude of models, including those with both random and fixed effects.

10. *IBNR techniques*

An important statistical problem for the practicing actuary is the forecasting of the total of the claims that are Incurred, But Not Reported, hence the acronym IBNR, or not fully settled. Most techniques to determine estimates for this total are based on so-called run-off triangles, in which claim totals are grouped by year of origin and development year. Many traditional actuarial reserving methods turn out to be maximum likelihood estimations in special cases of GLMs.

We describe the workings of the ubiquitous chain ladder method to predict future losses, as well as, briefly, the Bornhuetter-Ferguson method, which aims to incorporate actuarial knowledge about the portfolio. We also show how these methods can be implemented in R, using the glm function. In this same framework, many extensions and variants of the chain ladder method can easily be introduced. England and Verrall have proposed methods to describe the prediction error with the chain ladder method, both an analytical estimate of the variance and a bootstrapping method to obtain an estimate for the predictive distribution. We describe an R implementation of these methods.

11. *More on GLMs*

For the second edition, we extended the material in virtually all chapters, mostly involving the use of R, but we also add some more material on GLMs. We briefly recapitulate the Gauss-Markov theory of ordinary linear models found in many other texts on statistics and econometrics, and explain how the algorithm by Nelder and Wedderburn works, showing how it can be implemented in R. We also study the stochastic component of a GLM, stating that the observations are independent random variables with a distribution in a subclass of the exponential family. The well-known normal, Poisson and gamma families have a variance proportional to μ^p for $p = 0, 1, 2$, respectively, where μ is the mean (heteroskedasticity). The so-called Tweedie class contains random variables, in fact compound Poisson–gamma risks, having variance proportional to μ^p for some $p \in (1, 2)$. These mean-variance relations are interesting for actuarial purposes. Extensions to R contributed by Dunn and Smyth provide routines computing cdf, inverse cdf, pdf and random drawings of such random variables, as well as to estimate GLMs with Tweedie distributed risks.

Educational aspects

As this text has been in use for a long time now at the University of Amsterdam and elsewhere, we could draw upon a long series of exams, resulting in long lists of exercises. Also, many examples are given, making this book well-suited as a textbook. Some less elementary exercises have been marked by [♠], and these might be skipped.

The required mathematical background is on a level such as acquired in the first stage of a bachelors program in quantitative economics (econometrics or actuarial science), or mathematical statistics. To indicate the level of what is needed, the book by Bain and Engelhardt (1992) is a good example. So the book can be used either in the final year of such a bachelors program, or in a subsequent masters program in actuarial science proper or in quantitative financial economics with a strong insurance component. To make the book accessible to non-actuaries, notation and jargon from life insurance mathematics is avoided. Therefore also students in applied mathematics or statistics with an interest in the stochastic aspects of insurance will be able to study from this book. To give an idea of the mathematical rigor and statistical sophistication at which we aimed, let us remark that moment generating functions are used routinely, while characteristic functions and measure theory are avoided in general. Prior experience with regression models, though helpful, is not required.

As a service to the student help is offered, in Appendix B, with many of the exercises. It takes the form of either a final answer to check one's work, or a useful hint. There is an extensive index, and the tables that might be needed in an exam are printed in the back. The list of references is not a thorough justification with bibliographical data on every result used, but more a collection of useful books and papers containing more details on the topics studied, and suggesting further reading.

Ample attention is given to exact computing techniques, and the possibilities that R provides, but also to old fashioned approximation methods like the Central Limit Theorem (CLT). The CLT itself is generally too crude for insurance applications, but slight refinements of it are not only fast, but also often prove to be surprisingly accurate. Moreover they provide solutions of a parametric nature such that one does not have to recalculate everything after a minor change in the data. Also, we want to stress that 'exact' methods are as exact as their input. The order of magnitude of errors resulting from inaccurate input is often much greater than the one caused by using an approximation method.

The notation used in this book conforms to what is usual in mathematical statistics as well as non-life insurance mathematics. See for example the book by Bowers et al. (1986, 1997), the non-life part of which is similar in design to the first part of this book. In particular, random variables are capitalized, though not all capitals actually denote random variables.

Acknowledgments

First and most of all, the authors would like to thank David Vyncke for all the work he did on the first edition of this book. Many others have provided useful input. We thank (in pseudo-random order) Hans Gerber, Elias Shiu, Angela van Heerwaarden, Dennis Dannenburg, Richard Verrall, Klaus Schmidt, Bjørn Sundt, Vsevolod Malinovskii, who translated this textbook into Russian, Qihe Tang, who translated it into Chinese in a joint project with Taizhong Hu and Shixue Cheng, Roger Laeven, Ruud Koning, Pascal Schoenmaekers, Julien Tomas, Katrien Antonio as well as Steven Vanduffel for their comments, and Vincent Goulet for helping us with some R-problems. Jan Dhaene and Marc Goovaerts acknowledge the support of the Fortis Chair on Financial and Actuarial Risk Management.

World wide web support

The authors would like to keep in touch with the users of this text. On the internet page *http://www1.fee.uva.nl/ke/act/people/kaas/ModernART.htm* we maintain a list of all typos that have been found so far, and indicate how teachers may obtain solutions to the exercises as well as the slides used at the University of Amsterdam for courses based on this book.

To save users a lot of typing, and typos, this site also provides the R commands used for the examples in the book.

Amsterdam,   *Rob Kaas*
Leuven,   *Marc Goovaerts*
Louvain-la-Neuve,   *Jan Dhaene*
March 22, 2009   *Michel Denuit*

Contents

There are 10^{11} stars in the galaxy. That used to be a huge number. But it's only a hundred billion. It's less than the national deficit! We used to call them astronomical numbers. Now we should call them economical numbers —
Richard Feynman (1918–1988)

Chapter 1
Utility theory and insurance

> *The sciences do not try to explain, they hardly even try to interpret, they mainly make models. By a model is meant a mathematical construct which, with the addition of certain verbal interpretations, describes observed phenomena. The justification of such a mathematical construct is solely and precisely that it is expected to work —*
> John von Neumann (1903 - 1957)

1.1 Introduction

The insurance industry exists because people are willing to pay a price for being insured. There is an economic theory that explains why insureds are willing to pay a premium larger than the *net premium*, that is, the mathematical expectation of the insured loss. This theory postulates that a decision maker, generally without being aware of it, attaches a value $u(w)$ to his wealth w instead of just w, where $u(\cdot)$ is called his *utility function*. To decide between random losses X and Y, he compares $E[u(w-X)]$ with $E[u(w-Y)]$ and chooses the loss with the highest expected utility. With this model, the insured with wealth w is able to determine the maximum premium P^+ he is prepared to pay for a random loss X. This is done by solving the equilibrium equation $E[u(w-X)] = u(w-P)$. At the equilibrium, he does not care, in terms of utility, if he is insured or not. The model applies to the other party involved as well. The insurer, with his own utility function and perhaps supplementary expenses, will determine a minimum premium P^-. If the insured's maximum premium P^+ is larger than the insurer's minimum premium P^-, both parties involved increase their utility if the premium is between P^- and P^+.

Although it is impossible to determine a person's utility function exactly, we can give some plausible properties of it. For example, more wealth would imply a higher utility level, so $u(\cdot)$ should be a non-decreasing function. It is also logical that 'reasonable' decision makers are *risk averse*, which means that they prefer a fixed loss over a random loss with the same expected value. We will define some classes of utility functions that possess these properties and study their advantages and disadvantages.

Suppose that an insured can choose between an insurance policy with a fixed deductible and another policy with the same expected payment by the insurer and with the same premium. It can be shown that it is better for the insured to choose the former policy. If a reinsurer is insuring the total claim amount of an insurer's portfolio of risks, insurance with a fixed maximal retained risk is called a *stop-loss* reinsurance. From the theory of ordering of risks, we will see that this type of reinsurance is optimal for risk averse decision makers. In this chapter we prove that a stop-loss reinsurance results in the smallest variance of the retained risk. We also discuss a situation where the insurer prefers a *proportional* reinsurance, with a reinsurance payment proportional to the claim amount.

1.2 The expected utility model

Imagine that an individual runs the risk of losing an amount B with probability 0.01. He can insure himself against this loss, and is willing to pay a premium P for this insurance policy. If B is very small, then P will be hardly larger than $0.01B$. However, if B is somewhat larger, say 500, then P will be a little larger than 5. If B is very large, P will be a lot larger than $0.01B$, since this loss could result in bankruptcy. So the premium for a risk is not *homogeneous*, that is, not proportional to the risk.

Example 1.2.1 (St. Petersburg paradox)
For a price P, one may enter the following game. A fair coin is tossed until a head appears. If this takes n trials, the payment is an amount 2^n. Therefore, the expected gain from the game equals $\sum_{n=1}^{\infty} 2^n (\frac{1}{2})^n = \infty$. Still, unless P is small, it turns out that very few are willing to enter the game, which means no one merely looks at expected profits.   ▽

In economics, the model developed by Von Neumann & Morgenstern (1947) aims to describe how decision makers choose between uncertain prospects. If a decision maker is able to choose consistently between potential random losses X, then there exists a utility function $u(\cdot)$ to appraise the wealth w such that the decisions he makes are exactly the same as those resulting from comparing the losses X based on the expectation $E[u(w - X)]$. In this way, a complex decision is reduced to the comparison of real numbers.

For the comparison of X with Y, the utility function $u(\cdot)$ and its linear transform $au(\cdot) + b$ for some $a > 0$ are equivalent, since they result in the same decision:

$$E[u(w - X)] \leq E[u(w - Y)] \quad \text{if and only if}$$
$$E[au(w - X) + b] \leq E[au(w - Y) + b]. \tag{1.1}$$

So from each class of equivalent utility functions, we can select one, for example by requiring that $u(0) = 0$ and $u(1) = 1$. Assuming $u'(0) > 0$, we could also use the utility function $v(\cdot)$ with $v(0) = 0$ and $v'(0) = 1$:

$$v(x) = \frac{u(x) - u(0)}{u'(0)}. \tag{1.2}$$

It is impossible to determine which utility functions are used 'in practice'. Utility theory merely states the existence of a utility function. We could try to reconstruct a decision maker's utility function from the decisions he takes, by confronting him with a large number of questions like: "Which premium P are you willing to pay to avoid a loss 1 that could occur with probability q"? Without loss of generality, we take $u(0) = 0$, $u(-1) = -1$ and initial wealth $w = 0$, by shifting the utility function over a distance w. Then we learn for which value of P we have

$$u(-P) = (1 - q)u(0) + qu(-1) = -q. \tag{1.3}$$

In practice, we would soon experience the limitations of this procedure: the decision maker will grow increasingly irritated as the interrogation continues, and his decisions will become inconsistent, for example because he asks a larger premium for a smaller risk or a totally different premium for nearly the same risk. Such mistakes are inevitable unless the decision maker is using a utility function explicitly.

Example 1.2.2 (Risk loving versus risk averse)
Suppose that a person owns a capital w and that he values his wealth by the utility function $u(\cdot)$. He is given the choice of losing the amount b with probability $\frac{1}{2}$ or just paying a fixed amount $\frac{1}{2}b$. He chooses the former if $b = 1$, the latter if $b = 4$, and if $b = 2$ he does not care. Apparently the person likes a little gamble, but is afraid of a larger one, like someone with a fire insurance policy who takes part in a lottery. What can be said about the utility function $u(\cdot)$?

Choose again $w = 0$ and $u(0) = 0$, $u(-1) = -1$. The decision maker is indifferent between a loss 2 with probability $\frac{1}{2}$ and a fixed loss 1 ($b = 2$). This implies that

$$u(-1) = \frac{u(0) + u(-2)}{2}. \tag{1.4}$$

The function $u(\cdot)$ is neither convex nor concave, since for $b = 1$ and $b = 4$ we get

$$u(-\tfrac{1}{2}) < \frac{u(0) + u(-1)}{2} \quad \text{and} \quad u(-2) > \frac{u(0) + u(-4)}{2}. \tag{1.5}$$

Note that a function that is convex is often synonymously called 'concave upward', a concave function is 'concave downward'. The connection between convexity of a real function f and convexity of sets is that the so-called epigraph of f, that is, the set of points lying on or above its graph, is a convex set.

Since $u(0) = 0$ and $u(-1) = -1$, (1.4) and (1.5) yield

$$u(-2) = -2, \quad u(-\tfrac{1}{2}) < -\tfrac{1}{2} \quad \text{and} \quad u(-4) < -4. \tag{1.6}$$

Connecting these five points gives a graph that lies below the diagonal for $-1 < x < 0$ and $x < -2$, and above the diagonal for $x \in (-2, -1)$. ∇

We assume that utility functions are non-decreasing, although the reverse is conceivable, for example in the event of capital levy. Hence, the *marginal utility* is non-negative: $u'(x) \geq 0$. The *risk averse* decision makers are an important group. They have a *decreasing marginal utility*, so $u''(x) \leq 0$. Note that we will not be very rigorous in distinguishing between the notions increasing and non-decreasing. If needed, we will use the phrase 'strictly increasing'. To explain why such decision makers are called risk averse, we use the following fundamental theorem (for a proof, see Exercises 1.2.1 and 1.2.2):

Theorem 1.2.3 (Jensen's inequality)
If $v(x)$ is a convex function and Y is a random variable, then

$$E[v(Y)] \geq v(E[Y]),\qquad(1.7)$$

with equality if and only if $v(\cdot)$ is linear on the support of Y (the set of its possible values), including the case that $\mathrm{Var}[Y] = 0$.   ▽

From this inequality, it follows that for a concave utility function

$$E[u(w - X)] \leq u(E[w - X]) = u(w - E[X]).\qquad(1.8)$$

Apparently, decision makers with such utility functions prefer to pay a fixed amount $E[X]$ instead of a risky amount X, so they are indeed *risk averse*.

Now, suppose that a risk averse insured with capital w has the utility function $u(\cdot)$. Assuming he is insured against a loss X for a premium P, his expected utility will increase if

$$E[u(w - X)] \leq u(w - P).\qquad(1.9)$$

Since $u(\cdot)$ is a non-decreasing continuous function, this is equivalent to $P \leq P^+$, where P^+ denotes the maximum premium to be paid. This so-called *zero utility premium* is the solution to the following utility equilibrium equation

$$E[u(w - X)] = u(w - P^+).\qquad(1.10)$$

The insurer, say with utility function $U(\cdot)$ and capital W, will insure the loss X for a premium P if $E[U(W + P - X)] \geq U(W)$, hence $P \geq P^-$ where P^- denotes the minimum premium to be asked. This premium follows from solving the utility equilibrium equation reflecting the insurer's position:

$$U(W) = E[U(W + P^- - X)].\qquad(1.11)$$

A deal improving the expected utility for both sides will be possible if $P^+ \geq P^-$.

From a theoretical point of view, insurers are often considered to be virtually risk neutral. So for any risk X, disregarding additional costs, a premium $E[X]$ is sufficient. Therefore,

$$E[U(W + E[X] - X)] = U(W) \text{ for any risk } X.\qquad(1.12)$$

In Exercise 1.2.3 it is proved that this entails that the utility function $U(\cdot)$ must be linear.

Example 1.2.4 (Risk aversion coefficient)
Given the utility function $u(\cdot)$, how can we approximate the maximum premium P^+ for a risk X?

Let μ and σ^2 denote the mean and variance of X. Using the first terms in the series expansion of $u(\cdot)$ in $w - \mu$, we obtain

$$u(w - P^+) \approx u(w - \mu) + (\mu - P^+)u'(w - \mu);$$
$$u(w - X) \approx u(w - \mu) + (\mu - X)u'(w - \mu) + \tfrac{1}{2}(\mu - X)^2 u''(w - \mu). \tag{1.13}$$

Taking expectations on both sides of the latter approximation yields

$$E[u(w - X)] \approx u(w - \mu) + \tfrac{1}{2}\sigma^2 u''(w - \mu). \tag{1.14}$$

Substituting (1.10) into (1.14), it follows from (1.13) that

$$\tfrac{1}{2}\sigma^2 u''(w - \mu) \approx (\mu - P^+)u'(w - \mu). \tag{1.15}$$

Therefore, the maximum premium P^+ for a risk X is approximately

$$P^+ \approx \mu - \tfrac{1}{2}\sigma^2 \frac{u''(w - \mu)}{u'(w - \mu)}. \tag{1.16}$$

This suggests the following definition: the (absolute) *risk aversion coefficient* $r(w)$ of the utility function $u(\cdot)$ at wealth w is given by

$$r(w) = -\frac{u''(w)}{u'(w)}. \tag{1.17}$$

Then the maximum premium P^+ to be paid for a risk X is approximately

$$P^+ \approx \mu + \tfrac{1}{2}r(w - \mu)\sigma^2. \tag{1.18}$$

Note that $r(w)$ does not change when $u(\cdot)$ is replaced by $au(\cdot) + b$. From (1.18), we see that the risk aversion coefficient indeed reflects the degree of risk aversion: the more risk averse one is, the larger the premium one is willing to pay. $\quad\nabla$

1.3 Classes of utility functions

Besides the linear functions, other families of suitable utility functions exist that have interesting properties:

linear utility:	$u(w) = w$	
quadratic utility:	$u(w) = -(\alpha - w)^2$	$(w \le \alpha)$
logarithmic utility:	$u(w) = \log(\alpha + w)$	$(w > -\alpha)$
exponential utility:	$u(w) = -\alpha e^{-\alpha w}$	$(\alpha > 0)$
power utility:	$u(w) = w^c$	$(w > 0,\ 0 < c \le 1)$

(1.19)

These utility functions, and of course their linear transforms as well, have a non-negative and non-increasing marginal utility; for the quadratic utility function, we set $u(w) = 0$ if $w > \alpha$. The risk aversion coefficient for the linear utility function is 0, while for the exponential utility function, it equals α. For the other utility functions, it can be written as $(\gamma + \beta w)^{-1}$ for some γ and β, see Exercise 1.3.1.

Note that linear utility leads to the *principle of equivalence* and *risk neutrality* (net premiums), since if $E[w + P^- - X] = w$, then $P^- = E[X]$.

Example 1.3.1 (Exponential premium)
Suppose that an insurer has an exponential utility function with parameter α. What is the minimum premium P^- to be asked for a risk X?
Solving the equilibrium equation (1.11) with $U(x) = -\alpha e^{-\alpha x}$ yields

$$P^- = \frac{1}{\alpha} \log(m_X(\alpha)), \tag{1.20}$$

where $m_X(\alpha) = E[e^{\alpha X}]$ is the moment generating function of X at argument α. We observe that this *exponential premium* is independent of the insurer's current wealth W, in line with the risk aversion coefficient being a constant.

The expression for the maximum premium P^+ is the same as (1.20), see Exercise 1.3.3, but now of course α represents the risk aversion of the insured. Assume that the loss X is exponentially distributed with parameter β. Taking $\beta = 0.01$ yields $E[X] = \frac{1}{\beta} = 100$. If the insured's utility function is exponential with parameter $\alpha = 0.005$, then

$$P^+ = \frac{1}{\alpha} \log(m_X(\alpha)) = 200 \log\left(\frac{\beta}{\beta - \alpha}\right) = 200 \log(2) \approx 138.6, \tag{1.21}$$

so the insured is willing to accept a sizable loading on the net premium $E[X]$. ∇

The approximation (1.18) from Example 1.2.4 yields

$$P^+ \approx E[X] + \tfrac{1}{2}\alpha \mathrm{Var}[X] = 125. \tag{1.22}$$

Obviously, the approximation (1.22) is increasing with α, but also the premium (1.20) is increasing if X is a non-negative random variable with finite variance, as we will prove next.

Theorem 1.3.2 (Exponential premium increases with risk aversion)
The exponential premium (1.20) for a risk X is an increasing function of the risk aversion α.

Proof. For $0 < \alpha < \gamma$, consider the strictly concave function $v(\cdot)$ with

$$v(x) = x^{\alpha/\gamma}. \tag{1.23}$$

From Jensen's inequality, it follows that

$$v(E[Y]) > E[v(Y)], \tag{1.24}$$

for any random variable Y with $\text{Var}[Y] > 0$. Take $Y = e^{\gamma X}$, then $v(Y) = e^{\alpha X}$ and

$$\{\text{E}[e^{\gamma X}]\}^\alpha = \{(\text{E}[Y])^{\alpha/\gamma}\}^\gamma = \{v(\text{E}[Y])\}^\gamma > \{\text{E}[v(Y)]\}^\gamma = \{\text{E}[e^{\alpha X}]\}^\gamma. \quad (1.25)$$

Therefore,

$$\{m_X(\alpha)\}^\gamma < \{m_X(\gamma)\}^\alpha, \quad (1.26)$$

which implies that, for any $\gamma > \alpha$, for the exponential premiums we have

$$\frac{1}{\alpha} \log(m_X(\alpha)) < \frac{1}{\gamma} \log(m_X(\gamma)). \quad (1.27)$$

So the proof is completed. ∇

Just as for the approximation (1.18), the limit of (1.20) as $\alpha \downarrow 0$ is the net premium. This follows immediately from the series expansion of $\log(m_X(t))$, see also Exercise 1.3.4. But as $\alpha \uparrow \infty$, the exponential premium tends to $\max[X]$, while the approximation goes to infinity.

Example 1.3.3 (Quadratic utility)
Suppose that for $w < 5$, the insured's utility function is $u(w) = 10w - w^2$. What is the maximum premium P^+ as a function of w, $w \in [0,5]$, for an insurance policy against a loss 1 with probability $\frac{1}{2}$? What happens to this premium if w increases?

Again, we solve the equilibrium equation (1.10). The expected utility after a loss X equals

$$\text{E}[u(w - X)] = 11w - \tfrac{11}{2} - w^2, \quad (1.28)$$

and the utility after paying a premium P equals

$$u(w - P) = 10(w - P) - (w - P)^2. \quad (1.29)$$

By the equilibrium equation (1.10), the right hand sides of (1.28) and (1.29) should be equal, and after some calculations we find the maximum premium as

$$P = P(w) = \sqrt{\left(\tfrac{11}{2} - w\right)^2 + \tfrac{1}{4}} - (5 - w), \quad w \in [0,5]. \quad (1.30)$$

One may verify that $P'(w) > 0$, see also Exercise 1.3.2. We observe that a decision maker with quadratic utility is willing to pay larger premiums as his wealth increases toward the saturation point 5. Because of this property, quadratic utility is less appropriate to model the behavior of risk averse decision makers. The quadratic utility function still has its uses, of course, since knowing only the expected value and the variance of the risk suffices to do the calculations. ∇

Example 1.3.4 (Uninsurable risk)
A decision maker with an exponential utility function with risk aversion $\alpha > 0$ wants to insure a gamma distributed risk with shape parameter n and scale parameter 1. See Table A. Determine P^+ and prove that $P^+ > n$. When is $P^+ = \infty$ and what does that mean?

From formula (1.20), it follows that

$$P^+ = \frac{1}{\alpha} \log(m_X(\alpha)) = \begin{cases} -\frac{n}{\alpha} \log(1-\alpha) & \text{for } 0 < \alpha < 1 \\ \infty & \text{for } \alpha \geq 1. \end{cases} \qquad (1.31)$$

Since $\log(1+x) < x$ for all $x > -1$, $x \neq 0$, we have also $\log(1-\alpha) < -\alpha$ and consequently $P^+ > E[X] = n$. So the resulting premium is larger than the net premium. If $\alpha \geq 1$, then $P^+ = \infty$, which means that the decision maker is willing to pay any finite premium. An insurer with risk aversion $\alpha \geq 1$ insuring the risk will suffer a loss, in terms of utility, for any finite premium P, since also $P^- = \infty$. For such insurers, the risk is *uninsurable*.   ▽

Remark 1.3.5 (Allais paradox (1953), Yaari's dual theory (1987))
Consider the following possible capital gains:

$$X = \quad 1\,000\,000 \quad \text{with probability } 1$$

$$Y = \begin{cases} 5\,000\,000 & \text{with probability } 0.10 \\ 1\,000\,000 & \text{with probability } 0.89 \\ 0 & \text{with probability } 0.01 \end{cases}$$

$$V = \begin{cases} 1\,000\,000 & \text{with probability } 0.11 \\ 0 & \text{with probability } 0.89 \end{cases}$$

$$W = \begin{cases} 5\,000\,000 & \text{with probability } 0.10 \\ 0 & \text{with probability } 0.90 \end{cases}$$

Experimental economy has revealed that, having a choice between X and Y, many people choose X, but at the same time they prefer W over V. This result violates the expected utility hypothesis, since, assuming an initial wealth of 0, the latter preference $E[u(W)] > E[u(V)]$ is equivalent to $0.11\,u(1\,000\,000) < 0.1\,u(5\,000\,000) + 0.01\,u(0)$, but the former leads to exactly the opposite inequality. Apparently, expected utility does not always describe the behavior of decision makers adequately. Judging from this example, it would seem that the attraction of being in a completely safe situation is stronger than expected utility indicates, and induces people to make irrational decisions.

Yaari (1987) has proposed an alternative theory of decision making under risk that has a very similar axiomatic foundation. Instead of using a utility function, Yaari's dual theory computes 'certainty equivalents' not as expected values of transformed wealth levels (utilities), but with distorted probabilities of large gains and losses. It turns out that this theory leads to paradoxes that are very similar to the ones vexing utility theory.   ▽

1.4 Stop-loss reinsurance

Reinsurance treaties usually do not cover the risk fully. *Stop-loss (re)insurance* covers the top part. It is defined as follows: for a loss X, assumed non-negative, the payment is

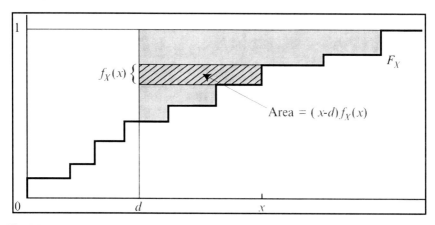

Fig. 1.1 Graphical derivation of (1.33) for a discrete cdf.

$$(X - d)_+ := \max\{X - d, 0\} = \begin{cases} X - d & \text{if } X > d; \\ 0 & \text{if } X \le d. \end{cases} \qquad (1.32)$$

The insurer retains a risk d (his *retention*) and lets the reinsurer pay for the remainder, so the insurer's loss stops at d. In the reinsurance practice, the retention equals the maximum amount to be paid out for every single claim and d is called the *priority*. We will prove that, regarding the variance of the insurer's retained loss, a stop-loss reinsurance is optimal. The other side of the coin is that reinsurers do not offer stop-loss insurance under the same conditions as other types of reinsurance.

Theorem 1.4.1 (Net stop-loss premium)
Consider the *stop-loss premium*, by which we mean the net premium $\pi_X(d) := \mathrm{E}[(X - d)_+]$ for a stop-loss contract. Both in the discrete case, where $F_X(x)$ is a step function with a step $f_X(x)$ in x, and in the continuous case, where $F_X(x)$ has $f_X(x)$ as its derivative, the stop-loss premium is given by

$$\pi_X(d) = \left\{ \begin{array}{l} \sum_{x>d}(x - d)f_X(x) \\ \int_d^\infty (x - d)f_X(x)\,dx \end{array} \right\} = \int_d^\infty [1 - F_X(x)]\,dx. \qquad (1.33)$$

Proof. This proof can be given in several ways. A graphical 'proof' for the discrete case is given in Figure 1.1. The right hand side of the equation (1.33), that is, the total shaded area enclosed by the graph of $F_X(x)$, the horizontal line at 1 and the vertical line at d, is divided into small bars with a height $f_X(x)$ and a width $x - d$. Their total area equals the left hand side of (1.33). The continuous case can be proved in the same way by taking limits, considering bars with an infinitesimal height.

To prove it in the continuous case by partial integration $\int Fg = FG - \int fG$, write

$$E[(X-d)_+] = \int_d^\infty (x-d)f_X(x)\,dx$$
$$= -(x-d)[1-F_X(x)]\Big|_d^\infty + \int_d^\infty [1-F_X(x)]\,dx. \tag{1.34}$$

Note that we do not just take $F_X(x)$ as an antiderivative of $f_X(x)$. The only choice $F_X(x)+C$ that is a candidate to produce finite terms on the right hand side is $F_X(x)-1$. With partial integration, the integrated term often vanishes, but this is not always easy to prove. In this case, for $x \to \infty$ this can be seen as follows: since $E[X] < \infty$, the integral $\int_0^\infty t f_X(t)\,dt$ is convergent, and hence the 'tails' tend to zero, so

$$x[1-F_X(x)] = x\int_x^\infty f_X(t)\,dt \le \int_x^\infty t f_X(t)\,dt \downarrow 0 \text{ for } x \to \infty. \tag{1.35}$$

In many cases, easier proofs result from using Fubini's theorem, that is, swapping the order of integration in double integrals, taking care that the integration is over the same set of values in $\mathbb{R}^2$. In this case, we have only one integral, but this can be fixed by writing $x-d = \int_d^x dy$. The pairs (x,y) over which we integrate satisfy $d < y < x < \infty$, so this leads to

$$\int_d^\infty (x-d)f_X(x)\,dx = \int_d^\infty \int_d^x dy\, f_X(x)\,dx = \int_d^\infty \int_y^\infty f_X(x)\,dx\,dy$$
$$= \int_d^\infty [1-F_X(y)]\,dy, \tag{1.36}$$

By the same device, the proof in the discrete case can be given as

$$\sum_{x>d} (x-d)f_X(x) = \sum_{x>d}\int_d^x dy\, f_X(x) = \int_d^\infty \sum_{x>y} f_X(x)\,dy = \int_d^\infty [1-F_X(y)]\,dy. \tag{1.37}$$

Note that in a Riemann integral, we may change the value of the integrand in a countable set of arguments without affecting the outcome of the integral, so it does not matter if we take $\sum_{x\ge y} f_X(x)$ or $\sum_{x>y} f_X(x)$. ∇

Theorem 1.4.2 (Stop-loss transform)
The function $\pi_X(d) = E[(X-d)_+]$, called the *stop-loss transform* of X, is a continuous convex function that is strictly decreasing in the retention d, as long as $F_X(d) < 1$. When $F_X(d) = 1$, $\pi_X(d) = 0$, and always $\pi_X(\infty) = 0$. If X is non-negative, then $\pi_X(0) = E[X]$, and for $d < 0$, $\pi_X(d)$ decreases linearly with slope -1.

Proof. From (1.33), it follows that:

$$\pi_X'(d) = F_X(d) - 1. \tag{1.38}$$

Since $F_X(x) = \Pr[X \le x]$, each cdf F_X is continuous from the right. Accordingly, the derivative in (1.38) is a right hand derivative. Since $F_X(x)$ is non-decreasing, $\pi_X(d)$ is a continuous and convex function that is strictly decreasing in d, as long as

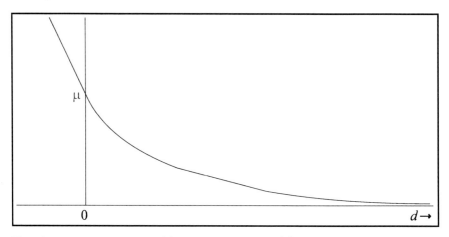

Fig. 1.2 A stop-loss transform $\pi_X(d)$ for a risk $X \geq 0$ with $E[X] = \mu$.

$F_X(d) < 1$. If X is non-negative, then $\pi_X(d) = E[(X-d)_+] = E[X-d] = E[X]-d$ for $d \leq 0$. That $\lim_{d \to \infty} \pi_X(d) = 0$ is evident. These properties are illustrated in Figure 1.2. ▽

In the next theorem, we prove the important result that a stop-loss insurance minimizes the variance of the retained risk.

Theorem 1.4.3 (Optimality of stop-loss reinsurance)

Let $I(X)$ be the payment on some reinsurance contract if the loss is X, with $X \geq 0$. Assume that $0 \leq I(x) \leq x$ holds for all $x \geq 0$. Then

$$E[I(X)] = E[(X-d)_+] \implies \text{Var}[X - I(X)] \geq \text{Var}[X - (X-d)_+]. \qquad (1.39)$$

Proof. Because of the previous theorem, for every $I(\cdot)$ we can find a retention d such that the expectations $E[I(X)]$ and $E[(X-d)_+]$ are equal. We write the retained risks as follows:

$$V(X) = X - I(X) \quad \text{and} \quad W(X) = X - (X-d)_+. \qquad (1.40)$$

Since $E[V(X)] = E[W(X)]$, it suffices to prove that

$$E[\{V(X)-d\}^2] \geq E[\{W(X)-d\}^2]. \qquad (1.41)$$

A sufficient condition for this to hold is that $|V(X)-d| \geq |W(X)-d|$ with probability one. This is trivial in the event $X \geq d$, since then $W(X) \equiv d$ holds. For $X < d$, we have $W(X) \equiv X$, and hence

$$V(X) - d = X - d - I(X) \leq X - d = W(X) - d < 0. \qquad (1.42)$$

This completes the proof. ▽

The essence of the proof is that stop-loss reinsurance makes the risk 'as close to d' as possible, under the retained risks with a fixed mean. So it is more attractive because it is less risky, more predictable. As stated before, this theorem can be extended: using the theory of ordering of risks, one can prove that stop-loss insurance not only minimizes the variance of the retained risk, but also maximizes the insured's expected utility, see Chapter 7.

In the above theorem, it is crucial that the premium for a stop-loss coverage is the same as the premium for another type of coverage with the same expected payment. Since the variance of the reinsurer's capital will be larger for a stop-loss coverage than for another coverage, the reinsurer, who is without exception at least slightly risk averse, in practice will charge a higher premium for a stop-loss insurance.

Example 1.4.4 (A situation in which proportional reinsurance is optimal)
To illustrate the importance of the requirement that the premium does not depend on the type of reinsurance, we consider a related problem: suppose that the insurer collects a premium $(1+\theta)E[X]$ and that he is looking for the most profitable reinsurance $I(X)$ with $0 \le I(X) \le X$ and prescribed variance

$$\text{Var}[X - I(X)] = V. \tag{1.43}$$

So the insurer wants to maximize his expected profit, under the assumption that the instability of his own financial situation is fixed in advance. We consider two methods for the reinsurer to calculate his premium for $I(X)$. In the first scenario (A), the reinsurer collects a premium $(1+\lambda)E[I(X)]$, structured like the direct insurer's. In the second scenario (B), the reinsurer determines the premium according to the *variance principle*, which means that he asks as a premium the expected value plus a loading equal to a constant, say α, times the variance of $I(X)$. Then the insurer can determine his expected profit, which equals the collected premium less the expected value of the retained risk and the reinsurance premium, as follows:

$$\begin{aligned} \text{A}: \quad & (1+\theta)E[X] - E[X-I(X)] - (1+\lambda)E[I(X)] \\ & = \theta E[X] - \lambda E[I(X)]; \\ \text{B}: \quad & (1+\theta)E[X] - E[X-I(X)] - (E[I(X)] + \alpha\text{Var}[I(X)]) \\ & = \theta E[X] - \alpha\text{Var}[I(X)]. \end{aligned} \tag{1.44}$$

As one sees, in both scenarios the expected profit equals the original expected profit $\theta E[X]$ reduced by the expected profit of the reinsurer. Clearly, we have to minimize the expected profit of the reinsurer, hence the following minimization problems A and B arise:

$$\begin{array}{llll} \text{A}: & \text{Min } E[I(X)] & \text{B}: & \text{Min Var}[I(X)] \\ & \text{s.t. Var}[X-I(X)] = V & & \text{s.t. Var}[X-I(X)] = V \end{array} \tag{1.45}$$

To solve Problem B, we write

$$\text{Var}[I(X)] = \text{Var}[X] + \text{Var}[I(X)-X] - 2\text{Cov}[X, X-I(X)]. \tag{1.46}$$

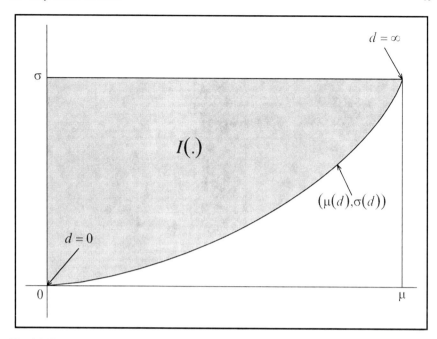

Fig. 1.3 Expected value and standard deviation of the retained risk for different reinsurance contracts. The boundary line constitutes the stop-loss contracts with $d \in [0, \infty)$. Other feasible reinsurance contracts are to be found in the shaded area.

Since the first two terms on the right hand side are given, the left hand side is minimal if the covariance term is maximized. This can be accomplished by taking X and $X - I(X)$ linearly dependent, choosing $I(x) = \gamma + \beta x$. From $0 \leq I(x) \leq x$, we find $\gamma = 0$ and $0 \leq \beta \leq 1$; from (1.43), it follows that $(1 - \beta)^2 = V/\text{Var}[X]$. So, if the variance V of the retained risk is given and the reinsurer uses the variance principle, then *proportional reinsurance* $I(X) = \beta X$ with $\beta = 1 - \sqrt{V/\text{Var}[X]}$ is optimal.

For the solution of problem A, we use Theorem 1.4.3. By calculating the derivatives with respect to d, we can prove that not just $\mu(d) = \text{E}[X - (X - d)_+]$, but also $\sigma^2(d) = \text{Var}[X - (X - d)_+]$ is continuously increasing in d. See Exercise 1.4.3. Notice that $\mu(0) = \sigma^2(0) = 0$ and $\mu(\infty) = \mu = \text{E}[X]$, $\sigma^2(\infty) = \sigma^2 = \text{Var}[X]$.

In Figure 1.3, we plot the points $(\mu(d), \sigma(d))$ for $d \in [0, \infty)$ for some loss random variable X. Because of Theorem 1.4.3, other reinsurance contracts $I(\cdot)$ can only have an expected value and a standard deviation of the retained risk above the curve in the μ, σ^2-plane, since the variance is at least as large as for the stop-loss reinsurance with the same expected value. This also implies that such a point can only be located to the left of the curve. From this we conclude that, just as in Theorem 1.4.3, the non-proportional stop-loss solution is optimal for problem A. The stop-loss contracts in this case are Pareto-optimal: there are no other solutions with both a smaller variance and a higher expected profit. ∇

1.5 Exercises

For *hints* with the exercises, consult Appendix B. Also, the index at the end of the book might be a convenient place to look for explanations.

Section 1.2

1. Prove Jensen's inequality: if $v(x)$ is convex, then $E[v(X)] \geq v(E[X])$. Consider especially the case $v(x) = x^2$.

2. Also prove the reverse of Jensen's inequality: if $E[v(X)] \geq v(E[X])$ for every random variable X, then v is convex.

3. Prove: if $E[v(X)] = v(E[X])$ for every random variable X, then v is linear.

4. A decision maker has utility function $u(x) = \sqrt{x}$, $x \geq 0$. He is given the choice between two random amounts X and Y, in exchange for his entire present capital w. The probability distributions of X and Y are given by $\Pr[X = 400] = \Pr[X = 900] = 0.5$ and $\Pr[Y = 100] = 1 - \Pr[Y = 1600] = 0.6$. Show that he prefers X to Y. Determine for which values of w he should decline the offer. Can you think of utility functions with which he would prefer Y to X?

5. Prove that $P^- \geq E[X]$ for risk averse insurers.

6. An insurer undertakes a risk X and after collecting the premium, he owns a capital $w = 100$. What is the maximum premium the insurer is willing to pay to a reinsurer to take over the complete risk, if his utility function is $u(w) = \log(w)$ and $\Pr[X = 0] = \Pr[X = 36] = 0.5$? Find not only the exact value, but also the approximation (1.18) of Example 1.2.4.

7. Assume that the reinsurer's minimum premium to take over the risk of the previous exercise equals 19 and that the reinsurer has the same utility function. Determine his capital W.

8. Describe the utility function of a person with the following risk behavior: after winning an amount 1, he answers 'yes' to the question 'double or quits?'; after winning again, he agrees only after a long huddle; the third time he says 'no'.

9. Verify that $P^+[2X] < 2P^+[X]$ when $w = 0$, $X \sim \text{Bernoulli}(1/2)$ and

$$u(x) = \begin{cases} 2x/3 & \text{for } x > -3/4, \\ 2x+1 & \text{for } -1 < x < -3/4, \\ 3x+2 & \text{for } x < -1. \end{cases}$$

Section 1.3

1. Prove that the utility functions in (1.19) have a non-negative and non-increasing marginal utility. Show how the risk aversion coefficient of all these utility functions can be written as $r(w) = (\gamma + \beta w)^{-1}$.

2. Show that, for quadratic utility, the risk aversion increases with the capital. Check (1.28)–(1.30) and verify that $P'(w) > 0$ in (1.30).

3. Prove the formula (1.20) for P^- for the case of exponential utility. Also show that (1.10) yields the same solution for P^+.

4. Prove that the exponential premium P^- in (1.20) decreases to the net premium if the risk aversion α tends to zero.

5. Show that the approximation in Example 1.2.4 is exact if $X \sim N(\mu, \sigma^2)$ and $u(\cdot)$ is exponential.

6. Using the exponential utility function with $\alpha = 0.001$, determine which premium is higher: the one for $X \sim N(400, 25\,000)$ or the one for $Y \sim N(420, 20000)$. Determine for which values of α the former premium is higher.

7. Assume that the marginal utility of $u(w)$ is proportional to $1/w$, that is, $u'(w) = k/w$ for some $k > 0$ and all $w > 0$. What is $u(w)$? With this utility function and wealth $w = 4$, show that only prices $P < 4$ in the St. Petersburg paradox of Example 1.2.1 make entering the game worthwhile.

8. It is known that the premium P that an insurer with exponential utility function asks for a $N(1000, 100^2)$ distributed risk satisfies $P \geq 1250$. What can be said about his risk aversion α? If the risk X has *dimension* 'money', then what is the dimension of α?

9. For a random variable X with mean $E[X] = \mu$ and variance $Var[X] = \sigma^2$ it is known that for every possible $\alpha > 0$, the zero utility premium with exponential utility with risk aversion α contains a relative safety loading $\frac{1}{2}\sigma^2\alpha/\mu$ on top of the net premium. What distribution can X have?

10. Which utility function results if in the class of power utility functions w^c with $0 < c < 1$ we let $c \downarrow 0$? [Look at the linear transformation $(w^c - 1)/c$.]

11. Which class of utility functions has constant relative risk aversion (CRRA) $\frac{-wu''(w)}{u'(w)} \equiv \rho$?

12. For an exponential premium (1.20), prove that $P^-[2X] > 2P^-[X]$.

13. Assume that the insurer, from vast experience, knows a particular insurance risk is distributed as $X \sim \text{gamma}(2, \beta)$ with mean 50, while the insured himself, with inside knowledge, knows it is distributed as $X^\bullet \sim \text{exponential}(\beta^\bullet)$ with mean 45. They have exponential utility functions with risk aversions $\alpha = 0.001$ and $\alpha^\bullet = 0.005$ respectively. Find the interval of premiums for which both parties involved can increase their perceived expected utility.

Section 1.4

1. Sketch the stop-loss transform corresponding to the following cdf:

$$F(x) = \begin{cases} 0 & \text{for } x < 2 \\ x/4 & \text{for } 2 \leq x < 4 \\ 1 & \text{for } 4 \leq x \end{cases}$$

2. Determine the distribution of S if $E[(S - d)_+] = \frac{1}{3}(1 - d)^3$ for $0 \leq d \leq 1$.

3. [♠] Prove that, for the optimization of problem A,

$$\mu'(d) = 1 - F_X(d) \quad \text{and} \quad (\sigma^2)'(d) = 2[1 - F_X(d)][d - \mu(d)].$$

Verify that both are non-negative.

4. [♠] What happens if we replace '=' by '≤' in (1.43), taking V to be an upper bound for the variance of the retained risk in the scenarios A and B?

5. Define the coefficient of variation $V[\cdot]$ for a risk X with an expected value μ and a variance σ^2 as σ/μ. By comparing the variance of the retained risk $W(X) = X - (X - d)_+$ resulting from a stop-loss reinsurance with the one obtained from a suitable proportional reinsurance, show that $V[W] \leq V[X]$. Also show that $V[\min\{X, d\}]$ is non-increasing in d, by using the following equality: if $d < t$, then $\min\{X, d\} = \min\{\min\{X, t\}, d\}$.

6. Suppose for the random loss $X \sim N(0, 1)$ an insurance of *franchise* type is in operation: the amount $I(x)$ paid in case the damage is x equals x when $x \geq d$ for some $d > 0$, and zero

otherwise. Show that the net premium for this type of insurance is $\varphi(x)$, where $\varphi(\cdot)$ is the standard normal density, see Table A. Compare this with the net stop-loss premium with a retention d.

7. Consider the following R-statements and their output:

```
set.seed(2525); n <- 1e6; X <- rnorm(n)
rbind(c(mean(X[X>1]), mean(X*(X>1)), mean(pmax(X-1,0))),
         c(dnorm(1)/pnorm(-1), dnorm(1), dnorm(1)-pnorm(-1)))
## Result:
##               [,1]       [,2]        [,3]
## [1,] 1.524587 0.2416943 0.08316332
## [2,] 1.525135 0.2419707 0.08331547
```

Coincidence?

Chapter 2
The individual risk model

If the automobile had followed the same development cycle as the computer, a Rolls-Royce would today cost $100, get a million miles per gallon, and explode once a year, killing everyone inside — Robert X. Cringely

2.1 Introduction

In this chapter we focus on the distribution function of the total claim amount S for the portfolio of an insurer. We are not merely interested in the expected value and the variance of the insurer's random capital, but we also want to know the probability that the amounts paid exceed a fixed threshold. The distribution of the total claim amount S is also necessary to be able to apply the utility theory of the previous chapter. To determine the value-at-risk at, say, the 99.5% level, we need also good approximations for the inverse of the cdf, especially in the far tail. In this chapter we deal with models that still recognize the individual, usually different, policies. As is done often in non-life insurance mathematics, the time aspect will be ignored. This aspect is nevertheless important in disability and long term care insurance. For this reason, these types of insurance are sometimes considered life insurances.

In the insurance practice, risks usually cannot be modeled by purely discrete random variables, nor by purely continuous random variables. For example, in liability insurance a whole range of positive amounts can be paid out, each of them with a very small probability. There are two exceptions: the probability of having no claim, that is, claim size 0, is quite large, and the probability of a claim size that equals the maximum sum insured, implying a loss exceeding that threshold, is also not negligible. For expectations of such mixed random variables, we use the Riemann-Stieltjes integral as a notation, without going too deeply into its mathematical aspects. A simple and flexible model that produces random variables of this type is a mixture model, also called an 'urn-of-urns' model. Depending on the outcome of one drawing, resulting in one of the events 'no claim or maximum claim' or 'other claim', a second drawing is done from either a discrete distribution, producing zero or the maximal claim amount, or a continuous distribution. In the sequel, we present some examples of mixed models for the claim amount per policy.

Assuming that the risks in a portfolio are independent random variables, the distribution of their sum can be calculated by making use of convolution. Even with the computers of today, it turns out that this technique is quite laborious, so there is a need for other methods. One of the alternative methods is to make use of moment generating functions (mgf) or of related transforms like characteristic functions, probability generating functions (pgf) and cumulant generating functions (cgf). Sometimes it is possible to recognize the mgf of a sum of independent random variables and consequently identify the distribution function. And in some cases we can fruitfully employ a technique called the *Fast Fourier Transform* to reconstruct the density from a transform.

A totally different approach is to compute approximations of the distribution of the total claim amount S. If we consider S as the sum of a 'large' number of random variables, we could, by virtue of the Central Limit Theorem, approximate its distribution by a normal distribution with the same mean and variance as S. We will show that this approximation usually is not satisfactory for the insurance practice. Especially in the tails, there is a need for more refined approximations that explicitly recognize the substantial probability of large claims. More technically, the third central moment of S is usually greater than 0, while for the normal distribution it equals 0. We present an approximation based on a translated gamma random variable, as well as the normal power (NP) approximation. The quality of these approximations is similar. The latter can be calculated directly by means of a $N(0,1)$ table, the former requires using a computer.

Another way to approximate the individual risk model is to use the collective risk models described in the next chapter.

2.2 Mixed distributions and risks

In this section, we discuss some examples of insurance risks, that is, the claims on an insurance policy. First, we have to slightly extend the set of distribution functions we consider, because purely discrete random variables and purely continuous random variables both turn out to be inadequate for modeling the risks.

From the theory of probability, we know that every function $F(\cdot)$ that satisfies

$$F(-\infty) = 0; \quad F(+\infty) = 1$$
$$F(\cdot) \text{ is non-decreasing and right-continuous} \tag{2.1}$$

is a cumulative distribution function (cdf) of some random variable, for example of $F^{-1}(U)$ with $U \sim \text{uniform}(0,1)$, see Section 3.9.1 and Definition 5.6.1. If $F(\cdot)$ is a step function, that is, a function that is constant outside a denumerable set of discontinuities (steps), then $F(\cdot)$ and any random variable X with $F(x) = \Pr[X \le x]$ are called *discrete*. The associated probability density function (pdf) represents the height of the step at x, so

$$f(x) = F(x) - F(x-0) = \Pr[X = x] \quad \text{for all } x \in (-\infty, \infty). \tag{2.2}$$

Here, $F(x-0)$ is shorthand for $\lim_{\varepsilon \downarrow 0} F(x-\varepsilon)$; $F(x+0) = F(x)$ holds because of right-continuity. For all x, we have $f(x) \geq 0$, and $\sum_x f(x) = 1$ where the sum is taken over the denumerable set of all x with $f(x) > 0$.

Another special case is when $F(\cdot)$ is *absolutely continuous*. This means that if $f(x) = F'(x)$, then

$$F(x) = \int_{-\infty}^{x} f(t)\,dt. \tag{2.3}$$

In this case $f(\cdot)$ is called the probability density function, too. Again, $f(x) \geq 0$ for all x, while now $\int f(x)\,dx = 1$. Note that, just as is customary in mathematical statistics, this notation without integration limits represents the *definite* integral of $f(x)$ over the interval $(-\infty, \infty)$, and not just an arbitrary antiderivative, that is, any function having $f(x)$ as its derivative.

In statistics, almost without exception random variables are either discrete or continuous, but this is definitely not the case in insurance. Many distribution functions to model insurance payments have continuously increasing parts, but also some positive steps. Let Z represent the payment on some contract. There are three possibilities:

1. The contract is claim-free, hence $Z = 0$.
2. The contract generates a claim that is larger than the maximum sum insured, say M. Then, $Z = M$.
3. The contract generates a 'normal' claim, hence $0 < Z < M$.

Apparently, the cdf of Z has steps in 0 and in M. For the part in-between we could use a discrete distribution, since the payment will be some integer multiple of the monetary unit. This would produce a very large set of possible values, each of them with a very small probability, so using a continuous cdf seems more convenient. In this way, a cdf arises that is neither purely discrete, nor purely continuous. In Figure 2.2 a diagram of a mixed continuous/discrete cdf is given, see also Exercise 1.4.1.

The following urn-of-urns model allows us to construct a random variable with a distribution that is a mixture of a discrete and a continuous distribution. Let I be an *indicator random variable*, with values $I = 1$ or $I = 0$, where $I = 1$ indicates that some event has occurred. Suppose that the probability of the event is $q = \Pr[I = 1]$, $0 \leq q \leq 1$. If $I = 1$, in the second stage the claim Z is drawn from the distribution of X, if $I = 0$, then from Y. This means that

$$Z = IX + (1-I)Y. \tag{2.4}$$

If $I = 1$ then Z can be replaced by X, if $I = 0$ it can be replaced by Y. Note that we may act as if not just I and X, Y are independent, but in fact the triple (X, Y, I); only the conditional distributions of $X \mid I = 1$ and of $Y \mid I = 0$ are relevant, so we can take for example $\Pr[X \leq x \mid I = 0] = \Pr[X \leq x \mid I = 1]$ just as well. Hence, the cdf of Z can be written as

$$
\begin{aligned}
F(z) &= \Pr[Z \le z] \\
&= \Pr[Z \le z, I = 1] + \Pr[Z \le z, I = 0] \\
&= \Pr[X \le z, I = 1] + \Pr[Y \le z, I = 0] \\
&= q \Pr[X \le z] + (1 - q) \Pr[Y \le z].
\end{aligned}
\tag{2.5}
$$

Now, let X be a discrete random variable and Y a continuous random variable. From (2.5) we get

$$
F(z) - F(z - 0) = q \Pr[X = z] \quad \text{and} \quad F'(z) = (1 - q) \frac{\mathrm{d}}{\mathrm{d}z} \Pr[Y \le z].
\tag{2.6}
$$

This construction yields a cdf $F(z)$ with steps where $\Pr[X = z] > 0$, but it is not a step function, since $F'(z) > 0$ on the support of Y.

To calculate the moments of Z, the moment generating function $\mathrm{E}[e^{tZ}]$ and the stop-loss premiums $\mathrm{E}[(Z - d)_+]$, we have to calculate the expectations of functions of Z. For that purpose, we use the iterative formula of conditional expectations, also known as the law of total expectation, the law of iterated expectations, the tower rule, or the smoothing theorem:

$$
\mathrm{E}[W] = \mathrm{E}[\mathrm{E}[W \mid V]].
\tag{2.7}
$$

We apply this formula with $W = g(Z)$ for an appropriate function $g(\cdot)$ and replace V by I. Then, introducing $h(i) = \mathrm{E}[g(Z) \mid I = i]$, we get, using (2.6) at the end:

$$
\begin{aligned}
\mathrm{E}[g(Z)] &= \mathrm{E}[\mathrm{E}[g(Z) \mid I]] = qh(1) + (1 - q)h(0) = \mathrm{E}[h(I)] \\
&= q\mathrm{E}[g(Z) \mid I = 1] + (1 - q)\mathrm{E}[g(Z) \mid I = 0] \\
&= q\mathrm{E}[g(X) \mid I = 1] + (1 - q)\mathrm{E}[g(Y) \mid I = 0] \\
&= q\mathrm{E}[g(X)] + (1 - q)\mathrm{E}[g(Y)] \\
&= q \sum_z g(z) \Pr[X = z] + (1 - q) \int_{-\infty}^{\infty} g(z) \frac{\mathrm{d}}{\mathrm{d}z} \Pr[Y \le z] \, \mathrm{d}z \\
&= \sum_z g(z)[F(z) - F(z - 0)] + \int_{-\infty}^{\infty} g(z) F'(z) \, \mathrm{d}z.
\end{aligned}
\tag{2.8}
$$

Remark 2.2.1 (Riemann-Stieltjes integrals)

The result in (2.8), consisting of a sum and an ordinary Riemann integral, can be written as a right hand Riemann-Stieltjes integral:

$$
\mathrm{E}[g(Z)] = \int_{-\infty}^{\infty} g(z) \, \mathrm{d}F(z).
\tag{2.9}
$$

The integrator is the differential $\mathrm{d}F(z) = F_Z(z) - F_Z(z - \mathrm{d}z)$. It replaces the probability of z, that is, the height of the step at z if there is one, or $F'(z) \, \mathrm{d}z$ if there is no step at z. Here, $\mathrm{d}z$ denotes a positive infinitesimally small number. Note that the cdf $F(z) = \Pr[Z \le z]$ is continuous from the right. In life insurance mathematics theory, Riemann-Stieltjes integrals were used as a tool to describe situations in which it is

vital which value of the integrand should be taken: the limit from the right, the limit
from the left, or the actual function value. Actuarial practitioners have not adopted
this convention. We avoid this problem altogether by considering continuous integrands only.   ∇

Remark 2.2.2 (Mixed random variables and mixed distributions)
We can summarize the above as follows: a mixed continuous/discrete cdf $F_Z(z) =$
$\Pr[Z \leq z]$ arises when a mixture of random variables

$$Z = IX + (1 - I)Y \tag{2.10}$$

is used, where X is a discrete random variable, Y is a continuous random variable
and I is a Bernoulli(q) random variable, with X, Y and I independent. The cdf of Z
is again a mixture, that is, a convex combination, of the cdfs of X and Y, see (2.5):

$$F_Z(z) = qF_X(z) + (1 - q)F_Y(z) \tag{2.11}$$

For expectations of functions $g(\cdot)$ of Z we get the same mixture of expectations of
$E[g(X)]$ and $E[g(Y)]$, see (2.8):

$$E[g(Z)] = qE[g(X)] + (1 - q)E[g(Y)]. \tag{2.12}$$

It is important to make a distinction between the urn-of-urns model (2.10) leading
to a convex combination of *cdfs*, and a convex combination of *random variables*
$T = qX + (1 - q)Y$. Although (2.12) is valid for $T = Z$ in case $g(z) = z$, the random
variable T does not have (2.11) as its cdf. See also Exercises 2.2.8 and 2.2.9.   ∇

Example 2.2.3 (Insurance against bicycle theft)
We consider an insurance policy against bicycle theft that pays b in case the bicycle
is stolen, upon which event the policy ends. Obviously, the number of payments
is 0 or 1 and the amount is known in advance, just as with life insurance policies.
Assume that the probability of theft is q and let $X = Ib$ denote the claim payment,
where I is a Bernoulli(q) distributed indicator random variable, with $I = 1$ if the
bicycle is stolen, $I = 0$ if not. In analogy to (2.4), we can rewrite X as $X = Ib + (1 -
I)0$. The distribution and the moments of X can be obtained from those of I:

$$\Pr[X = b] = \Pr[I = 1] = q; \qquad \Pr[X = 0] = \Pr[I = 0] = 1 - q;$$
$$E[X] = bE[I] = bq; \qquad \text{Var}[X] = b^2\text{Var}[I] = b^2q(1 - q). \tag{2.13}$$

Now suppose that only half the amount is paid out in case the bicycle was not locked.
Some bicycle theft insurance policies have a restriction like this. Insurers check this
by requiring that all the original keys have to be handed over in the event of a
claim. Then, $X = IB$, where B represents the random payment. Assuming that the
probabilities of a claim $X = 400$ and $X = 200$ are 0.05 and 0.15, we get

$$\Pr[I = 1, B = 400] = 0.05; \quad \Pr[I = 1, B = 200] = 0.15. \tag{2.14}$$

Hence, $\Pr[I = 1] = 0.2$ and consequently $\Pr[I = 0] = 0.8$. Also,

$$\Pr[B = 400 \,|\, I = 1] = \frac{\Pr[B = 400, I = 1]}{\Pr[I = 1]} = 0.25. \qquad (2.15)$$

This represents the conditional probability that the bicycle was locked given the fact that it was stolen.   ▽

Example 2.2.4 (Exponential claim size, if there is a claim)
Suppose that risk X is distributed as follows:

1. $\Pr[X = 0] = \frac{1}{2}$;
2. $\Pr[X \in [x, x + dx)] = \frac{1}{2}\beta e^{-\beta x} dx$ for $\beta = 0.1, x > 0$,

where dx denotes a positive infinitesimal number. What is the expected value of X, and what is the maximum premium for X that someone with an exponential utility function with risk aversion $\alpha = 0.01$ is willing to pay?

The random variable X is not continuous, because the cdf of X has a step in 0. It is also not a discrete random variable, since the cdf is not a step function; its derivative, which in terms of infinitesimal numbers equals $\Pr[x \le X < x + dx]/dx$, is positive for $x > 0$. We can calculate the expectations of functions of X by dealing with the steps in the cdf separately, see (2.9). This leads to

$$E[X] = \int_{-\infty}^{\infty} x \, dF_X(x) = 0 \, dF_X(0) + \int_0^{\infty} x F_X'(x) \, dx = \frac{1}{2} \int_0^{\infty} x \beta e^{-\beta x} dx = 5. \quad (2.16)$$

If the utility function of the insured is exponential with parameter $\alpha = 0.01$, then (1.21) yields for the maximum premium P^+:

$$
\begin{aligned}
P^+ &= \frac{1}{\alpha} \log(m_X(\alpha)) = \frac{1}{\alpha} \log\left(e^0 dF_X(0) + \frac{1}{2} \int_0^{\infty} e^{\alpha x} \beta e^{-\beta x} dx \right) \\
&= \frac{1}{\alpha} \log\left(\frac{1}{2} + \frac{1}{2}\frac{\beta}{\beta - \alpha} \right) = 100 \log\left(\frac{19}{18} \right) \approx 5.4.
\end{aligned}
\qquad (2.17)
$$

This same result can of course be obtained by writing X as in (2.10).   ▽

Example 2.2.5 (Liability insurance with a maximum coverage)
Consider an insurance policy against a liability loss S. We want to determine the expected value, the variance and the distribution function of the payment X on this policy, when there is a deductible of 100 and a maximum payment of 1000. In other words, if $S \le 100$ then $X = 0$, if $S \ge 1100$ then $X = 1000$, otherwise $X = S - 100$. The probability of a positive claim ($S > 100$) is 10% and the probability of a large loss ($S \ge 1100$) is 2%. Given $100 < S < 1100$, S has a uniform$(100, 1100)$ distribution. Again, we write $X = IB$ where I denotes the number of payments, 0 or 1, and B represents the amount paid, if any. Therefore,

$$
\begin{aligned}
&\Pr[B = 1000 \,|\, I = 1] = 0.2; \\
&\Pr[B \in (x, x + dx) \,|\, I = 1] = c \, dx \quad \text{for } 0 < x < 1000.
\end{aligned}
\qquad (2.18)
$$

Integrating the latter probability over $x \in (0, 1000)$ yields 0.8, so $c = 0.0008$.

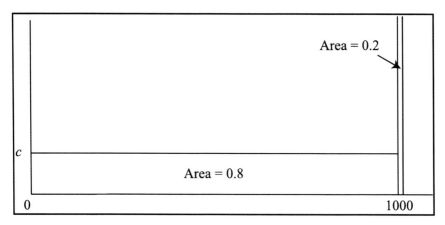

Fig. 2.1 'Probability density function' of B given $I = 1$ in Example 2.2.5.

The conditional distribution function of B, given $I = 1$, is neither discrete, nor continuous. In Figure 2.1 we attempt to depict a pdf by representing the probability mass at 1000 by a bar of infinitesimal width and infinite height such that the area equals 0.2. In actual fact we have plotted $f(\cdot)$, where $f(x) = 0.0008$ on $(0, 1000)$ and $f(x) = 0.2/\varepsilon$ on $(1000, 1000 + \varepsilon)$ with $\varepsilon > 0$ very small.

For the cdf F of X we have

$$
\begin{aligned}
F(x) = \Pr[X \le x] &= \Pr[IB \le x] \\
&= \Pr[IB \le x, I = 0] + \Pr[IB \le x, I = 1] \\
&= \Pr[IB \le x \mid I = 0] \Pr[I = 0] + \Pr[IB \le x \mid I = 1] \Pr[I = 1]
\end{aligned}
\tag{2.19}
$$

which yields

$$
F(x) = \begin{cases}
0 \times 0.9 + 0 \times 0.1 = 0 & \text{for } x < 0 \\
1 \times 0.9 + 1 \times 0.1 = 1 & \text{for } x \ge 1000 \\
1 \times 0.9 + cx \times 0.1 & \text{for } 0 \le x < 1000.
\end{cases}
\tag{2.20}
$$

A graph of the cdf F is shown in Figure 2.2. For the differential ('density') of F, we have

$$
dF(x) = \begin{cases}
0.9 & \text{for } x = 0 \\
0.02 & \text{for } x = 1000 \\
0 & \text{for } x < 0 \text{ or } x > 1000 \\
0.00008\, dx & \text{for } 0 < x < 1000.
\end{cases}
\tag{2.21}
$$

The moments of X can be calculated by using this differential. ∇

The variance of risks of the form IB can be calculated through the conditional distribution of B, given I, by use of the well-known *variance decomposition rule*, see (2.7), which is also known as the law of total variance:

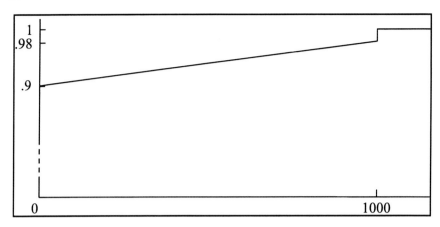

Fig. 2.2 Cumulative distribution function F of X in Example 2.2.5.

$$\text{Var}[W] = \text{Var}[\text{E}[W \,|\, V]] + \text{E}[\text{Var}[W \,|\, V]]. \tag{2.22}$$

In statistics, the first term is the component of the variance of W, not explained by knowledge of V; the second is the explained component of the variance. The conditional distribution of $B \,|\, I = 0$ is irrelevant, so for convenience, we let it be equal to the one of $B \,|\, I = 1$, meaning that we take I and B to be independent. Then, letting $q = \text{Pr}[I = 1]$, $\mu = \text{E}[B]$ and $\sigma^2 = \text{Var}[B]$, we have $\text{E}[X \,|\, I = 1] = \mu$ and $\text{E}[X \,|\, I = 0] = 0$. Therefore, $\text{E}[X \,|\, I = i] = \mu i$ for both values $i = 0, 1$, and analogously, $\text{Var}[X \,|\, I = i] = \sigma^2 i$. Hence,

$$\text{E}[X \,|\, I] \equiv \mu I \quad \text{and} \quad \text{Var}[X \,|\, I] \equiv \sigma^2 I, \tag{2.23}$$

from which it follows that

$$
\begin{aligned}
\text{E}[X] \ &= \text{E}[\text{E}[X \,|\, I]] = \text{E}[\mu I] = \mu q; \\
\text{Var}[X] \ &= \text{Var}[\text{E}[X \,|\, I]] + \text{E}[\text{Var}[X \,|\, I]] = \text{Var}[\mu I] + \text{E}[\sigma^2 I] \\
&= \mu^2 q(1 - q) + \sigma^2 q.
\end{aligned}
\tag{2.24}
$$

Notice that a continuous cdf F is not necessarily absolutely continuous in the sense of (2.3), as is demonstrated by the following example.

Example 2.2.6 ([♠] The Cantor cdf; continuous but not absolutely continuous) Let $X_1, X_2, \ldots$ be an infinite sequence of independent Bernoulli($1/2$) random variables. Define the following random variable:

$$W = \sum_{i=1}^{\infty} \frac{2X_i}{3^i} = \frac{2}{3}X_1 + \frac{1}{3}\sum_{i=1}^{\infty} \frac{2X_{i+1}}{3^i} \tag{2.25}$$

Then the possible values of W are, in the ternary system, $0.d_1 d_2 d_3 \ldots$ with $d_i \in \{0, 2\}$ for all $i = 1, 2, \ldots$, and with $d_i = 2$ occurring if $X_i = 1$. Obviously, all of these values have zero probability as they correspond to *all* X_i having specific outcomes, so F_W is continuous.

Also, all intervals of real numbers in $(0,1)$ having a ternary digit $d_i = 1$ on some place $i = 1,2,\ldots,n$ are not possible values of W, hence F_W is constant on the union B_n of all those intervals. But it is easy to see that the total length of these intervals tends to 1 as $n \to \infty$.

So we have constructed a continuous cdf F_W, known as the *Cantor distribution function*, that is constant except on a set of length 0 (known as the *Cantor set*). The cdf F_W cannot be equal to the integral over its derivative, since this is zero almost everywhere with respect to the Lebesgue measure ('interval length'). So though F_W is continuous, it is not absolutely continuous as in (2.3). ∇

2.3 Convolution

In the individual risk model we are interested in the distribution of the total S of the claims on a number of policies, with

$$S = X_1 + X_2 + \cdots + X_n, \tag{2.26}$$

where X_i, $i = 1,2,\ldots,n$, denotes the payment on policy i. The risks X_i are assumed to be independent random variables. If this assumption is violated for some risks, for example in case of fire insurance policies on different floors of the same building, then these risks could be combined into one term in (2.26).

The operation 'convolution' calculates the distribution function of $X + Y$ from the cdfs of two independent random variables X and Y as follows:

$$
\begin{aligned}
F_{X+Y}(s) &= \Pr[X + Y \le s] \\
&= \int_{-\infty}^{\infty} \Pr[X + Y \le s \mid X = x]\, dF_X(x) \\
&= \int_{-\infty}^{\infty} \Pr[Y \le s - x \mid X = x]\, dF_X(x) \\
&= \int_{-\infty}^{\infty} \Pr[Y \le s - x]\, dF_X(x) \\
&= \int_{-\infty}^{\infty} F_Y(s - x)\, dF_X(x) =: F_X * F_Y(s).
\end{aligned}
\tag{2.27}
$$

The cdf $F_X * F_Y(\cdot)$ is called the convolution of the cdfs $F_X(\cdot)$ and $F_Y(\cdot)$. For the density function we use the same notation. If X and Y are discrete random variables, we find for the cdf of $X + Y$ and the corresponding density

$$F_X * F_Y(s) = \sum_x F_Y(s - x) f_X(x) \quad \text{and} \quad f_X * f_Y(s) = \sum_x f_Y(s - x) f_X(x), \tag{2.28}$$

where the sum is taken over all x with $f_X(x) > 0$. If X and Y are continuous random variables, then

$$F_X * F_Y(s) = \int_{-\infty}^{\infty} F_Y(s-x) f_X(x) \, dx \qquad (2.29)$$

and, taking the derivative under the integral sign to find the density,

$$f_X * f_Y(s) = \int_{-\infty}^{\infty} f_Y(s-x) f_X(x) \, dx. \qquad (2.30)$$

Since $X + Y \equiv Y + X$, the convolution operator $*$ is *commutative*: $F_X * F_Y$ is identical to $F_Y * F_X$. Also, it is *associative*, since for the cdf of $X + Y + Z$, it does not matter in which order we do the convolutions, therefore

$$(F_X * F_Y) * F_Z \equiv F_X * (F_Y * F_Z) \equiv F_X * F_Y * F_Z. \qquad (2.31)$$

For the sum of n independent and identically distributed random variables with marginal cdf F, the cdf is the n-fold convolution power of F, which we write as

$$F * F * \cdots * F =: F^{*n}. \qquad (2.32)$$

Example 2.3.1 (Convolution of two uniform distributions)
Suppose that $X \sim \text{uniform}(0,1)$ and $Y \sim \text{uniform}(0,2)$ are independent. What is the cdf of $X + Y$?

The indicator function of a set A is defined as follows:

$$I_A(x) = \begin{cases} 1 & \text{if } x \in A \\ 0 & \text{if } x \notin A. \end{cases} \qquad (2.33)$$

Indicator functions provide us with a concise notation for functions that are defined differently on some intervals. For all x, the cdf of X can be written as

$$F_X(x) = x I_{[0,1)}(x) + I_{[1,\infty)}(x), \qquad (2.34)$$

while $F_Y'(y) = \frac{1}{2} I_{[0,2)}(y)$ for all y, which leads to the differential

$$dF_Y(y) = \frac{1}{2} I_{[0,2)}(y) \, dy. \qquad (2.35)$$

The convolution formula (2.27), applied to $Y + X$ rather than $X + Y$, then yields

$$F_{Y+X}(s) = \int_{-\infty}^{\infty} F_X(s-y) \, dF_Y(y) = \int_0^2 F_X(s-y) \frac{1}{2} \, dy, \; s \geq 0. \qquad (2.36)$$

The interval of interest is $0 \leq s < 3$. Subdividing it into $[0,1)$, $[1,2)$ and $[2,3)$ yields

$$F_{X+Y}(s) = \left\{ \int_0^s (s-y)\tfrac{1}{2}\,dy \right\} I_{[0,1)}(s)$$

$$+ \left\{ \int_0^{s-1} \tfrac{1}{2}\,dy + \int_{s-1}^s (s-y)\tfrac{1}{2}\,dy \right\} I_{[1,2)}(s) \qquad (2.37)$$

$$+ \left\{ \int_0^{s-1} \tfrac{1}{2}\,dy + \int_{s-1}^2 (s-y)\tfrac{1}{2}\,dy \right\} I_{[2,3)}(s)$$

$$= \tfrac{1}{4}s^2 I_{[0,1)}(s) + \tfrac{1}{4}(2s-1)I_{[1,2)}(s) + [1 - \tfrac{1}{4}(3-s)^2]I_{[2,3)}(s).$$

Notice that $X + Y$ is symmetric around $s = 1.5$. Although this problem could be solved graphically by calculating the probabilities by means of areas, see Exercise 2.3.5, the above derivation provides an excellent illustration that, even in simple cases, convolution can be a laborious process. $\quad \triangledown$

Example 2.3.2 (Convolution of discrete distributions)
Let $f_1(x) = \tfrac{1}{4}, \tfrac{1}{2}, \tfrac{1}{4}$ for $x = 0,1,2$, $f_2(x) = \tfrac{1}{2}, \tfrac{1}{2}$ for $x = 0,2$ and $f_3(x) = \tfrac{1}{4}, \tfrac{1}{2}, \tfrac{1}{4}$ for $x = 0,2,4$. Let f_{1+2} denote the convolution of f_1 and f_2 and let f_{1+2+3} denote the convolution of f_1, f_2 and f_3. To calculate F_{1+2+3}, we need to compute the values as shown in Table 2.1. In the discrete case, too, convolution is clearly a laborious exercise. Note that the more often we have $f_i(x) \neq 0$, the more calculations need to be done. $\quad \triangledown$

Table 2.1 Convolution computations for Example 2.3.2

x	$f_1(x)$	$* f_2(x)$	$= f_{1+2}(x)$	$* f_3(x)$	$= f_{1+2+3}(x)$	$\Rightarrow F_{1+2+3}(x)$
0	1/4	1/2	1/8	1/4	1/32	1/32
1	1/2	0	2/8	0	2/32	3/32
2	1/4	1/2	2/8	1/2	4/32	7/32
3	0	0	2/8	0	6/32	13/32
4	0	0	1/8	1/4	6/32	19/32
5	0	0	0	0	6/32	25/32
6	0	0	0	0	4/32	29/32
7	0	0	0	0	2/32	31/32
8	0	0	0	0	1/32	32/32

Example 2.3.3 (Convolution of iid uniform distributions)
Let X_i, $i = 1,2,\ldots,n$, be independent and identically uniform$(0,1)$ distributed. By using the convolution formula and induction, it can be shown that for all $x > 0$, the pdf of $S = X_1 + \cdots + X_n$ equals

$$f_S(x) = \frac{1}{(n-1)!} \sum_{h=0}^{[x]} \binom{n}{h} (-1)^h (x-h)^{n-1} \qquad (2.38)$$

where $[x]$ denotes the integer part of x. See also Exercise 2.3.4. $\quad \triangledown$

Example 2.3.4 (Convolution of Poisson distributions)
Let $X \sim \text{Poisson}(\lambda)$ and $Y \sim \text{Poisson}(\mu)$ be independent random variables. From (2.28) we have, for $s = 0, 1, 2, \ldots,$

$$f_{X+Y}(s) = \sum_{x=0}^{s} f_Y(s-x) f_X(x) = \frac{e^{-\mu-\lambda}}{s!} \sum_{x=0}^{s} \binom{s}{x} \mu^{s-x} \lambda^x$$

$$= e^{-(\lambda+\mu)} \frac{(\lambda+\mu)^s}{s!}, \tag{2.39}$$

where the last equality is the binomial theorem. Hence, $X + Y$ is $\text{Poisson}(\lambda + \mu)$ distributed. For a different proof, see Exercise 2.4.2. ∇

2.4 Transforms

Determining the distribution of the sum of independent random variables can often be made easier by using transforms of the cdf. The *moment generating function (mgf)* suits our purposes best. For a non-negative random variable X, it is defined as

$$m_X(t) = \text{E}\left[e^{tX}\right], \quad -\infty < t < h, \tag{2.40}$$

for some h. The mgf is going to be used especially in an interval around 0, which requires $h > 0$ to hold. Note that this is the case only for light-tailed risks, of which exponential moments $\text{E}[e^{\varepsilon x}]$ for some $\varepsilon > 0$ exist.

If X and Y are independent, then

$$m_{X+Y}(t) = \text{E}\left[e^{t(X+Y)}\right] = \text{E}\left[e^{tX}\right] \text{E}\left[e^{tY}\right] = m_X(t) m_Y(t). \tag{2.41}$$

So, the convolution of cdfs corresponds to simply multiplying the mgfs. Note that the mgf-transform is one-to-one, so every cdf has exactly one mgf. Also, it is continuous, in the sense that the mgf of the limit of a series of cdfs is the limit of the mgfs. See Exercises 2.4.12 and 2.4.13.

For random variables with a heavy tail, such as the Pareto distributions, the mgf does not exist. The *characteristic function*, however, always exists. It is defined as:

$$\phi_X(t) = \text{E}\left[e^{itX}\right] = \text{E}\left[\cos(tX) + i\sin(tX)\right], \quad -\infty < t < \infty. \tag{2.42}$$

A disadvantage of the characteristic function is the need to work with complex numbers, although applying the same function formula derived for real t to imaginary t as well produces the correct results most of the time, resulting for example in the $\text{N}(0, 2)$ distribution with mgf $\exp(t^2)$ having $\exp((it)^2) = \exp(-t^2)$ as its characteristic function.

As their name indicates, moment generating functions can be used to generate moments of random variables. The usual series expansion of e^x yields

$$m_X(t) = \mathrm{E}[e^{tX}] = \sum_{k=0}^{\infty} \frac{\mathrm{E}[X^k]t^k}{k!}, \tag{2.43}$$

so the k-th moment of X equals

$$\mathrm{E}[X^k] = \frac{\mathrm{d}^k}{\mathrm{d}t^k} m_X(t) \bigg|_{t=0}. \tag{2.44}$$

Moments can also be generated from the characteristic function in similar fashion.

The *probability generating function (pgf)* is reserved for random variables with natural numbers as values:

$$g_X(t) = \mathrm{E}[t^X] = \sum_{k=0}^{\infty} t^k \Pr[X = k]. \tag{2.45}$$

So, the probabilities $\Pr[X = k]$ in (2.45) are just the coefficients in the series expansion of the pgf. The series (2.45) converges absolutely if $|t| \le 1$.

The *cumulant generating function (cgf)* is convenient for calculating the third central moment; it is defined as:

$$\kappa_X(t) = \log m_X(t). \tag{2.46}$$

Differentiating (2.46) three times and setting $t = 0$, one sees that the coefficients of $t^k/k!$ for $k = 1, 2, 3$ are $\mathrm{E}[X]$, $\mathrm{Var}[X]$ and $\mathrm{E}[(X - \mathrm{E}[X])^3]$. The quantities generated this way are the *cumulants* of X, and they are denoted by κ_k, $k = 1, 2, \ldots$ One may also proceed as follows: let μ_k denote $\mathrm{E}[X^k]$ and let, as usual, the 'big O notation' $O(t^k)$ denote 'terms of order t to the power k or higher'. Then

$$m_X(t) = 1 + \mu_1 t + \tfrac{1}{2}\mu_2 t^2 + \tfrac{1}{6}\mu_3 t^3 + O(t^4), \tag{2.47}$$

which, using $\log(1+z) = z - \tfrac{1}{2}z^2 + \tfrac{1}{3}z^3 + O(z^4)$, yields

$$\begin{aligned}
\log m_X(t) &= \log\left(1 + \mu_1 t + \tfrac{1}{2}\mu_2 t^2 + \tfrac{1}{6}\mu_3 t^3 + O(t^4)\right) \\
&= \mu_1 t + \tfrac{1}{2}\mu_2 t^2 + \tfrac{1}{6}\mu_3 t^3 + O(t^4) \\
&\quad - \tfrac{1}{2}\left\{\mu_1^2 t^2 + \mu_1\mu_2 t^3 + O(t^4)\right\} \\
&\quad + \tfrac{1}{3}\left\{\mu_1^3 t^3 + O(t^4)\right\} + O(t^4) \\
&= \mu_1 t + \tfrac{1}{2}(\mu_2 - \mu_1^2)t^2 + \tfrac{1}{6}(\mu_3 - 3\mu_1\mu_2 + 2\mu_1^3)t^3 + O(t^4) \\
&= \mathrm{E}[X]t + \mathrm{Var}[X]\tfrac{1}{2}t^2 + \mathrm{E}[(X - \mathrm{E}[X])^3]\tfrac{1}{6}t^3 + O(t^4).
\end{aligned} \tag{2.48}$$

The *skewness* of a random variable X is defined as the following dimension-free quantity:

$$\gamma_X = \frac{\kappa_3}{\sigma^3} = \frac{\mathrm{E}[(X - \mu)^3]}{\sigma^3}, \tag{2.49}$$

with $\mu = E[X]$ and $\sigma^2 = \text{Var}[X]$. If $\gamma_X > 0$, large values of $X - \mu$ are likely to occur, hence the (right) tail of the cdf is heavy. A negative skewness $\gamma_X < 0$ indicates a heavy left tail. If X is symmetric then $\gamma_X = 0$, but having zero skewness is not sufficient for symmetry. For some counterexamples, see the exercises.

The cumulant generating function, the probability generating function, the characteristic function and the moment generating function are related by

$$\kappa_X(t) = \log m_X(t); \quad g_X(t) = m_X(\log t); \quad \phi_X(t) = m_X(it). \tag{2.50}$$

In Exercise 2.4.14 the reader is asked to examine the last of these equalities. Often the mgf can be extended to the whole complex plane in a natural way. The mgf operates on the real axis, the characteristic function on the imaginary axis.

2.5 Approximations

A well-known method to approximate a cdf is based on the Central Limit Theorem (CLT). We study this approximation as well as two more accurate ones that involve three moments rather than just two.

2.5.1 Normal approximation

Next to the Law of Large Numbers, the Central Limit Theorem is the most important theorem in statistics. It states that by adding up a large number of independent random variables, we get a normally distributed random variable in the limit. In its simplest form, the Central Limit Theorem (CLT) is as follows:

Theorem 2.5.1 (Central Limit Theorem)
If $X_1, X_2, \ldots, X_n$ are independent and identically distributed random variables with mean μ and variance $\sigma^2 < \infty$, then

$$\lim_{n \to \infty} \Pr\left[\sum_{i=1}^{n} X_i \le n\mu + x\sigma\sqrt{n}\right] = \Phi(x). \tag{2.51}$$

Proof. We restrict ourselves to proving the convergence of the sequence of cgfs. Let $S^* = (X_1 + \cdots + X_n - n\mu)/\sigma\sqrt{n}$, then for $n \to \infty$ and for all t:

$$\begin{aligned}
\log m_{S^*}(t) &= -\frac{\sqrt{n}\mu t}{\sigma} + n\left\{\log m_X\left(\frac{t}{\sigma\sqrt{n}}\right)\right\} \\
&= -\frac{\sqrt{n}\mu t}{\sigma} + n\left\{\mu\left(\frac{t}{\sigma\sqrt{n}}\right) + \tfrac{1}{2}\sigma^2\left(\frac{t}{\sigma\sqrt{n}}\right)^2 + O\left(\left(\frac{1}{\sqrt{n}}\right)^3\right)\right\} \quad (2.52) \\
&= \tfrac{1}{2}t^2 + O\left(\frac{1}{\sqrt{n}}\right),
\end{aligned}$$

which converges to the cgf of the $N(0,1)$ distribution, with mgf $\exp(\frac{1}{2}t^2)$. As a consequence, the cdf of S^* converges to the standard normal cdf Φ. ∇

As a result, if the summands are *independent* and have *finite* variance, we can approximate the cdf of $S = X_1 + \cdots + X_n$ by

$$F_S(s) \approx \Phi\left(s; \sum_{i=1}^{n} E[X_i], \sum_{i=1}^{n} \text{Var}[X_i]\right). \tag{2.53}$$

This approximation can safely be used if n is 'large'. But it is difficult to define 'large', as is shown in the following examples.

Example 2.5.2 (Generating approximately normal random deviates fast)
If pseudo-random numbers can be generated fast (using bit-manipulations), but computing logarithms and the inverse normal cdf takes a lot of time, approximately $N(0,1)$ distributed pseudo-random drawings numbers can conveniently be produced by adding up twelve uniform$(0,1)$ numbers and subtracting 6 from their sum. This technique is based on the CLT with $n = 12$. Comparing this cdf with the normal cdf, using (2.38), yields a maximum difference of 0.002. Hence, the CLT performs quite well in this case. See also Exercise 2.4.5. ∇

Example 2.5.3 (Illustrating the various approximations)
Suppose that $n = 1000$ young men take out a life insurance policy for a period of one year. The probability of dying within this year is 0.001 for everyone and the payment for every death is 1. We want to calculate the probability that the total payment is at least 4. This total payment is binomial$(1000, 0.001)$ distributed and since $n = 1000$ is large and $p = 0.001$ is small, we will approximate this probability by a Poisson(np) distribution. Calculating the probability at $3 + \frac{1}{2}$ instead of at 4, applying a continuity correction needed later on, we find

$$\Pr[S \geq 3.5] = 1 - e^{-1} - e^{-1} - \tfrac{1}{2}e^{-1} - \tfrac{1}{6}e^{-1} = 0.01899. \tag{2.54}$$

Note that the exact binomial probability is 0.01893. Although n is much larger than in the previous example, the CLT gives a poor approximation: with $\mu = E[S] = 1$ and $\sigma^2 = \text{Var}[S] = 1$, we find

$$\Pr[S \geq 3.5] = \Pr\left[\frac{S - \mu}{\sigma} \geq \frac{3.5 - \mu}{\sigma}\right] \approx 1 - \Phi(2.5) = 0.0062. \tag{2.55}$$

The CLT approximation is not very good because of the extreme skewness of the terms X_i and the resulting skewness of S, which is $\gamma_S = 1$. In the previous example, we started from symmetric terms, leading to a higher order of convergence, as can be seen from derivation (2.52). ∇

As an alternative for the CLT, we give two more refined approximations: the translated gamma approximation and the normal power approximation (NP). In numerical examples, they turn out to be much more accurate than the CLT approximation. As regards the quality of the approximations, there is not much to choose between the two. Their inaccuracies are minor compared with the errors that result from the lack of precision in the estimates of the first three moments that are involved.

2.5.2 Translated gamma approximation

Most total claim distributions are skewed to the right (skewness $\gamma > 0$), have a non-negative support and are unimodal. So they have roughly the shape of a gamma distribution. To gain more flexibility, apart from the usual parameters α and β we allow a shift over a distance x_0. Hence, we approximate the cdf of S by the cdf of $Z + x_0$, where $Z \sim \text{gamma}(\alpha, \beta)$ (see Table A). We choose α, β and x_0 in such a way that the approximating random variable has the same first three moments as S.

The translated gamma approximation can then be formulated as follows:

$$F_S(s) \approx G(s - x_0; \alpha, \beta),$$
$$\text{where } G(x; \alpha, \beta) = \frac{1}{\Gamma(\alpha)} \int_0^x y^{\alpha-1} \beta^\alpha e^{-\beta y} \mathrm{d}y, \quad x \geq 0. \tag{2.56}$$

Here $G(x; \alpha, \beta)$ is the gamma cdf. We choose α, β and x_0 such that the first three moments are the same, hence $\mu = x_0 + \frac{\alpha}{\beta}$, $\sigma^2 = \frac{\alpha}{\beta^2}$ and $\gamma = \frac{2}{\sqrt{\alpha}}$ (see Table A), so

$$\alpha = \frac{4}{\gamma^2}, \quad \beta = \frac{2}{\gamma \sigma} \quad \text{and} \quad x_0 = \mu - \frac{2\sigma}{\gamma}. \tag{2.57}$$

It is required that the skewness γ is strictly positive. In the limit $\gamma \downarrow 0$, the normal approximation appears. Note that if the first three moments of the cdf $F(\cdot)$ are equal to those of $G(\cdot)$, by partial integration it can be shown that the same holds for $\int_0^\infty x^j [1 - F(x)] \mathrm{d}x$, $j = 0, 1, 2$. This leaves little room for these cdfs to be very different from each other.

Example 2.5.4 (Illustrating the various approximations, continued)
If $S \sim \text{Poisson}(1)$, we have $\mu = \sigma = \gamma = 1$, and (2.57) yields $\alpha = 4$, $\beta = 2$ and $x_0 = -1$. Hence, $\Pr[S \geq 3.5] \approx 1 - G(3.5 - (-1); 4, 2) = 0.0212$. This value is much closer to the exact value than the CLT approximation. ∇

The translated gamma approximation leads to quite simple formulas to approximate the moments of a stop-loss claim $(S - d)_+$ or of the retained loss $S - (S - d)_+$. To evaluate the gamma cdf is easy in R, and in spreadsheet programs the gamma distribution is also included, although the accuracy sometimes leaves much to be desired. Note that in many applications, for example MS Excel, the parameter β should be replaced by $1/\beta$. In R, specify $\beta = 2$ by using `rate=2`, or by `scale=1/2`.

Example 2.5.5 (Translated gamma approximation)
A total claim amount S has expected value 10000, standard deviation 1000 and skewness 1. From (2.57) we have $\alpha = 4$, $\beta = 0.002$ and $x_0 = 8000$. Hence,

$$\Pr[S > 13000] \approx 1 - G(13000 - 8000; 4, 0.002) = 0.010. \qquad (2.58)$$

The regular CLT approximation is much smaller: 0.0013. Using the inverse of the gamma distribution function, the value-at-risk on a 95% level is found by reversing the computation (2.58), resulting in 11875.   ▽

2.5.3 NP approximation

Another approximation that uses three moments of the approximated random variable is the Normal Power approximation. It goes as follows.
  If $E[S] = \mu$, $Var[S] = \sigma^2$ and $\gamma_S = \gamma$, then, for $s \geq 1$,

$$\Pr\left[\frac{S - \mu}{\sigma} \leq s + \frac{\gamma}{6}(s^2 - 1)\right] \approx \Phi(s) \qquad (2.59)$$

or, equivalently, for $x \geq 1$,

$$\Pr\left[\frac{S - \mu}{\sigma} \leq x\right] \approx \Phi\left(\sqrt{\frac{9}{\gamma^2} + \frac{6x}{\gamma} + 1} - \frac{3}{\gamma}\right). \qquad (2.60)$$

The second formula can be used to approximate the cdf of S, the first produces approximate quantiles. If $s < 1$ (or $x < 1$), the correction term is negative, which implies that the CLT gives more conservative results.

Example 2.5.6 (Illustrating the various approximations, continued)
If $S \sim \text{Poisson}(1)$, then the NP approximation yields $\Pr[S \geq 3.5] \approx 1 - \Phi(2) = 0.0228$. Again, this is a better result than the CLT approximation.
  The R-calls needed to produce all the numerical values are the following:

```
x <- 3.5; mu <- 1; sig <- 1; gam <- 1; z <- (x-mu)/sig
1-pbinom(x, 1000, 0.001)                              ##   0.01892683
1-ppois(x,1)                                          ##   0.01898816
1-pnorm(z)                                            ##   0.00620967
1-pnorm(sqrt(9/gam^2 + 6*z/gam + 1) - 3/gam)          ##   0.02275013
1-pgamma(x-(mu-2*sig/gam), 4/gam^2, 2/gam/sig) ##   0.02122649
```

Equations (2.53), (2.60) and (2.56)–(2.57) were used.   ▽

Example 2.5.7 (Recalculating Example 2.5.5 by the NP approximation)
We apply (2.59) to determine the capital that covers loss S with probability 95%:

$$\Pr\left[\frac{S - \mu}{\sigma} \leq s + \frac{\gamma}{6}(s^2 - 1)\right] \approx \Phi(s) = 0.95 \quad \text{if } s = 1.645, \qquad (2.61)$$

hence for the desired 95% quantile of S we find

$$E[S] + \sigma_S\left(1.645 + \frac{\gamma}{6}(1.645^2 - 1)\right) = E[S] + 1.929\sigma_S = 11929. \qquad (2.62)$$

To determine the probability that capital 13000 will be insufficient to cover the losses S, we apply (2.60) with $\mu = 10000$, $\sigma = 1000$ and $\gamma = 1$:

$$\Pr[S > 13000] = \Pr\left[\frac{S - \mu}{\sigma} > 3\right] \approx 1 - \Phi(\sqrt{9 + 6 \times 3 + 1} - 3)$$

$$= 1 - \Phi(2.29) = 0.011.$$

(2.63)

Note that the translated gamma approximation gave 0.010, against only 0.0013 for the CLT.   ▽

Remark 2.5.8 (Justifying the NP approximation)
For $U \sim N(0,1)$, consider the random variable $Y = U + \frac{\gamma}{6}(U^2 - 1)$. It is easy to verify that (see Exercise 2.5.21), writing $w(x) = \sqrt{\left(\frac{9}{\gamma^2} + \frac{6x}{\gamma} + 1\right)_+}$, we have

$$F_Y(x) = \Phi\left(+w(x) - \frac{3}{\gamma}\right) - \Phi\left(-w(x) - \frac{3}{\gamma}\right) \approx \Phi\left(w(x) - \frac{3}{\gamma}\right).$$

(2.64)

The term $\Phi(-w(x) - 3/\gamma)$ accounts for small U leading to large Y. It is generally negligible, and vanishes as $\gamma \downarrow 0$.

Also, using $E[U^6] = 15$, $E[U^4] = 3$ and $E[U^2] = 1$, for small γ one can prove

$$F[Y] = 0; \quad E[Y^2] = 1 + O(\gamma^2); \quad E[Y^3] = \gamma(1 + O(\gamma^2)).$$

(2.65)

Therefore, the first three moments of $\frac{S-\mu}{\sigma}$ and Y as defined above are alike. This, with (2.64), justifies the use of formula (2.60) to approximate the cdf of $\frac{S-\mu}{\sigma}$.   ▽

Remark 2.5.9 ([♠] Deriving NP using the Edgeworth expansion)
Formula (2.59) can be derived by the use of a certain expansion for the cdf, though not in a mathematically rigorous way. Define $Z = (S - E[S])/\sqrt{\text{Var}[S]}$, and let $\gamma = E[Z^3]$ be the skewness of S (and Z). For the cgf of Z we have

$$\log m_Z(t) = \tfrac{1}{2}t^2 + \tfrac{1}{6}\gamma t^3 + \dots,$$

(2.66)

hence

$$m_Z(t) = e^{t^2/2} \cdot \exp\left\{\tfrac{1}{6}\gamma t^3 + \dots\right\} = e^{t^2/2} \cdot \left(1 + \tfrac{1}{6}\gamma t^3 + \dots\right).$$

(2.67)

The 'mgf' (generalized to functions that are not a density) of $\varphi^{(3)}(x)$, with $\varphi(x)$ the $N(0,1)$ density, can be found by partial integration:

$$\int_{-\infty}^{\infty} e^{tx} \varphi^{(3)}(x)\,dx = e^{tx} \varphi^{(2)}(x)\Big|_{-\infty}^{\infty} - \int_{-\infty}^{\infty} t e^{tx} \varphi^{(2)}(x)\,dx$$

$$= 0 - 0 + \int_{-\infty}^{\infty} t^2 e^{tx} \varphi^{(1)}(x)\,dx$$

(2.68)

$$= 0 - 0 + 0 - \int_{-\infty}^{\infty} t^3 e^{tx} \varphi(x)\,dx = -t^3 e^{t^2/2}.$$

Therefore we recognize the cdf corresponding to mgf (2.67) as:

$$F_Z(x) = \Phi(x) - \tfrac{1}{6}\gamma\Phi^{(3)}(x) + \ldots \tag{2.69}$$

Formula (2.69) is called the *Edgeworth expansion* for F_Z; leaving out the dots gives an *Edgeworth approximation* for it. There is no guarantee that the latter is an increasing function. To derive the NP approximation formula (2.59) from it, we try to find a correction $\delta = \delta(s)$ to the argument s such that

$$F_Z(s + \delta) \approx \Phi(s). \tag{2.70}$$

That means that we have to find a zero for the auxiliary function $g(\delta)$ defined by

$$g(\delta) = \Phi(s) - \left\{\Phi(s + \delta) - \tfrac{1}{6}\gamma\Phi^{(3)}(s + \delta)\right\}. \tag{2.71}$$

Using a Taylor expansion $g(\delta) \approx g(0) + \delta g'(0)$ we may conclude that $g(\delta) = 0$ for $\delta \approx -g(0)/g'(0)$, so

$$\delta \approx \frac{-\tfrac{1}{6}\gamma\Phi^{(3)}(s)}{-\Phi'(s) + \tfrac{1}{6}\gamma\Phi^{(4)}(s)} = \frac{-\tfrac{1}{6}\gamma(s^2 - 1)\varphi(s)}{\left(-1 + \tfrac{1}{6}\gamma(-s^3 + 3s)\right)\varphi(s)}. \tag{2.72}$$

Since the skewness γ is of order $\lambda^{-1/2}$, see for example (2.48), therefore small for large portfolios, we drop the term with γ in the denominator of (2.72), leading to

$$F_Z(s + \delta) \approx \Phi(s) \qquad \text{when} \qquad \delta = \tfrac{1}{6}\gamma(s^2 - 1). \tag{2.73}$$

This is precisely the NP approximation (2.59) given earlier.

The dots in formula (2.69) denote the inverse mgf-transform of the dots in (2.67). It is not possible to show that the terms replaced by dots in this formula are small, let alone their absolute sum. So it is an exaggeration to say that the approximations obtained this way, dropping terms of a possibly divergent series and then using an approximate inversion, are justified by theoretical arguments. ∇

2.6 Application: optimal reinsurance

An insurer is looking for an optimal reinsurance for a portfolio consisting of 20 000 one-year life insurance policies that are grouped as follows:

Insured amount b_k	Number of policies n_k
1	10 000
2	5 000
3	5 000

The probability of dying within one year is $q_k = 0.01$ for each insured, and the policies are independent. The insurer wants to optimize the probability of being able to meet his financial obligations by choosing the best retention, which is the maximum payment per policy. The remaining part of a claim is paid by the reinsurer. For example, if the retention is 1.6 and someone with insured amount 2 dies, then the insurer pays 1.6, the reinsurer pays 0.4. After collecting the premiums, the insurer holds a capital B from which he has to pay the claims and the reinsurance premium. This premium is assumed to be 120% of the net premium.

First, we set the retention equal to 2. From the point of view of the insurer, the policies are then distributed as follows:

Insured amount b_k	Number of policies n_k
1	10 000
2	10 000

The expected value and the variance of the insurer's total claim amount S are equal to

$$
\begin{aligned}
\mathrm{E}[S] &= n_1 b_1 q_1 + n_2 b_2 q_2 \\
&= 10000 \times 1 \times 0.01 + 10000 \times 2 \times 0.01 = 300, \\
\mathrm{Var}[S] &= n_1 b_1^2 q_1 (1-q_1) + n_2 b_2^2 q_2 (1-q_2) \\
&= 10000 \times 1 \times 0.01 \times 0.99 + 10000 \times 4 \times 0.01 \times 0.99 = 495.
\end{aligned}
\tag{2.74}
$$

By applying the CLT, we get for the probability that the costs S plus the reinsurance premium $1.2 \times 0.01 \times 5000 \times 1 = 60$ exceed the available capital B:

$$
\Pr[S + 60 > B] = \Pr\left[\frac{S - \mathrm{E}[S]}{\sigma_S} > \frac{B - 360}{\sqrt{495}}\right] \approx 1 - \Phi\left(\frac{B - 360}{\sqrt{495}}\right).
\tag{2.75}
$$

We leave it to the reader to determine this same probability for retentions between 2 and 3, as well as to determine which retention for a given B leads to the largest probability of survival. See the exercises with this section.

2.7 Exercises

Section 2.2

1. Determine the expected value and the variance of $X = IB$ if the claim probability equals 0.1. First, assume that B equals 5 with probability 1. Then, let $B \sim \text{uniform}(0, 10)$.

2. Throw a true die and let X denote the outcome. Then, toss a coin X times. Let Y denote the number of heads obtained. What are the expected value and the variance of Y?

3. In Example 2.2.4, plot the cdf of X. Also determine, with the help of the obtained differential, the premium the insured is willing to pay for being insured against an inflated loss $1.1X$. Do the same by writing $X = IB$. Has the zero utility premium followed inflation exactly?

4. Calculate $E[X]$, $\text{Var}[X]$ and the moment generating function $m_X(t)$ in Example 2.2.5 with the help of the differential. Also plot the 'density'.

5. If $X = IB$, what is $m_X(t)$?

6. Consider the following cdf F: $F(x) = \begin{cases} 0 & \text{for } x < 2, \\ \frac{x}{4} & \text{for } 2 \le x < 4, \\ 1 & \text{for } 4 \le x. \end{cases}$

   Determine independent random variables I, X and Y such that $Z = IX + (1 - I)Y$ has cdf F, $I \sim$ Bernoulli, X is a discrete and Y a continuous random variable.

7. The differential of cdf F is $dF(x) = \begin{cases} dx/3 & \text{for } 0 < x < 1 \text{ and } 2 < x < 3, \\ \frac{1}{6} & \text{for } x \in \{1, 2\}, \\ 0 & \text{elsewhere.} \end{cases}$

   Find a discrete cdf G, a continuous cdf H and a real constant c with the property that $F(x) = cG(x) + (1 - c)H(x)$ for all x.

8. Suppose that $T = qX + (1 - q)Y$ and $Z = IX + (1 - I)Y$ with $I \sim$ Bernoulli(q) and I, X and Y independent. Compare $E[T^k]$ with $E[Z^k]$, $k = 1, 2$.

9. In the previous exercise, assume additionally that X and Y are independent $N(0, 1)$. What distributions do T and Z have?

10. [♠] In Example 2.2.6, show that $E[W] = \frac{1}{2}$ and $\text{Var}[W] = \frac{1}{8}$.
    Also show that $m_W(t) = e^{t/2} \prod_{i=1}^{\infty} \cosh(t/3^i)$. Recall that $\cosh(t) = (e^t + e^{-t})/2$.

Section 2.3

1. Calculate $\Pr[S = s]$ for $s = 0, 1, \ldots, 6$ when $S = X_1 + 2X_2 + 3X_3$ and $X_j \sim$ Poisson(j).

2. Determine the number of multiplications of non-zero numbers that are needed for the calculation of all probabilities $f_{1+2+3}(x)$ in Example 2.3.2. How many multiplications are needed to calculate $F_{1+\cdots+n}(x)$, $x = 0, \ldots, 4n - 4$ if $f_k = f_3$ for $k = 4, \ldots, n$?

3. Prove by convolution that the sum of two independent normal random variables, see Table A, has a normal distribution.

4. [♠] Verify the expression (2.38) in Example 2.3.3 for $n = 1, 2, 3$ by using convolution. Determine $F_S(x)$ for these values of n. Using induction, verify (2.38) for arbitrary n.

5. Assume that $X \sim$ uniform($0, 3$) and $Y \sim$ uniform($-1, 1$). Calculate $F_{X+Y}(z)$ graphically by using the area of the sets $\{(x, y) \mid x + y \le z, x \in (0, 3) \text{ and } y \in (-1, 1)\}$.

Section 2.4

1. Determine the cdf of $S = X_1 + X_2$ where the X_k are independent and exponential(k) distributed. Do this both by convolution and by calculating the mgf, identifying the corresponding density using the method of partial fractions.

2. Same as Example 2.3.4, but now by making use of the mgfs.

3. What is the fourth cumulant κ_4 in terms of the central moments?

4. Prove that cumulants actually cumulate in the following sense: if X and Y are independent, then the kth cumulant of $X + Y$ equals the sum of the kth cumulants of X and Y.

5. Prove that the sum of twelve independent uniform$(0,1)$ random variables has variance 1 and expected value 6. Determine κ_3 and κ_4.
   Plot the difference between the cdf of this random variable and the N$(6,1)$ cdf, using the expression for $F_S(x)$ found in Exercise 2.3.4.

6. Determine the skewness of a Poisson(μ) distribution.

7. Determine the skewness of a gamma(α, β) distribution.

8. If S is symmetric, then $\gamma_S = 0$. Prove this, but also, for $S = X_1 + X_2 + X_3$ with $X_1 \sim$ Bernoulli(0.4), $X_2 \sim$ Bernoulli(0.7) and $X_3 \sim$ Bernoulli(p), all independent, find a value of p such that S has skewness $\gamma_S = 0$, and verify that S is not symmetric.

9. Determine the skewness of a risk of the form Ib where $I \sim$ Bernoulli(q) and b is a fixed amount. For which values of q and b is the skewness equal to zero, and for which of these values is I actually symmetric?

10. Determine the pgf of the binomial, the Poisson and the negative binomial distribution, see Table A.

11. Determine the cgf and the cumulants of the following distributions: Poisson, binomial, normal and gamma.

12. Show that X and Y are equal in distribution if they have the same support $\{0, 1, \ldots, n\}$ and the same pgf. If $X_1, X_2, \ldots$ are risks, again with range $\{0, 1, \ldots, n\}$, such that the pgfs of X_i converge to the pgf of Y for each argument t when $i \to \infty$, verify that also $\Pr[X_i = x] \to \Pr[Y = x]$ for all x.

13. Show that X and Y are equal in distribution if they have the same support $\{0, \delta, 2\delta, \ldots, n\delta\}$ for some $\delta > 0$ and moreover, they have the same mgf.

14. [♠] Examine the equality $\phi_X(t) = m_X(it)$ from (2.50), for the special case that $X \sim$ exponential(1). Show that the characteristic function is real-valued if X is symmetric around 0.

15. Show that the skewness of $Z = X + 2Y$ is 0 if $X \sim$ binomial$(8, p)$ and $Y \sim$ Bernoulli$(1 - p)$. For which values of p is Z symmetric?

16. For which values of δ is the skewness of $X - \delta Y$ equal to 0, if $X \sim$ gamma$(2, 1)$ and $Y \sim$ exponential(1)?

17. Can the pgf of a random variable be used to generate moments? Can the mgf of an integer-valued random variable be used to generate probabilities?

Section 2.5

1. What happens if we replace the argument 3.5 in Example 2.5.3 by $3 - 0, 3 + 0, 4 - 0$ and $4 + 0$? Is a correction for continuity needed here?

2. Prove that both versions of the NP approximation are equivalent.

3. If $Y \sim$ gamma(α, β) and $\gamma_Y = \frac{2}{\sqrt{\alpha}} \leq 4$, then $\sqrt{4\beta Y} - \sqrt{4\alpha - 1} \overset{\approx}{\sim} N(0,1)$. See ex. 2.5.14 for a comparison of the first four moments. So approximating a translated gamma approximation with parameters α, β and x_0, we also have $\Pr[S \leq s] \approx \Phi\left(\sqrt{4\beta(s - x_0)} - \sqrt{4\alpha - 1}\right)$.
   Show $\Pr[S \leq s] \approx \Phi\left(\sqrt{\frac{8}{\gamma} \frac{s - \mu}{\sigma} + \frac{16}{\gamma^2}} - \sqrt{\frac{16}{\gamma^2} - 1}\right)$ if $\alpha = \frac{4}{\gamma^2}, \beta = \frac{2}{\gamma \sigma}, x_0 = \mu - \frac{2\sigma}{\gamma}$.
   Inversely, show $\Pr\left[S \leq x_0 + \frac{1}{4\beta}(y + \sqrt{4\alpha - 1})^2\right] \approx 1 - \varepsilon$ if $\Phi(y) = 1 - \varepsilon$,
   as well as $\Pr\left[\frac{S - \mu}{\sigma} \leq y + \frac{\gamma}{8}(y^2 - 1) + y(\sqrt{1 - \gamma^2/16} - 1)\right] \approx \Phi(y)$.

4. Show that the translated gamma approximation as well as the NP approximation result in the normal approximation (CLT) if μ and σ^2 are fixed and $\gamma \downarrow 0$.

5. Approximate the critical values of a χ^2_{18} distribution for $\varepsilon = 0.05, 0.1, 0.5, 0.9, 0.95$ with the NP approximation and compare the results with the exact values.

6. In the previous exercise, what is the result if the translated gamma approximation is used?

7. Use the identity 'having to wait longer than x for the nth event' $\equiv$ 'at most $n-1$ events occur in $(0,x)$' in a Poisson process to prove that $\Pr[Z > x] = \Pr[N < n]$ if $Z \sim$ gamma$(n,1)$ and $N \sim$ Poisson(x). How can this fact be used to calculate the translated gamma approximation?

8. Compare the exact critical values of a χ^2_{18} distribution for $\varepsilon = 0.05, 0.1, 0.5, 0.9, 0.95$ with the approximations obtained in exercise 2.5.3.

9. An insurer's portfolio contains 2 000 one-year life insurance policies. Half of them are characterized by a payment $b_1 = 1$ and a probability of dying within 1 year of $q_1 = 1\%$. For the other half, we have $b_2 = 2$ and $q_2 = 5\%$. Use the CLT to determine the minimum safety loading, as a percentage, to be added to the net premium to ensure that the probability that the total payment exceeds the total premium income is at most 5%.

10. As the previous exercise, but now using the NP approximation. Employ the fact that the third cumulant of the total payment equals the sum of the third cumulants of the risks.

11. Show that the right hand side of (2.60) is well-defined for all $x \geq -1$. What are the minimum and the maximum values? Is the function increasing? What happens if $x = 1$?

12. Suppose that X has expected value $\mu = 1000$ and standard deviation $\sigma = 2000$. Determine the skewness γ if (i) X is normal, (ii) $X/\phi \sim$ Poisson(μ/ϕ), (iii) $X \sim$ gamma(α, β), (iv) $X \sim$ inverse Gaussian(α, β) or (v) $X \sim$ lognormal(ν, τ^2). Show that the skewness is infinite if (vi) $X \sim$ Pareto. See also Table A.

13. A portfolio consists of two types of contracts. For type k, $k = 1, 2$, the claim probability is q_k and the number of policies is n_k. If there is a claim, then its size is x with probability $p_k(x)$:

	n_k	q_k	$p_k(1)$	$p_k(2)$	$p_k(3)$
Type 1	1000	0.01	0.5	0	0.5
Type 2	2000	0.02	0.5	0.5	0

Assume that the contracts are independent. Let S_k denote the total claim amount of the contracts of type k and let $S = S_1 + S_2$. Calculate the expected value and the variance of a contract of type k, $k = 1, 2$. Then, calculate the expected value and the variance of S. Use the CLT to determine the minimum capital that covers all claims with probability 95%.

14. [♠] Let $U \sim$ gamma$(\alpha, 1)$, $Y \sim$ N$(\sqrt{4\alpha - 1}, 1)$ and $T = \sqrt{4U}$. Show that E$[U^t] = \Gamma(\alpha + t)/\Gamma(\alpha)$, $t > 0$. Then show that E$[Y^j] \approx$ E$[T^j]$, $j = 1, 3$, by applying $\Gamma(\alpha + 1/2)/\Gamma(\alpha) \approx \sqrt{\alpha - 1/4}$ and $\alpha\Gamma(\alpha) = \Gamma(\alpha + 1)$. Also, show that E$[Y^2] =$ E$[T^2]$ and E$[Y^4] =$ E$[T^4] - 2$.

15. [♠] A justification for the 'correction for continuity', see Example 2.5.3, used to approximate cdf of integer valued random variables by continuous ones, goes as follows. Let G be the continuous cdf of some non-negative random variable, and construct cdfs H by $H(k+\varepsilon) = G(k+0.5)$, $k = 0, 1, 2, \ldots, 0 \leq \varepsilon < 1$. Using the *midpoint rule* with intervals of length 1 to approximate the right hand side of (1.33) at $d = 0$, show that the means of G and H are about equal. Conclude that if G is a continuous cdf that is a plausible candidate for approximating the discrete cdf F and has the same mean as F, by taking $F(x) := G(x + 0.5)$ one gets an approximation with the proper mean value. [Taking $F(x) = G(x)$ instead, one gets a mean that is about $\mu + 0.5$ instead of μ. Thus very roughly speaking, each tail probability of the sum approximating (1.33) will be too big by a factor $\frac{1}{2\mu}$.]

16. To get a feel for the approximation error as opposed to the error caused by errors in the estimates of μ, σ and γ needed for the NP approximation and the gamma approximation,

recalculate Example 2.5.5 if the following parameters are changed: (i) $\mu = 10100$ (ii) $\sigma = 1020$ (iii) $\mu = 10100$ and $\sigma = 1020$ (iv) $\gamma = 1.03$. Assume that the remaining parameters are as they were in Example 2.5.5.

17. The function pNormalPower, when implemented carelessly, sometimes produces the value NaN (not a number). Why and when could that happen? Build in a test to cope with this situation more elegantly.

18. Compare the results of the translated gamma approximation with an exact Poisson(1) distribution using the calls pTransGam(0:10,1,1,1) and ppois(0:10,1).
    To see the effect of applying a correction for continuity, compare also with the result of pTransGam(0:10+0.5,1,1,1).

19. Repeat the previous exercise, but now for the Normal Power approximation.

20. Note that we have prefixed the (approximate) cdf with p, as is customary in R. Now write quantile functions qTransGam and qNormalPower, and do some testing.

21. Prove (2.64) and (2.65).

Section 2.6

1. In the situation of Section 2.6, calculate the probability that B will be insufficient for retentions $d \in [2,3]$. Give numerical results for $d = 2$ and $d = 3$ if $B = 405$.

2. Determine the retention $d \in [2,3]$ that minimizes this probability for $B = 405$. Which retention is optimal if $B = 404$?

3. Calculate the probability that B will be insufficient if $d = 2$ by using the NP approximation.

Chapter 3
Collective risk models

Any sufficiently advanced technology is indistinguishable from magic — A.C. Clarke's third law of prediction, 1973

3.1 Introduction

In this chapter, we introduce collective risk models. Just as in Chapter 2, we calculate the distribution of the total claim amount, but now we regard the portfolio as a collective that produces a random number N of claims in a certain time period. We write

$$S = X_1 + X_2 + \cdots + X_N, \qquad (3.1)$$

where X_i is the ith claim. Obviously, the total claims $S = 0$ if $N = 0$. The terms of S in (3.1) correspond to actual claims; in (2.26), there are many terms equal to zero, corresponding to the policies that do not produce a claim. We assume that the individual claims X_i are independent and identically distributed, and also that N and all X_i are independent. In the special case that N is Poisson distributed, S has a *compound Poisson distribution.* If N has a (negative) binomial distribution, then S has a *compound (negative) binomial distribution.*

In collective models, some policy information is ignored. If a portfolio contains only one policy that could generate a high claim amount, this amount will appear at most once in the individual model (2.26). In the collective model (3.1), however, it could occur several times. Moreover, in collective models we require the claim number N and the claim amounts X_i to be independent. This makes it somewhat less appropriate to model a car insurance portfolio, since for example bad weather conditions will cause a lot of small claim amounts. In practice, however, the influence of these phenomena appears to be small.

A collective risk model turns out to be both computationally efficient and rather close to reality. We give some algorithms to calculate the distribution of (3.1). An obvious but laborious method is convolution, conditioning on $N = n$ for all n. We also discuss the sparse vector algorithm. This can be used if $N \sim$ Poisson, and is based on the fact that the frequencies of the claim amounts can be proved to be independent Poisson random variables. For a larger class of distributions, we can use Panjer's recursion, which expresses the probability of $S = s$ recursively in terms of the probabilities of $S = k$, $k = 0, 1, \ldots, s-1$. Another approach is to use the *Fast Fourier Transform* to invert the characteristic function.

We can express the moments of S in terms of those of N and X_i. With this information we can again approximate the distribution of S using the CLT if $E[N]$ is large, as well as by the translated gamma approximation and the normal power approximation (NP) from the previous chapter.

Next, we look for appropriate distributions for N and X_i such that the collective model fits closely to a given individual model. It will turn out that the Poisson distribution and the negative binomial distribution are often appropriate choices for N. We will show some relevant relations between these distributions. We will also discuss some special properties of the compound Poisson distributions. Many parametric distributions are suitable to model insurance losses. We study their properties, including how to estimate the parameters by maximum likelihood and how to simulate random drawings from them.

Stop-loss insurance policies are not only in use for reinsurance treaties, but also for insuring absence due to illness, or if there is a deductible. We give a number of techniques to calculate stop-loss premiums for discrete distributions, but also for several continuous distributions. With the help of the approximations for distribution functions introduced in Chapter 2, we can also approximate stop-loss premiums.

3.2 Compound distributions

Assume that S is a compound random variable such as in (3.1), with terms X_i distributed as X. Further use the following notation:

$$\mu_k = E[X^k], \qquad P(x) = \Pr[X \le x], \qquad F(s) = \Pr[S \le s]. \qquad (3.2)$$

We can then calculate the expected value of S by using the conditional distribution of S, given N. First, we use the condition $N = n$ to substitute outcome n for the random variable N on the left of the conditioning bar below. Next, we use the independence of X_i and N to dispose of the condition $N = n$. This gives the following computation:

$$
\begin{aligned}
E[S] = E\big[E[S\,|\,N]\big] &= \sum_{n=0}^{\infty} E[X_1 + \cdots + X_N \,|\, N = n]\Pr[N = n] \\
&= \sum_{n=0}^{\infty} E[X_1 + \cdots + X_n \,|\, N = n]\Pr[N = n] \\
&= \sum_{n=0}^{\infty} E[X_1 + \cdots + X_n]\Pr[N = n] \\
&= \sum_{n=0}^{\infty} n\mu_1 \Pr[N = n] = \mu_1 E[N].
\end{aligned}
\qquad (3.3)
$$

Note that the expected claim total equals expected claim number times expected claim size.

The variance can be determined with the variance decomposition rule (2.22):

$$\begin{aligned}
\mathrm{Var}[S] &= \mathrm{E}\big[\mathrm{Var}[S\,|\,N]\big] + \mathrm{Var}\big[\mathrm{E}[S\,|\,N]\big] \\
&= \mathrm{E}\big[N\mathrm{Var}[X]\big] + \mathrm{Var}[N\mu_1] \\
&= \mathrm{E}[N]\mathrm{Var}[X] + \mu_1^2\mathrm{Var}[N] \quad \text{(for all } N) \\
&= \mathrm{E}[N]\mathrm{E}[X^2] \quad \text{(when especially } N \sim \text{ Poisson).}
\end{aligned} \tag{3.4}$$

The same technique as used in (3.3) yields for the mgf:

$$\begin{aligned}
m_S(t) &= \mathrm{E}\big[\mathrm{E}[e^{tS}\,|\,N]\big] \\
&= \sum_{n=0}^{\infty} \mathrm{E}\big[e^{t(X_1+\cdots+X_N)}\,|\,N=n\big]\,\mathrm{Pr}[N=n] \\
&= \sum_{n=0}^{\infty} \mathrm{E}\big[e^{t(X_1+\cdots+X_n)}\big]\,\mathrm{Pr}[N=n] \\
&= \sum_{n=0}^{\infty} \big\{m_X(t)\big\}^n\,\mathrm{Pr}[N=n] = \mathrm{E}\big[(e^{\log m_X(t)})^N\big] \\
&= m_N(\log m_X(t)).
\end{aligned} \tag{3.5}$$

Example 3.2.1 (A compound distribution with closed form cdf)
Let $N \sim$ geometric(p), $0 < p < 1$, and $X \sim$ exponential(1). What is the cdf of S?
    Write $q = 1 - p$. First, we compute the mgf of S, and then we try to identify it. For $qe^t < 1$, which means $t < -\log q$, we have

$$m_N(t) = \sum_{n=0}^{\infty} e^{nt} pq^n = \frac{p}{1 - qe^t}. \tag{3.6}$$

Since $X \sim$ exponential(1), so $m_X(t) = (1-t)^{-1}$, (3.5) yields

$$m_S(t) = m_N(\log m_X(t)) = \frac{p}{1 - qm_X(t)} = p + q\frac{p}{p-t}, \tag{3.7}$$

so the mgf of S is a mixture of the mgfs of the constant 0 and of the exponential(p) distribution. Because of the one-to-one correspondence of cdfs and mgfs, we may conclude that the cdf of S is the same mixture:

$$F(x) = p + q(1 - e^{-px}) = 1 - qe^{-px} \quad \text{for } x \geq 0. \tag{3.8}$$

This is a distribution with a jump of size p in 0, exponential otherwise. ∇

This example is unique in the sense that it presents the only non-trivial compound distribution with a closed form for the cdf.

3.2.1 Convolution formula for a compound cdf

The conditional distribution of S, given $N = n$, allows us to calculate F:

$$F(x) = \Pr[S \leq x] = \sum_{n=0}^{\infty} \Pr[X_1 + \cdots + X_N \leq x \mid N = n] \Pr[N = n], \qquad (3.9)$$

so

$$F(x) = \sum_{n=0}^{\infty} P^{*n}(x) \Pr[N = n], \quad f(x) = \sum_{n=0}^{\infty} p^{*n}(x) \Pr[N = n]. \qquad (3.10)$$

These expressions are the *convolution formulas* for a compound cdf.

Example 3.2.2 (Application of the convolution formula)
Let $\Pr[N = j - 1] = j/10$ for $j = 1,2,3,4$, and let $p(1) = 0.4$, $p(2) = 0.6$. By using (3.10), $F(x)$ can be calculated as follows:

x	$p^{*0}(x)$	$p^{*1}(x)$	$p^{*2}(x)$	$p^{*3}(x)$		$f(x)$	$F(x)$
0	1					0.1000	0.1000
1		0.4				0.0800	0.1800
2		0.6	0.16			0.1680	0.3480
3			0.48	0.064		0.1696	0.5176
4			0.36	0.288		:	:
5				0.432		:	.
:				:		:	:
	$\updownarrow \times$	$+\updownarrow \times$	$+\updownarrow \times$	$+\updownarrow \times$	$=$	$\uparrow$ $\Rightarrow$	$\uparrow$
$\Pr[N = n]$	0.1	0.2	0.3	0.4			

The probabilities $\Pr[N = n]$ in the bottom row are multiplied by the numbers in a higher row. Then, the sum of these results is put in the corresponding row in the column $f(x)$. For example: $0.2 \times 0.6 + 0.3 \times 0.16 = 0.168$. ▽

Note that if we attempt convolution in case of arbitrary discrete claim sizes rather than integer-valued ones such as here, the number of possible values and the required number of computations increase exponentially.

Example 3.2.3 (Compound distributions, exponential claim amounts)
From expression (3.10) for $F(x)$, we see that it is convenient to choose the distribution of X in such a way that the n-fold convolution is easy to calculate. This is the case for the normal and the gamma distribution: the sum of n independent $N(\mu, \sigma^2)$ random variables is $N(n\mu, n\sigma^2)$, while the sum of n gamma(α, β) random variables is a gamma$(n\alpha, \beta)$ random variable.

Suppose the claim amounts have an exponential(1) distribution, which is the same as gamma(α, β) with $\alpha = \beta = 1$. In Poisson waiting time processes, see also Exercise 2.5.7 and Chapter 4, the probability of waiting at least a time x for the

n-th event, which is at the same time the probability that at most $n - 1$ events have occurred at time x, is a Poisson(x) probability. Hence we have

$$1 - P^{*n}(x) = \int_x^\infty y^{n-1} \frac{e^{-y}}{(n-1)!} \, dy = e^{-x} \sum_{i=0}^{n-1} \frac{x^i}{i!}. \qquad (3.11)$$

This can also be proved with partial integration or by comparing the derivatives, see Exercise 3.2.7. So, for $x > 0$,

$$1 - F(x) = \sum_{n=1}^\infty \Pr[N = n] e^{-x} \sum_{i=0}^{n-1} \frac{x^i}{i!}. \qquad (3.12)$$

We can stop the outer summation as soon as $\Pr[N \geq n]$ is smaller than the required precision; also, two successive inner sums differ by the final term only, which implies that a single summation suffices.   ∇

Computing the distribution of the total claims is much easier if the terms are integer-valued, so we will often approximate X by rounding it to the nearest multiples of some discretization width.

3.3 Distributions for the number of claims

In practice, we will not have a lot of relevant data at our disposal to choose a distribution for N. To describe 'rare events', the Poisson distribution, having only one parameter to be estimated, is always the first choice. Also, its use can be justified if the underlying process can be described as a Poisson process, see Chapter 4. It is well-known that the expected value and the variance of a Poisson(λ) distribution are both equal to λ. If $\text{Var}[N]/\text{E}[N] > 1$, that is, there is *overdispersion*, one may use the negative binomial distribution instead. We consider two models in which the latter distribution is derived as a generalization of a Poisson distribution.

Example 3.3.1 (Poisson distribution, uncertainty about the parameter)
Assume that some car driver causes a Poisson(λ) distributed number of accidents in one year. The parameter λ is unknown and different for every driver. We assume that λ is the outcome of a random variable Λ. Then the conditional distribution of the number of accidents N in one year, given $\Lambda = \lambda$, is Poisson(λ). What is the marginal distribution of N?

Let $U(\lambda) = \Pr[\Lambda \leq \lambda]$ denote the distribution function of Λ. Then we can write the marginal probabilities of event $N = n$ as

$$\Pr[N = n] = \int_0^\infty \Pr[N = n \mid \Lambda = \lambda] \, dU(\lambda) = \int_0^\infty e^{-\lambda} \frac{\lambda^n}{n!} \, dU(\lambda), \qquad (3.13)$$

while for the unconditional mean and variance of N we have

$$E[N] = E[E[N|\Lambda]] = E[\Lambda];$$
$$\text{Var}[N] = E[\text{Var}[N|\Lambda]] + \text{Var}[E[N|\Lambda]] = E[\Lambda] + \text{Var}[\Lambda] \geq E[N]. \tag{3.14}$$

Now assume additionally that $\Lambda \sim \text{gamma}(\alpha, \beta)$, then, writing $p = \beta/(\beta+1)$,

$$m_N(t) = E[E[e^{tN}|\Lambda]] = E[\exp\{\Lambda(e^t - 1)\}] = m_\Lambda(e^t - 1)$$
$$= \left(\frac{\beta}{\beta - (e^t - 1)}\right)^\alpha = \left(\frac{p}{1 - (1-p)e^t}\right)^\alpha, \tag{3.15}$$

which from Table A we recognize as the mgf of a negative binomial$(\alpha, \beta/(\beta+1))$ distribution. It can be shown that the overdispersion $\text{Var}[N]/E[N]$ is $1/p = 1 + 1/\beta$.

Obviously, the value of Λ for a particular driver is a non-observable random variable. It is the 'long run claim frequency', the value to which the observed average number of accidents in a year would converge if the driver could be observed for a very long time, during which his claims pattern does not change. The distribution of Λ is called the structure distribution, see also Chapter 8.   ∇

Example 3.3.2 (Compound negative binomial is also compound Poisson)

At some intersection there are N traffic accidents with casualties in a year. There are L_i casualties in the ith accident, so $S = L_1 + L_2 + \cdots + L_N$ is the total number of casualties. Now assume $N \sim \text{Poisson}(\lambda)$ and $L_i \sim \text{logarithmic}(c)$ with $0 < c < 1$, so

$$\Pr[L_i = k] = \frac{c^k}{k\,h(c)}, \qquad k = 1, 2, \ldots \tag{3.16}$$

The division by the function $h(\cdot)$ serves to make the sum of the probabilities equal to 1, so from the usual series expansion of $\log(1+x)$, this function is $h(c) = -\log(1 - c)$, hence the name logarithmic distribution. What is the distribution of S?

The mgf of the terms L_i is given by

$$m_L(t) = \sum_{k=1}^{\infty} \frac{e^{tk}c^k}{k\,h(c)} = \frac{h(ce^t)}{h(c)}. \tag{3.17}$$

Then, for the mgf of S, we get

$$m_S(t) = m_N(\log m_L(t)) = \exp\lambda(m_L(t) - 1)$$
$$= \left(\exp\{h(ce^t) - h(c)\}\right)^{\lambda/h(c)} = \left(\frac{1-c}{1-ce^t}\right)^{\lambda/h(c)}, \tag{3.18}$$

which, see again Table A, we recognize as the mgf of a negative binomial distribution with parameters $\lambda/h(c) = -\lambda/\log(1-c)$ and $1 - c$.

On the one hand, the total payment Z for the casualties has a compound Poisson distribution since it is the sum of a Poisson(λ) number of payments per fatal accident (*cumulation*). On the other hand, summing over the casualties leads to a compound negative binomial distribution. It can be shown that if S_2 is compound negative binomial with parameters r and $p = 1 - q$ and claims distribution $P_2(\cdot)$, then S_2

has the same distribution as S_1, where S_1 is compound Poisson distributed with parameter λ and claims distribution $P_1(\cdot)$ given by:

$$\lambda = rh(q) \quad \text{and} \quad P_1(x) = \sum_{k=1}^{\infty} \frac{q^k}{kh(q)} P_2^{*k}(x). \qquad (3.19)$$

In this way, any compound negative binomial distribution can be written as a compound Poisson distribution.   ▽

Remark 3.3.3 (Compound Poisson distributions in probability theory)
The compound Poisson distributions are also object of study in probability theory. If we extend this class with its limits, to which the gamma and the normal distribution belong, then we have just the class of infinitely divisible distributions. This class consists of the random variables X with the property that for each n, a sequence of iid random variables $X_1, X_2, \ldots, X_n$ exists with $X \sim X_1 + X_2 + \cdots + X_n$.   ▽

3.4 Properties of compound Poisson distributions

In this section we prove some important theorems on compound Poisson distributions and use them to construct a better algorithm to calculate $F(\cdot)$ than the one given by (3.10). First, we show that the class of compound Poisson distributions is closed under convolution.

Theorem 3.4.1 (Sum of compound Poisson r.v.'s is compound Poisson)
If $S_1, S_2, \ldots, S_m$ are independent compound Poisson random variables with Poisson parameter λ_i and claims distribution P_i, $i = 1, 2, \ldots, m$, then $S = S_1 + S_2 + \cdots + S_m$ is compound Poisson distributed with specifications

$$\lambda = \sum_{i=1}^{m} \lambda_i \quad \text{and} \quad P(x) = \sum_{i=1}^{m} \frac{\lambda_i}{\lambda} P_i(x). \qquad (3.20)$$

Proof. Let m_i be the mgf of P_i. Then S has the following mgf:

$$m_S(t) = \prod_{i=1}^{m} \exp\left\{ \lambda_i [m_i(t) - 1] \right\} = \exp \lambda \left\{ \sum_{i=1}^{m} \frac{\lambda_i}{\lambda} m_i(t) - 1 \right\}. \qquad (3.21)$$

So S is a compound Poisson random variable with specifications (3.20).   ▽

Consequently, the total result of m independent compound Poisson portfolios is again compound Poisson distributed. The same holds if we observe the same portfolio in m years, assuming that the annual results are independent.

A special case is when the S_i have fixed claims x_i, hence $S_i = x_i N_i$ with $N_i \sim$ Poisson(λ_i). Assume the x_i to be all different. We get the random variable

$$S = x_1 N_1 + x_2 N_2 + \cdots + x_m N_m, \qquad (3.22)$$

which by Theorem 3.4.1 is compound Poisson with specifications:

$$\lambda = \lambda_1 + \cdots + \lambda_m \quad \text{and} \quad p(x_i) = \frac{\lambda_i}{\lambda}, i = 1, \ldots, m. \tag{3.23}$$

We can also prove the reverse statement, as follows:

Theorem 3.4.2 (Frequencies of claim sizes are independent Poisson)
Assume that S is compound Poisson distributed with parameter λ and with discrete claims distribution

$$\pi_i = p(x_i) = \Pr[X = x_i], \quad i = 1, 2, \ldots, m. \tag{3.24}$$

Suppose S is written as (3.22), where N_i denotes the frequency of the claim amount x_i, that is, the number of terms in S with value x_i. Then $N_1, \ldots, N_m$ are independent Poisson$(\lambda \pi_i)$ random variables, $i = 1, \ldots, m$.

Proof. Let $N = N_1 + \cdots + N_m$ and $n = n_1 + \cdots + n_m$. Conditionally on $N = n$, we have $N_1, \ldots, N_m \sim \text{Multinomial}(n, \pi_1, \ldots, \pi_m)$. Hence,

$$\begin{aligned}
&\Pr[N_1 = n_1, \ldots, N_m = n_m] \\
&= \Pr[N_1 = n_1, \ldots, N_m = n_m \mid N = n] \Pr[N = n] \\
&= \frac{n!}{n_1! n_2! \ldots n_m!} \pi_1^{n_1} \pi_2^{n_2} \ldots \pi_m^{n_m} e^{-\lambda} \frac{\lambda^n}{n!} \\
&= \prod_{i=1}^{m} e^{-\lambda \pi_i} \frac{(\lambda \pi_i)^{n_i}}{n_i!}.
\end{aligned} \tag{3.25}$$

By summing over all n_i, $i \neq k$, we see that N_k is marginally Poisson$(\lambda \pi_k)$ distributed. The N_i are independent since $\Pr[N_1 = n_1, \ldots, N_m = n_m]$ is the product of the marginal probabilities of $N_i = n_i$. ∇

Example 3.4.3 (Application: sparse vector algorithm)
If the claims X are integer-valued and non-negative, we can calculate the compound Poisson cdf F in an efficient way. We explain this by an example. Let $\lambda = 4$ and $\Pr[X = 1, 2, 3] = \frac{1}{4}, \frac{1}{2}, \frac{1}{4}$. Then, gathering together terms as we did in (3.22), we can write S as $S = 1N_1 + 2N_2 + 3N_3$ and calculate the distribution of S by convolution. We can compute $f(x) = \Pr[S = x]$ as follows:

x	$\Pr[N_1 = x]$ $(e^{-1}\times)$	$*$	$\Pr[2N_2 = x]$ $(e^{-2}\times)$	$=$	$\Pr[N_1 + 2N_2 = x]$ $(e^{-3}\times)$	$*$	$\Pr[3N_3 = x]$ $(e^{-1}\times)$	$=$	$\Pr[S = x]$ $(e^{-4}\times)$
0	1		1		1		1		1
1	1		–		1		–		1
2	1/2		2		5/2		–		5/2
3	1/6		–		13/6		1		19/6
4	1/24		2		:		–		:
:	:		:		:		:		:
	$\uparrow$		$\uparrow$				$\uparrow$		
	$1/x!$		$2^{x/2}/(x/2)!$				$1/(x/3)!$		

The density of the total amount of the claims of size $1, 2, \ldots, j-1$ is convoluted with the one of jN_j. In the column with probabilities of jN_j, only the rows $0, j, 2j, \ldots$ are filled, which is why this algorithm is called a 'sparse vector' algorithm. These probabilities are Poisson($\lambda \pi_j$) probabilities.

Implementing the Sparse vector algorithm in R is easy, since convolution of two vectors can be handled by the convolve function. It employs a technique called the *Fast Fourier Transform (FFT)*, the workings of which are explained in Section 3.6. An R-implementation of the sparse vector algorithm is as follows:

```
SparseVec <- function (freq)
{if (any(freq<0)) stop("negative frequency")
 M <- length(freq)
 mu <- sum((1:M)*freq); sigma2 <- sum((1:M)^2*freq)
 ##mean and variance of the compound r.v.; see (3.4)
 MM <- ceiling(mu + 10 * sqrt(sigma2)) + 6
 fs <- dpois(0:(MM-1), freq[1])   ##density of S_1 = 1*N_1
 for (j in 2:M)
 {MMM <- trunc((MM-1)/j)
  fj <- rep(0, MM)   ##construct the density of j*N_j
  fj[(0:MMM)*j+1] <- dpois(0:MMM, freq[j])
  fs <- convolve(fs, rev(fj), type="o") }
 ##fs is the density of S_j = 1*N_1 + ... + j*N_j, j=2..M
 return(fs)   }
f <- SparseVec(c(1,2,1)); f[1:7] * exp(4)
```

The last line reproduces the first seven numbers in the last column of the table in Example 3.4.3. The argument freq contains the expected frequencies λp_j of each claim amount $j = 1, 2, \ldots$, which should of course be non-negative. The vector length MM is taken to be the mean plus 10 standard deviations plus 7, ensuring that sum(fs[1:MM]) will always be virtually equal to 1. The vector fs is initialized to the density of $1N_1$, and convoluted with the one of jN_j in step j, $j = 2, \ldots, m$. Note that it is required that the second vector given as an argument to convolve is reversed and that the type given is "o", short for "open". The function result returned is the probability distribution of the compound random variable.

The algorithm given is fast since it uses the efficient FFT technique to do the convolutions. It is, however, not a proper sparse vector algorithm, since the fact that the vector fj has zeros at places that are non-multiples of j is never used. It can be shown that for large m and n, to compute probabilities of $0, \ldots, n$ with a maximal claim amount m by convolve takes $O(mn \log n)$ operations, while the sparse vector algorithm needs $O(n^2 \log m)$ (see Exercise 3.5.7).   ▽

3.5 Panjer's recursion

In 1981, Panjer described a method to calculate the probabilities $f(x)$ recursively. In fact, the method can be traced back to as early as Euler. As a result of Panjer's publication, a lot of other articles have appeared in the actuarial literature covering similar recursion relations. The recursion relation described by Panjer is as follows:

Theorem 3.5.1 (Panjer's recursion)
Consider a compound distribution with integer-valued non-negative claims with pdf
$p(x)$, $x = 0,1,2,\ldots$, for which, for some real a and b, the probability q_n of having n
claims satisfies the following recursion relation

$$q_n = \left(a + \frac{b}{n}\right)q_{n-1}, \quad n = 1,2,\ldots \tag{3.26}$$

Then the following relations for the probability of a total claim equal to s hold:

$$f(0) = \begin{cases} \Pr[N = 0] & \text{if } p(0) = 0; \\ m_N(\log p(0)) & \text{if } p(0) > 0; \end{cases} \tag{3.27}$$

$$f(s) = \frac{1}{1 - ap(0)}\sum_{h=1}^{s}\left(a + \frac{bh}{s}\right)p(h)f(s-h), \quad s = 1,2,\ldots$$

Proof. From $\Pr[S = 0] = \sum_{n=0}^{\infty}\Pr[N = n]p^n(0)$ we get the starting value $f(0)$. Write
$T_k = X_1 + \cdots + X_k$. First, note that because of symmetry:

$$\mathrm{E}\left[a + \frac{bX_1}{s}\,\Big|\,T_k = s\right] = a + \frac{b}{k}. \tag{3.28}$$

This expectation can also be determined in the following way:

$$\mathrm{E}\left[a + \frac{bX_1}{s}\,\Big|\,T_k = s\right] = \sum_{h=0}^{s}\left(a + \frac{bh}{s}\right)\Pr[X_1 = h\,|\,T_k = s]$$

$$= \sum_{h=0}^{s}\left(a + \frac{bh}{s}\right)\frac{\Pr[X_1 = h]\Pr[T_k - X_1 = s - h]}{\Pr[T_k = s]}. \tag{3.29}$$

Because of (3.26) and the previous two equalities, we have, for $s = 1,2,\ldots$,

$$f(s) = \sum_{k=1}^{\infty}q_k\Pr[T_k = s] = \sum_{k=1}^{\infty}q_{k-1}\left(a + \frac{b}{k}\right)\Pr[T_k = s]$$

$$= \sum_{k=1}^{\infty}q_{k-1}\sum_{h=0}^{s}\left(a + \frac{bh}{s}\right)\Pr[X_1 = h]\Pr[T_k - X_1 = s - h]$$

$$= \sum_{h=0}^{s}\left(a + \frac{bh}{s}\right)\Pr[X_1 = h]\sum_{k=1}^{\infty}q_{k-1}\Pr[T_k - X_1 = s - h] \tag{3.30}$$

$$= \sum_{h=0}^{s}\left(a + \frac{bh}{s}\right)p(h)f(s-h)$$

$$= ap(0)f(s) + \sum_{h=1}^{s}\left(a + \frac{bh}{s}\right)p(h)f(s-h),$$

from which the second relation of (3.27) follows immediately. ∇

Example 3.5.2 (Distributions suitable for Panjer's recursion)
Only the following distributions satisfy relation (3.26):

1. Poisson(λ) with $a = 0$ and $b = \lambda \geq 0$; in this case, (3.27) simplifies to:

$$f(0) = e^{-\lambda(1-p(0))};$$

$$f(s) = \frac{1}{s}\sum_{h=1}^{s}\lambda h p(h)f(s-h); \qquad (3.31)$$

2. Negative binomial(r,p) with $p = 1 - a$ and $r = 1 + \frac{b}{a}$; so $0 < a < 1$ and $a+b > 0$;
3. Binomial(k,p) with $p = \frac{a}{a-1}$ and $k = -\frac{b+a}{a}$; so $a < 0$, $b = -a(k+1)$.

If $a+b = 0$, then $q_0 = 1$ and $q_j = 0$ for $j = 1,2,\ldots$, so we get a Poisson(0) distribution. For other values of a and b than the ones used above, $q_n = \left(a + \frac{b}{n}\right)q_{n-1}$ for all $n = 1,2,\ldots$ cannot hold for a probability distribution:

- $q_0 \leq 0$ is not feasible, so assume $q_0 > 0$;
- $a+b < 0$ results in $q_1 < 0$;
- $a < 0$ and $b \neq a(n+1)$ for all n also results in negative probabilities;
- if $a \geq 1$ and $a+b > 0$, then from (3.26) we infer that $q_1 > 0$ and also that $nq_n = ((n-1)a+a+b)q_{n-1} \geq (n-1)q_{n-1}$, so $q_n \geq q_1/n$, $n = 1,2,\ldots$, so $\sum_n q_n = \infty$.

By allowing (3.26) to hold only for $n \geq 2$, hence admitting an arbitrary probability of no claims, we can find similar recursions for a larger group of counting distributions, which includes the logarithmic distributions. See Exercise 3.5.14 and 3.5.15. $\qquad \nabla$

Example 3.5.3 (Example 3.4.3 solved by Panjer's recursion)
As in Example 3.4.3, consider a compound Poisson distribution with $\lambda = 4$ and $\Pr[X = 1,2,3] = \frac{1}{4},\frac{1}{2},\frac{1}{4}$. Then (3.31) simplifies to

$$f(s) = \frac{1}{s}\left[f(s-1) + 4f(s-2) + 3f(s-3)\right], \quad s = 1,2,\ldots, \qquad (3.32)$$

and the starting value is $f(0) = e^{-4} \approx 0.0183$. We have

$$f(1) = f(0) = e^{-4},$$
$$f(2) = \frac{1}{2}\left[f(1) + 4f(0)\right] = \frac{5}{2}e^{-4}, \qquad (3.33)$$
$$f(3) = \frac{1}{3}\left[f(2) + 4f(1) + 3f(0)\right] = \frac{19}{6}e^{-4},$$

and so on. $\qquad \nabla$

Example 3.5.4 (Panjer's recursion and stop-loss premiums)
For an integer-valued S, we can write the stop-loss premium in an integer retention d as follows, see Section 1.4:

$$E[(S-d)_+] = \sum_{x=d}^{\infty}(x-d)f(x) = \sum_{x=d}^{\infty}[1-F(x)]. \qquad (3.34)$$

The stop-loss premium is piecewise linear in the retention on the intervals where the cdf remains constant, since for the right hand derivative we have by (1.38):

$$\frac{d}{dt}E[(S-t)_+] = F(t) - 1. \tag{3.35}$$

So the stop-loss premiums for non-integer d follow by linear interpolation.

With Panjer's recursion the stop-loss premiums can be calculated recursively, too, since from the last relation in (3.34), we have for integer d

$$\pi(d) := E[(S-d)_+] = \pi(d-1) - [1 - F(d-1)]. \tag{3.36}$$

As an example, take $S \sim$ compound Poisson(1) with $p(1) = p(2) = \frac{1}{2}$. Then, Panjer's recursion relation (3.31) simplifies to

$$f(x) = \frac{1}{x}\left[\frac{1}{2}f(x-1) + f(x-2)\right], \quad x = 1, 2, \ldots \tag{3.37}$$

with starting values

$$f(0) = e^{-1} \approx 0.368, \quad F(0) = f(0), \quad \pi(0) = E[S] = \lambda \mu_1 = \frac{3}{2}. \tag{3.38}$$

This leads to the following calculations:

x	$f(x) = (3.37)$	$F(x) = F(x-1) + f(x)$	$\pi(x) = \pi(x-1) - 1 + F(x-1)$
0	0.368	0.368	1.500
1	0.184	0.552	0.868
2	0.230	0.782	0.420
3	0.100	0.881	0.201
4	0.070	0.951	0.083
5	0.027	0.978	0.034

The advantage of computing the cdf and the stop-loss premiums simultaneously with the recursion is that there is no need to store the whole array of n values of $f(x)$, which might make a difference if the maximum claim m is much smaller than n. When in R the $f(x)$ values are stored in f, cumsum(f) produces the values of the cdf. For how to compute the successive stop-loss premiums, see Exercise 3.5.15. ∇

Remark 3.5.5 (Proof of Panjer's recursion through pgfs)
Panjer's recursion can also be derived from the probability generating functions. For the compound Poisson distribution, this goes as follows. First write

$$\frac{dg_S(t)}{dt} = \frac{d}{dt}\sum_{s=0}^{\infty} t^s \Pr[S=s] = \sum_{s=1}^{\infty} st^{s-1}\Pr[S=s]. \tag{3.39}$$

Just as in (3.5), we have

$$g_S(t) = g_N(g_X(t)) = \exp \lambda (g_X(t) - 1), \qquad (3.40)$$

so the derivative also equals $g'_S(t) = \lambda g_S(t) g'_X(t)$. For other distributions, similar expressions can be derived from (3.26). Now for $g_S(\cdot)$ and $g'_X(\cdot)$, substitute their series expansions:

$$
\begin{aligned}
\lambda g_S(t) g'_X(t) &= \lambda \left(\sum_{s=0}^{\infty} t^s \Pr[S = s] \right) \left(\sum_{x=1}^{\infty} x t^{x-1} \Pr[X = x] \right) \\
&= \sum_{x=1}^{\infty} \sum_{s=0}^{\infty} \lambda x t^{s+x-1} \Pr[S = s] \Pr[X = x] \\
&= \sum_{x=1}^{\infty} \sum_{v=x}^{\infty} \lambda x t^{v-1} \Pr[S = v - x] \Pr[X = x] \\
&= \sum_{v=1}^{\infty} \sum_{x=1}^{v} \lambda x t^{v-1} \Pr[S = v - x] \Pr[X = x].
\end{aligned}
\qquad (3.41)
$$

Comparing the coefficients of t^{s-1} in (3.39) and (3.41) yields

$$s \Pr[S = s] = \sum_{x=1}^{s} \lambda x \Pr[S = s - x] \Pr[X = x]. \qquad (3.42)$$

Dividing by s, one sees that this relation is equivalent with Panjer's recursion relation for the Poisson case (3.31).   ∇

Remark 3.5.6 (Convolution using Panjer's recursion)
How can we calculate the n-fold convolution of a distribution on $0, 1, 2, \ldots$ with Panjer's recursion?

Assume that $p(0) > 0$. If we replace X_i by $I_i Y_i$ where $\Pr[I_i = 1] = \Pr[X_i > 0] =: p$ and $Y_i \sim X_i | X_i > 0$, then $\sum_i X_i$ has the same distribution as $\sum_i I_i Y_i$, which gives us a compound binomial distribution with $p < 1$ as required in Example 3.5.2. Another method is to take limits for $p \uparrow 1$ in (3.27) for those values of a and b that produce a binomial(n, p) distribution.   ∇

Remark 3.5.7 (Implementing Panjer's recursion)
To compute the sum in the Panjer recursion (3.31), we might use R's loop mechanisms, but because R is an interpreted language it is worthwhile to 'vectorize' the computations, replacing the innermost loop by a call of sum. For this, note that to compute $f(s)$, the terms to be added in (3.31) are λ/s times the products of successive elements of the three vectors $(1, \ldots, m)$, $(p(1), \ldots, p(m))$ and $(f(s-1), \ldots, f(s-m))$. Here $m = \min\{s, r\}$ with r the maximal index for which $p_r > 0$. The second vector is the head of the p-vector, the third is the reverse of the tail part of the f-vector. An R program to implement Panjer's recursion and to reproduce the results of Example 3.5.3 is as follows:

```
Panjer.Poisson <- function (p, lambda)
{ if (sum(p)>1||any(p<0)) stop("p parameter not a density")
  if (lambda * sum(p) > 727) stop("Underflow")
```

```
cumul <- f <- exp(-lambda * sum(p))
r <- length(p)
s <- 0
repeat
{ s <- s+1
  m <- min(s, r)
  last <- lambda / s * sum(1:m * head(p,m) * rev(tail(f,m)))
  f <- c(f,last)
  cumul <- cumul + last
  if (cumul > 0.99999999) break  }
return(f)  }
Panjer.Poisson(c(0.25,0.5,0.25), 4) * exp(4)
```

The parameter p must contain the values of $p(1), p(2), \dots$, and it is checked if this, combined with $p(0) = 1 - \sum_h p(h)$, is indeed a density.

The parameter lambda representing λ should not be too big; in a standard Windows system problems arise if $\lambda(1 - p(0)) > 727$ holds, because in that case $f(0)$ is too small. R uses *double* precision (64-bit reals), but in programming environments employing *extended* precision (80-bit reals), one can easily cope with portfolios having $\lambda \approx 11340$, starting from $\Pr[S = 0] \approx 10^{-5000}$. In some older languages, a 48-bit real data type was used, leading to underflow already for $\lambda(1 - p(0)) \geq 88$. So for a portfolio of n life insurance policies with probabilities of claim equal to 0.5%, the calculation of $\Pr[S = 0]$ already experienced underflow for $n = 17600$. This underflow problem cannot be easily resolved in R itself, but it is possible to call compiled external code, using extended precision calculations, from R. This also reduces the running time, and in case of compound binomial probabilities, it might help remedy the numerical instability sometimes encountered.

The result of calling pp <- Panjer.Poisson(...) is a vector pp of probabilities $f(0), f(1), \dots, f(n)$ with the upper bound n such that $f(0) + f(1) + \cdots + f(n) > 1 - 10^{-8}$. Recall that in R, all arrays start with index 1, so pp[1] stores $f(0)$, and so on. ∇

3.6 Compound distributions and the Fast Fourier Transform

Another method to compute the probabilities of a compound distribution is based on inversion of the characteristic function. Let $X_1, X_2, \cdots \sim X$ be random variables with values in $\{0, 1, 2, \dots\}$, independent of each other as well as of the claim number N. Let $S = X_1 + \cdots + X_N$, and denote the probabilities by

$$p_h = \Pr[X = h]; \qquad q_n = \Pr[N = n]; \qquad f_s = \Pr[S = s]. \qquad (3.43)$$

Now let m be a number sufficiently large to let $\Pr[S \leq m] \approx 1$. For the Poisson and (negative) binomial case, Panjer's recursion requires $O(m^2)$ steps to compute $f_0, \dots, f_m$, or $O(m)$ if X is bounded. In this section we will introduce the Fast Fourier Transform method and show that it requires $O(m \log m)$ steps, for all compound distributions with a easily computable expression for the generating function of the number of claims.

The characteristic function of the compound random variable S is $\phi_S(t) \stackrel{\text{def}}{=} E[e^{itS}]$. Along the lines of (3.5) it can be proved that $\phi_S(t) = g_N(\phi_X(t))$, with g_N the pgf of N. The probabilities f_s can be found back from the characteristic function as follows:

$$f_s = \frac{1}{2\pi} \int_0^{2\pi} e^{-its} \phi_S(t) \, dt. \tag{3.44}$$

To prove this inversion formula is easy:

$$\frac{1}{2\pi} \int_0^{2\pi} e^{-its} \phi_S(t) \, dt = \frac{1}{2\pi} \int_0^{2\pi} \left\{ f_s + \sum_{k \neq s} f_k e^{-it(k-s)} \right\} dt = f_s + 0, \tag{3.45}$$

since for all $k \neq s$, as is readily verified by substituting $u = t(k-s)$,

$$\int_0^{2\pi} e^{-it(k-s)} \, dt = \int_0^{2\pi} \left\{ \cos(-t(k-s)) + i\sin(-t(k-s)) \right\} dt = 0. \tag{3.46}$$

Applying the trapezoidal rule to (3.44) with intervals of length $2\pi/n$, we see that an approximation for f_s is given by

$$f_s \approx r_s := \frac{1}{n} \sum_{h=0}^{n-1} e^{-i2\pi sh/n} \phi_S(2\pi h/n). \tag{3.47}$$

Note that (3.47) applies only if $s = 0, 1, \ldots, n-1$; for example for $s = n, 2n, \ldots$ we get the same approximation as for $s = 0$.

Now introduce the *discrete Fourier Transform* of a vector $\vec{f} = (f_0, \ldots, f_{n-1})$ as the vector $\vec{y} = \mathbf{T}^- \vec{f}$, with the matrix $\mathbf{T}^-$ defined as follows:

$$T_{jk}^- = e^{-i2\pi jk/n}, \qquad j, k = 0, 1, \ldots, n-1. \tag{3.48}$$

This means that every element t occurring in $\mathbf{T}^-$ is a unit root, with $t^n = 1$. Also define $\mathbf{T}^+$ in the same way, but with a plus sign in the exponent. Then approximation (3.47) can be written as

$$\vec{f} \approx \frac{1}{n} \mathbf{T}^- g_N(\mathbf{T}^+ \vec{p}). \tag{3.49}$$

All this would not be very useful but for two things. First, it is possible to compute approximation (3.49) very fast by an algorithm called the *Fast Fourier Transform*. It takes time $O(n \log n)$ and memory $O(n)$ only. If implemented in the standard way, it would require $O(n^2)$ operations and memory. Second, using another interpretation for the right hand side of (3.47), we will show how to make the error of the approximation negligible by taking n large enough.

The matrices $\mathbf{T}^+$ and $\mathbf{T}^-$ are in fact each other's inverse in the sense that $\frac{1}{n}\mathbf{T}^- = (\mathbf{T}^+)^{-1}$, because for fixed j, k, writing $\omega = e^{i2\pi(j-k)/n}$, we get

$$(\mathbf{T}^+\mathbf{T}^-)_{jk} = \sum_{h=0}^{n-1} e^{+i2\pi jh/n} e^{-i2\pi hk/n} = \sum_{h=0}^{n-1} \omega^h = \begin{cases} n & \text{if } j = k \\ \frac{1-\omega^n}{1-\omega} = 0 & \text{if } j \neq k. \end{cases} \tag{3.50}$$

If $z = x + \mathrm{i}y = r\mathrm{e}^{\mathrm{i}\phi}$, then $\bar{z} = x - \mathrm{i}y = r\mathrm{e}^{-\mathrm{i}\phi}$ is its complex conjugate. So

$$(\mathbf{T}^+\vec{g})_j = \sum_{h=0}^{n-1} e^{+i2\pi jh/n} g_h = \sum_{h=0}^{n-1} \overline{e^{-i2\pi jh/n} \overline{g_h}} = \overline{(\mathbf{T}^-\overline{\vec{g}})_j}. \tag{3.51}$$

Therefore the inverse operation of an FFT can be handled by the same algorithm, apart from taking complex conjugates and a division by the length of the vector.

To show that a Fast Fourier Transform algorithm can be constructed taking only time $O(n \log n)$, we use a 'divide and conquer' approach. Assume n even, then for $m = 0, 1, \ldots, n-1$, substituting $k = 2h + j$:

$$y_m = \sum_{h=0}^{n-1} e^{i2\pi mk/n} g_k = \sum_{j=0}^{1} \underbrace{\sum_{h=0}^{n/2-1} e^{i2\pi m 2h/n} g_{2h+j}} \tag{3.52}$$

For $j = 0$, the underbraced sum involves an FFT of length $\frac{n}{2}$ on $(g_0, g_2, \ldots, g_{n-2})$, for $j = 1$, on $(g_1, g_3, \ldots, g_{n-1})$. Therefore, using an induction assumption, to compute the FFT of length n takes time $2 \times \alpha \frac{1}{2} n \log \frac{1}{2} n + \beta n$ for some α and β, since two FFTs of length $n/2$ must be computed, plus a summation over j for each n. This adds up to $\alpha n \log n + n(\beta - \alpha \log 2)$, which is less than $\alpha n \log n$ in total provided $\alpha > \beta / \log 2$ is taken. Iterating this proves that FFT can be done using only $O(n \log n)$ operations.

If $\mathrm{Re}(g_j) = \mathrm{Pr}[Z = j]$, $\mathrm{Im}(g_j) = 0$, and $\mathrm{Pr}[Z \in \{0, 1, \ldots, n-1\}] = 1$ for a random variable Z, then for the characteristic function $\phi_Z(t) = \mathrm{E}[e^{\mathrm{i}tZ}]$ we have

$$\phi_Z(2\pi j/n) = \sum_{k=0}^{n-1} e^{i2\pi jk/n} g_k = (\mathbf{T}^+\vec{g})_j, \quad j = 0, 1, \ldots, n-1. \tag{3.53}$$

Since Z is integer-valued, $\phi_Z(t + 2\pi) = \mathrm{E}[e^{\mathrm{i}tZ} e^{\mathrm{i}2\pi Z}] = \phi_Z(t)$ for all real t, so ϕ_Z is periodical with period 2π.

By the above, for a random variable Z with support $\{0, 1, \ldots, n-1\}$ we have

$$y_j = \phi_Z(2\pi j/n) \quad \Longrightarrow \quad \mathbf{T}^+\vec{g} = \vec{y} \quad \Longrightarrow \quad \vec{g} = \frac{1}{n}\mathbf{T}^-\vec{y}. \tag{3.54}$$

To apply this to $S = X_1 + \cdots + X_N$, write $Z \equiv S \bmod n$: the remainder when S is divided by n. Then Z has support $\{0, 1, \ldots, n-1\}$, and $\mathrm{Pr}[Z \in \{S, S \pm n, S \pm 2n, \ldots\}] = 1$. Therefore their discrete Fourier transforms coincide:

$$\phi_S(2\pi j/n) = \phi_Z(2\pi j/n) \quad \text{for all integer } j. \tag{3.55}$$

For large n, $\Pr[Z = k] \approx \Pr[S = k]$. To compute the characteristic function of S from transforms of N and X, use the relation $\phi_S(t) = g_N(\phi_X(t))$, see also (3.5).

We summarize the discussion above in the following theorem.

Theorem 3.6.1 (Computing compound cdf by FFT)
If S has a compound distribution with specifications (3.43), we can approximate the probabilities f_s of $S = s$, $s = 0, \ldots, n-1$ by the exact probabilities r_s of $Z = s$, where $Z \equiv S \bmod n$. These probabilities can be computed, in time $O(n \log n)$, as:

$$\vec{f} \approx \vec{r} := \frac{1}{n} \mathbf{T}^- g_N(\mathbf{T}^+ \vec{q}). \tag{3.56}$$

The error $r_s - f_s = \Pr[S \in \{s+n, s+2n, \ldots\}]$, $s = 0, \ldots, n-1$. $\qquad \nabla$

Example 3.6.2 (Example 3.5.4 using FFT)
R has a built-in function `fft` to do the calculations. Using it, the probabilities of a compound Poisson($\lambda = 1$) random variable with claims distribution $\Pr[X = 1] = \Pr[X = 2] = \frac{1}{2}$ (see Example 3.5.4) can be reproduced as follows:

```
n <- 64; p <- rep(0, n); p[2:3] <- 0.5; lab <- 1
f <- Re(fft(exp(lab*(fft(p)-1)), inverse=TRUE))/n
```

Note that the R-function does not automatically do the division by n. Also note that we need to pad the p vector with enough zeros to let the resulting total probability of $S \in \{0, 1, \ldots, n-1\}$ be near enough to one. For the Fast Fourier Transform to live up to its name, the number of elements in the vector should preferably be a power of two, but in any case have a lot of factors. $\qquad \nabla$

In case $N \sim$ binomial or negative binomial, all one has to do is plug in the appropriate generating function. Also, for example a logarithmic number of claims can be handled easily. Moreover, the FFT-technique can be adapted to deal with negative claim amounts.

When the number n of conceivable total claim sizes is large, Panjer's recursion requires time $O(n^2)$ if the individual claim sizes are unbounded. In that case, FFT provides an easy to use and fast alternative, not quite exact but with a controllable error, and taking only $O(n \log n)$ time. For Panjer's recursion, one would typically have to compute the probabilities up to either the retention d, or to a certain quantile like the 75% quantile, but with FFT, it is mandatory to take n large enough to let $\Pr[S > n]$ be negligible. This does not make a lot of difference, asymptotically.

Note that for FFT, no for-loops were needed such as with Panjer's recursion. Therefore using FFT in many cases will be a lot faster than a recursive method.

3.7 Approximations for compound distributions

In the previous chapter, approximations were given that were refinements of the CLT, in which the distribution of a sum of a large number of random variables is approximated by a normal distribution. These approximations can also be used if the number of terms in a sum is a random variable with large values. For example, for the compound Poisson distribution with large λ we have the following counterpart of the CLT; similar results can be derived for other compound distributions.

Theorem 3.7.1 (CLT for compound Poisson distributions)
Let S be compound Poisson distributed with parameter λ and general claims cdf $P(\cdot)$ with finite variance. Then, with $\mu = \mathrm{E}[S]$ and $\sigma^2 = \mathrm{Var}[S]$,

$$\lim_{\lambda \to \infty} \mathrm{Pr}\left[\frac{S-\mu}{\sigma} \le x\right] = \Phi(x). \tag{3.57}$$

Proof. If $N_1, N_2, \ldots$ is a series of independent Poisson(1) random variables and if X_{ij}, $i = 1, 2, \ldots$, $j = 1, 2, \ldots$ are independent random variables with cdf $P(\cdot)$, then for integer-valued λ, we have

$$S \sim \sum_{j=1}^{\lambda}\sum_{i=1}^{N_j} X_{ij}, \quad \text{since } \sum_{j=1}^{\lambda} N_j \sim N. \tag{3.58}$$

As S in (3.58) is the sum of λ independent and identically distributed random variables, the CLT can be applied directly. Note that taking λ to be an integer presents no loss of generality, since the influence of the fractional part vanishes for large λ.

In this proof, we have reduced the situation to the Central Limit theorem. A proof along the lines of the one of Theorem 2.5.1 is asked in Exercise 3.7.3.   ∇

To use the CLT, translated gamma approximation and normal power approximation (NP) one needs the cumulants of S. Again, let μ_k denote the kth moment of the claims distribution. Then, for the compound Poisson distribution, we have

$$\kappa_S(t) = \lambda(m_X(t) - 1) = \lambda \sum_{k=1}^{\infty} \mu_k \frac{t^k}{k!}. \tag{3.59}$$

From (2.46) we know that the coefficients of $\frac{t^k}{k!}$ are the cumulants. Hence mean, variance and third central moment of a compound Poisson(λ, x) random variable with *raw moments* $\mu_j = \mathrm{E}[X^j]$ are given by

$$\mathrm{E}[S] = \lambda\mu_1, \quad \mathrm{Var}[S] = \lambda\mu_2 \quad \text{and} \quad \mathrm{E}[(S - \mathrm{E}[S])^3] = \lambda\mu_3. \tag{3.60}$$

The skewness is proportional to $\lambda^{-1/2}$:

$$\gamma_S = \frac{\mu_3}{\mu_2^{3/2}\sqrt{\lambda}}. \tag{3.61}$$

Remark 3.7.2 (Asymptotics and underflow)
There are certain situations in which one would have to resort to approximations. First of all, if the calculation time is uncomfortably long: for the calculation of $f(s)$

in (3.31) for large s, we need a lot of multiplications, see Exercise 3.5.4. Second, the recursion might not 'get off the ground'; see Remark 3.5.7.

Fortunately, the approximations improve with increasing λ; they are asymptotically exact, since in the limit they coincide with the usual normal approximation based on the CLT.   ∇

3.8 Individual and collective risk model

In the preceding sections we have shown that replacing the individual model by the collective risk model has distinct computational advantages. In this section we focus on the question which collective model should be chosen. We consider a situation from life insurance, but the same situation occurs in non-life insurance, for example when fines are imposed (malus) if an employee gets disabled.

Consider n one-year life insurance policies. At death, which happens with probability q_i, the claim amount on policy i is b_i, assumed positive, otherwise it is 0. We want to approximate the total amount of the claims on all policies using a collective model. For that purpose, we replace the I_i payments of size b_i for policy i, where $I_i \sim$ Bernoulli(q_i), by a Poisson(λ_i) distributed number of payments b_i. Instead of the cdf of the total payment in the individual model

$$\tilde{S} = \sum_{i=1}^{n} I_i b_i, \quad \text{with} \quad \Pr[I_i = 1] = q_i = 1 - \Pr[I_i = 0], \quad (3.62)$$

we consider the cdf of the following approximating random variable:

$$S = \sum_{i=1}^{n} Y_i, \quad \text{with} \quad Y_i = N_i b_i = \sum_{j=1}^{N_i} b_i \quad \text{and} \quad N_i \sim \text{Poisson}(\lambda_i). \quad (3.63)$$

If we choose $\lambda_i = q_i$, the expected number of payments for policy i is equal in both models. To stay on the safe side, we could also choose $\lambda_i = -\log(1 - q_i) > q_i$. With this choice, the probability of 0 claims on policy i is equal in both the collective and the individual model. This way, we incorporate implicit margins by using a larger total claim size than the original one. See also Section 7.4.1 and Remark 3.8.2.

Although (3.63) still has the form of an individual model, S is a compound Poisson distributed random variable because of Theorem 3.4.1, so it is indeed a collective model as in (3.1). The specifications are:

$$\lambda = \sum_{i=1}^{n} \lambda_i \quad \text{and} \quad P(x) = \sum_{i=1}^{n} \frac{\lambda_i}{\lambda} I_{[b_i, \infty)}(x), \quad (3.64)$$

with the indicator function $I_A(x) = 1$ if $x \in A$ and 0 otherwise. From this it is clear that the expected numbers of payments are equal if $\lambda_i = q_i$ is taken:

$$\lambda = \sum_{i=1}^{n} \lambda_i = \sum_{i=1}^{n} q_i. \tag{3.65}$$

Also, by (3.62) and (3.63), the expectations of $\widetilde{S}$ and S are then equal:

$$E[\widetilde{S}] = \sum_{i=1}^{n} q_i b_i = E[S]. \tag{3.66}$$

For the variances of S and $\widetilde{S}$ we have

$$\mathrm{Var}[S] = \sum_{i=1}^{n} q_i b_i^2; \quad \mathrm{Var}[\widetilde{S}] = \sum_{i=1}^{n} q_i(1 - q_i)b_i^2 = \mathrm{Var}[S] - \sum_{i=1}^{n}(q_i b_i)^2. \tag{3.67}$$

We see that S has a larger variance. If $\lambda_i = q_i$ then using a collective model results in risk averse decision makers tending to take more conservative decisions, see further Chapter 7. Also notice that the smaller $\sum_{i=1}^{n}(q_i b_i)^2$ is, the less the collective model will differ from the individual model.

Remark 3.8.1 (*The* collective model)
By *the* collective model for a portfolio, we mean a compound Poisson distribution as in (3.64) with $\lambda_i - q_i$. We also call it the *canonical* collective approximation.

In Exercise 3.8.3 we show that in the situation (3.62), *the* collective model can be obtained as well by replacing each claim X_i by a Poisson(1) number of independent claims with the same distribution as X_i. We can also do this if the random variables X_i are more general than those in (3.62). For example, assume that contract i produces claims $b_0 = 0, b_1, b_2, \ldots, b_n$ with probabilities $p_0, p_1, \ldots, p_n$. Since X_i equals exactly one of these values, we can write

$$X_i \equiv I_0 b_0 + I_1 b_1 + \cdots + I_n b_n, \tag{3.68}$$

with $I_j = 1$ if $X_i = b_j$, zero otherwise. So $\mathrm{Pr}[I_j = 1] = p_j$ for the marginal distributions of I_j, and their joint distribution is such that $I_0 + I_1 + \cdots + I_n \equiv 1$. One can show that if we choose the canonical collective model, we actually replace X_i by the compound Poisson distributed random variable Y_i, with

$$Y_i = N_0 b_0 + N_1 b_1 + \cdots + N_n b_n, \tag{3.69}$$

where the N_j are independent Poisson(p_j) random variables. In this way, the expected frequencies of all claim sizes remain unchanged.   ∇

Remark 3.8.2 (Model for an *open* portfolio)
The second proposed model with $\lambda_i = -\log(1 - q_i)$ can be used to model an open portfolio, with entries and exits not on renewal times. Assume that in a certain policy the waiting time W until death has an exponential(β) distribution. For the probability of no claims to be $1 - q$, we must have $\mathrm{Pr}[W > 1] = 1 - q$, so $\beta = -\log(1 - q)$. Now assume that, at the moment of death, each time we replace this policy by an identical one. Thus, we have indeed an open model for our portfolio. The waiting

times until death are always exponentially(β) distributed. But from the theory of Poisson processes, see also Exercise 2.5.7, we know that the number of deaths before time 1 is Poisson(β) distributed. In this model, we in fact replace, for each i, the ith policy by a Poisson($-\log(1-q_i)$) distributed number of copies. Since $I_i \sim \min\{N_i, 1\}$, the *open* collective model we get this way is a safe approximation to the individual model, as it allows for more claims per policy than one. See also Section 7.4.1.   ▽

Remark 3.8.3 (Negative risk amounts)
If we assume that the b_i are positive integers, then we can quickly calculate the probabilities for S, and consequently quickly approximate those for $\tilde{S}$, with Panjer's recursion. But if the b_i can be negative as well as positive, we cannot use this recursion. In that case, we can split up S in two parts $S = S^+ - S^-$ where S^+ is the sum of the terms Y_i in (3.63) with $b_i \geq 0$. By Theorem 3.4.2, S^+ and S^- are independent compound Poisson random variables with non-negative terms. The cdf of S can then be found by convolution of those of S^+ and S^-.

To find the stop-loss premium $E[(S-d)_+]$ for only one value of d, the convolution of S^+ and S^- is not needed. Conditioning on the total S^- of the negative claims, we can rewrite the stop-loss premium as follows:

$$E[(S-d)_+] = \sum_{x \geq 0} E[(S^+ - (x+d))_+] \Pr[S^- = x]. \tag{3.70}$$

To calculate this we only need the stop-loss premiums of S^+, which follow as a by-product of Panjer's recursion, see Example 3.5.4. Then the desired stop-loss premium can be calculated with a simple summation. For the convolution, a double summation is necessary, or it could be handled through the use of `convolve`.

Note that the FFT-technique, see Example 3.6.2, is not restricted to non-negative claim amounts.   ▽

3.9 Loss distributions: properties, estimation, sampling

In a compound model for losses, we have to specify both the claim number distribution and the claim severity distribution. For the former we often take the Poisson distribution, such as in the canonical or the open collective model, or when the assumptions of a Poisson process apply, see Chapter 4. In case of overdispersion, due to parameter uncertainty or cumulation of events, see Examples 3.3.1 and 3.3.2, we might use the negative binomial distribution. For some purposes, for example to compute premium reductions in case of a deductible, it is convenient to use a parametric distribution that fits the observed severity distribution well. Depending on the type of insurance at hand, candidates may vary from light tailed (for example the Gaussian distribution) to very heavy-tailed (Pareto). In this section we will present some severity distributions, explain their properties and suggest when to use them. We use maximum likelihood to estimate parameters. Often it is useful to generate pseudo-random samples from the loss distribution, for example if we want to compute the financial consequences of applying some risk management instrument like a complicated reinsurance scheme.

3.9.1 Techniques to generate pseudo-random samples

For many distributions, pseudo-random samples may be drawn by using standard R functions. They often use the *inversion method*, also known as the probability integral transform. It is based on the fact that if $U \sim \text{uniform}(0,1)$, then $F^{-1}(U) \sim F$ because $\Pr[F^{-1}(U) \le x] = \Pr[U \le F(x)] = F(x)$. For example the function rnorm in its standard mode applies the inverse normal cdf qnorm to results of runif, see also Appendix A.

The function runif to generate uniform pseudo-random numbers in R is state-of-the-art. Its default method is Mersenne-Twister, described in R's help-files as a "twisted generalized feedback shift register algorithm with period $2^{19937} - 1$ and equidistribution in 623 consecutive dimensions (over the whole period)".

Another sampling method is the *rejection method*. Suppose that it is hard to sample from density $f(\cdot)$, but that an easier to handle distribution $g(\cdot)$ exists satisfying $f(x) \le kg(x) \; \forall x$ for some appropriate bound $k \ge 1$. We get a random outcome from $f(\cdot)$ by sampling a point uniformly from the area below the graph of $f(x)$, and taking its x-coordinate; the probability of an outcome x or less is then just $F(x)$. John von Neumann's (1951) idea was not to do this directly, but to sample a random point below the graph of $kg(x)$, drawing its x-coordinate using, in most cases, the inversion method, and then its y-coordinate uniformly from $(0, kg(x))$. The x-coordinate is accepted as an outcome from $f(\cdot)$ if this random point happens to be under $f(x)$ as well. If rg() produces a random drawing from $g(x)$ and $f(x)/g(x) \le k$, an R-program to draw x randomly from $f(\cdot)$ could be as follows:

```
repeat {x <- rg(); if (runif(1) < f(x)/k/g(x)) break}
```

The number of points rejected is a geometric($1/k$) random variable, so the smaller k, the faster the random number generation.

In many cases, we can construct drawings from given distributions by using the fact that they are a simple transformation of other random variables for which a standard R-function r... exists to produce pseudo-random values. See Exercise 3.9.24 for some applications of this, including sampling from (log)normal, Pareto, Erlang, Weibull or Gompertz distributions.

In mixed models, one may first draw from the distribution of the conditioning random variable (structure variable), and next from the conditional distribution of the random variable of interest.

When the cdf $F(x) = G(x)H(x)$ for simple G and H, or the survival function is $1 - F(x) = (1 - G(x))(1 - H(x))$, a random value from F is produced by taking the maximum (minimum) of independent random variables distributed as G and

H. See Exercise 3.9.16 where this device is employed to sample from a Makeham distributed lifetime random variable.

3.9.2 Techniques to compute ML-estimates

We aim to obtain estimates for $\alpha, \beta, \ldots$ by maximizing the (log-)likelihood

$$\ell(\alpha, \beta, \ldots; \vec{y}) = \log \prod f_{Y_i}(y_i; \alpha, \beta, \ldots). \tag{3.71}$$

A a first step, we inspect the *normal equations* $\partial\ell/\partial\alpha = 0, \partial\ell/\partial\beta = 0, \ldots$ to see if they admit an explicit solution. This is the case for normal, lognormal, Poisson and inverse Gaussian samples. For Pareto, the optimal solution is explicit but it is not the solution to the normal equations. It may also happen that the normal equations provide a partial solution, often in the sense that the optimal parameters must be related in such a way that the fitted mean coincides with the sample mean. The advantage of this is that it reduces the dimension of the maximization problem to be solved; one of the parameters may be substituted away. This occurs with the negative binomial and the gamma distributions. If only one normal equation remains to be solved, this can be done using the R function `uniroot`, which is both reliable and fast. Alternatively, the optimization can be done using `optimize` to search an interval for a minimum or maximum of a real function. If needed, one can use `optim` for optimization in more dimensions.

3.9.3 Poisson claim number distribution

Not just by the Poisson process often underlying rare events, see Chapter 4, but also by the mere fact that it has only one parameter to estimate, the Poisson distribution is an attractive candidate for the claim number of the compound total loss. For a Poisson(λ) sample $Y_1 = y_1, \ldots, Y_n = y_n$, the loglikelihood is

$$\ell(\lambda; \vec{y}) = \log \prod f_{Y_i}(y_i; \lambda) = -n\lambda + \sum y_i \log \lambda - \sum \log y_i!, \tag{3.72}$$

which gives $\widehat{\lambda} = \bar{Y}$ as the maximum likelihood estimator of λ.

In insurance situations, often the numbers of claims pertain to policies that were not in force during a full calendar year, but only a known fraction of it. We denote this exposure for policy i by w_i. In that case, it follows from the properties of a Poisson process, see also Chapter 4, that the number of claims for policy i has a Poisson(λw_i) distribution. Therefore the loglikelihood is

$$\begin{aligned}\ell(\lambda; \vec{y}, \vec{w}) &= \log \prod f_{Y_i}(y_i; \lambda w_i) \\ &= -\sum \lambda w_i + \sum y_i \log \lambda + \sum y_i \log w_i - \sum \log y_i!,\end{aligned} \tag{3.73}$$

so in this case $\hat{\lambda} = \sum Y_i / \sum w_i$ is the maximum likelihood estimator of λ, that is, the number of claims divided by the total exposure.

If we consider Y_i/w_i, the number of claims per unit of exposure, one claim in a contract that was insured for only nine months counts as $\frac{4}{3}$. Then $\hat{\lambda}$ is the *weighted* average of these quantities, with weights $w_i / \sum w_i$. In practice, however, often simply a straight average of the number of claims per unit of exposure is taken.

Random sampling from the Poisson distribution is achieved by calling rpois. See Exercise 3.9.17 for a way to do this random sampling by the use of (3.11), which relates numbers of events to exponential waiting times.

3.9.4 Negative binomial claim number distribution

We have seen in Section 3.3 that sometimes it is proper to use a claim number distribution more spread than the Poisson, the variance of which equals the mean. In fact, both parameter uncertainty and cumulation in a Poisson process may lead to a negative binomial(r, p) claim number, see Examples 3.3.1 and 3.3.2, with overdispersion factor $\text{Var}[N]/\text{E}[N] = 1/p$. On the basis of a sample of outcomes $Y_1 = y_1, \ldots, Y_n = y_n$, we want to estimate the parameters r, p by maximum likelihood. In Table A one sees that the corresponding density equals

$$f_Y(y; r, p) = \binom{r+y-1}{y} p^r (1-p)^y, \quad y = 0, 1, \ldots \tag{3.74}$$

For non-integer r, the binomial coefficient is defined using gamma-functions, with $x! = \Gamma(x+1)$ for all real $x > 0$.

Now, just as in Theorem 3.4.2, let N_j, $j = 0, 1, \ldots$, count the number of times a sample element Y_i in the sample equals the value j. Observe that $\sum N_j \equiv n$ holds, while $\sum Y_i \equiv \sum j N_j$. In the loglikelihood of the total sample we find logarithms of the factorials in the density (3.74), as well as of the factorials arising because the places i in the sample where $Y_i = j$ occurs may be chosen arbitrarily, see (3.25). The latter are constant with respect to r and p, so we ignore them, and get

$$\ell(r, p; \vec{y}) = \log \prod_{j=0}^{\infty} \{f_Y(j; r, p)\}^{n_j} + \ldots$$

$$= \sum_j n_j \{\log(r+j-1) + \cdots + \log r - \log j!\} \tag{3.75}$$

$$+ rn \log p + \sum_j j n_j \log(1-p) + \ldots$$

Note that the outcomes $y_1, \ldots, y_n$ do not carry more information about the parameters r, p than do their frequencies $n_0, n_1, \ldots$ Apart from the order in which the iid observations occurred, the sample can be reconstructed from the frequencies. The conditional joint density of the sample, given the frequencies, no longer depends on the parameters r and p, or in other words, these frequencies are sufficient statistics. The iterative computation of the ML-estimates using the frequencies is of course faster for large integer-valued samples.

We first look at the partial derivative $\partial\ell/\partial p$. It must be zero for those r, p for which the likelihood is maximal:

$$0 = \frac{\partial\ell}{\partial p} = \frac{rn}{p} - \frac{\sum jn_j}{1-p} \iff \frac{r(1-p)}{p} = \bar{y} \iff p = \frac{r}{r+\bar{y}}, \tag{3.76}$$

This equation expresses that the mean of the ML-estimated density equals the observed sample mean.

The second ML-equation $\partial\ell/\partial r = 0$ results in

$$0 = \sum_{j=1}^{\infty} n_j \left(\frac{1}{r} + \cdots + \frac{1}{r+j-1} \right) + n\log p. \tag{3.77}$$

From this, no explicit expression for the ML-estimators can be derived. But substituting $p = r/(r+\bar{y})$ from (3.76) into (3.77) results in a one-dimensional equation for r that can be solved numerically using R. One can also use R's function `optimize` to do the maximization over r, or simply use `optim`; see below.

About generating negative binomial(r, p) random samples, see Table A, we first remark that in elementary probability texts, such random variables are introduced as the number of failures before the rth success in a sequence of Bernoulli trials, or sometimes as the number of trials needed, successes and failures combined. This requires r to be integer. For the general case, instead of the standard function `rnbinom` we can use the mixed model of Example 3.3.1 to generate negative binomial(r, p) outcomes, by first drawing the parameter from a suitable gamma distribution and next drawing from a Poisson distribution with that parameter. The following R commands draw a negative binomial sample, count the frequencies of each outcome and compare these with the theoretical frequencies:

```
set.seed(1); n <- 2000; r <- 2; p <- 0.5
hh <- rpois(n, rgamma(n,r,p/(1-p)))
n.j <- tabulate(1+hh); j <- 0:max(hh)
rbind(n.j, round(dnbinom(j,r,p)*n))
```

In the first line, we initialize the random number generator so as to be able to reproduce the results. The second line draws a sample using the method suggested. The function `tabulate` counts the frequencies of the numbers $1, 2, \ldots$ in its argument. We added 1 to each element of the sample to include the frequency of 0 as well. Running this script one sees that the observed frequencies match the theoretical ones, computed using `dnbinom`, quite closely.

To get initial estimates for the parameters, we use the method of moments:

$$\frac{r(1-p)}{p} = \bar{y}; \quad \frac{r(1-p)}{p^2} = \overline{y^2} - \bar{y}^2. \tag{3.78}$$

Solving these equations for r and p is done by dividing the first by the second:

```
y.bar <- sum(j*n.j/n); y2.bar <- sum(j^2*n.j/n)
p0 <- y.bar/(y2.bar-y.bar^2); r0 <- p0 * y.bar/(1-p0)
```

For how to solve the second ML-equation (3.77) using `uniroot` and inserting (3.76), see Exercise 3.9.26. To maximize the loglikelihood using the `optimize` function of R, the loglikelihood is computed by specifying `log=T` in the density and taking a weighted sum with as weights the frequencies n_j for all $j = 0, 1, \ldots$:

```
g <- function (r) {sum(dnbinom(j,r,r/(r+y.bar),log=T)*n.j)}
r <- optimize(g, c(r0/2, 2*r0), maximum=T, tol=1e-12)$maximum
p <- r/(r+y.bar)
```

We get $\hat{r} = 1.919248$ and $\hat{p} = 0.4883274$.

By using the general R function `optim`, we do not have to rely on the fact that in the optimum, $p = r/(r+\bar{y})$ must hold because of (3.76):

```
h <- function (x) {-sum(dnbinom(j,x[1],x[2],log=T)*n.j)}
optim(c(r0,p0), h, control=list(reltol=1e-14))
```

The first argument of `optim` is a vector of starting values, the second the function `h(x)` with x the vector of parameters. The first element of x represents r, the second is p. The `control` argument contains a list of refinements of the optimization process; to get the exact same results as before, we have set the relative tolerance to 10^{-14}. Note the minus-sign in the definition of h, needed because the standard mode of `optim` is to minimize a function

3.9.5 Gamma claim severity distributions

The gamma(α, β) distribution can be used to model non-negative losses if the tail of the cdf is not too 'heavy', such as in motor insurance for damage to the own vehicle. Density, cdf, quantiles and random deviates for this distribution are given by the standard R functions `dgamma`, `pgamma`, `qgamma` and `rgamma`, respectively. Note that the parameter α corresponds to the `shape` parameter of these functions, while β corresponds to `rate`, $1/\beta$ to `scale`. See Table A.

To find maximum likelihood estimators for the parameters α, β on the basis of a random sample $Y_1 = y_1, \ldots, Y_n = y_n$ from a gamma(α, β) distribution, proceed as follows. The loglikelihood $\ell(\alpha, \beta)$ of the parameters is given by

$$\ell(\alpha, \beta; \vec{y}) = \log \prod_{i=1}^{n} f_Y(y_i; \alpha, \beta). \qquad (3.79)$$

Filling in the gamma density $f_Y(y) = \frac{1}{\Gamma(\alpha)} \beta^\alpha y^{\alpha-1} e^{-\beta y}$, we get

$$\ell(\alpha, \beta) = n\alpha \log \beta - n \log \Gamma(\alpha) + (\alpha - 1) \log \prod y_i - \beta \sum y_i. \qquad (3.80)$$

One of the ML-equations ensures again that the fitted mean is the observed mean:

$$\frac{\partial \ell}{\partial \beta} = \frac{n\alpha}{\beta} - \sum y_i = 0 \iff \widehat{\beta} = \frac{\widehat{\alpha}}{\bar{y}}. \tag{3.81}$$

Writing $\overline{\log y}$ for the mean of the logarithms of the observations, we see from the other one that there is no explicit solution:

$$\frac{\partial \ell}{\partial \alpha} = n\log\beta - n\frac{\Gamma'(\alpha)}{\Gamma(\alpha)} + \sum \log y_i = 0 \iff$$
$$\log\widehat{\alpha} - \frac{\Gamma'(\widehat{\alpha})}{\Gamma(\widehat{\alpha})} - \log\bar{y} + \overline{\log y} = 0. \tag{3.82}$$

Using the R-function `digamma()` to compute the digamma (psi) function defined as $\Gamma'(\alpha)/\Gamma(\alpha)$, the solution can be found like this:

```
set.seed(2525); y <- rgamma(2000, shape=5, rate=1)
aux <- log(mean(y))  - mean(log(y))
f <- function(x) log(x) - digamma(x) - aux
alpha <- uniroot(f, c(1e-8,1e8))$root    ##  5.049
beta <- alpha/mean(y)                    ##  1.024
```

The interval $(10^{-8}, 10^8)$ in which a zero for f above is sought covers skewnesses from $+2 \times 10^{-4}$ to $+2 \times 10^4$. The function $f(x)$ decreases from $f(0) = +\infty$ to $f(\infty) < 0$. This is because $\log(x) - \Gamma'(x)/\Gamma(x)$ decreases to 0, and, as is proved in Exercise 9.3.12, aux is strictly positive unless all y_i are equal.

Just as for the negative binomial distribution, optimal $\widehat{\alpha}$ and $\widehat{\beta}$ can also be found by using `optimize` and $\widehat{\beta} = \widehat{\alpha}/\bar{y}$, or `optim`.

3.9.6 Inverse Gaussian claim severity distributions

A distribution that sometimes appears in the actuarial literature, for several purposes, is the inverse Gaussian (IG). Its properties resemble those of the gamma and lognormal distributions. Its name derives from the inverse relationship that exists between the cumulant generating functions of these distributions and those of Gaussian distributions, see (3.85) below. Various parameterizations are in use. Just like with the gamma distribution, we will use a shape parameter α and a scale parameter β. See also Table A.

The probability density function of the IG distribution is:

$$f(x; \alpha, \beta) = \frac{\alpha}{\sqrt{2\pi\beta}} x^{-\frac{3}{2}} e^{-\frac{(\alpha - \beta x)^2}{2\beta x}}, \quad x > 0. \tag{3.83}$$

The main reason the IG distribution has never gained much popularity is because it is not easy to manage mathematically. Indeed to prove that the density integrates to one is not at all trivial without knowing the corresponding cdf, which is:

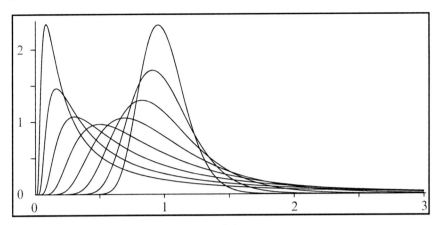

Fig. 3.1 Inverse Gaussian densities for $\alpha = \beta = \frac{1}{4},\frac{1}{2},1,2,4,8,16,32$ (tops from left to right).

$$F(x;\alpha,\beta) = \Phi\left(\frac{-\alpha}{\sqrt{\beta x}} + \sqrt{\beta x}\right) + e^{2\alpha}\Phi\left(\frac{-\alpha}{\sqrt{\beta x}} - \sqrt{\beta x}\right), \quad x > 0. \qquad (3.84)$$

This function has $\lim_{x\downarrow 0} F(x;\alpha,\beta) = 0$, $\lim_{x\to\infty} F(x;\alpha,\beta) = 1$ and its derivative is (3.83), which is non-negative on $(0,\infty)$. So (3.84) is an absolutely continuous cdf, and (3.83) is its density. Detailed proofs are asked in Exercise 3.9.3.

Using the fact that (3.83) is a density to compute the resulting integral, we can prove that the mgf equals

$$m(t;\alpha,\beta) = \exp\left\{\alpha\left[1 - \sqrt{1 - 2t/\beta}\right]\right\}, \quad t \le \frac{\beta}{2}. \qquad (3.85)$$

Notice that the mgf is finite for $t = \beta/2$, but not for $t > \beta/2$.

The special case with $\alpha = \beta$ is also known as the Wald distribution. From the mgf one easily sees that β is indeed a scale parameter, since βX is inverse Gaussian$(\alpha, 1)$ if $X \sim$ inverse Gaussian(α,β). We also see that adding two independent inverse Gaussian distributed random variables, with parameters α_1,β and α_2,β, yields an inverse Gaussian random variable with parameters $\alpha_1 + \alpha_2,\beta$. The expected value and the variance are α/β and α/β^2 respectively, just as for the gamma distribution. The easiest way to show this is by taking a series expansion of the cgf. The skewness is $3/\sqrt{\alpha}$, as opposed to $2/\sqrt{\alpha}$ for a gamma distribution with the same mean and variance. The flexibility of the inverse Gaussian distributions, from very skew to almost normal, is illustrated in Figure 3.1. All depicted distributions have the same mean $\alpha/\beta = 1$, and a strictly positive mode.

For the inverse Gaussian(α,β) distribution, the loglikelihood is

$$\ell(\alpha,\beta;\vec{y}) = \sum\left(\log\frac{\alpha}{\sqrt{2\pi\beta}} - \frac{3}{2}\log y_i - \frac{\alpha^2}{2\beta y_i} + \alpha - \frac{\beta y_i}{2}\right). \qquad (3.86)$$

Setting the partial derivatives equal to zero gives

$$\frac{\partial \ell}{\partial \alpha} = \sum \left(\frac{1}{\alpha} - \frac{\alpha}{\beta y_i} + 1 \right) = 0;$$

$$\frac{\partial \ell}{\partial \beta} = \sum \left(\frac{-1}{2\beta} + \frac{\alpha^2}{2\beta^2 y_i} - \frac{y_i}{2} \right) = 0. \tag{3.87}$$

Writing $\overline{1/y}$ for the average of the numbers $1/y_i$, we can rewrite this as

$$\frac{1}{\alpha} + 1 = \frac{\alpha}{\beta} \overline{1/y} \quad \text{and} \quad \frac{1}{\alpha} + \frac{\beta}{\alpha} \overline{y} = \frac{\alpha}{\beta} \overline{1/y}, \tag{3.88}$$

so we get very simple explicit expressions for the ML parameter estimates:

$$\widehat{\alpha} = \frac{1}{\overline{y}\,\overline{1/y} - 1} \quad \text{and} \quad \widehat{\beta} = \frac{\widehat{\alpha}}{\overline{y}}. \tag{3.89}$$

The second equation ensures that, again, the mean of the fitted distribution equals the sample mean. It is an easy exercise to show that in case of a sample $Y_1, \ldots, Y_n$, the quantities $\sum Y_i$ and $\sum 1/Y_i$ are sufficient statistics.

In the package statmod extending R one finds functions d/p/q/rinvgauss for density, cdf, quantiles and random number generation with this distribution. Similar functions can be found in SuppDists (then with capital G). The parameters used there are not the same we use; see also Exercise 3.9.12. They are the mean $\mu = \alpha/\beta$ and a precision parameter λ. The latter is taken in such a way that the variance α/β^2 equals μ^3/λ, therefore $\lambda = \alpha^2/\beta$. Conversely, $\alpha = \lambda/\mu$ and $\beta = \lambda/\mu^2$. Generating a random sample and estimating the parameters from it by maximum likelihood, using (3.89), then goes as follows:

```
library(statmod); set.seed(2525)
y <- rinvgauss(2000, mu=5, lambda=3)
alpha <- 1/(mean(y)*mean(1/y)-1); beta <- alpha/mean(y)
```

We get $\widehat{\alpha} = 0.626$; $\widehat{\beta} = 0.128$. So, $\widehat{\mu} = \widehat{\alpha}/\widehat{\beta} = 4.89$ and $\widehat{\lambda} = \widehat{\alpha}^2/\widehat{\beta} = 3.06$. The true values $\mu = \alpha/\beta = 5$, $\lambda = \alpha^2/\beta = 3$ give $\alpha = 0.6$, $\beta = 0.12$. In Exercise 3.9.12, the reader is asked to verify that we always get feasible estimates $\widehat{\alpha} > 0$ and $\widehat{\beta} > 0$ this way, in other words, that $\overline{y}\,\overline{1/y} > 1$ must hold in (3.89).

3.9.7 Mixtures/combinations of exponential distributions

Another useful class of parametric claim severity distributions, especially in the context of ruin theory (Chapter 4), consists of mixtures/combinations of exponential distributions. A *mixture* arises if the parameter of an exponential distribution is a random variable that is α with probability q and β with probability $1 - q$. The density is then given by

$$p(x) = q\alpha e^{-\alpha x} + (1-q)\beta e^{-\beta x}, \quad x > 0. \tag{3.90}$$

For each q with $0 \le q \le 1$, the function $p(\cdot)$ is a probability density function. But also for $q < 0$ or $q > 1$, $p(\cdot)$ in (3.90) is sometimes a pdf. Since $\int p(x)\, dx = 1$ always holds, we only have to check if $p(x) \ge 0$ for all x. From Exercise 3.9.4, we learn that it suffices to check $p(0) \ge 0$ and $p(\infty) \ge 0$. Assuming $\alpha < \beta$, this holds if $1 < q \le \beta/(\beta - \alpha)$, and in this case (3.90) is called a *combination* of exponential distributions.

An example of a proper combination of exponential distributions is given by

$$p(x) = 2(e^{-x} - e^{-2x}) = 2 \times 1e^{-1x} - 1 \times 2e^{-2x}, \tag{3.91}$$

which has $q = 2$, $\alpha = 1$ and $\beta = 2$. A second example is the function

$$p(x) = \frac{4}{3}(e^{-x} - \frac{1}{2}e^{-2x}) = \frac{4}{3} \times 1e^{-1x} - \frac{1}{3} \times 2e^{-2x}. \tag{3.92}$$

If $X \sim$ exponential(α) and $Y \sim$ exponential(β), with $\alpha \ne \beta$, then

$$m_{X+Y}(t) = \frac{\alpha\beta}{(\alpha - t)(\beta - t)} = \frac{\beta}{\beta - \alpha}\frac{\alpha}{\alpha - t} - \frac{\alpha}{\beta - \alpha}\frac{\beta}{\beta - t}. \tag{3.93}$$

This is the mgf of density (3.90) with $q = \frac{\beta}{\beta - \alpha}$. So a sum of independent exponential random variables has a combination of exponential distributions as its density. The reverse is not always true: (3.91) is the pdf of the convolution of an exponential(1) and an exponential(2) distribution, since $q = \beta/(\beta - \alpha) = 2$, but the pdf (3.92) cannot be written as such a convolution.

If $\alpha \uparrow \beta$, then $\beta/(\beta - \alpha) \to \infty$, and $X + Y$ tends to a gamma$(2, \beta)$ random variable. Hence, the gamma distributions with $r = 2$ are limits of densities that are combinations of exponential distributions, and the same holds for all gamma distributions with an integer shape parameter (so-called Erlang distributions).

A mixture of exponential distributions with parameters $0 < \alpha < \beta$ and $0 \le q \le 1$ can be generated using the urn-of-urns model $Z = IX/\alpha + (1 - I)Y/\beta$, with X, Y and I independent, X and $Y \sim$ exponential(1) and $I \sim$ Bernoulli(q). There is also a two-stage model that produces *all* random variables with pdf (3.90). For this, let $I \sim$ Bernoulli(γ) with $0 \le \gamma \le 1$, and let $0 < \alpha < \beta$. Then

$$Z = I\frac{X}{\alpha} + \frac{Y}{\beta} \tag{3.94}$$

has as its mgf

$$m_Z(t) = \left(1 - \gamma + \gamma\frac{\alpha}{\alpha - t}\right)\frac{\beta}{\beta - t} = \frac{\alpha\beta - t\beta(1 - \gamma)}{(\alpha - t)(\beta - t)}. \tag{3.95}$$

To show that this is the mgf of a combination or a mixture of exponential distributions, it suffices to find q, using partial fractions, such that (3.95) equals the mgf of

(3.90), which is

$$q\frac{\alpha}{\alpha-t} + (1-q)\frac{\beta}{\beta-t}. \tag{3.96}$$

Comparing (3.95) and (3.96) we see that $q\alpha + (1-q)\beta = \beta(1-\gamma)$, hence

$$q = \frac{\beta\gamma}{\beta-\alpha}. \tag{3.97}$$

Since $0 < \alpha < \beta$, we have that $0 \le q \le 1$ if $0 \le \gamma \le 1 - \alpha/\beta$, and then Z is a mixture of exponential distributions. If $1 - \alpha/\beta < \gamma \le 1$, then $q > 1$, and Z is a combination of exponential distributions.

The loss Z in (3.94) can be viewed as the result of an experiment where one suffers a loss Y/β in any case and where it is decided by a trial with probability γ of success whether one loses an additional amount X/α. Another interpretation is that the loss is drawn from either Y/β or $X/\alpha + Y/\beta$, since $Z = I(X/\alpha + Y/\beta) + (1 - I)Y/\beta$. If $\gamma = 1$, again a sum of two exponential distributions arises.

Writing R-functions to compute the cdf $P(x)$ and the density $p(x)$ is trivial; quantiles $x = P^{-1}(u)$ with $0 < u < 1$ follow by (numerical) inversion of the cdf, hence solving $P(x) = u$ for x. For this, one may call the R-function `uniroot`. But to generate random values, simply use model (3.94):

```
set.seed(1); n <- 2000; q <- 1.5; alpha <- 1; beta <- 2
gam <- (beta-alpha)/beta * q
y <- rbinom(n,1,gam) * rexp(n)/alpha + rexp(n)/beta
```

One way to estimate the parameters of the joint likelihood $\prod p(y_i; q, \alpha, \beta)$, with $p(x)$ as in (3.90), is by using the method of moments. Not for every combination of first three sample moments feasible parameters of (3.90) can be found leading to the right mean, variance and skewness; for details, consult Babier and Chan (1992), and see also Example 4.9.1. ML-optimization requires a three-dimensional maximization that cannot be easily reduced in dimension such as the ones we encountered before. But it is easy to simply let `optim` do the work, for example by:

```
f <- function (y, q, alpha, beta){
  q * alpha * exp(-alpha*y) + (1-q) * beta * exp(-beta*y)}
h <- function (x) {-sum(log(f(y, x[1], x[2], x[3])))}
optim(c(0.8, 0.9, 1.8), h)
```

The resulting parameter estimates are $\hat{q} = 1.285$, $\hat{\alpha} = 0.941$, $\hat{\beta} = 2.362$.

3.9.8 Lognormal claim severities

Using that $X \sim \text{lognormal}(\mu, \sigma^2)$ if and only if $\log X \sim N(\mu, \sigma^2)$, for the cdf and the density of this claim severity distribution, by the chain rule we get

$$F_X(x) = \Phi(\log x; \mu, \sigma^2) \quad \text{and} \quad f_X(x) = \frac{1}{x}\varphi(\log x; \mu, \sigma^2), \tag{3.98}$$

with φ and Φ the normal density and cdf. See also Table A.

It is easy to write R functions `d/p/q/rlnorm` based on these relations with the normal distribution, but such functions are also included in the package `stats`.

To compare the fatness of the tail with the one of the inverse Gaussian(α, β) distribution with the same mean α/β and variance α/β^2, note that the lognormal skewness is $\gamma = (3 + 1/\alpha)/\sqrt{\alpha}$, see Table A, while the inverse Gaussian skewness is $\gamma = 3/\sqrt{\alpha}$. The lognormal distribution is suitable as a severity distribution for branches with moderately heavy-tailed claims, like fire insurance.

Maximum likelihood estimation in case of a lognormal(μ, σ^2) random sample is simple by reducing it to the normal case. If the sample is $Y_1 = y_1, \ldots, Y_n = y_n$, let $X_i = \log Y_i$, then $X_i \sim N(\mu, \sigma^2)$, so as is well-known, $\widehat{\mu} = \overline{X}$ and $\widehat{\sigma}^2 = \overline{(X - \widehat{\mu})^2}$.

3.9.9 Pareto claim severities

The Pareto(α, x_0) distribution, see Table A, can be used for branches with high probability of large claims, notably liability insurance. In Exercise 3.9.1 it is proved that $Y \sim \text{Pareto}(\alpha, x_0)$ is equivalent to $\log(Y/x_0) \sim \text{exponential}(\alpha)$. With this property, it is easy to write the function `rpareto`; the other ones are trivial.

To compute ML-estimates for a Pareto(α, x_0) sample is slightly different from what we saw before, because in this case the optimal estimates cannot be produced by simply solving the ML equations. The loglikelihood with a Pareto(α, x_0) sample $Y_1 = y_1, \ldots, Y_n = y_n$ is, if $y_{(1)} := \min(y_i)$ denotes the sample minimum:

$$\ell(\alpha, x_0; \vec{y}) = \begin{cases} n\log \alpha + \alpha n \log x_0 - (\alpha + 1)\sum \log y_i & \text{if } x_0 \leq y_{(1)}; \\ -\infty & \text{if } x_0 > y_{(1)}. \end{cases} \qquad (3.99)$$

For each choice of α, we have $\ell(\alpha, x_0) \leq \ell(\alpha, y_{(1)})$, so the ML estimate for x_0 must be $\widehat{x}_0 = y_{(1)}$. Further,

$$\frac{\partial \ell(\alpha, \widehat{x}_0)}{\partial \alpha} = 0 \iff \frac{n}{\alpha} + n\log\widehat{x}_0 - \sum \log y_i = 0 \qquad (3.100)$$

$$\iff \widehat{\alpha} = \left(\frac{1}{n}\sum \log(y_i/\widehat{x}_0)\right)^{-1}.$$

The shape of the ML estimate $\widehat{\alpha}$ is not surprising, since the transformed sample $X_i = \log(Y_i/x_0)$ is distributed as exponential(α), for which case the ML estimate of α is known to be $1/\overline{X}$. For the same reason, a random Pareto(α, x_0) sample can be generated by multiplying the exponents of an exponential(α) random sample by x_0. So to draw a sample and find ML estimates of α, x_0, do

```
set.seed(2525); x0 <- 100; alpha <- 2; n <- 2000
y <- x0*exp(rexp(n)/alpha)
x0.hat <- min(y); alpha.hat <- 1/mean(log(y/x0.hat))
```

The resulting ML-estimates are $\widehat{x}_0 = 100.024$ and $\widehat{\alpha} = 2.0014$. Determining estimates for Pareto samples by the method of moments might present problems since the population moments $E[Y^j]$ only exist for powers $j < \alpha$.

3.10 Stop-loss insurance and approximations

The payment by a reinsurer in case of a stop-loss reinsurance with retention d for a loss S is equal to $(S - d)_+$. In this section we look for analytical expressions for the net stop-loss premium for some distributions. Note that expressions for stop-loss premiums can also be used to calculate net excess of loss premiums.

If $\pi(d)$ denotes the stop-loss premium for a loss with cdf $F(\cdot)$ as a function of d, then $\pi'(d + 0) = F(d) - 1$. This fact can be used to verify the expressions for stop-loss premiums. For the necessary integrations, we often use partial integration.

Example 3.10.1 (Stop-loss premiums for the normal distribution)
If $X \sim N(\mu, \sigma^2)$, what is the stop-loss premium for X if the retention is d?

As always for non-standard normal distributions, it is convenient to consider the case $\mu = 0$ and $\sigma^2 = 1$ first, and then use the fact that if $U \sim N(0,1)$, then $X = \sigma U + \mu \sim N(\mu, \sigma^2)$. The required stop-loss premium follows from

$$E[(X - d)_+] = E[(\sigma U + \mu - d)_+] = \sigma E\left[\left(U - \frac{d - \mu}{\sigma}\right)_+\right]. \tag{3.101}$$

Since $\varphi'(u) = -u\varphi(u)$, we have the following relation

$$\int_t^\infty u\varphi(u)\, du = \int_t^\infty [-\varphi'(u)]\, du = \varphi(t). \tag{3.102}$$

It immediately follows that

$$\pi(t) = E[(U - t)_+] = \varphi(t) - t[1 - \Phi(t)], \tag{3.103}$$

and hence

$$E[(X - d)_+] = \sigma\varphi\left(\frac{d - \mu}{\sigma}\right) - (d - \mu)\left[1 - \Phi\left(\frac{d - \mu}{\sigma}\right)\right]. \tag{3.104}$$

For a table with a number of stop-loss premiums for the standard normal distribution, we refer to Example 3.10.5 below. See also Table C at the end of this book. ∇

Example 3.10.2 (Gamma distribution)
Another distribution that has a rather simple expression for the stop-loss premium is the gamma distribution. If $S \sim \text{gamma}(\alpha, \beta)$ and $G(\cdot; \alpha, \beta)$ denotes the cdf of S, then it can be shown that

$$E[(S - d)_+] = \frac{\alpha}{\beta}[1 - G(d; \alpha + 1, \beta)] - d[1 - G(d; \alpha, \beta)]. \tag{3.105}$$

We can also derive expressions for the higher moments of the stop-loss payment $E[(S-d)_+^k]$, $k = 2,3,\ldots$. Even the mgf can be calculated analogously, and consequently also exponential premiums for the stop-loss payment. $\qquad\qquad\qquad\nabla$

Remark 3.10.3 (Moments of the retained loss)
Since either $S \le d$, so $(S-d)_+ = 0$, or $S > d$, so $(S-d)_+ = S-d$, the following equivalence holds in general:

$$[(S-d)-(S-d)_+][(S-d)_+] \equiv 0. \qquad (3.106)$$

With this, we can derive the moments of the retained loss $S - (S-d)_+$ from those of the stop-loss payment, using the equivalence

$$\begin{aligned} \{S-d\}^k &\equiv \{[S-d-(S-d)_+] + (S-d)_+\}^k \\ &\equiv \{S-d-(S-d)_+\}^k + \{(S-d)_+\}^k. \end{aligned} \qquad (3.107)$$

This holds since, due to (3.106), the remaining terms in the binomial expansion vanish. $\qquad\qquad\qquad\nabla$

In this way, if the loss approximately follows a translated gamma distribution, one can approximate the expected value, the variance and the skewness of the retained loss. See Exercise 3.10.4.

Example 3.10.4 (Stop-loss premiums approximated by NP)
The probabilities of $X > y$ for some random variable can be approximated quite well with the NP approximation. Is it possible to derive an approximation for the stop-loss premium for X too?
    Define the following auxiliary functions for $u \ge 1$ and $y \ge 1$:

$$q(u) = u + \frac{\gamma}{6}(u^2 - 1) \quad \text{and} \quad w(y) = \sqrt{\frac{9}{\gamma^2} + \frac{6y}{\gamma} + 1} - \frac{3}{\gamma}. \qquad (3.108)$$

From Section 2.5 we recall that $w(q(u)) = u$ and $q(w(y)) = y$. Furthermore, $q(\cdot)$ and $w(\cdot)$ are monotonically increasing, and $q(u) \ge y$ if and only if $w(y) \le u$. Let Z be a random variable with expected value 0, standard deviation 1 and skewness $\gamma > 0$. We will derive the stop-loss premiums of random variables X with $E[X] = \mu$, $\text{Var}[X] = \sigma^2$ and skewness γ from those of Z with the help of (3.101).
    The NP approximation (2.59) states that

$$\Pr[Z > q(u)] = \Pr[w(Z) > u] \approx 1 - \Phi(u) \quad \text{if } u \ge 1. \qquad (3.109)$$

Assume that $U \sim N(0,1)$ and define $V = q(U)$ if $U \ge 1$, $V = 1$ otherwise, so $V = q(\max\{U,1\})$. Then

$$\Pr[V > q(u)] = \Pr[U > u] = 1 - \Phi(u), \quad u \ge 1. \qquad (3.110)$$

Hence

$$\Pr[Z > y] \approx \Pr[V > y] = 1 - \Phi(w(y)), \quad y \ge 1. \qquad (3.111)$$

The stop-loss premium of Z in $d > 1$ can be approximated through the stop-loss premium of V, since

$$\int_d^\infty \Pr[Z > y]\,dy \approx \int_d^\infty \Pr[V > y]\,dy = \mathrm{E}[(V - d)_+]$$

$$= \int_{-\infty}^\infty (q(\max\{u, 1\}) - d)_+ \varphi(u)\,du \qquad (3.112)$$

$$= \int_{w(d)}^\infty (q(u) - d)\varphi(u)\,du.$$

To calculate this integral, we use that $\frac{d}{du}[u\varphi(u)] = (1 - u^2)\varphi(u)$, so

$$\int_t^\infty [u^2 - 1]\varphi(u)\,du = t\varphi(t). \qquad (3.113)$$

Substituting (3.102) and (3.113) and the function $q(\cdot)$ into (3.112) yields

$$\mathrm{E}[(Z - d)_+] \approx \int_{w(d)}^\infty \left(u + \frac{\gamma}{6}(u^2 - 1) - d\right)\varphi(u)\,du$$

$$\qquad (3.114)$$

$$= \varphi(w(d)) + \frac{\gamma}{6}w(d)\varphi(w(d)) - d[1 - \Phi(w(d))]$$

as an approximation for the net stop-loss premium for any risk Z with mean 0, variance 1 and skewness γ. $\qquad\qquad \triangledown$

Example 3.10.5 (Comparing various approximations of stop-loss premiums)
What are approximately the stop-loss premiums for X with $\mathrm{E}[X] = \mu = 0$, $\mathrm{Var}[X] = \sigma^2 = 1$ and skewness $\gamma = 0, 1, 2$, for retentions $d = 0, \frac{1}{2}, \dots, 4$?

To get NP-approximations we apply formula (3.104) if $\gamma = 0$, (3.114) otherwise. The parameters of a translated gamma distributed random variable with expected value 0, variance 1 and skewness γ are $\alpha = 4/\gamma^2$, $\beta = 2/\gamma$ and $x_0 = -2/\gamma$. For $\gamma \downarrow 0$, (3.105) yields the stop-loss premiums for a $N(0, 1)$ distribution. All gamma stop-loss premiums are somewhat smaller than the NP approximated ones.

The NP-approximation (3.114) yields plausible results for $d = 0$, but the results in Table 3.1 for gamma are surprising, in that the stop-loss premiums decrease with increasing skewness. From (3.116) below, it immediately follows that if all stop-loss premiums for one distribution are larger than those of another distribution with the same expected value, then the former has a larger variance. Since in this case the variances are equal, as well as larger stop-loss premiums of the translated gamma, there have to be smaller ones. With NP, lower stop-loss premiums for higher skewness might occur, for example, in case $d < 0$. Note that the translated gamma approximation gives the stop-loss premium for a risk with the right expected value and variance. On the other hand, NP gives approximate stop-loss premiums for a random variable with almost the same moments. Obviously, random variables exist having the NP tail probabilities in the area $d \in (0, \infty)$, as well as the correct first three moments.

Table 3.1 Approximate stop-loss premiums at various retentions for a standardized random variable with skewness 0, 1 and 2, using the CLT, the NP and the gamma approximations.

	CLT	Normal Power		gamma	
d	$\gamma = 0$	$\gamma = 1$	$\gamma = 2$	$\gamma = 1$	$\gamma = 2$
0.0	.3989	.4044	.4195	.3907	.3679
0.5	.1978	.2294	.2642	.2184	.2231
1.0	.0833	.1236	.1640	.1165	.1353
1.5	.0293	.0637	.1005	.0598	.0821
2.0	.0085	.0316	.0609	.0297	.0498
2.5	.0020	.0151	.0365	.0144	.0302
3.0	.0004	.0070	.0217	.0068	.0183
3.5	.0001	.0032	.0128	.0032	.0111
4.0	.0000	.0014	.0075	.0015	.0067

For arbitrary μ and σ, simply use a translation like the one given in (3.101). In that case, first determine $d = (t - \mu)/\sigma$, then multiply the corresponding stop-loss premium in the above table by σ, and if necessary, use interpolation. $\qquad \nabla$

3.10.1 Comparing stop-loss premiums in case of unequal variances

In this subsection we compare the stop-loss premiums of two risks with equal expected value, but with unequal variance. It is impossible to formulate an exact general rule, but we can state some useful approximating results.

Just as one gets the expected value by integrating the distribution function over $(0, \infty)$, one can in turn integrate the stop-loss premiums. In Exercise 3.10.1, the reader is invited to prove that, if $U \geq 0$ with probability 1,

$$\frac{1}{2}\text{Var}[U] = \int_0^\infty \left\{ \text{E}[(U - t)_+] - (\mu - t)_+ \right\} dt. \tag{3.115}$$

The integrand in this equation is always non-negative. From (3.115), it follows that if U and W are risks with equal expectation μ, then

$$\int_0^\infty \left\{ \text{E}[(U - t)_+] - \text{E}[(W - t)_+] \right\} dt = \frac{1}{2} \left\{ \text{Var}[U] - \text{Var}[W] \right\}. \tag{3.116}$$

By approximating the integral in (3.116) with the trapezoidal rule with interval width 1, see also (3.47), we can say the following about the total of all differences in the stop-loss premiums of U and W (notice that we do not use absolute values):

$$\sum_{i=1}^\infty \left\{ \text{E}[(U - i)_+] - \text{E}[(W - i)_+] \right\} \approx \frac{1}{2} \left\{ \text{Var}[U] - \text{Var}[W] \right\}. \tag{3.117}$$

So, if we replace the actual stop-loss premiums of U by those of W, then (3.117) provides an approximation for the total error in all integer-valued arguments. In Chapter 7 we examine conditions for $E[(U-d)_+] \geq E[(W-d)_+]$ to hold for all d. If that is the case, then all terms in (3.117) are positive and consequently, the maximum error in all of these terms will be less than the right-hand side.

If two integrands are approximately proportional, their ratio is about equal to the ratio of the corresponding integrals. So from (3.115) we get:

$$\frac{E[(U-t)_+] - (\mu-t)_+}{E[(W-t)_+] - (\mu-t)_+} \approx \frac{Var[U]}{Var[W]}. \qquad (3.118)$$

The approximation is exact if $\mu = E[U]$ and $W = (1-I)\mu + IU$ with $I \sim \text{Bernoulli}(\alpha)$ independent of U and $\alpha = Var[W]/Var[U]$, see Exercise 3.10.2.

If $t \geq \mu$, then $(\mu-t)_+ = 0$, so the approximation (3.118) simplifies to the following rule of thumb:

Rule of thumb 3.10.6 (Ratio of stop-loss premiums)
For retentions t larger than the expectation $\mu = E[U] = E[W]$, we have for the stop-loss premiums of risks U and W:

$$\frac{E[(U-t)_+]}{E[(W-t)_+]} \approx \frac{Var[U]}{Var[W]}. \qquad (3.119)$$

This rule works best for intermediate values of t, see below. ▽

Example 3.10.7 ('Undefined wife')
Exercise 3.7.4 deals with the situation where it is unknown for which of the insureds a widow's benefit might have to be paid. If the frequency of being married is 80%, we can either multiply all risk amounts by 0.8 and leave the probability of dying within one year as it is, or we can multiply the mortality probability by 0.8 and leave the payment as it is. We derived that the resulting variance of the total claim amount in the former case is approximately 80% of the variance in the latter case. So, if we use the former method to calculate the stop-loss premiums instead of the correct method, then the resulting stop-loss premiums for retentions that are larger than the expected claim cost are approximately 20% too small. ▽

Example 3.10.8 (Numerical evaluation of the Rule of thumb)
We calculated the stop-loss premiums for a $N(0,1.01)$ and a $N(0,1.25)$ distribution at retentions $d = 0, \frac{1}{2}, 1, \ldots, 3$, to compare them with those of a $N(0,1)$ distribution. According to Rule of thumb 3.10.6, these should be 1.01 and 1.25 times as big respectively. Table 3.2 gives the factor by which that factor should be multiplied to get the real error. For example, for $d = 0$ the quotient $\pi(d;0,1.01)/\pi(d;0,1)$ equals 1.005 instead of 1.01, so the error is only 50% of the one predicted by the Rule of thumb. As can be seen, the Rule of thumb correction factor is too large for retentions close to the expected value, too small for large retentions and approximately correct for retentions equal to the expected value plus 0.6 standard deviation. The Rule of thumb correction factor has a large error for retentions in the far tail where the

Table 3.2 Factors by which the N(0, 1.01) and N(0, 1.25) stop-loss premiums deviate from those of N(0, 1), expressed in terms of the Rule of thumb correction factor

d	$\pi(d;0,1)$	Correction factors	
		$1+0.01\times$	$1+0.25\times$
0.0	0.39894	0.50	0.47
0.5	0.19780	0.89	0.85
1.0	0.08332	1.45	1.45
1.5	0.02931	2.22	2.35
2.0	0.00849	3.20	3.73
2.5	0.00200	4.43	5.84
3.0	0.00038	5.92	9.10

stop-loss premiums of the distribution with the smaller variance are negligible but those of the distribution with the larger variance are not.   ∇

3.11 Exercises

Section 3.2

1. Calculate (3.3), (3.4) and (3.5) in case N has the following distribution: a) Poisson(λ), b) binomial(n,p) and c) negative binomial(r,p).
2. Give the counterpart of (3.5) for the cumulant generating function.
3. Assume that the number of eggs in a bird's nest is a Poisson(λ) distributed random variable, and that the probability that a female hatches out equals p. Determine the distribution of the number of female hatchlings in a bird's nest.
4. Let S be compound Poisson distributed with $\lambda = 2$ and $p(x) = x/10, x = 1,2,3,4$. Apply (3.10) to calculate the probabilities of $S = s$ for $s \leq 4$.
5. Complete the table in Example 3.2.2 for $x = 0,\dots,6$. Determine the expected value and the variance of N, X and S.
6. Determine the expected value and the variance of S, where S is defined as in Example 3.2.2, except that N is Poisson distributed with $\lambda = 2$.
7. Prove relation (3.11) by partial integration. Do the same by differentiating both sides of the equation and examining one value, either $x = 0$ or $x \to \infty$.

Section 3.3

1. Show that the Poisson distribution also arises as the limit of the negative binomial(r,p) distribution if $r \to \infty$ and $p \to 1$ such that $r(1 - p) = \lambda$ remains constant.
2. Under which circumstances does the usual Poisson distribution arise instead of the negative binomial in Examples 3.3.1 and 3.3.2?
3. [♠] Prove (3.19).

Section 3.4

1. The same as Exercise 3.2.4, but now with the sparse vector algorithm.

2. What happens with (3.23) if some x_i are equal in (3.22)?

3. Assume that S_1 is compound Poisson with $\lambda_1 = 4$ and claims $p_1(j) = \frac{1}{4}, j = 0, 1, 2, 3$, and S_2 is also compound Poisson with $\lambda_2 = 2$ and $p_2(j) = \frac{1}{2}, j = 2, 4$. If S_1 and S_2 are independent, then what is the distribution of $S_1 + S_2$?

4. In Exercise 3.2.3, prove that the number of males is independent of the number of females.

5. Let N_j, $j = 1, 2$, denote the number of claims of size j in Example 3.2.2. Are N_1 and N_2 independent?

6. Assume that S is compound Poisson distributed with parameter λ and with discrete claims distribution $p(x)$, $x > 0$. Consider S_0, a compound Poisson distribution with parameter $\lambda_0 = \lambda/\alpha$ for some α with $0 < \alpha < 1$, and with claims distribution $p_0(x)$ where $p_0(0) = 1 - \alpha$ and $p_0(x) = \alpha p(x)$ for $x > 0$. Prove that S and S_0 have the same distribution by comparing their mgfs. Also show that $S \sim S_0$ holds because the frequencies of the claim amounts $x \neq 0$ in (3.22) have the same distribution.

7. How many multiplications *with non-zero numbers* does the sparse vector algorithm of Example 3.4.3 take to compute all probabilities $\Pr[S = x], x = 0, 1, \ldots, n-1$? Assume the claim sizes to be bounded by T, and remember that $1 + 1/2 + 1/3 + \cdots + 1/T \approx \log T + 0.5772$ (the Euler-Mascheroni constant).

8. Redo Exercise 3.4.1 using R.

Section 3.5

1. The same as Exercise 3.2.4, but now with Panjer's recursion relation.

2. The same as Exercise 3.4.6, first part, but now by proving with induction that Panjer's recursion yields the same probabilities $f(s)$.

3. Verify Example 3.5.2.

4. In case of a compound Poisson distribution for which the claims have mass points $1, 2, \ldots, m$, determine how many multiplications have to be done to calculate the probability $F(t)$ using Panjer's recursion. Distinguish the cases $m < t$ and $m \geq t$.

5. Prove that $E[N] = (a+b)/(1-a)$ if $q_n = \Pr[N = n]$ satisfies (3.26).

6. In Example 3.5.4, determine the retention d for which $\pi(d) = 0.3$.

7. Let N_1, N_2 and N_3 be independent and Poisson(1) distributed. For the retention $d = 2.5$, determine $E[(N_1 + 2N_2 + 3N_3 - d)_+]$.

8. Assume that S_1 is compound Poisson distributed with parameter $\lambda = 2$ and claim sizes $p(1) = p(3) = \frac{1}{2}$. Let $S_2 = S_1 + N$, where N is Poisson(1) distributed and independent of S_1. Determine the mgf of S_2. What is the corresponding distribution? Determine $\Pr[S_2 \leq 2.4]$.

9. Determine the parameters of an integer-valued compound Poisson distributed Z if for some $\alpha > 0$, Panjer's recursion relation equals $\Pr[Z = s] = f(s) = \frac{\alpha}{s}[f(s-1) + 2f(s-2)]$, $s = 1, 2, 3, \ldots$ [Don't forget the case $p(0) \neq 0!$]

10. Assume that S is compound Poisson distributed with parameter $\lambda = 3$, $p(1) = \frac{5}{6}$ and $p(2) = \frac{1}{6}$. Calculate $f(x)$, $F(x)$ and $\pi(x)$ for $x = 0, 1, 2, \ldots$. Also calculate $\pi(2.5)$.

11. Derive formulas from (3.34) for the stop-loss premium that only use $f(0), f(1), \ldots, f(d-1)$ and $F(0), F(1), \ldots, F(d-1)$ respectively.

12. Give a formula, analogous to (3.36), to calculate $E[(S-d)_+^2]$.

13. [♠] Write functions `Panjer.NegBin` and `Panjer.Bin` to handle (negative) binomial claim numbers.

14. Let $M \sim$ negative binomial(r, p) for some r, p with $0 < p < 1$. Which distribution does L have, if $\Pr[L = m] = \lim_{r \to 0} \Pr[M = m \,|\, M > 0]$, $m = 1, 2, \dots$? Show that, for this random variable, relation (3.26) holds for all $n = 2, 3, \dots$

15. [♠] By a slight extension of the proof of Theorem 3.5.1, it can be shown that if (3.26) holds for all $n = 2, 3, \dots$, the probability of a claim total s satisfies the following relation:

$$f(s) = \frac{[q_1 - (a+b)q_0]f(s) + \sum_{h=1}^{s}(a + \frac{bh}{s})p(h)f(s-h)}{1 - ap(0)}, \quad s = 1, 2, \dots$$

Here q_0 is the arbitrary amount of probability at zero given to the frequency distribution. If $q_1 = (a+b)q_0$, the first term in the numerator vanishes, resulting in (3.27). The class of counting distributions satisfying (3.26) for $n > k$ is known as the (a, b, k) class, $k = 0, 1, \dots$. The $(a, b, 1)$ class includes zero-modified or zero-truncated distributions. By the previous exercise, the logarithmic distribution is in the $(a, b, 1)$ class, but not in the $(a, b, 0)$ class. Write a Panjering function for the $(a, b, 1)$ class.

16. Using the R- functions `cumsum` and `rev`, show how the vector of stop-loss premiums at retention $d = 0, 1, \dots$ can be obtained from a vector of probabilities p with p[1] = $\Pr[S = 0]$, p[2] = $\Pr[S = 1]$, and so on. Use relation (3.34).

Section 3.6

1. Describe what is produced by the R-calls:

```
y <- rep(0,64); y[2:7] <- 1/6; Re(fft(fft(y)^10,T))/64
```

2. Use the `fft` function to compute the stop-loss premium at retention $d = 10$ for a compound distribution with claim sizes 1 and 3 each with probability $\frac{1}{2}$ and as claim number a logarithmic random variable L with $E[L] = 3$, see Example 3.3.2. The same when $\Pr[L = 0, 1, 3, 7, 8, 9, 10] = .3, .2, .1, .1, .1, .1, .1$.

Section 3.7

1. Assume that S is compound Poisson distributed with parameter $\lambda = 12$ and uniform$(0, 1)$ distributed claims. Approximate $\Pr[S < 10]$ with the CLT approximation, the translated gamma approximation and the NP approximation.

2. Assume that S is compound Poisson distributed with parameter $\lambda = 10$ and $\chi^2(4)$ distributed claims. Approximate the distribution function of S with the translated gamma approximation. With the NP approximation, estimate the quantile s such that $F_S(s) \approx 0.95$, as well as the probability $F_S(E[S] + 3\sqrt{\text{Var}[S]})$.

3. Prove Theorem 3.7.1 by proving, just as in the proof of Theorem 3.5.1, that as $\lambda \to \infty$, the mgf of a standardized compound Poisson(λ, X) random variable $(S - \mu_S)/\sigma_S$ converges to $e^{t^2/2}$ for all t.

Section 3.8

1. Show that $\lambda_j = -\log(1 - q_j)$ yields both a larger expectation and a larger variance of S in (3.63) than $\lambda_j = q_j$ does. For both cases, compare $\Pr[I_i = j]$ and $\Pr[N_i = j]$, $j = 0, 1, 2, \ldots$ in (3.62) and (3.63), as well as the cdfs of I_i and N_i.

2. Consider a portfolio of 100 one-year life insurance policies that are evenly divided between the insured amounts 1 and 2 and probabilities of dying within this year 0.01 and 0.02. Determine the expectation and the variance of the total claims $\widetilde{S}$. Choose an appropriate compound Poisson distribution S to approximate $\widetilde{S}$ and compare the expectations and the variances. Determine for both S and $\widetilde{S}$ the parameters of a suitable approximating translated gamma distribution.

3. Show, by comparing the respective mgfs, that the following representations of *the* collective model are equivalent:

   1. The compound Poisson distribution specified in (3.64) with $\lambda_i = q_i$.
   2. The random variable $\sum_i N_i b_i$ from (3.63) with $\lambda_i = q_i$.
   3. The random variable $Z_1 + \cdots + Z_n$ where the Z_i are compound Poisson distributed with claim number parameter 1 and claims distribution equal to those of $I_i b_i$.
   4. The compound Poisson distribution with parameter $\lambda = n$ and claims distribution $Q(x) = \frac{1}{n} \sum_j \Pr[X_j \leq x]$. [Hence $Q(\cdot)$ is the arithmetic mean of the cdfs of the claims. It can be interpreted as the cdf of a claim from a *randomly* chosen policy, where each policy has probability $\frac{1}{n}$.]

4. In a portfolio of n one-year life insurance policies for men, the probability of dying in this year equals q_i for the ith policyholder. In case of death, an amount b_i has to be paid out, but only if it turns out that the policy holder leaves a widow behind. This information is not known to the insurer in advance ('undefined wife'), but it is known that this probability equals 80% for each policy. In this situation, we can approximate the individual model by a collective one in two ways: by replacing the insured amount for policy i by $0.8b_i$, or by replacing the claim probability for policy i by $0.8q_i$. Which method is correct? Determine the variance of the total claims for both methods. Show how we can proceed in both cases, if we have a program at our disposal that calculates stop-loss premiums from a mortality table and an input file containing the sex, the age and the risk amount.

5. [♠] At what value of x in (3.70) may we stop the summation if an absolute precision ε is required?

6. Consider a portfolio with 2 classes of policies. Class i contains 1000 policies with claim size $b_i = i$ and claim probability 0.01, for $i = 1, 2$. Let B_i denote the number of claims in class i. Write the total claims S as $S = B_1 + 2B_2$ and let $N = B_1 + B_2$ denote the number of claims. Consider the compound binomial distributed random variable $T = X_1 + X_2 + \cdots + X_N$ with $\Pr[X_i = 1] = \Pr[X_i = 2] = 1/2$. Compare S and T as regards the maximum value, the expected value, the variance, the claim number distribution and the distribution. Do the same for B_1 and $B_2 \sim \text{Poisson}(10)$.

7. Consider an excess of loss reinsurance on some portfolio. In case of a claim x, the reinsurer pays out an amount $h(x) = (x - \beta)_+$. The claims process is compound Poisson with claim number parameter 10 and uniform(1000,2000) distributed claim sizes. For $\beta \in [1000, 2000]$, determine the distribution of the total amount to be paid out by the reinsurer in a year.

8. Consider two portfolios P_1 and P_2 with the following characteristics:

	Risk amount	Number of policies	Claim probability
P_1	z_1	n_1	q_1
	z_2	n_2	q_2
P_2	z_1	$2n_1$	$\frac{1}{3}q_1$
	z_2	$2n_2$	$\frac{1}{2}q_2$

For the *individual* risk models for P_1 and P_2, determine the difference of the variance of the total claims amount. Check if the *collective* approximation of P_1 equals the one of P_2, both constructed with the recommended methods.

9. A certain portfolio contains two types of contracts. For type k, $k = 1,2$, the claim probability equals q_k and the number of policies equals n_k. If there is a claim, then with probability $p_k(x)$ it equals x, as follows:

	n_k	q_k	$p_k(1)$	$p_k(2)$	$p_k(3)$
Type 1	1000	0.01	0.5	0	0.5
Type 2	2000	0.02	0.5	0.5	0

Assume that all policies are independent. Construct a collective model T to approximate the total claims. Make sure that both the expected number of positive claims and the expected total claims agree. Give the simplest form of Panjer's recursion relation in this case; also give a starting value. With the help of T, approximate the capital that is required to cover all claims in this portfolio with probability 95%. Use an approximation based on three moments, and compare the results with those of Exercise 2.5.13.

10. Consider a portfolio containing n contracts that all produce a claim 1 with probability q. What is the distribution of the total claims according to the individual model, *the* collective model and the *open* collective model? If $n \to \infty$, with q fixed, does the individual model S converge to the collective model T, in the sense that the difference of the probabilities $\Pr[(S - \mathrm{E}[S])/\sqrt{\mathrm{Var}[S]} \le x] - \Pr[(T - \mathrm{E}[S])/\sqrt{\mathrm{Var}[S]} \le x]$ converges to 0?

Section 3.9

1. Determine the mean and the variance of the lognormal and the Pareto distribution, see also Table A. Proceed as follows: $Y \sim$ lognormal(μ, σ^2) means $\log Y \sim \mathrm{N}(\mu, \sigma^2)$; if $Y \sim$ Pareto(α, x_0), then $Y/x_0 \sim$ Pareto$(\alpha, 1)$ and $\log(Y/x_0) \sim$ exponential(α).

2. Determine which parameters of the distributions in this section are scale parameters, in the sense that λX, or more general $f(\lambda)X$ for some function f, has a distribution that does not depend on λ. Show that neither the skewness γ_X nor the coefficient of variation σ_X/μ_X depend on such parameters. Determine these two quantities for the given distributions.

3. [♠] Prove that the expression in (3.84) is indeed a cdf that is 0 in $x = 0$, tends to 1 for $x \to \infty$ and has a positive derivative (3.83). Also verify that (3.85) is the mgf, and confirm the other statements about the inverse Gaussian distributions.

4. Show that the given conditions on q in (3.90) are sufficient for $p(\cdot)$ to be a pdf.

5. Determine the cdf $\Pr[Z \le d]$ and the stop-loss premium $\mathrm{E}[(Z - d)_+]$ for a mixture or combination Z of exponential distributions as in (3.90). Also determine the conditional distribution of $Z - z$, given $Z > z$.

6. Determine the mode of mixtures and combinations of exponential distributions. Also determine the mode and the median of the lognormal distribution.

7. Determine the mode of the inverse Gaussian(α, α) distribution. For the parameter values of Figure 3.1, use your computer to determine the median of this distribution.

8. Write R-functions `d/p/q/rCombExp` for mixtures/combinations of exponential distributions.

9. If $E[X] = 1$ and $\sqrt{\mathrm{Var}[X]}/E[X] = 2$ is the coefficient of variation, find the 97.5% quantile if $X \sim$ gamma, inverse Gaussian, lognormal, and normal.

10. [♠] A small insurer has three branches: Fire, Liability and Auto. The total claims on the branches are all independent compound Poisson random variables. The specifications are:

   - Fire: $\lambda_1 = 10$ is the mean claim number; claim sizes are inverse Gaussian with mean $\mu_1 = 10000$ and coefficient of variation $\sigma_1/\mu_1 = 0.5$;
   - Liability: $\lambda_2 = 3$; Pareto claim sizes with mean 30000 and coefficient of variation 0.5;
   - Auto: $\lambda_3 = 500$; gamma claim sizes with mean 1000 and coefficient of variation 0.3.

   Using the Normal Power approximation, compute the Value-at-Risk (quantiles s such that $\Pr[S > s] = q$) for the combined portfolio at the levels $q = 0.5, 0.75, 0.9, 0.95, 0.99$. The premium income $P = P_1 + P_2 + P_3$ contains a safety loading 30% for Fire, 50% for Liability and 10% for Auto. It is invested in such a way that its value after a year is lognormal with mean $1.1P$ and standard deviation $0.3P$. Find the probability that the invested capital is enough to cover the claims. Find the initial capital that is needed to ensure that with probability 99%, the insurer does not get broke by the claims incurred and/or the investment losses.

11. [♠] In the situation of Remark 3.8.3, find the probability vectors of a compound Poisson(50, $p(7) = p(10) = p(19) = 1/3$) distribution for S^+ and compound Poisson(3, $p(1) = p(4) = 1/2$) for S^-. Use either FFT or Panjer's recursion. Then, compute the stop-loss premiums at $d = \lfloor E[S] + k\sigma_S \rfloor$ for $k = 1, 2, 3$ using (3.70). Also, compute the whole probability vector of $S = S^+ - S^-$ using `convolve`. Finally, use the fact that S itself is also a compound Poisson random variable (with possible claim sizes $-1, -4, +7, +10, +19$) to compute the vector of probabilities $\Pr[S = s]$ using the FFT-technique.

12. Prove that (3.89) produces positive estimates $\hat{\alpha}$ and $\hat{\beta}$. Hint: apply Jensen's inequality to the r.v. Z with $\Pr[Z = y_i] = \frac{1}{n}, i = 1, \dots, n$.

13. Prove that `aux > 0`, so $\overline{\log y} < \log \bar{y}$, holds when computing ML-estimates for a gamma sample using the R program involving `uniroot` given.

14. [♠] For range $y > 0$, shape parameter $\alpha > 0$ and scale parameter $\beta > 0$, a Weibull(α, β) random variable Y has density $f(y; \alpha, \beta) = \alpha\beta(\beta y)^{\alpha-1}\exp\left(-(\beta y)^\alpha\right)$. Find the cdf of Y. Show that $X = (\beta Y)^\alpha$ has an exponential(1) distribution. Also show that the ML equations for estimating α and β using a sample $Y_1 = y_1, \dots, Y_n = y_n$ can be written as $g(\alpha) := 1/\alpha + \overline{\log y} - \overline{y^\alpha \log y}/\overline{y^\alpha} = 0$ and $\beta = (1/\overline{y^\alpha})^{1/\alpha}$. Using `rweibull`, generate a Weibull sample of size 2000 with parameters $\alpha = 5$ and $\beta = 1$, and compute the ML estimates for the parameters from this sample.

15. Write a function `rTransGam` to generate random deviates from translated gamma distributions. Draw a large sample and verify if the first three sample moments are 'close' to the theoretical values.
   Identify the problem with programming `rNormalPower`.

16. In life actuarial science, it is sometimes useful to be able to generate samples from Gompertz and Makeham lifetime distributions. The mortality rate of Makeham's law equals

$$\mu_X(x) \stackrel{\text{def}}{=} \frac{f_X(x)}{1 - F_X(x)} = -\frac{\mathrm{d}}{\mathrm{d}x}\log(1 - F_X(x)) = a + bc^x,$$

while Gompertz' law is the special case with $a = 0$. Assuming $c > 1$, the second component of the mortality rate increases exponentially with age; the first is a constant. Since the minimum $X = \min(Y, Z)$ of two independent random variables is easily seen to have $\mu_X(x) = \mu_Y(x) +$

$\mu_Z(x)$ as its mortality rate, under Makeham's law people either die from 'senescence' after time $Y \sim$ Gompertz(b,c), or from some 'accident' occurring independently after time $Z \sim$ exponential(a), whichever happens first. In the parameterization given, X has the following survival function:

$$1 - F_X(x) = \Pr[Z > x]\Pr[Y > x] = \exp\left(-ax - \frac{b}{\log c}(c^x - 1)\right), \quad x > 0.$$

If $Y \sim$ Gompertz(b,c), show that $c^Y - 1$ is exponential$(\frac{b}{\log c})$. Use this fact to write a function rGompertz to generate random deviates from the Gompertz distribution. Use this function to write a function rMakeham. Verify if your results make sense using, e.g., $a = 0.0005$, $b = 0.00007$, $c = 1.1$.

17. Use relation (3.11) to construct an R function to do one drawing from a Poisson(λ) distribution.

18. [♠] If U_1, U_2 are independent uniform$(0,1)$ random variables, show that $\tan(\pi(U_1 - \frac{1}{2})) \sim$ Cauchy$(0,1)$. Also show that $\Phi^{-1}(U_1)/\Phi^{-1}(U_2)$ has this same distribution. [Because it has fat tails and an easy form, this distribution is often used to find a majorizing function in the rejection method.]

19. Apply the rejection method of Section 3.9.1 to construct an R-function to draw from the triangular distribution $f(x) = x$ on $(0,1)$, $2 - x$ on $(1,2)$. Use the uniform$(0,2)$ density to get an upper bound for $f(x)$.
    Also, draw from $f(x)$ by using the fact that $U_1 + U_2 \sim f(x)$ if the U_i are iid uniform$(0,1)$.

20. The rejection method can also applied to draw from integer-valued random variables. In fact, to generate values from N, generate a value from $N + U$ and round down. Here $U \sim$ uniform$(0,1)$, independent of N. Sketch the density of $N + U$. Apply to a binomial$(3, p)$ distribution.

21. What is produced by sum(runif(10)<1/3)? What by sum(runif(12))-6?

22. [♠] For the inverse Gaussian distributions, show that $\sum Y_i$ and $\sum \frac{1}{Y_i}$ are jointly sufficient. Recall that this may be proved by using the factorization criterion, that is, by showing that g and h exist such that the joint density of $Y_1, \ldots, Y_n$ can be factorized as $f_{Y_1,\ldots,Y_n}(y_1,\ldots,y_n; \alpha, \beta) = g\left(\sum y_i, \sum \frac{1}{y_i}; \alpha, \beta\right) h(y_1, \ldots, y_n)$.

23. [♠] Let $X \sim$ gamma$(\alpha + 1, 1)$, $\alpha > 0$, and $U \sim$ uniform$(0,1)$ be independent. Prove that $m_{\log X}(t) = \frac{\Gamma(\alpha + t + 1)}{\Gamma(\alpha + 1)}$ and $m_{\frac{1}{\alpha}\log U}(t) = \frac{1}{1 + t/\alpha}$. What is the distribution of $XU^{1/\alpha}$?

24. Let $U \sim$ uniform$(0,1)$. Name the distribution of the following transformed random variables:

    a) $B = I_{(0,p)}(U)$ and $N = \sum_1^n B_i$ with $B_i \sim B$ iid (see Exercise 3.9.21);
    b) $X = \Phi^{-1}(U)$ and $\mu + \sigma X$;
    c) $e^{\mu + \sigma X}$;
    d) $Y = -\log U$ and Y/β;
    e) $\sum_1^n Y_i/\beta$ with $Y_i \sim Y$ iid;
    f) $e^Y (= 1/U)$ and $x_0 e^{Y/\beta}$;
    g) $Y^{1/\alpha}/\beta$ (see Exercise 3.9.14);
    h) $\log(1 + Y\log(c)/b)/\log(c)$ (see Exercise 3.9.16).

25. [♠] Assume we have observations of the number of claims $N_i = n_i$, $i = 1,\ldots,n$ in a portfolio of risks. It is known that, conditionally given that the value of a structure variable $\Lambda_i = \lambda_i$, this number has a Poisson$(\lambda_i w_i)$ distribution, where the $w_i \in [0,1]$ are known exposures. Also assume $\Lambda_i \sim$ gamma(α, β). Find the marginal distribution of the N_i. Draw a random sample of size $n = 10000$ from the N_i, assuming the w_i are 0.6 for the first 2000 policies, 1.0 for the remainder. Take α, β such that the mean number of claims per unit of exposure is 0.1, and the coefficient of variation of the risk parameters $\sigma_\Lambda/\mu_\Lambda = 1.3$. From this sample, find the ML-estimates for α, β.

26. [♠] Solve the second ML-equation (3.77) using `uniroot` and inserting (3.76). For this, write a function of x delivering the sum of $n_1, n_2, \ldots$ times the cumulative sums of $\frac{1}{x}, \frac{1}{x+1}, \ldots$, less $n \log \frac{x}{x+\bar{y}}$.

27. [♠] How does one generate pseudo-random samples from the following distributions: Student, chi-square, beta, Fisher-Snedecor? Use their well-known relations to normal random samples.

28. Write a function to generate pseudo-random samples from the logistic distribution, see Example 11.5.6.

Section 3.10

1. Assume that X is normally distributed with expectation 10000 and standard deviation 1000. Determine the stop-loss premium for a retention 13000. Do the same for a random variable Y that has the same first two moments as X, but skewness 1.

2. Show that $E[(S - d)_+] = E[S] - d + \int_0^d (d - x) \, dF(x) = E[S] - \int_0^d [1 - F(x)] \, dx$.

3. If $X \sim N(\mu, \sigma^2)$, show that $\int_\mu^\infty E[(X - t)_+] \, dt = \frac{1}{4}\sigma^2$ and determine $E[(X - \mu)_+]$.

4. Verify (3.105). Also verify (3.106) and (3.107), and show how these can be used to approximate the variance of the retained loss.

5. Give an expression for the net premium if the number of claims is Poisson(λ) distributed and the claim size is Pareto distributed. Assume that there is a deductible d.

6. [♠] Let $X \sim \text{lognormal}(\mu, \sigma^2)$. Show that for $d > 0$, the stop-loss premium is

$$E[(X - d)_+] = e^{\mu + \sigma^2/2} \Phi\left(\frac{-\log d + \mu + \sigma^2}{\sigma}\right) - d\Phi\left(\frac{-\log d + \mu}{\sigma}\right).$$

Compare your result with the Black-Scholes option pricing formula, and explain.

7. In Table 3.1, does using linear interpolation to calculate the stop-loss premium in for example $d = 0.4$ for one of the given values for γ yield a result that is too high or too low?

8. Assume that N is an integer-valued risk with $E[(N - d)_+] = E[(U - d)_+]$ for $d = 0, 1, 2, \ldots$, where $U \sim N(0, 1)$. Determine $\Pr[N = 1]$.

9. Let $\pi(t) = E[(U - t)_+]$ denote the stop-loss premium for $U \sim N(0, 1)$ and retention t, $-\infty < t < \infty$. Show that $\pi(-t), t \geq 0$ satisfies $\pi(-t) = t + \pi(+t)$. Sketch $\pi(t)$.

10. In Sections 3.9 and 3.10, the retention is written as $\mu + k\sigma$, so it is expressed in terms of a number of standard deviations above the expected loss. However, in the insurance practice, the retention is always expressed as a percentage of the expected loss. Consider two companies for which the risk of absence due to illness is to be covered by stop-loss insurance. This risk is compound Poisson distributed with parameter λ_i and exponentially distributed individual losses X with $E[X] = 1000$. Company 1 is small: $\lambda_1 = 3$; company 2 is large: $\lambda_2 = 300$. What are the net stop-loss premiums for both companies in case the retention d equals 80%, 100% and 120% of the expected loss respectively? Express these amounts as a percentage of the expected loss and use the normal approximation.

11. For the normal, lognormal and gamma distributions, as well as mixtures/combinations of exponentials, write functions like `slnorm` giving the stop-loss premiums of the corresponding random variables.
    Also, give a function yielding an approximate stop-loss premium for an r.v. having mean μ, variance σ^2 and skewness γ, based on the Normal Power approximation, see (3.114). Do the same for the translated gamma approximation, see (2.57) and (3.105).

12. Using R, verify Tables 3.1 and 3.2.

13. Prove (3.115) and (3.116) and verify that the integrand in (3.115) is non-negative.

14. Show that (3.118) is exact if $W = (1-I)\mu + IU$ with $\mu = E[U]$ and $I \sim$ Bernoulli(α), for $\alpha = \text{Var}[W]/\text{Var}[U]$.

15. Verify Rule of thumb 3.10.6 for the case $U \sim$ Poisson(1) and $V \sim$ binomial(10, $\frac{1}{10}$).

16. Assume that $X_1, X_2, \ldots$ are independent and identically distributed risks that represent the loss on a portfolio in consecutive years. We could insure these risks with separate stop-loss contracts for one year with a retention d, but we could also consider only one contract for the whole period of n years with a retention nd. Show that $E[(X_1 - d)_+] + \cdots + E[(X_n - d)_+] \geq E[(X_1 + \cdots + X_n - nd)_+]$. If $d \geq E[X_i]$, examine how the total net stop-loss premium for the one-year contracts $E[(X_1 - d)_+]$ relates to the stop-loss premium for the n-year period $E[(X_1 + \cdots + X_n - nd)_+]$.

17. Let $B_1 \sim$ binomial(4,0.05), $B_2 \sim$ binomial(2,0.1), $S = B_1 + B_2$ and $T \sim$ Poisson(0.4). For the retentions $d = \frac{1}{2}, 1, \frac{3}{2}$, use the Rule of thumb 3.10.6 and discuss the results.

18. Derive (3.117) from the trapezoidal rule $\int_0^\infty f(x)\,dx \approx \frac{1}{2\delta} \sum_1^\infty [f(i\delta) + f((i-1)\delta)]$ with interval width $\delta = 1$.

Chapter 4
Ruin theory

Survival Probabilities, The Goal of Risk Theory —
Hilary Seal 1978

4.1 Introduction

In this chapter we focus again on collective risk models, but now in the long term. We consider the development in time of the capital $U(t)$ of an insurer. This is a stochastic process that increases continuously because of earned premiums, and decreases stepwise at times that claims occur. When the capital becomes negative, we say that ruin occurs. Let $\psi(u)$ denote the probability that this ever happens, under the assumption that the annual premium and the claims process remain unchanged. This probability is a useful management tool since it serves as an indication of the soundness of the insurer's combination of premiums and claims process, in relation to the available initial capital $u = U(0)$. A high probability of ultimate ruin indicates instability: measures such as reinsurance or raising premiums should be considered, or the insurer should attract extra working capital.

The probability of ruin enables one to compare portfolios, but we cannot attach any absolute meaning to the probability of ruin, as it does not actually represent the probability that the insurer will go bankrupt in the near future. First of all, it might take centuries for ruin to actually happen. Second, obvious interventions in the process such as paying out dividends or raising the premium for risks with an unfavorable claims performance are ruled out in the definition of the probability of ruin. Furthermore, the effects of inflation and return on capital are supposed to cancel each other out exactly. The ruin probability only accounts for the insurance risk, not for possible mismanagement. Finally, the state of ruin is merely a mathematical abstraction: with a capital of -1, the insurer is not broke in practice, and with a capital of $+1$, the insurer can hardly be called solvent. As a result, the exact value of a ruin probability is not of vital importance; a good approximation is just as useful.

Nevertheless, the calculation of the probability of ruin is one of the central problems in actuarial science. The classical ruin model assumes that iid claims arrive according to a Poisson process. In this setting it is possible to determine the moment generating function with the probability $1 - \psi(u)$ of not getting ruined (the non-ruin or *survival probability*), but only two types of claim severities are known for which the probability of ruin can easily be calculated. These are the exponential random variables and sums, minima, maxima, mixtures and combinations of such random variables, as well as random variables with only a finite number of values. But for other distributions, an elegant and usually sufficiently tight bound $\psi(u) \leq e^{-Ru}$ can be found, the so-called *Lundberg upper bound*. The real number R in this expression is called the *adjustment coefficient*; it can be calculated by solving an equation that contains the mgf of the claims, their expectation, and the ratio of premium and expected claims. This bound can often be used instead of the actual ruin probability: the higher R, the lower the upper bound for the ruin probability and, hence, the safer the situation. Also, in most cases it actually represents an asymptotic expression for $\psi(u)$, so $\psi(u)$ is proportional to e^{-Ru} for large u. This means that e^{-R} can be interpreted as the factor by which the ruin probability decreases if the initial capital in the ruin process is increased from u to $u + 1$, for large u. Another interpretation for R is that it is precisely that risk aversion coefficient that would have led to the current annual premium keeping exponential utility at the same level. The fact that $\psi(u)$ is proportional to e^{-Ru} for large u means that R is the asymptotic hazard rate $-\psi'(u)/\psi(u)$ with $\psi(u)$. For the special case of exponential claim sizes, in fact $\psi(u) = \psi(0)e^{-Ru}$ holds for all u.

Multiplying both the premium rate and the expected claim frequency by the same factor does not change the probability of eventual ruin: it does not matter if we make the clock run faster. There have been attempts to replace the ruin probability by a more 'realistic' quantity, for example the *finite time* ruin probability, which is the probability of ruin before time t_0. But this quantity behaves somewhat less orderly and introduces the choice of the length of the time interval as an extra problem. Another alternative risk measure arises if we consider the capital in discrete time points $0, 1, 2, \ldots$ only, for example at the time of the closing of the books. For this *discrete time model*, we will derive some results.

First, we will discuss the Poisson process as a model to describe the development in time of the number of claims. A characteristic feature of the Poisson process is that it is *memoryless*, in the sense that the probability of the event of a claim occurring in the next second is independent of the history of the process. The advantage of a process being memoryless is the mathematical simplicity; the disadvantage is that it is often not realistic. The total of the claims paid in a Poisson process constitutes a compound Poisson process. Based on the convenient mathematical properties of this process, we will be able to derive in a simple way Lundberg's exponential upper bound for the ruin probability, as well as the formula $\psi(u) = \psi(0)e^{-Ru}$ for the ruin probability with initial capital u if the claim sizes are exponential. Here $\psi(0)$ turns out to be simply the ratio of annual expected claims over annual premiums. For non-exponential claim sizes, this same formula is a useful approximation, having the correct value at zero, as well as the correct asymptotic hazard rate.

In the second part of this chapter, we will derive the mgf of the non-ruin probability by studying the maximal aggregate loss, which represents the maximum difference over time between the earned premiums and the total payments up to any moment. Using this mgf, we will determine the value of the ruin probability in case the claims are distributed according to variants of the exponential distribution. Next, we will consider some approximations for the ruin probability.

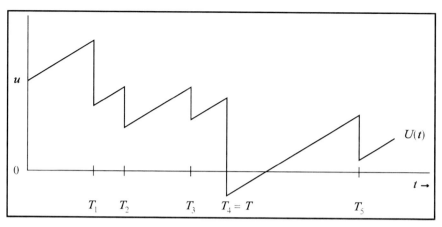

Fig. 4.1 A realization of the risk process $U(t)$.

4.2 The classical ruin process

A stochastic process consists of related random variables, indexed by the time t. We define the *surplus process* or *risk process* as follows:

$$U(t) = u + ct - S(t), \quad t \geq 0, \tag{4.1}$$

where

$U(t) = $ the insurer's capital at time t;

$u = U(0) = $ the initial capital;

$c = $ the (constant) premium income per unit of time;

$S(t) = X_1 + X_2 + \cdots + X_{N(t)},$

with

$N(t) = $ the number of claims up to time t, and

$X_i = $ the size of the ith claim, assumed non-negative.

A typical realization of the risk process is depicted in Figure 4.1. The random variables $T_1, T_2, \ldots$ denote the time points at which a claim occurs. The slope of the process is c if there are no claims; at times $t = T_j$ for some j, the capital drops by X_j, which is the size of the jth claim. Since in Figure 4.1, at time T_4 the total of the incurred claims $X_1 + X_2 + X_3 + X_4$ is larger than the initial capital u plus the earned premium cT_4, the remaining surplus $U(T_4)$ is less than 0. This state of the process is called *ruin*, and the point in time at which this occurs for the first time is denoted by T. So

$$T = \begin{cases} \min\{t \,|\, t \geq 0 \,\&\, U(t) < 0\}; \\ \infty \quad \text{if } U(t) \geq 0 \text{ for all } t. \end{cases} \tag{4.2}$$

In case the probability of $T = \infty$ is positive, the random variable T is called *defective*. The probability that ruin ever occurs, that is, the probability that T is finite, is called the ruin probability. It is written as follows:

$$\psi(u) = \Pr[T < \infty]. \tag{4.3}$$

Before we turn to the claim process $S(t)$, representing the total claims up to time t, we first look at the process $N(t)$ of the number of claims up to t. We will assume that $N(t)$ is a so-called Poisson process:

Definition 4.2.1 (Poisson process)
The process $N(t)$ is a *Poisson process* if for some $\lambda > 0$ called the intensity of the process, the increments of the process have the following property:

$$N(t+h) - N(t) \sim \text{Poisson}(\lambda h) \tag{4.4}$$

for all $t > 0$, $h > 0$ and each history $N(s)$, $s \le t$.   ∇

As a result, the increments in a Poisson process have the following properties:

Independence: the increments $N(t_i + h_i) - N(t_i)$ are independent for disjoint intervals $(t_i, t_i + h_i)$, $i = 1, 2, \ldots$;
Stationarity: $N(t+h) - N(t)$ is Poisson(λh) distributed for every value of t.

Next to this global definition of the claim number process, we can also consider increments $N(t+dt) - N(t)$ in *infinitesimally* small intervals $(t, t+dt)$. For the Poisson process we have by (4.4):

$$
\begin{aligned}
\Pr[N(t+dt) - N(t) = 1 \mid N(s), 0 \le s \le t] &= e^{-\lambda dt} \lambda \, dt = \lambda \, dt, \\
\Pr[N(t+dt) - N(t) = 0 \mid N(s), 0 \le s \le t] &= e^{-\lambda dt} = 1 - \lambda \, dt, \\
\Pr[N(t+dt) - N(t) \ge 2 \mid N(s), 0 \le s \le t] &= 0.
\end{aligned}
\tag{4.5}
$$

Actually, as is common when differentials are involved, the final equalities in (4.5) are not really quite equalities, but they are only valid up to terms of order $(dt)^2$.

A third way to define such a process is by considering the *waiting times*

$$W_1 = T_1, \quad W_j = T_j - T_{j-1}, \; j = 2, 3, \ldots \tag{4.6}$$

Because Poisson processes are memoryless, these waiting times are independent exponential(λ) random variables, and they are also independent of the history of the process. This can be shown as follows: if the history $\mathcal{H}$ represents an arbitrary realization of the process up to time t with the property that $T_{i-1} = t$, then

$$\Pr[W_i > h \mid \mathcal{H}] = \Pr[N(t+h) - N(t) = 0 \mid \mathcal{H}] = e^{-\lambda h}. \tag{4.7}$$

If $N(t)$ is a Poisson process, then $S(t)$ is a compound Poisson process; for a fixed $t = t_0$, the aggregate claims $S(t_0)$ have a compound Poisson distribution with parameter λt_0.

We need some more notation: the cdf and the moments of the individual claims X_i are denoted by

$$P(x) = \Pr[X_i \le x]; \quad \mu_j = \mathrm{E}[X_i^j], \ i,j = 1,2,\dots \qquad (4.8)$$

The *loading factor* or *safety loading* θ is defined by $c = (1+\theta)\lambda\mu_1$, hence

$$\theta = \frac{c}{\lambda\mu_1} - 1. \qquad (4.9)$$

4.3 Some simple results on ruin probabilities

In this section we give a short and elegant proof of F. Lundberg's exponential upper bound for the ruin probability. Also, we derive the ruin probability for a Poisson ruin process with exponential claims. First we introduce the adjustment coefficient.

Definition 4.3.1 (Adjustment coefficient)
In a ruin process with claims distributed as $X \ge 0$ having $\mathrm{E}[X] = \mu_1 > 0$, the *adjustment coefficient R* is the positive solution of the following equation in r:

$$1 + (1+\theta)\mu_1 r = m_X(r). \qquad (4.10)$$

See also Figure 4.2. $\qquad\qquad \triangledown$

In general, the adjustment coefficient equation (4.10) has one positive solution: $m_X(t)$ is strictly convex since $m_X''(t) = \mathrm{E}[X^2 e^{tX}] > 0$, $m_X'(0) < (1+\theta)\mu_1$, and, with few exceptions, $m_X(t) \to \infty$ continuously. Note that for $\theta \downarrow 0$, the limit of R is 0, while for $\theta \uparrow \infty$, we see that R tends to the asymptote of $m_X(r)$, or to ∞.

The adjustment coefficient equation (4.10) is equivalent to $\lambda + cR = \lambda m_X(R)$. Now write S for the total claims in an interval of length 1; consequently $c - S$ is the profit in that interval. Then S is compound Poisson with parameter λ, so $m_S(r) = \exp\{\lambda(m_X(r)-1)\}$. Then R can also be found as the positive solution to any of the following equivalent equations, see Exercise 4.3.1:

$$e^{Rc} = \mathrm{E}[e^{RS}] \iff m_{c-S}(-R) = 1 \iff c = \frac{1}{R}\log m_S(R). \qquad (4.11)$$

Remark 4.3.2 (Adjustment coefficient and risk aversion)
From the last equation and (1.20) we see that, in case of an exponential utility function, R is just that risk aversion α that leads to an annual premium c. $\qquad \triangledown$

Example 4.3.3 (Adjustment coefficient for an exponential distribution)
Assume that X is exponentially distributed with parameter $\beta = 1/\mu_1$. The corresponding adjustment coefficient is the positive solution of

$$1 + (1+\theta)\mu_1 r = m_X(r) = \frac{\beta}{\beta - r}. \qquad (4.12)$$

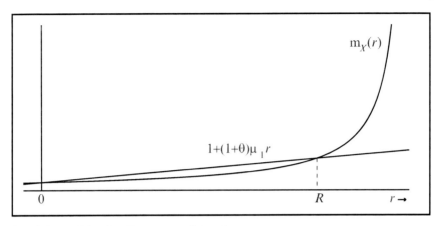

Fig. 4.2 Determining the adjustment coefficient R.

The solutions of this equation are the trivial excluded solution $r = 0$ and

$$r = R = \frac{\theta \beta}{1 + \theta}. \tag{4.13}$$

So this situation admits an explicit expression for the adjustment coefficient.   ∇

For most distributions, there is no explicit expression for the adjustment coefficient. To facilitate solving (4.10) numerically, one can use the fact that $R \in [0, 2\theta\mu_1/\mu_2]$, see Exercise 4.3.2, but $R < \beta$ must also hold in (4.12).

In the next theorem, we prove F. Lundberg's famous exponential inequality for the ruin probability. Surprisingly, the proof involves mathematical induction.

Theorem 4.3.4 (Lundberg's exponential bound for the ruin probability)
For a compound Poisson risk process with an initial capital u, a premium per unit of time c, claims with cdf $P(\cdot)$ and mgf $m_X(t)$, and an adjustment coefficient R that satisfies (4.10), we have the following inequality for the ruin probability:

$$\psi(u) \le e^{-Ru}. \tag{4.14}$$

Proof. Define $\psi_k(u)$, $-\infty < u < \infty$ and $k = 0, 1, 2, \ldots$, as the probability that ruin occurs at or before the kth claim. Since for $k \to \infty$, $\psi_k(u)$ increases to its limit $\psi(u)$ for all u, it suffices to prove that $\psi_k(u) \le e^{-Ru}$ for each k. For $k = 0$ the inequality holds, since $\psi_0(u) = 1$ if $u < 0$, and $\psi_0(u) = 0$ if $u \ge 0$. We are going to split up the event 'ruin at or before the kth claim' as regards time and size of the first claim. Assume that it occurs between time t and $t + dt$. This event has a 'probability' $\lambda e^{-\lambda t} dt$. Also assume it has a size between x and $x + dx$, which has a probability $dP(x)$. Then the capital right after time t equals $u + ct - x$. Integrating over x and t yields

$$\psi_k(u) = \int_0^\infty \int_0^\infty \psi_{k-1}(u + ct - x) \, dP(x) \, \lambda e^{-\lambda t} \, dt. \tag{4.15}$$

Now assume that the induction hypothesis holds for $k - 1$, that is, $\psi_{k-1}(u) \leq e^{-Ru}$ for all real u. Then, (4.15) leads to

$$
\begin{aligned}
\psi_k(u) &\leq \int_0^\infty \int_0^\infty \exp\{-R(u + ct - x)\}\, dP(x)\, \lambda e^{-\lambda t}\, dt \\
&= e^{-Ru} \int_0^\infty \lambda \exp\{-t(\lambda + Rc)\}\, dt \int_0^\infty e^{Rx}\, dP(x) \qquad (4.16) \\
&= e^{-Ru} \frac{\lambda}{\lambda + cR} m_X(R) = e^{-Ru},
\end{aligned}
$$

where the last equality follows from (4.10). $\triangledown$

Corollary 4.3.5 (Positive adjustment coefficient means ruin is not certain)
Since there is a positive probability $e^{-\lambda}$ of having no claims until time 1, and $\psi(c) \leq e^{-Rc}$ by the above theorem, for any non-negative initial capital $u \geq 0$ we have $1 - \psi(u) \geq 1 - \psi(0) \geq e^{-\lambda}(1 - e^{-Rc})$. So the probability of non-ruin, also known as the survival probability, is strictly positive when $R > 0$ holds, that is, when the mgf of the claim severity distribution is finite on an interval containing zero. $\triangledown$

Next, we derive an explicit expression for the ruin probability in a compound Poisson risk process with claims that are exponential. It turns out that, except for a factor $\psi(0)$, Lundberg's exponential upper bound (4.14) is an equality here.

Theorem 4.3.6 (Ruin probability with exponential claims)
For a Poisson ruin process with claims $\sim$ exponential$(1/\mu_1)$, intensity λ and premium rate $c = (1 + \theta)\lambda\mu_1$ with $\theta > 0$, the ruin probability with initial capital u is

$$
\psi(u) = \psi(0)e^{-Ru}, \qquad (4.17)
$$

where $R = \frac{\theta}{(1+\theta)\mu_1}$ is the adjustment coefficient solving (4.12), and $\psi(0) = \frac{1}{1+\theta}$.

Proof. To simplify notation, we assume that both the claims and the interarrival times are exponential(1), so $\mu_1 = \lambda = 1$; the general case with arbitrary time and money units follows easily. For this situation, we want to find the ruin probability $\psi(u)$ with initial capital $u \geq 0$. Our proof has two steps. In the first, we derive a simple equation involving differentials that the ruin probability must satisfy. From this, we get an expression for the ruin probability in which the ruin probability $\psi(0)$ without initial capital still is unknown. In the second step, to find $\psi(0)$ we use a formula similar to (4.15).

  Assume we have a realization of a risk process with initial capital u in which ruin occurs. In that case, there is a ruining claim, of size $X \sim$ exponential(1). It is known that $X > v$, where v represents the capital present just before ruin in this realization of the process. Now assume that the initial capital was not u, but infinitesimally larger, so $u + du$. There are two cases in which our particular realization of the process also results in ruin starting from $u + du$. First, the ruining claim may be so big that an extra du of initial capital would not have made a difference, in other words, $X > v + du$. Since X is exponential, we have $\Pr[X > v + du \mid X > v] = e^{-v-du}/e^{-v} = 1 - du$. Note that this probability is not dependent on v. Second, the extra amount

du might for the moment save us from ruin. This event has conditional probability du. But then, we might get ruined at a later time, and because the Poisson process is memoryless, the probability of that happening is just $\psi(0)$. Therefore we have the following differential equation:

$$\psi(u+du) = \psi(u)\big(1 - du + \psi(0)\,du\big). \tag{4.18}$$

Rearranging this gives

$$\frac{\psi'(u)}{\psi(u)} = \psi(0) - 1. \tag{4.19}$$

The left hand side is $d\log\psi(u)/du$, so necessarily

$$\log\psi(u) = u(\psi(0) - 1) + \log\psi(0). \tag{4.20}$$

Therefore, for a classical risk process with unit exponential claims, the ruin probability, as a function of the initial capital u, must be of the following form:

$$\psi(u) = \begin{cases} \psi(0)e^{-u(1-\psi(0))} & \text{for } u \ge 0; \\ 1 & \text{for } u < 0. \end{cases} \tag{4.21}$$

The second step in the proof is to find the value of $\psi(0)$ as a function of c. For this, assume the first claim equals x and occurs at time t; this same device was used to derive (4.15). The 'probability' of this event equals $e^{-t}\,dt\,e^{-x}\,dx$. After that claim, the capital is $u + ct - x$, so the ruin probability at u can be expressed as

$$\psi(u) = \int_0^\infty \int_0^\infty \psi(u + ct - x)\,e^{-t}\,dt\,e^{-x}\,dx, \tag{4.22}$$

which, using (4.21), for the special case $u = 0$ can be rewritten as

$$\psi(0) = \int_0^\infty \int_{x/c}^\infty \psi(0)e^{-(1-\psi(0))(ct-x)}\,e^{-t}\,dt\,e^{-x}\,dx$$
$$+ \int_0^\infty \int_0^{x/c} e^{-t}\,dt\,e^{-x}\,dx. \tag{4.23}$$

Using $\int_0^\infty e^{-\alpha x}\,dx = 1/\alpha$ and substituting $y = ct - x$, this results in:

$$\psi(0) = \frac{\psi(0)}{c}\int_0^\infty e^{-x/c}\int_0^\infty e^{-y(1-\psi(0)+1/c)}\,dy\,e^{-x}\,dx$$
$$+ \int_0^\infty (1 - e^{-x/c})e^{-x}\,dx$$
$$= \frac{\psi(0)}{c}\int_0^\infty e^{-x/c}\frac{1}{1-\psi(0)+1/c}\,e^{-x}\,dx + 1 - \frac{1}{1+1/c}$$
$$= \frac{\psi(0)}{c}\frac{1}{1-\psi(0)+1/c}\frac{1}{1+1/c} + \frac{1}{c+1}, \tag{4.24}$$

which can be simplified to the following quadratic equation expressing $\psi(0)$ in c:

$$\psi(0) = \frac{1}{c} \frac{1}{1 - \psi(0) + 1/c}. \tag{4.25}$$

There are two solutions. One is $\psi(0) = 1$, which is excluded by Corollary 4.3.5. The other solution is $\psi(0) = \frac{1}{c}$, so (4.17) is proved.   ▽

The equality $\psi(0) = \frac{1}{1+\theta}$, proved in (4.25) for exponential claims, actually holds for all distributions of the claim sizes, see (4.59) in Corollary 4.7.2. Since $\psi(0) \uparrow 1$ if $\theta \downarrow 0$, by Theorem 4.3.6 we have $\psi(u) \equiv 1$ for exponential claims when there is no safety loading, that is, when $\theta = 0$.

Example 4.3.7 (Simulating a Poisson ruin process)
The following R program counts how often ruin occurs in the first $n = 400$ claims of a simulated ruin process. This way we get an estimate of $\psi_n(u)$, the probability of ruin at or before the nth claim, as defined in the proof of Theorem 4.3.4. The claims are gamma(α, α) distributed with $\alpha = 2$. Because the claims are not very heavy-tailed, it is to be expected that if ruin occurs, it will occur early. This is because the surplus process will drift away from zero, since $E[U(t)] = u + \theta \lambda t E[X]$ and $\text{Var}[U(t)] = \lambda t E[X^2]$. So this program gives a good approximation for $\psi(u)$:

```
lab <- 1; EX <- 1; theta <- 0.3; u <- 7.5; alpha <- 2
n <- 400; nSim <- 10000; set.seed(2)
c. <- (1+theta)*lab*EX
N <- rep(Inf, nSim)
for (k in 1:nSim){
    Wi <- rexp(n)/lab; Ti <- cumsum(Wi)
    Xi <- rgamma(n, shape=alpha)/alpha ## severity has mean EX=1
    Si <- cumsum(Xi); Ui <- u + Ti*c. - Si
    ruin <- !all(Ui>=0)
    if (ruin) N[k] <- min(which(Ui<0))}
N <- N[N<Inf]; length(N); mean(N); sd(N); max(N)
##   745  30.78792  24.71228  255
```

The vector `Ui` contains the surplus after the ith claim has been paid. The 745 finite ruin claim numbers `N[N<Inf]` have a maximum of 255, mean 30.8 and s.d. 24.7. So it is highly improbable that ruin will occur after the 400th claim, therefore $\psi(u) \approx \psi_{400}(u)$. Hence a good estimate of the ultimate ruin probability is $\psi(u) \approx 7.45\%$.

As the reader may verify, the approximation $\psi(u) \approx \psi(0)e^{-Ru}$ gives 7.15%. The exact ruin probability, computed using the technique of Section 4.8, is 7.386%. Lundberg's exponential upper bound $e^{-Ru} = 9.29\%$ for this case.   ▽

4.4 Ruin probability and capital at ruin

To be able to give more results about the ruin process, we have to derive some more theory. First we give an expression for the ruin probability that involves the mgf of

$U(T)$, which is the capital at the moment of first ruin, conditionally given the event that ruin occurs in a finite time period.

Theorem 4.4.1 (Ruin probability and capital at ruin)
The ruin probability for $u \geq 0$ satisfies

$$\psi(u) = \frac{e^{-Ru}}{E[e^{-RU(T)} \,|\, T < \infty]}. \tag{4.26}$$

Proof. Let $R > 0$ and $t > 0$. Then,

$$E[e^{-RU(t)}] = E[e^{-RU(t)} \,|\, T \leq t]\Pr[T \leq t] + E[e^{-RU(t)} \,|\, T > t]\Pr[T > t]. \tag{4.27}$$

The left-hand side equals e^{-Ru}. This can be shown as follows: since $U(t) = u + ct - S(t)$ and $S(t) \sim$ compound Poisson with parameter λt, we have, using equation (4.10) to prove that the expression in square brackets equals 1:

$$\begin{aligned}
E[e^{-RU(t)}] &= E[e^{-R\{u+ct-S(t)\}}] \\
&= e^{-Ru}[e^{-Rc}\exp\{\lambda(m_X(R) - 1)\}]^t \\
&= e^{-Ru}.
\end{aligned} \tag{4.28}$$

For the first conditional expectation in (4.27) we take $v \in [0,t]$ and write, using $U(t) = U(v) + c(t - v) - [S(t) - S(v)]$, see also (4.28):

$$\begin{aligned}
E[e^{-RU(t)} \,|\, T = v] &= E[e^{-R\{U(v)+c(t-v)-[S(t)-S(v)]\}} \,|\, T = v] \\
&= E[e^{-RU(v)} \,|\, T = v]e^{-Rc(t-v)}E[e^{R\{S(t)-S(v)\}} \,|\, T = v] \\
&= E[e^{-RU(v)} \,|\, T = v]\{e^{-Rc}\exp[\lambda(m_X(R) - 1)]\}^{t-v} \\
&= E[e^{-RU(T)} \,|\, T = v].
\end{aligned} \tag{4.29}$$

The total claims $S(t) - S(v)$ between v and t has again a compound Poisson distribution. What happens after v is independent of what happened before v, so $U(v)$ and $S(t) - S(v)$ are independent. The term in curly brackets equals 1, again by equation (4.10). Equality (4.29) holds for all $v \leq t$, and therefore $E[e^{-RU(t)} \,|\, T \leq t] = E[e^{-RU(T)} \,|\, T \leq t]$ also holds.

Since $\Pr[T \leq t] \uparrow \Pr[T < \infty]$ for $t \to \infty$, it suffices to show that the last term in (4.27) vanishes for $t \to \infty$. For that purpose, we split the event $T > t$ according to the size of $U(t)$. More precisely, we consider the cases $U(t) \leq u_0(t)$ and $U(t) > u_0(t)$ for some suitable function $u_0(t)$. Notice that $T > t$ implies that we are not in ruin at time t, which means that $U(t) \geq 0$, so $e^{-RU(t)} \leq 1$. We have

$$\begin{aligned}
&E[e^{-RU(t)} \,|\, T > t]\Pr[T > t] \\
&= E[e^{-RU(t)} \,|\, T > t, 0 \leq U(t) \leq u_0(t)]\Pr[T > t, 0 \leq U(t) \leq u_0(t)] \\
&\quad + E[e^{-RU(t)} \,|\, T > t, U(t) > u_0(t)]\Pr[T > t, 0 \leq U(t) > u_0(t)] \\
&\leq \Pr[U(t) \leq u_0(t)] + E[\exp(-Ru_0(t))].
\end{aligned} \tag{4.30}$$

The second term vanishes if $u_0(t) \to \infty$. For the first term, note that $U(t)$ has an expected value $\mu(t) = u + ct - \lambda t \mu_1$ and a variance $\sigma^2(t) = \lambda t \mu_2$. Because of Chebyshev's inequality, it suffices to choose the function $u_0(t)$ such that $(\mu(t) - u_0(t))/\sigma(t) \to \infty$. We can for example take $u_0(t) = t^{2/3}$. $\triangledown$

Corollary 4.4.2 (Consequences of Theorem 4.4.1)
The following observations are immediate from the previous theorem:

1. If $\theta \downarrow 0$, then the chord in Figure 4.2 tends to a tangent line and, because of Theorem 4.4.1, $\psi(u) \to 1$; if $\theta \leq 0$ then $\psi(u) \equiv 1$, see Exercise 4.4.1.
2. If $T < \infty$, then $U(T) < 0$. Hence, the denominator in (4.26) is greater than 1, so $\psi(u) < e^{-Ru}$; this is yet another proof of Theorem 4.3.4.
3. If b is an upper bound for the claims, then $\Pr[U(T) > -b] = 1$, so we have an exponential lower bound for the ruin probability: $\psi(u) > e^{-R(u+b)}$.
4. As the size of the initial capital does not greatly affect the probability distribution of the capital at ruin, it is quite plausible that the denominator of (4.26) has a finite limit, say c_0, as $u \to \infty$. Then $c_0 > 1$. This yields the following asymptotic approximation for $\psi(\cdot)$: for large u, we have $\psi(u) \approx \frac{1}{c_0} e^{-Ru}$.
5. If $R > 0$, then $1 - \psi(u) > 0$ for all $u \geq 0$. As a consequence, if $1 - \psi(u_0) = 0$ for some $u_0 \geq 0$, then $R = 0$ must hold, so $1 - \psi(u) = 0$ for all $u \geq 0$. $\triangledown$

Remark 4.4.3 (Interpretation of the adjustment coefficient; martingale)
In (4.28), we saw that the adjustment coefficient R has the property that $E[e^{-RU(t)}]$ is constant in t. In other words, $e^{-RU(t)}$ is a *martingale*: it can be interpreted as the fortune of a gambler who is involved in a sequence of fair games. Note that if R is replaced by any other non-zero real number, the expression in square brackets in (4.28) is not equal to 1, so in fact the adjustment coefficient is the unique positive number R with the property that $e^{-RU(t)}$ is a martingale. $\triangledown$

Example 4.4.4 (Ruin probability, exponential claims)
From (4.26), we can verify expression (4.17) for the ruin probability if the claims have an exponential(β) distribution. For this purpose, assume that the history of the process up to time t is given by $\mathcal{H}$, that ruin occurs at time $T = t$ and that the capital $U(T - 0)$ just before ruin equals v. So the ruining claim, of size X, say, is given to be larger than v. Then, for each $\mathcal{H}$, v, t and y, we have:

$$\Pr[-U(T) > y \mid \mathcal{H}] = \Pr[X > v + y \mid X > v] = \frac{e^{-\beta(v+y)}}{e^{-\beta v}} = e^{-\beta y}. \quad (4.31)$$

Apparently, the deficit $-U(T)$ at ruin also has an exponential(β) distribution, so the denominator of (4.26) equals $\beta/(\beta - R)$. With $\beta = 1/\mu_1$ and $R = \theta\beta/(1+\theta)$, see (4.13), and thus $\beta/(\beta - R) = 1 + \theta$, we find again the ruin probability (4.17) in case of exponential($\beta = 1/\mu_1$) claims:

$$\psi(u) = \frac{1}{1+\theta} \exp\left(-\frac{\theta\beta u}{1+\theta}\right) = \frac{1}{1+\theta} \exp\left(-\frac{\theta}{1+\theta}\frac{u}{\mu_1}\right) = \psi(0)e^{-Ru}. \quad (4.32)$$

In this specific case, the denominator of (4.26) does not depend on u. $\triangledown$

4.5 Discrete time model

In the discrete time model, we consider more general risk processes $U(t)$ than the compound Poisson process from the previous sections, but now only on the time points $n = 0, 1, 2, \ldots$ Write G_n for the profit in the interval $(n - 1, n]$, therefore

$$U(n) = u + G_1 + G_2 + \cdots + G_n, \quad n = 0, 1, \ldots \tag{4.33}$$

Later on, we will discuss what happens if we assume that $U(t)$ is a compound Poisson process, but for the moment we only assume that the profits $G_1, G_2, \ldots$ are independent and identically distributed, with $\Pr[G_n < 0] > 0$, but $E[G_n] = \mu > 0$. We define a discrete time version of the ruin time $\widetilde{T}$, the ruin probability $\widetilde{\psi}(u)$ and the adjustment coefficient $\widetilde{R} > 0$, as follows:

$$\widetilde{T} = \min\{n : U(n) < 0\}; \quad \widetilde{\psi}(u) = \Pr[\widetilde{T} < \infty]; \quad m_G(-\widetilde{R}) = 1. \tag{4.34}$$

The last equation has a unique solution. This can be seen as follows: since $E[G] > 0$ and $\Pr[G < 0] > 0$, we have $m'_G(0) > 0$ and $m_G(-r) \to \infty$ for $r \to \infty$, while $m''_G(-r) = E[G^2 e^{-Gr}] > 0$, so $m_G(\cdot)$ is a convex function.

Example 4.5.1 (Compound Poisson distributed annual claims)
In the special case that $U(t)$ is a compound Poisson process, we have $G_n = c - Z_n$ where Z_n denotes the compound Poisson distributed total claims in year n. From (4.11), we know that R satisfies the equation $m_{c-Z}(-R) = 1$. Hence, $\widetilde{R} = R$. ∇

Example 4.5.2 (Normally distributed annual claims)
If $G_n \sim N(\mu, \sigma^2)$, with $\mu > 0$, then $\widetilde{R} = 2\mu/\sigma^2$ follows from:

$$\log(m_G(-r)) = 0 = -\mu r + \frac{1}{2}\sigma^2 r^2. \tag{4.35}$$

Combining this with the previous example, we observe the following. If we consider a compound Poisson process with a large Poisson parameter, therefore with many claims between the time points $0, 1, 2, \ldots$, then S_n will approximately follow a normal distribution. Hence the adjustment coefficients will be close to each other, so $R \approx 2\mu/\sigma^2$. But if we take $\mu = c - \lambda\mu_1 = \theta\lambda\mu_1$ and $\sigma^2 = \lambda\mu_2$ in Exercise 4.3.2, we see that $2\mu/\sigma^2$ is an upper bound for R. ∇

Analogously to Theorem 4.4.1, the following equality relates the ruin probability and the capital at ruin:

$$\widetilde{\psi}(u) = \frac{e^{-\widetilde{R}u}}{E[e^{-\widetilde{R}U(\widetilde{T})} \mid \widetilde{T} < \infty]}. \tag{4.36}$$

So in the discrete time model one can give an exponential upper bound for the ruin probability, too, which is

$$\widetilde{\psi}(u) \le e^{-\widetilde{R}u}. \tag{4.37}$$

4.6 Reinsurance and ruin probabilities

In the economic environment we postulated in Chapter 1, reinsurance contracts should be compared by their expected utility. In practice, however, this method is not applicable. Alternatively, we can compare the ruin probabilities under various reinsurance policies. This too is quite difficult. Therefore we will concentrate on the adjustment coefficient and try to obtain a more favorable one by reinsurance. It is exactly from this possibility that the adjustment coefficient takes its name.

In reinsurance we transfer some of our expected profit to the reinsurer, in exchange for more stability in our position. These two conflicting criteria cannot be optimized at the same time. A similar problem arises in statistics where one finds a trade-off between the power and the size of a test. In our situation, we can follow the same procedure used in statistics, that is, maximizing one criterion while restricting the other. We could, for example, maximize the expected profit subject to the condition that the adjustment coefficient R is larger than some R_0.

We consider two situations. In the discrete time ruin model, we reinsure the total claims in one year and then examine the discrete adjustment coefficient $\tilde{R}$. In the continuous time model, we compare R for proportional reinsurance and excess of loss reinsurance, with a retention for each claim.

Example 4.6.1 (Discrete time compound Poisson claim process)
Consider the compound Poisson distribution with $\lambda = 1$ and $p(1) = p(2) = \frac{1}{2}$ from Example 3.5.4. What is the discrete adjustment coefficient $\tilde{R}$ for the total claims S in one year, if the loading factor θ equals 0.2, meaning that the annual premium c equals 1.8?

The adjustment coefficient $\tilde{R}$ is calculated by solving the equation:

$$\lambda + cr = \lambda m_X(r) \iff 1 + 1.8r = \frac{1}{2}e^r + \frac{1}{2}e^{2r}. \tag{4.38}$$

Using R's numerical routine `uniroot` for one-dimensional root (zero) finding, we find the adjustment coefficient in (4.38) by

```
f <- function (r) exp(r)/2 + exp(2*r)/2 - 1 - 1.8*r
R <- uniroot(f, lower=0.00001, upper=1)$root ## 0.2105433
```

The first parameter of `uniroot` is the function f of which the zero R is sought. The `lower` and `upper` parameter give an interval in which the root of the equation $f(R) = 0$ is known to lie; the signs of the function f must be opposite at both ends of that interval. Note that the function is computed at both endpoints first, so the upper bound must be chosen with some care in this case. Of course one can use the upper bound $R \le 2\theta\mu_1/\mu_2$, see Exercise 4.3.2, if this is smaller than the asymptote of m_X. Apart from the `root`, the function `uniroot` provides more information about the zero, including an indication of its precision.

Now assume that we take out a stop-loss reinsurance with $d = 3$. The reinsurance premium loading is $\xi = 0.8$, so if $d = 3$, from Example 3.5.4 we see that the reinsurance premium is $(1+\xi)\mathrm{E}[(S-d)_+] = 1.8\pi(3) = 0.362$. To determine the adjustment coefficient, we calculate the distribution of the profit in one year G_i, which consists of the premium income less the reinsurance premium and the retained loss. Hence,

$$G_i = \begin{cases} 1.8 - 0.362 - S_i & \text{if } S_i = 0,1,2,3; \\ 1.8 - 0.362 - 3 & \text{if } S_i > 3. \end{cases} \tag{4.39}$$

The corresponding discrete adjustment coefficient $\widetilde{R}$, which is the solution of (4.34), is approximately 0.199.

Because of the reinsurance, our expected annual profit is reduced. It is equal to our original expected profit less the one of the reinsurer. For example, for $d = 3$ it equals $1.8 - 1.5 - \xi\pi(3) = 0.139$. In the following table, we show the results for different values of the retention d:

Retention d	$\widetilde{R}$	Expected profit
3	0.199	0.139
4	0.236	0.234
5	0.230	0.273
∞	0.211	0.300

We see that the decision $d = 3$ is not rational: it is dominated by $d = 4$, $d = 5$ as well as $d = \infty$, that is, no reinsurance, since they all yield both a higher expected profit and more stability in the sense of a larger adjustment coefficient $\widetilde{R}$.   ∇

Example 4.6.2 (Reinsurance, individual claims)
Reinsurance may also affect each individual claim, instead of only the total claims in a period. Assume that the reinsurer pays $h(x)$ if the claim amount is x. In other words, the retained loss equals $x - h(x)$. We consider two special cases:

$$h(x) = \begin{cases} \alpha x & 0 \le \alpha \le 1 & \text{proportional reinsurance} \\ (x-\beta)_+ & 0 \le \beta & \text{excess of loss reinsurance} \end{cases} \tag{4.40}$$

Obviously, proportional reinsurance can be considered as a reinsurance on the total claims just as well. We will examine the usual adjustment coefficient R_h, which is the root of

$$\lambda + (c - c_h)r = \lambda \int_0^\infty e^{r[x-h(x)]}\, \mathrm{d}P(x), \tag{4.41}$$

where c_h denotes the reinsurance premium. The reinsurer uses a loading factor ξ on the net premium. Assume that $\lambda = 1$, and $p(x) = \frac{1}{2}$ for $x = 1$ and $x = 2$. Furthermore, let $c = 2$, so $\theta = \frac{1}{3}$, and consider two values $\xi = \frac{1}{3}$ and $\xi = \frac{2}{5}$.

In case of proportional reinsurance $h(x) = \alpha x$, the premium equals

$$c_h = (1+\xi)\lambda\mathrm{E}[h(X)] = (1+\xi)\frac{3}{2}\alpha, \tag{4.42}$$

so, because of $x - h(x) = (1-\alpha)x$, (4.41) leads to the equation

$$1 + [2 - (1 + \xi)\frac{3}{2}\alpha]r = \frac{1}{2}e^{r(1-\alpha)} + \frac{1}{2}e^{2r(1-\alpha)}. \qquad (4.43)$$

For $\xi = \frac{1}{3}$, we have $c_h = 2\alpha$ and $R_h = \frac{0.325}{1-\alpha}$; for $\xi = \frac{2}{5}$, we have $c_h = 2.1\alpha$.

Next, we consider the excess of loss reinsurance $h(x) = (x - \beta)_+$, with $0 \le \beta \le 2$. The reinsurance premium equals

$$c_h = (1 + \xi)\lambda[\frac{1}{2}h(1) + \frac{1}{2}h(2)] = \frac{1}{2}(1 + \xi)[(1 - \beta)_+ + (2 - \beta)_+], \qquad (4.44)$$

while $x - h(x) = \min\{x, \beta\}$, and therefore R_h is the root of

$$1 + (2 - \frac{1}{2}(1 + \xi)[(1 - \beta)_+ + (2 - \beta)_+])r = \frac{1}{2}[e^{\min\{\beta,1\}r} + e^{\min\{\beta,2\}r}]. \qquad (4.45)$$

In the table below, we give the results R_h for different values of β, compared with the same results in case of proportional reinsurance with the same expected payment by the reinsurer: $\frac{3}{2}\alpha = \frac{1}{2}(1 - \beta)_+ + \frac{1}{2}(2 - \beta)_+$.

For $\xi = \frac{1}{3}$, the loading factors of the reinsurer and the insurer are equal, and the more reinsurance we take, the larger the adjustment coefficient is. If the reinsurer's loading factor equals $\frac{2}{5}$, then for $\alpha \ge \frac{5}{6}$ the expected retained loss $\lambda E[X - h(X)] = \frac{3}{2}(1 - \alpha)$ is not less than the retained premium $c - c_h = 2 - 2.1\alpha$. Consequently, the resulting retained loading factor is not positive, and eventual ruin is a certainty. The same phenomenon occurs in case of excess of loss reinsurance with $\beta \le \frac{1}{4}$. In the table below, this situation is denoted by the symbol $*$.

	β	2.0	1.4	0.9	0.6	0.3	0.15	0.0
	α	0.0	0.2	0.4	0.6	0.8	0.9	1.0
$\xi = \frac{1}{3}$	XL	.325	.444	.611	.917	1.83	3.67	∞
	Prop.	.325	.407	.542	.813	1.63	3.25	∞
$\xi = \frac{2}{5}$	XL	.325	.425	.542	.676	.426	$*$	$*$
	Prop.	.325	.390	.482	.602	.382	$*$	$*$

From the table we see that all adjustment coefficients for excess of loss coverage (XL) are at least as large as those for proportional reinsurance (Prop.) with the same expected payment. This is not a coincidence: by using the theory on ordering of risks, it can be shown that XL coverage always yields the best R-value as well as the smallest ruin probability among all reinsurance contracts with the same expected value of the retained risk; see Section 7.4.4. $\triangledown$

4.7 Beekman's convolution formula

In this section we show that the non-ruin probability can be written as a compound geometric distribution function with known specifications. For this purpose, we con-

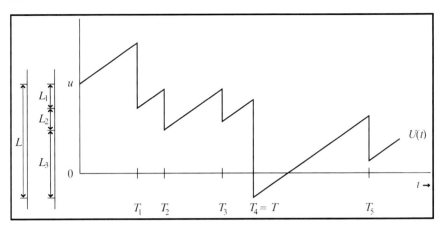

Fig. 4.3 The quantities $L, L_1, L_2, \ldots$

sider the *maximal aggregate loss*, in other words the maximal difference over t between the payments and the earned premium up to time t:

$$L = \max\{S(t) - ct \mid t \geq 0\}. \tag{4.46}$$

Since $S(0) = 0$, we have $L \geq 0$. The event $L > u$ occurs if, and only if, a finite point in time t exists for which $U(t) < 0$. In other words, the events $L > u$ and $T < \infty$ are equivalent and consequently

$$\psi(u) = 1 - F_L(u). \tag{4.47}$$

Next, we consider the points where the surplus process reaches a new record low. This happens necessarily at points in time when a claim is paid. Let the random variables $L_j, j = 1, 2, \ldots$ denote the amounts by which the jth record low is less than the $j - 1$st one. See Figure 4.3 where there are three new record lows, assuming that the process drifts away to ∞ in the time period not shown. Let M be the random number of new records. We have

$$L = L_1 + L_2 + \cdots + L_M. \tag{4.48}$$

From the fact that a Poisson process is memoryless, it follows that the probability that a particular record low is the last one is the same every time. Hence, M follows a geometric distribution. For the same reason, the amounts of the improvements $L_1, L_2, \ldots$ are independent and identically distributed. The parameter of M, that is, the probability that the previous record is the last one, equals the probability to avoid ruin starting with initial capital 0, hence it equals $1 - \psi(0)$.

So L has a compound geometric distribution. Its specifications, that is, the value of the geometric parameter $1 - \psi(0)$ and the distribution of L_1, conditionally given $M \geq 1$, both follow from the following theorem:

Theorem 4.7.1 (Distribution of the capital at time of ruin)
If the initial capital u equals 0, then for all $y > 0$ we have:

$$\Pr[U(T) \in (-y - \mathrm{d}y, -y), T < \infty] = \frac{\lambda}{c}[1 - P(y)]\,\mathrm{d}y. \tag{4.49}$$

Proof. In a compound Poisson process, the probability of having a claim in the interval $(t, t + \mathrm{d}t)$ equals $\lambda\,\mathrm{d}t$, which is independent of t and of the history of the process up to that time. So, between 0 and $\mathrm{d}t$ there is either no claim (with probability $1 - \lambda\,\mathrm{d}t$), and the capital increases from u to $u + c\,\mathrm{d}t$, or one claim with random size X. In the latter case, there are two possibilities. If the claim size is less than u, then the process continues with capital $u + c\,\mathrm{d}t - X$. Otherwise ruin occurs, but the capital at ruin is only larger than y if $X > u + y$. Defining

$$G(u,y) = \Pr[U(T) \in (-\infty, -y), T < \infty \,|\, U(0) = u], \tag{4.50}$$

we can write

$$G(u,y) = (1 - \lambda\,\mathrm{d}t)G(u + c\,\mathrm{d}t, y)$$
$$+ \lambda\,\mathrm{d}t \left\{ \int_0^u G(u - x, y)\,\mathrm{d}P(x) + \int_{u+y}^\infty \mathrm{d}P(x) \right\}. \tag{4.51}$$

If G' denotes the partial derivative of G with respect to u, then

$$G(u + c\,\mathrm{d}t, y) = G(u, y) + c\,\mathrm{d}t\, G'(u, y). \tag{4.52}$$

Substitute (4.52) into (4.51) and divide by $c\,\mathrm{d}t$. Then we get

$$G'(u,y) = \frac{\lambda}{c}\left\{ G(u,y) - \int_0^u G(u - x, y)\,\mathrm{d}P(x) - \int_{u+y}^\infty \mathrm{d}P(x) \right\}. \tag{4.53}$$

Integrating this over $u \in [0, z]$ yields

$$G(z,y) - G(0,y) = \frac{\lambda}{c}\left\{ \int_0^z G(u,y)\,\mathrm{d}u - \int_0^z \int_0^u G(u - x, y)\,\mathrm{d}P(x)\,\mathrm{d}u \right.$$
$$\left. - \int_0^z \int_{u+y}^\infty \mathrm{d}P(x)\,\mathrm{d}u \right\}. \tag{4.54}$$

The double integrals in (4.54) can be reduced to single integrals as follows. For the first double integral, exchange the order of integration using Fubini, substitute $v = u - x$ and swap back the integration order. This leads to

$$\int_0^z \int_0^u G(u - x, y)\,\mathrm{d}P(x)\,\mathrm{d}u = \int_0^z \int_0^{z-v} G(v, y)\,\mathrm{d}P(x)\,\mathrm{d}v$$
$$= \int_0^z G(v, y)P(z - v)\,\mathrm{d}v. \tag{4.55}$$

In the second double integral in (4.54), we substitute $v = u + y$, leading to

$$\int_0^z \int_{u+y}^\infty dP(x)\,du = \int_y^{z+y} [1 - P(v)]\,dv. \tag{4.56}$$

Hence,

$$\begin{aligned} G(z,y) - G(0,y) = \frac{\lambda}{c}\Bigg\{ &\int_0^z G(u,y)[1 - P(z-u)]\,du \\ &- \int_y^{z+y} [1 - P(u)]\,du \Bigg\}. \end{aligned} \tag{4.57}$$

For $z \to \infty$, the first term on both sides of (4.57) vanishes, leaving

$$G(0,y) = \frac{\lambda}{c} \int_y^\infty [1 - P(u)]\,du, \tag{4.58}$$

which completes the proof. $\triangledown$

Corollary 4.7.2 (Consequences of Theorem 4.7.1)

1. The ruin probability at 0 depends on the safety loading only. Integrating (4.49) for $y \in (0,\infty)$ yields $\Pr[T < \infty]$, so for every $P(\cdot)$ with mean μ_1 we have

$$\psi(0) = \frac{\lambda}{c} \int_0^\infty [1 - P(y)]\,dy = \frac{\lambda}{c}\mu_1 = \frac{1}{1+\theta}. \tag{4.59}$$

2. Assuming that there is at least one new record low, L_1 has the same distribution as the amount with which ruin occurs starting from $u = 0$ (if ruin occurs). So we have the following expression for the density function of the record improvements:

$$f_{L_1}(y) = \frac{1 - P(y)}{(1+\theta)\mu_1}\frac{1}{\psi(0)} = \frac{1 - P(y)}{\mu_1}, \quad y > 0. \tag{4.60}$$

3. Let $H(x)$ denote the cdf of L_1 and p the parameter of M. Then, since L has a compound geometric distribution, the non-ruin probability of a risk process is given by Beekman's convolution formula:

$$1 - \psi(u) = \sum_{m=0}^\infty p(1-p)^m H^{m*}(u), \tag{4.61}$$

where

$$p = \frac{\theta}{1+\theta} \quad \text{and} \quad H(x) = 1 - \frac{1}{\mu_1} \int_x^\infty [1 - P(y)]\,dy. \tag{4.62}$$

4. The mgf of the maximal aggregate loss L, which because of (4.47) is also the mgf of the non-ruin probability $1 - \psi(u)$, is given by

$$m_L(r) = \frac{\theta}{1+\theta} + \frac{1}{1+\theta}\frac{\theta(m_X(r) - 1)}{1 + (1+\theta)\mu_1 r - m_X(r)}. \tag{4.63}$$

Proof. Only the last assertion requires a proof. Since $L = L_1 + \cdots + L_M$ with $M \sim$ geometric(p) for $p = \frac{\theta}{1+\theta}$, we have

$$m_L(r) = m_M(\log m_{L_1}(r)) = \frac{p}{1 - (1-p)m_{L_1}(r)}. \tag{4.64}$$

The mgf of L_1 follows from its density (4.60):

$$
\begin{aligned}
m_{L_1}(r) &= \frac{1}{\mu_1} \int_0^\infty e^{ry}(1 - P(y)) \, dy \\
&= \frac{1}{\mu_1} \left\{ \frac{1}{r}[e^{ry} - 1][1 - P(y)] \Big|_0^\infty + \int_0^\infty \frac{1}{r}[e^{ry} - 1] \, dP(y) \right\} \tag{4.65} \\
&= \frac{1}{\mu_1 r} [m_X(r) - 1],
\end{aligned}
$$

since the integrated term disappears at ∞ because for $t \to \infty$:

$$
\begin{aligned}
\int_0^\infty e^{ry} \, dP(y) < \infty &\Longrightarrow \int_t^\infty e^{ry} \, dP(y) \downarrow 0 \\
&\Longrightarrow e^{rt} \int_t^\infty dP(y) = e^{rt}(1 - P(t)) \downarrow 0. \tag{4.66}
\end{aligned}
$$

Substituting (4.65) into (4.64) then yields (4.63). Note that, as often, a proof using Fubini instead of partial integration is somewhat easier; see Theorem 1.4.3.   ▽

Remark 4.7.3 (Recursive formula for ruin probabilities)
The ruin probability in u can be expressed in the ruin probabilities at smaller initial capitals, as follows:

$$\psi(u) = \frac{\lambda}{c} \int_0^u [1 - P(y)]\psi(u - y) \, dy + \frac{\lambda}{c} \int_u^\infty [1 - P(y)] \, dy. \tag{4.67}$$

To prove this, note that $T < \infty$ implies that the surplus eventually will drop below the initial level, so, using (4.60):

$$
\begin{aligned}
\psi(u) &= \Pr[T < \infty] = \Pr[T < \infty, M > 0] \\
&= \Pr[T < \infty | M > 0] \Pr[M > 0] \\
&= \frac{1}{1+\theta} \int_0^\infty \Pr[T < \infty | L_1 = y] f_{L_1}(y) \, dy \tag{4.68} \\
&= \frac{\lambda}{c} \left(\int_0^u \psi(u - y)(1 - P(y)) \, dy + \int_u^\infty (1 - P(y)) \, dy \right),
\end{aligned}
$$

where we have substituted $c = (1 + \theta)\lambda \mu_1$.   ▽

4.8 Explicit expressions for ruin probabilities

Two situations exist for which we can easily compute the ruin probabilities. We already derived the analytical expression for the ruin probability in case of exponential claim size distributions, but such expressions can also be found for mixtures or combinations of exponentials. For discrete distributions, there is an algorithm.

In the previous section, we derived the mgf with the non-ruin probability $1 - \psi(u)$. To find an expression for the ruin probability, we only need to identify this mgf. We will describe how this can be done for mixtures and combinations of two exponential distributions, see Section 3.8. Since $1 - \psi(0) = \theta/(1+\theta)$ and

$$m_L(r) = \int_0^\infty e^{ru}d[1-\psi(u)] = 1-\psi(0) + \int_0^\infty e^{ru}(-\psi'(u))\,du, \qquad (4.69)$$

it follows from (4.63) that the 'mgf' of the function $-\psi'(u)$ equals

$$\int_0^\infty e^{ru}(-\psi'(u))\,du = \frac{1}{1+\theta}\frac{\theta(m_X(r)-1)}{1+(1+\theta)\mu_1 r - m_X(r)}. \qquad (4.70)$$

Note that, except for a constant, $-\psi'(u)$ is a density function, see Exercise 4.8.1. Now let X be a combination or a mixture of two exponential distributions such as in (3.90), so for some $\alpha < \beta$ and $0 \le q \le \frac{\beta}{\beta-\alpha}$ it has density function

$$p(x) = q\alpha e^{-\alpha x} + (1-q)\beta e^{-\beta x}, \quad x > 0. \qquad (4.71)$$

Then the right-hand side of (4.70), after multiplying both the numerator and the denominator by $(r-\alpha)(r-\beta)$, can be written as the ratio of two polynomials in r. By using partial fractions, this can be written as a sum of two terms of the form $\delta\gamma/(\gamma-r)$, corresponding to δ times the mgf with an exponential(γ) distribution. We give two examples to clarify this method.

Example 4.8.1 (Ruin probability for exponential distributions)

In (4.71), let $q = 0$ and $\beta = 1$, hence the claims distribution is exponential(1). Then, for $\delta = 1/(1+\theta)$ and $\gamma = \theta/(1+\theta)$, the right-hand side of (4.70) leads to

$$\frac{1}{1+\theta}\frac{\theta\left(\frac{1}{1-r}-1\right)}{1+(1+\theta)r-\frac{1}{1-r}} = \frac{\theta}{(1+\theta)[\theta-(1+\theta)r]} = \frac{\delta\gamma}{\gamma-r}. \qquad (4.72)$$

Except for the constant δ, this is the mgf of an exponential(γ) distribution. We conclude from (4.70) that $-\psi'(u)/\delta$ is equal to the density function of this distribution. By using the boundary condition $\psi(\infty) = 0$, we see that for the exponential(1) distribution

$$\psi(u) = \frac{1}{1+\theta}\exp\left(\frac{-\theta u}{1+\theta}\right), \qquad (4.73)$$

which corresponds to (4.32) in Section 4.4 for $\beta = \mu_1 = 1$, and to (4.17). $\qquad \triangledown$

Example 4.8.2 (Ruin probability, mixtures of exponential distributions)
Choose $\theta = 0.4$ and $p(x) = \frac{1}{2} \times 3e^{-3x} + \frac{1}{2} \times 7e^{-7x}$, $x > 0$. Then

$$m_X(r) = \frac{1}{2}\frac{3}{3-r} + \frac{1}{2}\frac{7}{7-r}; \quad \mu_1 = \frac{1}{2}\frac{1}{3} + \frac{1}{2}\frac{1}{7} = \frac{5}{21}. \tag{4.74}$$

So, after some calculations, the right-hand side of (4.70) leads to

$$\frac{6(5-r)}{7(6-7r+r^2)} = \frac{\delta}{1-r} + \frac{6\varepsilon}{6-r} \quad \text{when } \delta = \frac{24}{35} \text{ and } \varepsilon = \frac{1}{35}. \tag{4.75}$$

The ruin probability for this situation is given by

$$\psi(u) = \frac{24}{35}e^{-u} + \frac{1}{35}e^{-6u}. \tag{4.76}$$

Notice that $\psi(0) = \frac{1}{1+\theta}$ indeed holds. ∇

This method works fine for combinations of exponential distributions, too, and also for the limiting case gamma$(2,\beta)$, see Exercises 4.8.5–7. It is possible to generalize the method to mixtures/combinations of more than two exponential distributions, but then roots of polynomials of order three and higher have to be determined. For this, R's function `polyroot` could be used.

To find the coefficients in the exponents of expressions like (4.76) for the ruin probability, that is, the asymptotes of (4.75), we need the roots of the denominator of the right-hand side of (4.70). Assume that, in the density (4.71), $\alpha < \beta$ and $q \in (0,1)$. We have to solve the following equation:

$$1 + (1+\theta)\mu_1 r = q\frac{\alpha}{\alpha-r} + (1-q)\frac{\beta}{\beta-r}. \tag{4.77}$$

Notice that the right-hand side of this equation corresponds to the mgf of the claims only if r is to the left of the asymptotes, so if $r < \alpha$. If r is larger, then this mgf is $+\infty$; hence we write "$m_X(r)$" instead of $m_X(r)$ for these branches in Figure 4.4. From this figure, one sees immediately that the positive roots r_1 and r_2 of (4.77) are real numbers that satisfy

$$r_1 = R < \alpha < r_2 < \beta. \tag{4.78}$$

Remark 4.8.3 (Ruin probability for discrete distributions)
If the claims X can have only a finite number of positive values $x_1, x_2, \ldots, x_m$, with probabilities $p_1, p_2, \ldots, p_m$, the ruin probability equals

$$\psi(u) = 1 - \frac{\theta}{1+\theta}\sum_{k_1,\ldots,k_m}(-z)^{k_1+\cdots+k_m}e^z\prod_{j=1}^{m}\frac{p_j^{k_j}}{k_j!} \tag{4.79}$$

where $z = \frac{\lambda}{c}(u - k_1 x_1 - \cdots - k_m x_m)_+$. The summation extends over all values of $k_1, \ldots, k_m = 0, 1, 2, \ldots$ leading to $z > 0$, and hence is finite. For a proof of (4.79), see Gerber (1989). ∇

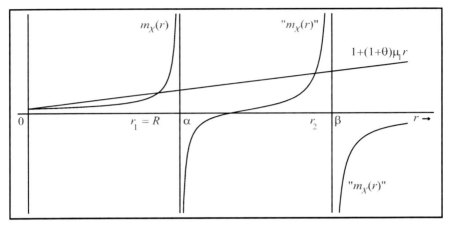

Fig. 4.4 Solutions of (4.77). Only the left branch of the graph is actually the mgf of X.

4.9 Approximation of ruin probabilities

For other distributions than the ones in the previous section, it is difficult to find the exact value of the ruin probability $\psi(u)$. Anyway, one may argue that this exact value is not very important, since in case of doubt, other factors will be decisive. So there is a need for good and simple approximations for the ruin probability.

We give some global properties of the ruin probability that should preferably be satisfied by the approximations. First, equation (4.59) yields $\psi(0) = \frac{1}{1+\theta}$. Next, we know that $\psi(u) = 1 - F_L(u)$, and thus, with partial integration,

$$
\begin{aligned}
\mathrm{E}[L] &= \int_0^\infty u\,\mathrm{d}[1 - \psi(u)] = \int_0^\infty \psi(u)\,\mathrm{d}u, \\
\mathrm{E}[L^2] &= \int_0^\infty u^2\,\mathrm{d}[1 - \psi(u)] = \int_0^\infty 2u\psi(u)\,\mathrm{d}u.
\end{aligned}
\tag{4.80}
$$

These moments of the maximal aggregate loss L can easily be computed since $L = L_1 + \cdots + L_M$ has a compound geometric distribution, with the distribution of M and L_j given in Section 4.7. The required moments of L_j are

$$
\mathrm{E}[L_j^k] = \frac{\mu_{k+1}}{\mu_1(k+1)}, \quad k = 1, 2, \ldots
\tag{4.81}
$$

Since $\mathrm{E}[M] = \frac{1}{\theta}$, we have

$$
\int_0^\infty \psi(u)\,\mathrm{d}u = \frac{\mu_2}{2\theta\mu_1}.
\tag{4.82}
$$

It can also be shown that

$$
\mathrm{Var}[L] = \frac{\mu_3}{3\theta\mu_1} + \frac{\mu_2^2}{4\theta^2\mu_1^2},
\tag{4.83}
$$

hence

$$\int_0^\infty u\psi(u)\,du = \frac{1}{2}E[L^2] = \frac{\mu_3}{6\theta\mu_1} + \frac{\mu_2^2}{4\theta^2\mu_1^2}. \tag{4.84}$$

After this groundwork, we introduce a number of possible approximations.

1. We already saw in Section 4.3 that one possible choice for the approximating function is to take

$$\psi(u) \approx \frac{1}{1+\theta}e^{-Ru}, \tag{4.85}$$

   with R the adjustment coefficient of the claim size distribution. This function has the correct value $\psi(0)$, and also the correct asymptotic hazard rate R.

2. In the previous approximation, we might also take the exact ruin probability of an exponential claim size distribution with adjustment coefficient R, that is

$$\psi(u) \approx (1 - R\mu_1)e^{-Ru}, \tag{4.86}$$

   This way, the hazard rate is correct, but the ruin probability at zero might differ.

3. Replacing the claims distribution by an exponential distribution with the same expected value, we get, see (4.32):

$$\psi(u) \approx \frac{1}{1+\theta}\exp\left(-\frac{\theta}{1+\theta}\frac{u}{\mu_1}\right). \tag{4.87}$$

   For $u = 0$, the approximation is correct, but in general, the integrals over the left-hand side and the right-hand side are different.

4. Approximating $\psi(u)$ by $\psi(0)e^{-ku}$ with k chosen such that (4.82) holds yields as an approximation

$$\psi(u) \approx \frac{1}{1+\theta}\exp\left(\frac{-2\theta\mu_1 u}{(1+\theta)\mu_2}\right). \tag{4.88}$$

   Note that if the claims are exponential($1/\mu_1$) random variables, then $\mu_2 = 2\mu_1^2$, so (4.87) and (4.88) both give the correct ruin probability.

5. We can approximate the ruin probability by a gamma distribution:

$$\psi(u) \approx \frac{1}{1+\theta}(1 - G(u,\alpha,\beta)), \quad u \geq 0. \tag{4.89}$$

To fit the first two moments, the parameters α and β of the gamma cdf $G(\cdot;\alpha,\beta)$ must meet the following conditions:

$$E[L] = \frac{1}{1+\theta}\frac{\alpha}{\beta}; \quad E[L^2] = \frac{1}{1+\theta}\left(\frac{\alpha^2}{\beta^2} + \frac{\alpha}{\beta^2}\right). \tag{4.90}$$

6. Just as in the first approximation, one can replace the claims distribution by another with a few moments in common, for which the corresponding ruin probability can be easily calculated. A suitable candidate for such a replacement is a mixture or combination of exponential distributions.

7. Another possible replacement is a discrete distribution. The ruin probabilities can easily be computed from (4.79). For each claims distribution, one can find a two-point distribution with the same first three moments. This is not always possible in case of a mixture/combination of two exponential distributions. Both methods yield good approximations. See Example 4.9.1 below.

8. From the theory of ordering of risks, it follows that one gets a lower bound for the ruin probability if one replaces the claims distribution with expectation μ by a one-point distribution on μ. A simple upper bound can be obtained if one knows the maximum value b of the claims. If one takes a claims distribution with probability μ/b of b and $1 - \mu/b$ of 0, a Poisson process arises that is equivalent to a Poisson process with claims always equal to b and claim number parameter $\lambda\mu/b$ instead of λ. So, both the lower bound and the upper bound can be calculated by using (4.79) with $m = 1$.

9. The geometric distribution allows the use of Panjer's recursion, provided the individual terms are integer-valued. This is not the case for the terms L_j of L, see (4.60). But we can easily derive lower and upper bounds this way, by simply rounding the L_j down to an integer multiple of δ to get a random variable L' that is suitable for Panjer's recursion, and gives an upper bound for $F_L(u)$ since $\Pr[L \le u] \le \Pr[L' \le u]$. Rounding up leads to a lower bound for $F_L(u)$. By taking δ small, we get quite good upper and lower bounds with little computational effort.

Example 4.9.1 (Approximating a ruin probability by using diatomic claims)
For the purpose of approximating a ruin probability by using an m-point claim size distribution with the same first few moments, we will confine ourselves to the first three moments of the distribution. So we will only use (4.79) in the special case $m = 2$, since a two-point distribution has three free parameters, the location of both mass points and one mass, and thus allows us to fit the first three moments of the distribution. An R-implementation of algorithm (4.79) for an $m = 2$ point claim size distribution is given below; it is not trivial to generalize it to the case of general m. Here the claims are x with probability px, and y with probability 1-px.

```
psi <- function (u, theta, x, y, px)
{ if (px<0||px>1||theta<=0||x<=0||y<=0||u<0) stop("bad params")
  mu <- x*px + y*(1-px)
  ss <- 0
  for (k in 0:(u/x))
  { n <- 0:((u-k*x)/y)
    h <- 1/mu/(1+theta) * (u - k*x - n*y)
    tt <- sum((-h)^(k+n)*exp(h)*(1-px)^n/factorial(n))
    ss <- ss + px^k / factorial(k) * tt }
  return(1 - theta/(1+theta)*ss) }
psi(2.5,0.5,1,2,0.5) ## 0.2475216
```

To determine a random variable X with a two-point distribution with x, y as mass points and $p_x = \Pr[X = x]$, three non-linear equations with as many unknowns must be solved. Analytical expressions exist, but here we will proceed differently. From relations (7.28) and (7.30) we deduce that $y = \mu + \sigma^2/(\mu - x)$ and $p_x = \sigma^2/(\sigma^2 +$

$(\mu - x)^2)$ characterize a two-point distribution with mass points equal to x and y, mean μ and variance σ^2. Then by varying x, we ensure that also the skewness is right. The last part is achieved by a call of the function `uniroot`. To fit the third moment $E[X^3] = x^3 p_x + y^3(1 - p_x)$, we use the identity

$$E[X^3] = E[((X - \mu) + \mu)^3] = \gamma\sigma^3 + 3\mu\sigma^2 + \mu^3. \tag{4.91}$$

The mass points with given μ, σ^2, γ, and the approximate $\psi(u)$ are found by the following R code:

```
mu <- 1.5; sig2 <- 0.25; gam <- 0;
mu3 <- gam * sig2^1.5 + 3 * mu * sig2 + mu^3
mu3.diff <- function(x)
{   y <- mu + sig2/(mu-x); px <- sig2 / (sig2+(x-mu)^2)
    px*x^3 + (1-px)*y^3 - mu3   }
x <- uniroot(mu3.diff, lower=0, upper=mu*0.9999999)$root
psi(2.5, 0.5, x, mu + sig2/(mu-x), sig2 / (sig2+(x-mu)^2))
```

The smaller of x and y obviously must be less than μ. That is why we look for a zero of `mu3.diff` in the interval $(0, \mu)$. The last line produces the approximate ruin probability for the given first three moments. ∇

4.10 Exercises

Section 4.2

1. Assume that the waiting times $W_1, W_2, \ldots$ are independent and identically distributed random variables with cdf $F(x)$ and density function $f(x)$, $x \geq 0$. Given $N(t) = i$ and $T_i = s$ for some $s \leq t$, what is the *conditional* probability of a claim occurring between points in time t and $t + dt$? (This generalization of a Poisson process is called a *renewal process*.)

2. Let $\{N(t), t \geq 0\}$ be a Poisson process with parameter λ, and let $p_n(t) = \Pr[N(t) = n]$ and $p_{-1}(t) \equiv 0$. Show that $p'_n(t) = -\lambda p_n(t) + \lambda p_{n-1}(t)$, $n = 0, 1, 2, \ldots$, and interpret these formulas by comparing $p_n(t)$ with $p_n(t + dt)$.

Section 4.3

1. Prove that the expressions in (4.11) are indeed equivalent to (4.10).
   Also prove (by partial integration or using Fubini), that

   $$1 + (1 + \theta)\mu r = 0 \text{ and } r > 0 \iff \int_0^\infty [e^{rx} - (1 + \theta)][1 - P(x)]\,dx = 0.$$

2. Use $e^{rx} > 1 + rx + \frac{1}{2}(rx)^2$ for $r > 0$ and $x > 0$ to prove that $R < 2\theta\mu_1/\mu_2$.

3. For $\theta = 0.4$ and $p(x) = \frac{1}{2}(3e^{-3x} + 7e^{-7x})$, determine the values of t for which $m_X(t)$ is finite, and also determine R.

4. If the claims distribution is discrete with $p(1) = p(2) = \frac{1}{2}$, then find θ if $R = \log 3$.

5. Which premium yields $e^{-Ru} = \varepsilon$?

6. If $\Pr[X = 0, 1, 2, 4] = \frac{1}{4}$, then determine R with R, for $\lambda = 1$ and $c = 3$.

7. Assume that the claims X in a ruin process with $\theta = \frac{2}{5}$ arise as follows: first, a value Y is drawn from two possible values 3 and 7, each with probability $\frac{1}{2}$. Next, conditionally on $Y = y$, the claim X is drawn from an exponential(y) distribution. Determine the adjustment coefficient R. If $R = 2$ for the same distribution, is θ larger or smaller than $\frac{2}{5}$?

8. In some ruin process, the individual claims have a gamma$(2, 1)$ distribution. Determine the loading factor θ as a function of the adjustment coefficient R. Also, determine $R(\theta)$. If the adjustment coefficient equals $\frac{4}{3}$, does $\theta = 2$ hold? Using a sketch of the graph of the mgf of the claims, discuss the behavior of R as a function of θ.

9. [♠] Discuss the determination of the adjustment coefficient R if in a ruin process the claims are lognormally distributed. Also, if the claims are inverse Gaussian.

10. Argue that $dc/dR \geq 0$. Use the relation $c = \frac{1}{R} \log(m_S(R))$, where S denotes the total claims in some period of length 1, to give an alternative proof that exponential premiums increase with the parameter (risk aversion) α. See also Theorem 11.3.2.

11. In Example 4.3.7, replace the claim severity distribution with a Pareto distribution with the same mean and variance, and replace $n = 400$ by $n = 1000$. What happens to the ruin probability? What happens to the adjustment coefficient? Does ruin still happen in the early stages of the process only?

12. In the same example, replace the severity distribution by a gamma(α, α) distribution with $\alpha = 1$ and $\alpha = 3$. Observe that the ruin probability decreases with α. Use uniroot to compute the adjustment coefficient when $\alpha = 3$.

Section 4.4

1. From Corollary 4.4.2, we know that $\psi(u) \to 1$ if $\theta \downarrow 0$. Why does this imply that $\psi(u) = 1$ if $\theta < 0$?

2. Which compound Poisson processes have a ruin probability $\psi(u) = \frac{1}{2}e^{-u}$?

3. For a compound Poisson process, it is known that the continuous ruin probability depends on the initial capital u in the following way: $\psi(u) = \alpha e^{-u} + \beta e^{-2u}$. Determine the adjustment coefficient for this process. Can anything be said about the Poisson parameter in this risk process? What is $E[\exp(-RU(T)) | T < \infty]$?

4. Assume that $\psi(\varepsilon) < 1$. By looking at the event "non-ruin and no claim before ε/c", with c denoting the premium income per unit of time, show that $\psi(0) < 1$ must hold.

5. For a certain risk process, it is given that $\theta = 0.4$ and $p(x) = \frac{1}{2}(3e^{-3x} + 7e^{-7x})$. Which of the numbers 0, 1 and 6 are roots of the adjustment coefficient equation $1 + (1 + \theta)\mu_1 R = m_X(R)$? Which one is the real adjustment coefficient?

   One of the four expressions below is the ruin probability for this process; determine which expression is the correct one, and argue why the other expressions cannot be the ruin probability.

   1. $\psi(u) = \frac{24}{35}e^{-u} + \frac{1}{35}e^{-6u}$;
   2. $\psi(u) = \frac{24}{35}e^{-u} + \frac{11}{35}e^{-6u}$;
   3. $\psi(u) = \frac{24}{35}e^{-0.5u} + \frac{1}{35}e^{-6.5u}$;
   4. $\psi(u) = \frac{24}{35}e^{-0.5u} + \frac{11}{35}e^{-6.5u}$.

6. The ruin probability for some ruin process equals $\psi(u) = \frac{1}{5}e^{-u} + \frac{2}{5}e^{-0.5u}$, $u \geq 0$. By using the fact that for ruin processes, in general, $\lim_{u \to \infty} \psi(u)/e^{-Ru} = c$ for some $c \in (0, 1)$, determine the adjustment coefficient R and the appropriate constant c in this case.

Section 4.5

1. Assume that the distribution of G_i satisfies $\Pr[G_i = +1] = p$ and $\Pr[G_i = -1] = q = 1 - p$. Further, $p > \frac{1}{2}$ and u is an integer. Determine $U(\widetilde{T})$ if $\widetilde{T} < \infty$. Express $\widetilde{\psi}(u)$ in terms of $\widetilde{R}$, and both $\widetilde{R}$ and $\widetilde{\psi}(u)$ in terms of p and q.

2. [♠] Assume that an insurer uses an exponential utility function $w(\cdot)$ with risk aversion α. Prove that $\mathrm{E}[w(U_{n+1}) \,|\, U_n = x] \geq w(x)$ if and only if $\alpha \leq \widetilde{R}$, and interpret this result.

3. Show that $\widetilde{T} \geq T$ with probability 1, as well as $\widetilde{\psi}(u) \leq \psi(u)$ for all u, if both are determined for a compound Poisson risk process.

4. Assume that the continuous infinite ruin probability for a compound Poisson process equals αe^{-u} in case of an initial capital u, for some constant α. Furthermore, the claims follow an exponential distribution with parameter 2 and the expected number of claims a year is 50. Determine the safety loading for this process. Also determine an upper bound for the discrete infinite ruin probability.

Section 4.6

1. The claim process on some insurance portfolio is compound Poisson with $\lambda = 1$ and $p(x) = e^{-x}, x > 0$. The loading factor is θ. Calculate the adjustment coefficient in case one takes out a proportional reinsurance $h(x) = \alpha x$ with a loading factor $\xi > 0$. Calculate the relative loading factor after this reinsurance. Which restrictions apply to α?

2. For the same situation as in the previous exercise, but now with excess of loss coverage $h(x) = (x - \beta)_+$, write down the adjustment coefficient equation, and determine the loading factor after reinsurance.

3. Assume that the claims per year W_i, $i = 1, 2, \ldots$, are $N(5, 1)$ distributed and that $\theta = 0.25$. A reinsurer covers a fraction α of each risk, applying a premium loading factor $\xi = 0.4$. Give the adjustment coefficient $\widetilde{R}$ for the reinsured portfolio, as a function of α. Which value optimizes the security of the insurer?

4. A total claims process is compound Poisson with $\lambda = 1$ and $p(x) = e^{-x}, x \geq 0$. The relative loading factor is $\theta = \frac{1}{2}$. One takes out a proportional reinsurance $h(x) = \alpha x$. The relative loading factor of the reinsurer equals 1. Determine the adjustment coefficient R_h. For which values of α is ruin not a certainty?

5. Reproduce the adjustment coefficients in both tables in this section.

Section 4.7

1. What is the mgf of L_1 if the claims (a) are equal to b with probability 1, and (b) have an exponential distribution?

2. Prove that $\Pr[L = 0] = 1 - \psi(0)$.

3. In Exercises 4.4.3 and 4.4.6, what is θ?

4. Verify equality (4.67) for the situation of Example 4.8.1 with $\theta = 1$.

5. [♠] Using equality (4.67), find the ruin probability for Example 4.8.2 recursively, for $u = 0, 0.1, 0.2, \ldots$. Check with the analytical outcomes.

6. Find $m_{L_1}(R)$.

Section 4.8

1. For which constant γ is $\gamma(-\psi'(u))$, $u > 0$ a density?

2. Make sketches like in Figure 4.4 to determine the asymptotes of (4.70), for a proper combination of exponential distributions and for a gamma$(2, \beta)$ distribution.

3. Calculate θ and R if $\psi(u) = 0.3e^{-2u} + 0.2e^{-4u} + \alpha e^{-6u}$, $u \geq 0$. Which values of α are possible taking into account that $\psi(u)$ decreases in u and that the safety loading θ is positive?

4. If $\lambda = 3$, $c = 1$ and $p(x) = \frac{1}{3}e^{-3x} + \frac{16}{3}e^{-6x}$, then determine μ_1, θ, m_X, (4.70), and an explicit expression for $\psi(u)$.

5. Determine $E[e^{-RU(t)}]$ in the previous exercise, with the help of (4.28). Determine independent random variables X, Y and I such that $IX + (1-I)Y$ has density $p(\cdot)$.

6. Just as Exercise 4.8.4, but now $p(x)$ is a gamma$(2, \frac{3}{5})$ density, $\lambda = 1$ and $c = 10$.

7. Determine $\psi(u)$ if $\theta = \frac{5}{7}$ and the claims X_i are equal to $X_i = Y_i/4 + Z_i/3$, with Y_i and $Z_i \sim$ exponential(1) and independent.

8. Sketch the density of L_j in case of a discrete claims distribution.

9. [♠] Prove (4.79) in case of $m = 1$ and $m = 2$.

10. Assume that the individual claims in a ruin process are equal to the maximum of two independent exponential(1) random variables, that is, $X_i = \max\{Y_{i1}, Y_{i2}\}$ with $Y_{ik} \sim$ exponential(1). Determine the cdf of X_i, and use this to prove that the corresponding density $p(x)$ is a combination of exponential distributions. Determine the loading factor θ in the cases that for the adjustment coefficient, we have $R = 0.5$ and $R = 2.5$.

11. Two companies have independent compound Poisson ruin processes with intensities $\lambda_1 = 1$ and $\lambda_2 = 8$, claims distributions exponential(3) and exponential(6), initial capitals u_1 and u_2 and loading factors $\theta_1 = 1$ and $\theta_2 = \frac{1}{2}$. These companies decide to merge, without changing their premiums. Determine the intensity, claims distribution, initial capital and loading factor of the ruin process for the merged company. Assuming $u_1 = u_2 = 0$, compare the probabilities of the following events (*continuous infinite ruin probabilities*): "both companies never go bankrupt" with "the merged company never goes bankrupt". Argue that for any choice of u_1 and u_2, the former event has a smaller probability than the latter.

12. [♠] Implement the method of Example 4.8.2 in R. As input, take u, θ and q, α, β as in (4.76), as output, $\psi(u)$ as well as $\delta, \varepsilon, r_1, r_2$ of (4.75).

13. [♠] Generalize the previous exercise for a mixture/combination of m exponentials.

Section 4.9

1. Verify (4.80), (4.81), (4.83)/(4.84), and (4.90). Solve α and β from (4.90).

2. Work out the details of the final approximation. [♠] Implement it using Panjer's recursion. Test and compare with the approximations above for some non-negative claim size distributions.

3. [♠] To be able to approximate the ruin probability, find the parameters of a mixture/combination of exponential distributions with matching μ, σ^2 and γ. Also give bounds for the skewness γ for this to be possible.

Chapter 5
Premium principles and Risk measures

Actuaries have long been tasked with tackling difficult quantitative problems. Loss estimates, reserves requirements, and other quantities have been their traditional domain, but the actuary of the future has an opportunity to broaden his/her scope of knowledge to include other risks facing corporations around the globe. While this may seem like a daunting task at first, the reality is that the skills required to analyze business risks are not a significant stretch of the traditional actuary's background — Timothy Essaye, `imageoftheactuary.org`

5.1 Introduction

The activities of an insurer can be described as a system in which the acquired capital increases because of (earned) premiums and interest, and decreases because of claims and costs. See also the previous chapter. In this chapter we discuss some mathematical methods to determine the premium from the distribution of the claims. The actuarial aspect of a premium calculation is to calculate a minimum premium, sufficient to cover the claims and, moreover, to increase the expected surplus sufficiently for the portfolio to be considered stable.

Bühlmann (1985) described a top-down approach for the premium calculation. First we look at the premium required by the total portfolio. Then we consider the problem of spreading the total premium over the policies in a fair way. To determine the minimum annual premium, we use the ruin model as introduced in the previous chapter. The result is an exponential premium (see Chapter 1), where the risk aversion parameter α follows from the maximal ruin probability allowed and the available initial capital. Assuming that the suppliers of the initial capital are to be rewarded with a certain annual dividend, and that the resulting premium should be as low as possible, therefore as competitive as possible, we can derive the optimal initial capital. Furthermore we show how the total premium can be spread over the policies in a fair way, while the total premium keeps meeting our objectives.

For the policy premium, a lot of premium principles can be justified. Some of them can be derived from models like the zero utility model, where the expected utility before and after insurance is equal. Other premium principles can be derived as an approximation of the exponential premium principle. We will verify to which extent these premium principles satisfy some reasonable requirements. We will also consider some characterizations of premium principles. For example, it turns out that the only utility preserving premium principles for which the total premium for independent policies equals the sum of the individual premiums are the net premium and the exponential premium.

As an application, we analyze how insurance companies can optimally form a 'pool'. Assuming exponential utility, it turns out that the most competitive total premium is obtained when the companies each take a fixed part of the pooled risk (coinsurance), where the proportion is inversely proportional to their risk aversion, hence proportional to their risk tolerance. See also Gerber (1979).

Mathematically speaking, a risk measure for a random variable S is just a functional connecting a real number to S. A prime example is of course a premium principle, another is the ruin probability for some given initial capital. The risk measure most often used in practice is the *Value-at-Risk* (VaR) at a certain (confidence) level p with $0 < p < 1$, which is the amount that will maximally be lost with probability p, therefore the inverse of the cdf. It is also called the quantile risk measure. Note the difference between the variance, written as $\text{Var}[X]$, and the value-at-risk, written in CamelCase as $\text{VaR}[X; p]$. Fortunately, Vector Autoregressive (VAR) models are outside the scope of this book.

The VaR is not ideal as a risk measure. One disadvantage is that it only looks at the probability of the shortfall of claims over capital being positive. But the size of the shortfall certainly matters; someone will have to pay for the remainder. Risk measures accounting for the size of the shortfall $(X - d)_+$ when capital d is available include the Tail-Value-at-Risk, the Expected Shortfall and the Conditional Tail Expectation.

Another possible disadvantage of VaR is that it is not subadditive: the sum of the VaRs for X and Y may be larger than the one for $X + Y$. Because of this, the VaR is not a *coherent* risk measure. But insisting on coherence is only justified when looking at complete markets, which the insurance market is not, since it is not always possible to diversify a risk.

5.2 Premium calculation from top-down

As argued in Chapter 4, insuring a certain portfolio of risks leads to a surplus that increases because of collected premiums and decreases in the event of claims. The following recurrent relations hold in the ruin model between the surpluses at integer times:

$$U_t = U_{t-1} + c - S_t, \quad t = 1, 2, \ldots \tag{5.1}$$

Ruin occurs if $U_\tau < 0$ for some real τ. We assume that the annual total claims S_t, $t = 1, 2, \ldots$, are independent and identically compound Poisson random variables, say $S_t \sim S$. The following question then arises: how large should the initial capital $U_0 = u$ and the premium $c = \pi[S]$ be to remain solvent at all times with a prescribed probability? The probability of ruin at integer times only is less than $\psi(u)$, which in turn is bounded from above by e^{-Ru}. Here R denotes the adjustment coefficient, which is the root of the equation $e^{Rc} = E[e^{RS}]$, see (4.11). If we set the upper bound equal to ε, then $R = |\log \varepsilon|/u$. Hence, we get a ruin probability bounded by ε by choosing the premium c as

$$c = \frac{1}{R}\log\left(E[e^{RS}]\right), \quad \text{where} \quad R = \frac{1}{u}|\log\varepsilon|. \tag{5.2}$$

This premium is the exponential premium (1.20) with parameter R. From Example 1.3.1, we know that the adjustment coefficient can be interpreted as a measure for the risk aversion: for the utility function $-\alpha e^{-\alpha x}$ with risk aversion α, the utility preserving premium is $c = \frac{1}{\alpha}\log(E[e^{\alpha X}])$.

A characteristic of the exponential premium is that choosing this premium for each policy also yields the right total premium for S. The reader may verify that if the payments X_j on policy j, $j = 1,\ldots,n$, are independent, then

$$S = X_1 + \cdots + X_n \quad \Longrightarrow \quad \frac{1}{R}\log(E[e^{RS}]) = \sum_{j=1}^{n}\frac{1}{R}\log(E[e^{RX_j}]). \tag{5.3}$$

Another premium principle that is additive in this sense is the variance principle, where for a certain parameter $\alpha \geq 0$ the premium is determined by

$$\pi[S] = E[S] + \alpha \text{Var}[S]. \tag{5.4}$$

In fact, every premium that is a linear combination of cumulants is additive. Premium (5.4) can also be obtained as an approximation of the exponential premium by considering only two terms of the Taylor expansion of the cgf, assuming that the risk aversion R is small, since

$$\pi[S] = \frac{1}{R}\kappa_S(R) = \frac{1}{R}\left(E[S]R + \text{Var}[S]\frac{R^2}{2} + \cdots\right) \approx E[S] + \frac{1}{2}R\text{Var}[S]. \tag{5.5}$$

So to approximate (5.2) by (5.4), α should be taken equal to $\frac{1}{2}R$. In view of (5.2) and $\tilde{\psi}(u) \leq e^{-Ru}$, we can roughly state that:

- doubling the loading factor α in (5.4) decreases the upper bound for the ruin probability from ε to ε^2;
- halving the initial capital requires the loading factor to be doubled if one wants to keep the same maximal ruin probability.

We will introduce a new aspect in the discrete time ruin model (5.1): how large should u be, if the premium c is to contain a yearly dividend iu for the shareholders who have supplied the initial capital? A premium at the portfolio level that ensures ultimate survival with probability $1 - \varepsilon$ and also incorporates this dividend is

$$\pi[S] = E[S] + \frac{|\log\varepsilon|}{2u}\text{Var}[S] + iu, \tag{5.6}$$

that is, the premium according to (5.2) and (5.5), plus the dividend iu. We choose u such that the premium is as competitive as possible, therefore as low as possible. By setting the derivative equal to zero, we see that a minimum is reached for $u = \sigma[S]\sqrt{|\log\varepsilon|/2i}$. Substituting this value into (5.6), we see that (see Exercise 5.2.1) the optimal premium is a *standard deviation premium*:

$$\pi[S] = E[S] + \sigma[S]\sqrt{2i|\log \varepsilon|}. \tag{5.7}$$

In the optimum, the loading $\pi[S] - E[S] - iu$ equals the dividend iu; notice that if i increases, then u decreases, but iu increases.

Finally, we have to determine which premium should be asked at the down level. We cannot just use a loading proportional to the standard deviation. The sum of these premiums for independent risks does not equal the premium for the sum, and consequently the top level would not be in balance: if we add a contract, the total premium no longer satisfies the specifications. On the other hand, as stated before, the variance principle is additive, just like the exponential and the net premium. Hence, (5.6) and (5.7) lead to Bühlmann's recommendation for the premium calculation:

1. Compute the optimal initial capital $u = \sigma[S]\sqrt{|\log \varepsilon|/2i}$ for S, i and ε;
2. Spread the total premium over the individual risks X_j by charging the following premium:

$$\pi[X_j] = E[X_j] + R\mathrm{Var}[X_j], \quad \text{where } R = |\log \varepsilon|/u. \tag{5.8}$$

Note that in this case the loading factor $R = \alpha$ of the variance premium, incorporating both a loading to avoid ruin and an equal loading from which to pay dividends, is twice as large as it would be without dividend, see (5.4) and (5.5). The total dividend and the necessary contribution to the expected growth of the surplus that is required to avoid ruin are spread over the policies in a similar way.

Bühlmann gives an example of a portfolio consisting of two kinds (A and B) of exponential risks with mean values 5 and 1:

Type	Number of risks	Expected value	Variance	Exponential premium	Variance premium	Stand. dev. premium
A	5	5	25	$-\frac{1}{R}\log(1-5R)$	$5 + \frac{R}{2}25$	
B	20	1	1	$-\frac{1}{R}\log(1-R)$	$1 + \frac{R}{2}1$	
Total	25	45	145			$45 + (2i|\log \varepsilon|145)^{\frac{1}{2}}$

Choose $\varepsilon = 1\%$, hence $|\log \varepsilon| = 4.6052$. Then, for the model with dividend, we have the following table of variance premiums for different values of i.

	Portfolio premium	Optimal u	Optimal R	Premium for A	Premium for B
$i = 2\%$	50.17	129.20	0.0356	5.89	1.0356
$i = 5\%$	53.17	81.72	0.0564	6.41	1.0564
$i = 10\%$	56.56	57.78	0.0797	6.99	1.0797

The portfolio premium and the optimal u follow from (5.7), R from (5.2), and the premiums for A and B are calculated according to (5.8). We observe that:

- the higher the required return i on the supplied initial capital u, the lower the optimal value for u;

- the loading is far from proportional to the risk premium: the loading as a percentage for risks of type A is 5 times the one for risks of type B;
- the resulting exponential premiums are close to the variance premiums given: if $i = 2\%$, the premium with parameter $2R$ is 6.18 for risks of type A and 1.037 for risks of type B.

5.3 Various premium principles and their properties

In this section, we give a list of premium principles that can be applied at the policy level as well as at the portfolio level. We also list some mathematical properties that one might argue a premium principle should have. Premium principles depend exclusively on the marginal distribution function of the random variable. Consequently, we will use both notations $\pi[F]$ and $\pi[X]$ for the premium of X, if F is the cdf of X. We will assume that X is a *bounded* non-negative random variable. Most premium principles can also be applied to unbounded and possibly negative claims. When the result is an infinite premium, the risk at hand is uninsurable.

We have encountered the following five premium principles in Section 5.2:

a) **Net premium:** $\pi[X] = E[X]$
   Also known as the equivalence principle; this premium is sufficient for a risk neutral insurer only.
b) **Expected value principle:** $\pi[X] = (1 + \alpha)E[X]$
   The loading equals $\alpha E[X]$, where $\alpha > 0$ is a parameter.
c) **Variance principle:** $\pi[X] = E[X] + \alpha \text{Var}[X]$
   The loading is proportional to $\text{Var}[X]$, and again $\alpha > 0$.
d) **Standard deviation principle:** $\pi[X] = E[X] + \alpha \sigma[X]$
   Here also $\alpha > 0$ should hold, to avoid getting ruined with probability 1.
e) **Exponential principle:** $\pi[X] = \frac{1}{\alpha} \log(m_X(\alpha))$
   The parameter $\alpha > 0$ is called the risk aversion. We already showed in Chapter 1 that the exponential premium increases if α increases. For $\alpha \downarrow 0$, the net premium arises; for $\alpha \to \infty$, the resulting premium equals the maximal value of X, see Exercise 5.3.11.

In the following two premium principles, the 'parameter' is a function.

f) **Zero utility premium:** $\pi[X]$ is such that $u(0) = E[u(\pi[X] - X)]$
   This concept was introduced in Chapter 1. The function $u(x)$ represents the utility a decision maker attaches to his present capital plus x. So, $u(0)$ is the utility of the present capital and $u(\pi[X] - X)$ is the utility after insuring a risk X against premium $\pi[X]$. The premium solving the utility equilibrium equation is called the zero utility premium. Each linear transform of $u(\cdot)$ yields the same premium. The function $u(\cdot)$ is usually assumed to be non-decreasing and concave. Accordingly it has positive but decreasing marginal utility $u'(x)$. The special choice $u(x) = \frac{1}{\alpha}(1 - e^{-\alpha x})$ leads to exponential utility; the net premium results for linear $u(\cdot)$.

g) **Mean value principle:** $\pi[X] = v^{-1}(\mathrm{E}[v(X)])$
The function $v(\cdot)$ is a convex and increasing valuation function. Again, the net premium and the exponential premium are special cases with $v(x) = x$ and $v(x) = e^{\alpha x}$, $\alpha > 0$.

The following premium principles are chiefly of theoretical importance:

h) **Percentile principle:** $\pi[X] = \min\{p \mid F_X(p) \geq 1 - \varepsilon\}$
The probability of a loss on contract X is at most ε, $0 \leq \varepsilon \leq 1$.
i) **Maximal loss principle:** $\pi[X] = \min\{p \mid F_X(p) = 1\}$
This premium arises as a limiting case of other premiums: (e) for $\alpha \to \infty$ and (h) for $\varepsilon \downarrow 0$. A 'practical' example: a pregnant woman pays some premium for an insurance contract that guarantees that the baby will be a girl; if it is a boy, the entire premium is refunded.
j) **Esscher principle:** $\pi[X] = \mathrm{E}[X e^{hX}]/\mathrm{E}[e^{hX}]$
Here, h is a parameter with $h > 0$. This premium is actually the net premium for a risk $Y = X e^{hX}/m_X(h)$. As one sees, Y results from X by enlarging the large values of X, while reducing the small values. The Esscher premium can also be viewed as the expectation with the so-called Esscher transform of $\mathrm{d}F_X(x)$, which has as a 'density':

$$\mathrm{d}\tilde{G}(x) = \frac{e^{hx}\,\mathrm{d}F_X(x)}{\int e^{hy}\,\mathrm{d}F_X(y)}. \tag{5.9}$$

This is the differential of a cdf with the same support as X, but for which the probabilities of small values are reduced in favor of the probabilities of large values. The net premium for Y gives a loaded premium for X.

5.3.1 Properties of premium principles

Below, we give five desirable properties for premium principles $\pi[X]$. Some other useful properties such as order preserving, which means that premiums for smaller risks should also be less, will be covered in Chapter 7.

1) **Non-negative loading:** $\pi[X] \geq \mathrm{E}[X]$
A premium without a positive loading will lead to ruin with certainty.
2) **No rip-off:** $\pi[X] \leq \min\{p \mid F_X(p) = 1\}$
The maximal loss premium (i) is a boundary case. If X is unbounded, this premium is infinite.
3) **Consistency:** $\pi[X + c] = \pi[X] + c$ for each c
If we raise the claim by some fixed amount c, then the premium should also be higher by the same amount. Synonyms for consistency are *cash invariance*, *translation invariance*, and more precisely, *translation equivariance*. Note that in general, a 'risk' need not be a non-negative random variable, though to avoid certain technical problems sometimes it is convenient to assume it is bounded from below.

4) **Additivity:** $\pi[X+Y] = \pi[X] + \pi[Y]$ for independent X, Y
Pooling independent risks does not affect the total premium needed.

5) **Iterativity:** $\pi[X] = \pi\big[\pi[X\,|\,Y]\big]$ for all X, Y
The premium for X can be calculated in two steps. First, apply $\pi[\cdot]$ to the conditional distribution of X, given $Y = y$. The resulting premium is a function $h(y)$, say, of y. Then, apply the same premium principle to the random variable $\pi[X\,|\,Y] := h(Y)$.

For the net premium, iterativity is the corresponding property for expected values (2.7). Otherwise, the iterativity criterion is rather artificial. As an example, assume that a certain driver causes a Poisson number X of accidents in one year, where the parameter λ is drawn from the distribution of the structure variable Λ. The number of accidents varies because of the Poisson deviation from the expectation λ, and because of the variation of the structure distribution. In case of iterativity, if we set premiums for both sources of variation one after another, we get the same premium as when we determine the premium for X directly.

Example 5.3.1 (Iterativity of the exponential principle)
The exponential premium principle is iterative. This can be shown as follows:

$$
\begin{aligned}
\pi[\pi[X\,|\,Y]] &= \frac{1}{\alpha}\log \mathrm{E}\big[e^{\alpha\pi[X\,|\,Y]}\big] = \frac{1}{\alpha}\log \mathrm{E}\big[\exp(\alpha\frac{1}{\alpha}\log \mathrm{E}[e^{\alpha X}\,|\,Y])\big] \\
&= \frac{1}{\alpha}\log \mathrm{E}\big[\mathrm{E}[e^{\alpha X}\,|\,Y]\big] = \frac{1}{\alpha}\log \mathrm{E}\big[e^{\alpha X}\big] = \pi[X].
\end{aligned}
\tag{5.10}
$$

After taking the expectation in an exponential premium, the transformations that were done before are successively undone. $\quad\triangledown$

Example 5.3.2 (Compound distributions)
Assume that $\pi[\cdot]$ is additive as well as iterative, and that S is a compound random variable with N terms distributed as X. The premium for S then equals

$$
\pi[S] = \pi[\pi[S\,|\,N]] = \pi[N\pi[X]].
\tag{5.11}
$$

Furthermore, if $\pi[\cdot]$ is also *proportional*, (or *homogeneous*), which means that $\pi[\alpha N] = \alpha\pi[N]$ for all $\alpha \geq 0$, then $\pi[S] = \pi[X]\pi[N]$. In general, proportionality does not hold, see for example Section 1.2. However, this property is used as a local working hypothesis for the calculation of the premium for similar contracts; without proportionality, the use of a tariff is meaningless. $\quad\triangledown$

In Table 5.1, we summarize the properties of our various premium principles. A "+" means that the property holds in general, a "−" that it does not, while especially an "e" means that the property only holds in case of an exponential premium (including the net premium). We assume that S is bounded from below. The proofs of these properties are asked in the exercises, but for the proof of most of the characterizations that zero utility and mean value principles with a certain additional property must be exponential, we refer to the literature. See also the following section.

Summarizing, one may state that only the exponential premium, the maximal loss principle and the net premium principle satisfy all these properties. The last

Table 5.1 Various premium principles and their properties

Principle ⟶	a) μ	b) $1+\lambda$	c) σ^2	d) σ	e) exp	f) $u(\cdot)$	g) $v(\cdot)$	h) %	i) max	j) Ess
Property ↓										
1) $\pi[X] \geq E[X]$	+	+	+	+	+	+	+	−	+	+
2) $\pi[X] \leq \max[X]$	+	−	−	−	+	+	+	+	+	+
3) $\pi[X+c] = \pi[X]+c$	+	−	+	+	+	+	e	+	+	+
4) $\pi[X+Y] = \pi[X]+\pi[Y]$	+	+	+	−	+	e	e	−	+	+
5) $\pi\big[\pi[X\,|\,Y]\big] = \pi[X]$	+	−	−	−	+	e	+	−	+	−

two are irrelevant in practice, so only the exponential premium principle survives this selection. See also Section 5.2. A drawback of the exponential premium has already been mentioned: it has the property that a decision maker's decisions do not depend on the capital he has acquired to date. On the other hand, this is also a strong point of this premium principle, since it is very convenient not to have to know one's current capital, which is generally either random or simply not precisely known at each point in time.

5.4 Characterizations of premium principles

In this section we investigate the properties marked with "e" in Table 5.1, and also some more characterizations of premium principles. Note that linear transforms of the functions $u(\cdot)$ and $v(\cdot)$ in (f) and (g) yield the same premiums. A technique to prove that only exponential utility functions $u(\cdot)$ have a certain property consists of applying this property to risks with a simple structure, and derive a differential equation for $u(\cdot)$ that holds only for exponential and linear functions. Since the linear utility functions are a limit of the exponential utility functions, we will not mention them explicitly in this section. For full proofs of the theorems in this section, we refer to Gerber (1979, 1985) as well as Goovaerts et al. (1984).

The entries "e" in Table 5.1 are studied in the following theorem.

Theorem 5.4.1 (Characterizing exponential principles)

1. A consistent mean value principle is exponential.
2. An additive mean value principle is exponential.
3. An additive zero utility principle is exponential.
4. An iterative zero utility principle is exponential.

Proof. Since for a mean value principle we have $\pi[X] = c$ if $\Pr[X = c] = 1$, consistency is just additivity with the second risk degenerate, so the second assertion follows from the first. The proof of the first, which will be given below, involves applying consistency to risks that are equal to x plus some Bernoulli(q) random variable, and computing the second derivative at $q = 0$ to show that a valuation function $v(\cdot)$ with the required property necessarily satisfies the differential equation $\frac{v''(x)}{v'(x)} = c$ for some constant c. The solutions are the linear and exponential valuation functions. The final assertion is proved in much the same way. The proof that an additive zero utility principle is exponential proceeds by deriving a similar equation, for which it turns out to be considerably more difficult to prove that the exponential utility functions are the unique solutions.

To prove that a consistent mean value principle is exponential, assume that $v(\cdot)$, which is a convex increasing function, yields a consistent mean value principle. Let $P(q)$ denote the premium, considered as a function of q, for a Bernoulli(q) risk S_q. Then, by definition,

$$v(P(q)) = qv(1) + (1-q)v(0). \tag{5.12}$$

The right-hand derivative of this equation in $q = 0$ yields

$$P'(0)v'(0) = v(1) - v(0), \tag{5.13}$$

so $P'(0) > 0$. The second derivative in $q = 0$ gives

$$P''(0)v'(0) + P'(0)^2 v''(0) = 0. \tag{5.14}$$

Because of the consistency, the premium for $S_q + x$ equals $P(q) + x$ for each constant x, and therefore

$$v(P(q) + x) = qv(1 + x) + (1-q)v(x). \tag{5.15}$$

The second derivative at $q = 0$ of this equation yields

$$P''(0)v'(x) + P'(0)^2 v''(x) = 0, \tag{5.16}$$

and, since $P'(0) > 0$, we have for all x that

$$\frac{v''(x)}{v'(x)} = \frac{v''(0)}{v'(0)}. \tag{5.17}$$

Consequently, $v(\cdot)$ is linear if $v''(0) = 0$, and exponential if $v''(0) > 0$. $\triangledown$

Remark 5.4.2 (Continuous and mixable premiums)
Another interesting characterization is the following one. A premium principle $\pi[\cdot]$ is continuous if $F_n \to F$ in distribution implies $\pi[F_n] \to \pi[F]$. If furthermore $\pi[\cdot]$ admits mixing, which means that $\pi[tF + (1-t)G] = t\pi[F] + (1-t)\pi[G]$ for cdfs F and G, as well as $\pi[c] = (1 + \lambda)c$ for all real c and some fixed λ, then it can be shown that $\pi[\cdot]$ must be the expected value principle $\pi[X] = (1 + \lambda)E[X]$. Note that the last condition can be replaced by additivity. To see this, observe that additivity implies that $\pi[n/m] = n/m\,\pi[1]$ for all natural n and m, and then use continuity. $\triangledown$

Finally, the Esscher premium principle can be justified as follows.

Theorem 5.4.3 (Characterization of Esscher premiums)
Assume an insurer has an exponential utility function with risk aversion α. If he charges a premium of the form $E[\varphi(X)X]$ where $\varphi(\cdot)$ is a continuous increasing

function with $E[\varphi(X)] = 1$, his utility is maximized if $\varphi(x)$ is proportional to $e^{\alpha x}$, hence if he uses the Esscher premium principle with parameter α.

Proof. The proof of this statement is based on the technique of variational calculus and adapted from Goovaerts et al. (1984). Let $u(\cdot)$ be a convex increasing utility function, and introduce $Y = \varphi(X)$. Then, because $\varphi(\cdot)$ increases continuously, we have $X = \varphi^{-1}(Y)$. Write $f(y) = \varphi^{-1}(y)$. To derive a condition for $E[u(-f(Y) + E[f(Y)Y])]$ to be maximal for all choices of continuous increasing functions when $E[Y] = 1$, consider a function $f(y) + \varepsilon g(y)$ for some arbitrary continuous function $g(\cdot)$. A little reflection will lead to the conclusion that the fact that $f(y)$ is optimal, and this new function is not, must mean that

$$\frac{d}{d\varepsilon} E\big[u\big(-f(Y) + E[f(Y)Y] + \varepsilon\{-g(Y) + E[g(Y)Y]\}\big)\big]\Big|_{\varepsilon=0} = 0. \qquad (5.18)$$

But this derivative is equal to

$$E\Big[u'\big(-f(Y) + E[f(Y)Y] + \varepsilon\{-g(Y) + E[g(Y)Y]\}\big) \\ \{-g(Y) + E[g(Y)Y]\}\Big]. \qquad (5.19)$$

For $\varepsilon = 0$, this derivative equals zero if

$$E\big[u'\big(-f(Y) + E[f(Y)Y]\big)g(Y)\big] = \\ E\big[u'\big(-f(Y) + E[f(Y)Y]\big)\big]E[g(Y)Y]. \qquad (5.20)$$

Writing $c = E[u'(-f(Y) + E[f(Y)Y])]$, this can be rewritten as

$$E\big[\{u'\big(-f(Y) + E[f(Y)Y]\big) - cY\}\{g(Y)\}\big] = 0. \qquad (5.21)$$

Since the function $g(\cdot)$ is arbitrary, by a well-known theorem from variational calculus we find that necessarily

$$u'\big(-f(y) + E[f(Y)Y]\big) - cy = 0. \qquad (5.22)$$

Using $x = f(y)$ and $y = \varphi(x)$, we see that ($\propto$ denoting 'is proportional to')

$$\varphi(x) \propto u'\big(-x + E[X\varphi(X)]\big). \qquad (5.23)$$

Now, if $u(x)$ is exponential(α), so $u(x) = -\alpha e^{-\alpha x}$, then

$$\varphi(x) \propto e^{-\alpha(-x + E[X\varphi(X)])} \propto e^{\alpha x}. \qquad (5.24)$$

Since $E[\varphi(X)] = 1$, we obtain $\varphi(x) = e^{\alpha x}/E[e^{\alpha X}]$ for the optimal standardized weight function. The resulting premium is an Esscher premium with parameter $h = \alpha$. $\triangledown$

Notice that the insurer uses a different weighting function for risks having different values of $E[\varphi(X)]$; these functions differ only by a constant factor.

5.5 Premium reduction by coinsurance

Consider n cooperating insurers that individually have exponential utility functions with parameter α_i, $i = 1, 2, \ldots, n$. Together, they want to insure a risk S by defining random variables $S_1, \ldots, S_n$ with

$$S \equiv S_1 + \cdots + S_n, \tag{5.25}$$

with S_i denoting the risk insurer i faces. S might for example be a new risk they want to take on together, or it may be their combined insurance portfolios that they want to redistribute. The total premium they need is

$$P = \sum_{i=1}^{n} \frac{1}{\alpha_i} \log E\left[e^{\alpha_i S_i}\right]. \tag{5.26}$$

This total premium depends on the choice of the S_i. How should the insurers split up the risk S in order to make the pool as competitive as possible, hence to minimize the total premium P?

It turns out that the optimal choice $\widetilde{S}_i$ for the insurers is when each of them insures a fixed part of S, to be precise

$$\widetilde{S}_i = \frac{\alpha}{\alpha_i} S \quad \text{with} \quad \frac{1}{\alpha} = \sum_{i=1}^{n} \frac{1}{\alpha_i}. \tag{5.27}$$

So, the optimal allocation is to let each insurer cover a fraction of the pooled risk that is proportional to the reciprocal of his risk aversion, hence to his risk tolerance. By (5.26) and (5.27), the corresponding total minimum premium is

$$\widetilde{P} = \sum_{i=1}^{n} \frac{1}{\alpha_i} \log E\left[e^{\alpha_i \widetilde{S}_i}\right] = \frac{1}{\alpha} \log E\left[e^{\alpha S}\right]. \tag{5.28}$$

This shows that the pool of optimally cooperating insurers acts as one insurer with an exponential premium principle with as risk tolerance the total risk tolerance of the companies involved.

The proof that $\widetilde{P} \le P$ for all other appropriate choices of $S_1 + \cdots + S_n \equiv S$ goes as follows. We have to prove that (5.28) is smaller than (5.26), so

$$\frac{1}{\alpha} \log E\left[\exp\left(\alpha \sum_{i=1}^{n} S_i\right)\right] \le \sum_{i=1}^{n} \frac{1}{\alpha_i} \log E\left[e^{\alpha_i S_i}\right], \tag{5.29}$$

which can be rewritten as

$$\mathrm{E}\left[\prod_{i=1}^{n} e^{\alpha S_i}\right] \leq \prod_{i=1}^{n}\left(\mathrm{E}\left[e^{\alpha_i S_i}\right]\right)^{\alpha/\alpha_i}. \qquad (5.30)$$

This in turn is equivalent to

$$\mathrm{E}\left[\prod_{i=1}^{n} \frac{e^{\alpha S_i}}{\mathrm{E}\left[e^{\alpha_i S_i}\right]^{\alpha/\alpha_i}}\right] \leq 1, \qquad (5.31)$$

or

$$\mathrm{E}\left[\exp\sum_i \frac{\alpha}{\alpha_i} T_i\right] \leq 1, \quad \text{with } T_i = \log\frac{e^{\alpha_i S_i}}{\mathrm{E}\left[e^{\alpha_i S_i}\right]}. \qquad (5.32)$$

We can prove inequality (5.32) as follows. Note that $\mathrm{E}[\exp(T_i)] = 1$ and that by definition $\sum_i \alpha/\alpha_i = 1$. Since e^x is a convex function, we have for all real $t_1, \ldots, t_n$

$$\exp\left(\sum_i \frac{\alpha}{\alpha_i} t_i\right) \leq \sum_i \frac{\alpha}{\alpha_i} \exp(t_i), \qquad (5.33)$$

and this implies that

$$\mathrm{E}\left[\exp\left(\sum_i \frac{\alpha}{\alpha_i} T_i\right)\right] \leq \sum_i \frac{\alpha}{\alpha_i} \mathrm{E}\left[e^{T_i}\right] = \sum_i \frac{\alpha}{\alpha_i} = 1. \qquad (5.34)$$

Hölder's inequality, which is well-known, arises by choosing $X_i = \exp(\alpha S_i)$ and $r_i = \alpha/\alpha_i$ in (5.30). See the exercises for the case $n = 2$, $r_1 = p$, $r_2 = q$.

5.6 Value-at-Risk and related risk measures

A risk measure for a random variable S is mathematically nothing but a functional connecting a real number to S. There are many possibilities. One is a premium, the price for taking over a risk. A ruin probability for some given initial capital measures the probability of becoming insolvent in the near or far future, therefore associates a real number with the random variable S representing the annual claims. It might be compound Poisson(λ, X), or N(μ, σ^2). The risk measure most often used in practice is simply the *Value-at-Risk* at a certain (confidence) level q with $0 < q < 1$, which is the amount that will maximally be lost with probability q, therefore the argument x where $F_S(x)$ crosses level q. It is also called the *quantile risk measure*. Some examples of the practical use of the VaR are the following.

• To prevent insolvency, the available economic capital must cover unexpected losses to a high degree of confidence. Banks often choose their confidence level according to a standard of solvency implied by a credit rating of A or AA. These target ratings require that the institution have sufficient equity to buffer losses over a one-year period with probability 99.90% and 99.97%, respectively, based

on historical one-year default rates. Typically, the reference point for the allocation of economic capital is a target rating of AA.
- For a portfolio of risks assumed to follow a compound Poisson risk process, the
  initial capital corresponding to some fixed probability ε of ultimate ruin is to
  be computed. This involves computing $\text{VaR}[L; 1 - \varepsilon]$, which is the VaR of the
  maximal aggregate loss L, see Section 4.7.
- Set the IBNR-reserve to be sufficient with a certain probability: the regulatory au-
  thorities prescribe, for example, the 75% quantile for the projected future claims
  as a capital to be held.
- Find out how much premium should be paid on a contract to have 60% certainty
  that there is no loss on that contract (*percentile premium principle*).

The definition of VaR is as follows:

Definition 5.6.1 (Value-at-Risk)
For a risk S, the *Value-at-Risk* (VaR) at (confidence) level p is defined as

$$\text{VaR}[S; p] \overset{\text{def}}{=} F_S^{-1}(p) \overset{\text{not}}{=} \inf\{s : F_S(s) \geq p\}. \qquad (5.35)$$

So the VaR is just the inverse cdf of S computed at p. $\qquad\qquad \triangledown$

Remark 5.6.2 (VaR is cost-optimal)
Assume that an insurer has available an economic capital d to pay claims from. He
must pay an annual compensation $i \cdot d$ to the shareholders. Also, there is the shortfall
over d, which is valued as its expected value $\text{E}[(S-d)_+]$ (this happens to be the net
premium a reinsurer might ask to take over the top part of the risk, but there need
not be reinsurance in force at all). So the total costs amount to

$$i \cdot d + \text{E}[(S-d)_+]. \qquad (5.36)$$

From the insurer's point of view, the optimal d turns out to be $\text{VaR}[S; 1-i]$. This
is easily seen by looking at the derivative of the cost (5.36), using the fact that
$\frac{\text{d}}{\text{d}t}\text{E}[(S-t)_+] = F_S(t) - 1$. So the VaR is the cost minimizing capital to be held in
this scenario. $\qquad\qquad \triangledown$

Since the typical values of i are around 10%, from this point of view the VaR-
levels required for a company to get a triple A rating (99.975% and over, say) are
suboptimal. On the other hand, maintaining such a rating might lead to a lower cost
of capital i and increase production of the company. Also, it is a matter of prestige.

Note that the cdf F_S, by its definition as $F_S(s) = \text{Pr}[S \leq s]$, is continuous from
the right, but F_S^{-1} as in (5.35) is left-continuous, with $\text{VaR}[S; p] = \text{VaR}[S; p - 0]$.
It has jumps at levels p where F_S has a constant segment. In fact, any number s
with $\text{VaR}[S; p] \leq s \leq \text{VaR}[S; p + 0]$ may serve as the VaR. See Figure 5.1 for an
illustration. We see that $F_S(s)$ has a horizontal segment at level 0.2; at this level, the
VaR jumps from the lower endpoint of the segment to the upper endpoint. Where F_S
has a jump, the VaR has a constant segment (between $p = 0.5$ and $p = 0.7$).

Remark 5.6.3 (Premiums are sometimes incoherent risk measures)
Relying on the VaR is often snubbed upon in financial circles these days, not as

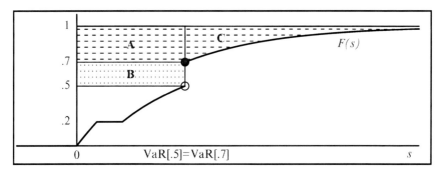

Fig. 5.1 Illustrating VaR and related risk measures for a cdf F_S

much because there is a danger in doing so, see Exercise 5.6.3, but because it is not a *coherent* risk measure in the sense of Artzner et al. (1997). To be called coherent, a risk measure must have respect for stochastic order, be positive homogeneous and translative, see Exercise 5.6.2, but it must also be *subadditive*. This last property means that the sum of the risk measures for a split-up portfolio is automatically an upper bound for the risk in the total portfolio. From Exercise 1.2.9 it is known that a risk-averse individual aiming to keep his utility at the same level or better is sometimes prepared to pay a premium for $2S$ that is strictly larger than twice the one for S. This is for example the case for exponential premiums, see Exercise 1.3.12, and the same superadditivity holds for Esscher premiums, see Exercise 5.6.11. Therefore, determining zero utility premiums may measure risks in a non-subadditive way, but to label this procedure incoherent is improper. Requiring subadditivity makes sense in complete markets where it is always possible to diversify a risk, but the insurance market simply is incomplete.   ▽

Example 5.6.4 (VaR is not subadditive)
The following is a counterexample for the subadditivity of VaR: if S and $T \sim$ Pareto$(1,1)$ independent, then for *all* $p \in (0,1)$ we have

$$\text{VaR}[S+T;p] > \text{VaR}[S;p] + \text{VaR}[T;p]. \tag{5.37}$$

To see that this is true, first verify that, since $F_S(x) = 1 - 1/x$, $x > 1$, we have $\text{VaR}[S;p] = \frac{1}{1-p}$. Next, using convolution (see Exercise 5.6.10), check that

$$\Pr[S+T \leq t] = 1 - \frac{2}{t} - 2\frac{\log(t-1)}{t^2}, \quad t > 2. \tag{5.38}$$

Now

$$\Pr\left[S+T \leq 2\text{VaR}[S;p]\right] = p - \frac{(1-p)^2}{2} \log\left(\frac{1+p}{1-p}\right) < p, \tag{5.39}$$

so for every p, we have $\text{VaR}[S;p] + \text{VaR}[T;p] < \text{VaR}[S+T;p]$. This gives a counterexample for VaR being subadditive; in fact we have proved that for this pair (S,T), VaR is *superadditive*.

Note that from (1.33) with $d = 0$ one might infer that $S+T$ and $2S$ should have cdfs that cross, so the same holds for the VaRs, being the inverse cdfs. Since we have $\text{VaR}[2S;p] = \text{VaR}[S;p] + \text{VaR}[T;p]$ for all p, this contradicts the above. But this reasoning is invalid because $E[S] = \infty$. To find counterexamples of subadditivity for VaR, it suffices to take $S \sim T$ with finite means. Then $S+T$ and $2S$ cannot have uniformly ordered quantile functions, so VaR must be subadditive at some levels, superadditive at others. ∇

The VaR itself does not tell the whole story about the risk, since if the claims S exceed the available funds d, someone still has to pay the remainder $(S-d)_+$. The Expected Shortfall measures 'how bad is bad':

Definition 5.6.5 (Expected shortfall)
For a risk S, the *Expected Shortfall* (ES) at level $p \in (0,1)$ is defined as

$$\text{ES}[S;p] = E[(S - \text{VaR}[S;p])_+]. \tag{5.40}$$

Thus, ES can be interpreted as the net stop-loss premium in the hypothetical situation that the excess over $d = \text{VaR}[S;p]$ is reinsured. ∇

In Figure 5.1, for levels $p \in [0.5, 0.7]$, $\text{ES}[S;p]$ is the stop-loss premium at retention $\text{VaR}[S;0.7]$, therefore just the area C; see also (1.33).

As we noted, VaR is not subadditive. Averaging high level VaRs, however, does produce a subadditive risk measure (see Property 5.6.10).

Definition 5.6.6 (Tail-Value-at-Risk (TVaR))
For a risk S, the *Tail-Value-at-Risk* (TVaR) at level $p \in (0,1)$ is defined as:

$$\text{TVaR}[S;p] = \frac{1}{1-p} \int_p^1 \text{VaR}[S;t]\, dt. \tag{5.41}$$

So the TVaR is just the arithmetic average of the VaRs of S from p on. ∇

Remark 5.6.7 (Other expressions for TVaR)
The Tail-Value-at-Risk can also be expressed as

$$\begin{aligned}
\text{TVaR}[S;p] &= \text{VaR}[S;p] + \frac{1}{1-p} \int_p^1 \{\text{VaR}[S;t] - \text{VaR}[S;p]\}\, dt \\
&= \text{VaR}[S;p] + \frac{1}{1-p} \text{ES}[S;p].
\end{aligned} \tag{5.42}$$

It is easy to see that the integral in (5.42) equals $\text{ES}[S;p]$, in similar fashion as in Figure 1.1.

Just as VaR, the TVaR arises naturally from the cost optimization problem considered in Remark 5.6.2. In fact, TVaR is the optimal value of the cost function, divided by i. To see why this holds, simply fill in $d = \text{VaR}[S;p = 1-i]$ in the cost function $i \cdot d + \mathrm{E}[(S-d)_+]$, see (5.36). As a consequence, we have the following characterization of TVaR:

$$\mathrm{TVaR}[S;p] = \inf_{d} \left\{ d + \frac{1}{1-p}\pi_S(d) \right\}. \tag{5.43}$$

Filling in other values than $d = \mathrm{VaR}[S;p]$ gives an upper bound for $\mathrm{TVaR}[S;p]$. ∇

In Figure 5.1, in view of (5.42) above, for level $p = \frac{1}{2}$, $(1-p)\mathrm{TVaR}[S;p]$ equals the total of the areas marked A, B and C. Therefore, the TVaR is just the average length of the dashed and the dotted horizontal lines in Figure 5.1.

TVaR resembles but is not always identical to the following risk measure.

Definition 5.6.8 (Conditional Tail Expectation)
For a risk S, the *Conditional Tail Expectation* (CTE) at level $p \in (0,1)$ is defined as

$$\mathrm{CTE}[S;p] = \mathrm{E}\left[S \,|\, S > \mathrm{VaR}[S;p]\right]. \tag{5.44}$$

So the CTE is the 'average loss in the worst $100(1-p)\%$ cases'.   ∇

Writing $d = \mathrm{VaR}[S;p]$ we have

$$\mathrm{CTE}[S;p] = \mathrm{E}[S \,|\, S > d] = d + \mathrm{E}[(S-d)_+ \,|\, S > d] = d + \frac{\mathrm{E}[(S-d)_+]}{\Pr[S > d]}. \tag{5.45}$$

Therefore we have for all $p \in (0,1)$

$$\mathrm{CTE}[S;p] = \mathrm{VaR}[S;p] + \frac{1}{1-F_S(\mathrm{VaR}[S;p])}\mathrm{ES}[S;p]. \tag{5.46}$$

So by (5.42), the CTE differs from the TVaR only when $p < F_S(\mathrm{VaR}[S;p])$, which means that F_S jumps over level p. In fact we have

$$\mathrm{CTE}[S;p] = \mathrm{TVaR}[S;F_S(F_S^{-1}(p))] \geq \mathrm{TVaR}[S;p] \geq \mathrm{VaR}[S;p]. \tag{5.47}$$

In Figure 5.1, levels p outside $(0.5,0.7)$ lead to TVaR and CTE being identical. In case $p \in [0.5,0.7]$, the CTE is just the total of the areas A and C, divided by $1 - F_S(\mathrm{VaR}[S;0.7])$. So the CTE at these levels p is the average length of the dashed horizontal lines in Figure 5.1 only. Diagrams of the VaR, ES, TVaR and CTE for risk S in Figure 5.1 are given in Figure 5.2. Observe that

- the VaR jumps where F_S is constant, and it is horizontal where F_S jumps;
- the ES, being a stop-loss premium at levels increasing with p, is decreasing in p, but not convex or even continuous;
- TVaR and CTE coincide except for p-values in the vertical part of F_S;
- TVaR is increasing and continuous.

Remark 5.6.9 (Terminology about ES, TVaR and CTE)
Some authors use the term Expected Shortfall to refer to CTE (or TVaR). As we define it, the ES is an *unconditional* mean of the shortfall, while TVaR and CTE

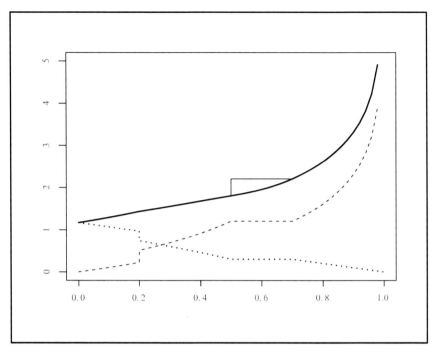

Fig. 5.2 TVaR (thick), CTE (thin), ES (dotted) and VaR (dashed) for the cdf F_S in Fig. 5.1, plotted against the level $p \in (0,1)$

involve the expected shortfall, given that it is positive. CTE and TVaR are also often used interchangeably. Defined as above, these quantities are the same only in case of continuous distributions. Another synonym for CTE that one may encounter in the literature is Conditional Value-at-Risk (CVaR).   ∇

Unlike the VaR, see Exercise 5.6.9 and Example 5.6.4, the Tail-Value-at-Risk is subadditive.

Property 5.6.10 (TVaR is subadditive)
TVaR$[S+T;p] \leq$ TVaR$[S;p] +$ TVaR$[T;p]$ holds for all pairs of risks (S,T) and for all $p \in (0,1)$.

Proof. We mentioned that in the characterization of TVaR (5.43) above, an upper bound for TVaR$[S;p]$ is obtained by filling in any other value for d. Now let especially $d = d_1 + d_2$, with $d_1 =$ VaR$[S;p]$ and $d_2 =$ VaR$[T;p]$. Then we have

$$\text{TVaR}[S+T;p] \le d + \frac{1}{1-p}\pi_{S+T}(d)$$

$$= d_1 + d_2 + \frac{1}{1-p}\text{E}[(S+T-d_1-d_2)_+]$$

$$\le d_1 + d_2 + \frac{1}{1-p}\text{E}[(S-d_1)_+ + (T-d_2)_+]$$ \hfill (5.48)

$$= \text{TVaR}[S;p] + \text{TVaR}[T;p].$$

Here we used that $s_+ + t_+ = (s_+ + t_+)_+ \ge (s+t)_+$ for all real s,t. ∇

In Exercise 5.6.5, we will see an example where CTE is not subadditive. For continuous risks, CTE and TVaR coincide, but there is also a neat more direct proof that CTE is subadditive in this case.

Property 5.6.11 (CTE is subadditive for continuous risks)
If S and T are random variables with continuous marginal cdfs and joint cdf, CTE is subadditive.

Proof. If $s = \text{VaR}[S;p]$, $t = \text{VaR}[T;p]$ and $z = \text{VaR}[S+T;p]$, by the assumed continuity we have $1 - p = \Pr[S+T>z] = \Pr[S>s] = \Pr[Y>t]$. It is not hard to see that if for some s, the events A and $S>s$ are equally likely, for the conditional means we have $\text{E}[S|A] \le \text{E}[S|S>s]$. Apply this with $A \equiv S+T>z$, then we get

$$\text{E}[S+T|A] = \text{E}[S|A] + \text{E}[T|A] \le \text{E}[S|S>s] + \text{E}[T|T>t]. \hfill (5.49)$$

This is tantamount to

$$\text{CTE}[S+T;p] \le \text{CTE}[S;p] + \text{CTE}[T;p], \hfill (5.50)$$

so subadditivity for the CTEs of continuous risks is proved. ∇

R provides nine ways to estimate a VaR at level p from a sample S of size n, differing subtly in the way the interpolation between the order statistics close to np of the sample is performed. See ?quantile for more information on this. Leaving out the type parameter is equivalent to choosing the default method type=7 in R. An example for a lognormal$(0, 1)$ random sample:

```
set.seed(1); S <- exp(rnorm(1000,0,1))
levels <- c(0.5,0.95,0.975,0.99,0.995)
quantile(S, levels, type=7)
##      50%        95%      97.5%        99%      99.5%
##0.9652926  5.7200919  7.4343584 10.0554305 11.5512498
```

The reader is asked to computes estimates of the other risk measures in this section in Exercise 5.6.4.

5.7 Exercises

Section 5.2

1. Show that (5.7) is valid.

2. What are the results in the table in case of a dividend $i = 2\%$ and $\varepsilon = 5\%$? Calculate the variance premium as well as the exponential premium.

Section 5.3

1. Let $X \sim$ exponential(1). Determine the premiums (a)–(e) and (h)–(j).

2. [♠] Prove that $\pi[X;\alpha] = \log(E[e^{\alpha X}])/\alpha$ is an increasing function of α, by showing that the derivative with respect to α is positive (see also Example 1.3.1).

3. Assume that the total claims for a car portfolio has a compound Poisson distribution with gamma distributed claims per accident. Determine the expected value premium if the loading factor equals 10%.

4. Determine the exponential premium for a compound Poisson risk with gamma distributed individual claims.

5. Calculate the variance premium for the claims distribution as in Exercise 5.3.3.

6. Show that the Esscher premium equals $\kappa'_X(h)$, where κ_X is the cgf of X.

7. What is the Esscher transformed density with parameter h for the following densities: exponential(α), binomial(n,p) and Poisson(λ)?

8. Show that the Esscher premium for X increases with the parameter h.

9. Calculate the Esscher premium for a compound Poisson distribution.

10. Show that the Esscher premium for small values of α boils down to a variance premium principle.

11. Assume that X is a finite risk with maximal value b, hence $\Pr[X \le b] = 1$ but $\Pr[X \ge b - \varepsilon] > 0$ for all $\varepsilon > 0$. Let π_α denote the exponential premium for X. Show that $\lim_{\alpha \to \infty} \pi_\alpha = b$.

12. Show that the exponential premium $\pi[X;\alpha]$ with risk aversion α is the difference quotient $(\kappa_X(\alpha) - \kappa_X(0))/\alpha$. Prove that it also can be written as a uniform mixture of Esscher premiums $\int_0^\alpha \pi[X;h]\,dh/\alpha$. From the fact that Esscher premiums increase with h, what can be concluded about the Esscher($h = \alpha$) premium compared with the exponential(α) premium?

13. In Table 5.1, prove the properties that are marked "+".

14. Construct counterexamples for the first 4 rows and the second column for the properties that are marked "−".

15. Investigate the additivity of a mixture of Esscher principles of the following type: $\pi[X] = p\pi[X;h_1] + (1 - p)\pi[X;h_2]$ for some $p \in [0,1]$, where $\pi[X;h]$ is the Esscher premium for risk X with parameter h.

16. Formulate a condition for dependent risks X and Y that implies that $\pi[X + Y] \le \pi[X] + \pi[Y]$ for the variance premium (subadditivity). Also show that this property holds for the standard deviation principle, no matter what the joint distribution of X and Y is.

Section 5.5

1. For a proof of Hölder's inequality in case of $n = 2$, let $p > 1$ and $q > 1$ satisfy $\frac{1}{p} + \frac{1}{q} = 1$. Successively prove that

    - if $u > 0$ and $v > 0$, then $uv \leq \frac{u^p}{p} + \frac{v^q}{q}$; (write $u = e^{s/p}$ and $v = e^{t/q}$);
    - if $E[U^p] = E[V^q] = 1$ and $\Pr[U > 0] = \Pr[V > 0] = 1$, then $E[UV] \leq 1$;
    - $|E[XY]| \leq E[|X|^p]^{1/p} E[|Y|^q]^{1/q}$.

2. Whose inequality arises for $p = q = 2$ in the previous exercise?

Section 5.6

1. Compute estimates for the TVaR and ES based on the same lognormal sample S as given in the R-example at the end of this section and at level 0.95.

2. Prove that the following properties hold for TVaR and VaR:

    - no rip-off $(T)\mathrm{VaR}[S;p] \leq \max[S]$
    - no unjustified loading $(T)\mathrm{VaR}[S;p] = c$ if $S \equiv c$
    - non-decreasing in p if $p < q$, then $(T)\mathrm{VaR}[S;p] \leq (T)\mathrm{VaR}[S;q]$
    - non-negative safety loading $\mathrm{TVaR}[S;p] \geq E[S]$
    - translative $(T)\mathrm{VaR}[S+c;p] = (T)\mathrm{VaR}[S;p] + c$
    - positively homogeneous $(T)\mathrm{VaR}[\alpha S] = \alpha \times (T)\mathrm{VaR}[S,p]$ for all $\alpha > 0$
    - TVaR is continuous in p but VaR and CTE are not

3. Let X_i, $i = 1, \ldots, 100$ be iid risks with $\Pr[X_i = -100] = 0.01$, $\Pr[X_i = +2] = 0.99$. Compare the VaRs of $S_1 = \sum_{i=1}^{100} X_i$ (diversified) and $S_2 = 100X_1$ (non-diversified). The diversified risk is obviously safer, but the VaRs cannot be uniformly smaller for two random variables with the same finite mean. Plot the cdfs of both these risks. Investigate which of these risks has a smaller VaR at various levels p.

4. Write R-functions computing TVaRs for (translated) gamma and lognormal distributions. Make plots of the TVaRs at levels $1\%, 3\%, \ldots, 99\%$.

5. Prove that CTE is not subadditive by looking at level $p = 0.9$ for the pair of risks (X,Y) with:

$$X \sim \text{uniform}(0,1); \qquad Y = \begin{cases} 0.95 - X & \text{if } X \leq 0.95 \\ 1.95 - X & \text{if } X > 0.95 \end{cases}$$

    Note that $X \sim Y \sim \text{uniform}(0,1)$ are continuous, but $X + Y$ is discrete.

6. Give expressions for ES, TVaR and CTE in case $S \sim \text{uniform}(a,b)$.

7. As the previous exercise, but now for $S \sim N(\mu, \sigma^2)$.

8. As the previous exercise, but now for $S \sim \text{lognormal}(\mu, \sigma^2)$.

9. Show that VaR is not subadditive using the example $X, Y \sim \text{Bernoulli}(0.02)$ iid and $p = 0.975$. Consider also the case that (X,Y) is bivariate normal.

10. Derive (5.38).

11. If $h > 0$, prove that $\pi[2S; h] > 2\pi[S; h]$ for Esscher premiums.

Chapter 6
Bonus-malus systems

It's a dangerous business going out your front door —
J.R.R. Tolkien (1892 - 1973)

6.1 Introduction

This chapter presents the theory behind bonus-malus methods for automobile insurance. This is an important branch of non-life insurance, in many countries even the largest in total premium income. A special feature of automobile insurance is that quite often and to everyone's satisfaction, a premium is charged that for a large part depends on the claims filed on the policy in the past. In experience rating systems such as these, bonuses can be earned by not filing claims, and a malus is incurred when many claims have been filed. Experience rating systems are common practice in reinsurance, but in this case, it affects the consumer directly. Actually, in case of a randomly fluctuating premium, the ultimate goal of insurance, that is, being in a completely secure financial position as regards this particular risk, is not reached. But in this type of insurance, the uncertainty still is greatly reduced. This same phenomenon occurs in other types of insurance, for example when part of the claims is not reimbursed by the insurer because there is a deductible.

That lucky policyholders pay for the damages caused by less fortunate insureds is the essence of insurance (*probabilistic solidarity*). But in private insurance, solidarity should not lead to inherently good risks paying for bad ones. An insurer trying to impose such *subsidizing solidarity* on his customers will see his good risks take their business elsewhere, leaving him with the bad risks. This may occur in the automobile insurance market when there are regionally operating insurers. Charging the same premiums nationwide will cause the regional risks, which for automobile insurance tend to be good risks because traffic is not so heavy there, to go to the regional insurer, who with mainly good risks in his portfolio can afford to charge lower premiums.

There is a psychological reason why experience rating is broadly accepted with car insurance, and not, for example, with health insurance. Bonuses are seen as rewards for careful driving, premium increases as an additional and well-deserved fine for the accident-prone. But someone who is ill is generally not to blame, and does not deserve to suffer in his pocket as well.

Traditionally, car insurance covers third party liability, as well as the damage to one's own vehicle. The latter is more relevant for rather new cars, since for reasons of moral hazard, insurers do not reimburse more than the current value of the car.

In Section 6.2, we describe the Dutch bonus-malus system, which is typical for such systems. Also, we briefly describe the reasons for adopting this new system. Bonus-malus systems are suitable for analysis by Markov chains, see Section 6.3. In this way, we will be able to determine the Loimaranta efficiency of such systems, that is, the elasticity of the mean asymptotic premium with respect to the claim frequency. In Chapter 8, we present a bonus-malus system that is a special case of a so-called credibility method. In Chapter 9, we study among other things some venerable non-life actuarial methods for automobile premium rating in the light of generalized linear models. Also, we present a case study on a portfolio not unlike the one used to construct the Dutch bonus-malus system. See also Appendix A.

6.2 A generic bonus-malus system

Every country has his own bonus-malus system, the wheel having been reinvented quite a few times. The system employed by many insurance companies in The Netherlands is the result of a large-scale investigation performed in 1982. It was prompted by the fact that the market was chaotic and in danger of collapsing. The data consisted of about 700 000 policies of which 50 particulars were known, and that produced 80 000 claims. Both claim frequency and average claim size were studied. Many Dutch insurers still use variants of the proposed system.

First, a basic premium is determined using rating factors like weight, list-price or capacity of the car, type of use of the car (private or for business), and the type of coverage (comprehensive, third party only, or a mixture). This is the premium that drivers without a known claims history have to pay. The bonus and malus for good and bad claims experience are implemented through the use of a so-called bonus-malus scale. One ascends one step, getting a greater bonus, after a claim-free year, and descends several steps after having filed one or more claims. The bonus-malus scale, including the percentages of the basic premium to be paid and the transitions made after 0, 1, 2, and 3 or more claims, is depicted in Table 6.1. In principle, new insureds enter at the step with premium level 100%. Other countries might use different rating factors and a different bonus-malus scale.

Not all relevant risk factors were usable as rating factors. Driving capacity, swiftness of reflexes, aggressiveness behind the wheel and knowledge of the highway code are hard to measure, while mileage is prone to deliberate misspecification. For some of these relevant factors, proxy measures can be found. One can get a good idea about mileage by looking at factors like weight and age of the car, as well as the type of fuel used, or type of usage (private or professional). In the Netherlands, diesel engines, for example, are mainly used by drivers with high mileage. Traffic density can be deduced from region of residence, driving speed from horse power and weight of the car. But it will remain impossible to assess the average future

Table 6.1 Transition rules and premium percentages for the Dutch bonus-malus system

Step	1	2	3	4	5	6	7	8	9	10	11	12	13	14
Percentage	120	100	90	80	70	60	55	50	45	40	37.5	35	32.5	30
0 claims $\rightarrow$	2	3	4	5	6	7	8	9	10	11	12	13	14	14
1 claim $\rightarrow$	1	1	1	1	2	3	4	5	6	7	7	8	8	9
2 claims $\rightarrow$	1	1	1	1	1	1	1	1	2	3	3	4	4	5
3^+ claims $\rightarrow$	1	1	1	1	1	1	1	1	1	1	1	1	1	1

claim behavior completely using data known in advance, therefore there is a need to use the actual claims history as a rating factor. Claims history is an ex post factor, which becomes fully known only just before the next policy year. Hence one speaks of ex post premium rating, where generally premiums are fixed ex ante.

In the investigation, the following was found. Next to car weight, cylinder capacity and horse power of the car provided little extra predicting power. It proved that car weight correlated quite well with the total claim size, which is the product of claim frequency and average claim size. Heavier cars tend to be used more often, and also tend to produce more damage when involved in accidents. Car weight is a convenient rating factor, since it can be found on official papers. In many countries, original list-price is used as the main rating factor for third party damage. This method has its drawbacks, however, because it is not reasonable to assume that someone would cause a higher third-party claim total if he has a metallic finish on his car or a more expensive audio system. It proved that when used next to car weight, catalogue price also did not improve predictions about third party claims. Of course for damage to the own vehicle, it remains the dominant rating factor. Note that the premiums proposed were not just any function of car weight and catalogue price, but they were directly proportional to these figures.

The factor 'past claims experience', implemented as 'number of claim-free years', proved to be a good predictor for future claims, even when used in connection with other rating factors. After six claim-free years, the risk still diminishes, although slower. This is reflected in the percentages in the bonus-malus scale given in Table 6.1. Furthermore, it proved that drivers with a bad claims history are worse than beginning drivers, justifying the existence of a malus class with a premium percentage of more than the 100% charged in the standard entry class.

An analysis of the influence of the region of residence on the claims experience proved that in less densely populated regions, fewer claims occurred, although somewhat larger. It appeared that the effect of region did not vanish with an increasing number of claim-free years. Hence the region effect was incorporated by a fixed discount percentage, in fact enabling the large companies to compete with the regionally operating insurers on an equal footing.

The age of the policyholder is very important for his claim behavior. The claim frequency at age 18 is about four times the one drivers of age 30–70 have. Part of this bad claim behavior can be ascribed to lack of experience, because after some years, the effect slowly vanishes. That is why it was decided not to let the basic premium vary by age, but merely to let young drivers enter at a less favorable step in the bonus-malus scale.

For commercial reasons, the profession of the policy holder as well as the make of the car were not incorporated in the rating system, even though these factors did have a noticeable influence.

Note that in the bonus-malus system, only the number of claims filed counts, not their size. Although it is clear that a bonus-malus system based on claim sizes is possible, such systems are hardly ever used with car insurance. This is because in general the unobserved driving qualities affect the claim number more than the claim severity distribution.

6.3 Markov analysis

Bonus-malus systems can be considered as special cases of Markov processes. In such processes, one jumps from one state to another in time. The Markov property says that the process is memoryless, as the probability of such transitions does not depend on how one arrived in a particular state. Using Markov analysis, one may determine which proportion of the drivers will eventually be on each step of the bonus-malus scale. Also, it gives a means to find out how effective the bonus-malus system is in determining adjusted premiums representing the driver's actual risk.

To fix ideas, let us look at a simple example. In a particular bonus-malus system, a driver pays premium c if he files one or more claims in the preceding two-year period, otherwise he pays a, with $a < c$. To describe this system by a bonus-malus scale, notice first that there are two groups of drivers paying the high premium, the ones who claimed last year, and the ones that filed a claim only in the year before that. So we have three states (steps):

1. Claim in the previous policy year; paid c at the previous policy renewal;
2. No claim in the previous policy year, claim in the year before; paid c;
3. Claim-free in the two most recent policy years; paid a.

First we determine the transition probabilities for a driver with probability p of having one or more claims in a policy year. In the event of a claim, he falls to state 1, otherwise he goes one step up, if possible. We get the following matrix $\mathbf{P}$ of transition probabilities p_{ij} to go from state i to state j:

$$\mathbf{P} = \begin{pmatrix} p & q & 0 \\ p & 0 & q \\ p & 0 & q \end{pmatrix}. \tag{6.1}$$

The matrix $\mathbf{P}$ is a stochastic matrix: every row represents a probability distribution over states to be entered, so all elements of it are non-negative. All row sums $\sum_j p_{ij}$ are equal to 1, since from any state i, one has to go to some state j. Therefore

$$\mathbf{P} \begin{pmatrix} 1 \\ 1 \\ 1 \end{pmatrix} = \begin{pmatrix} 1 \\ 1 \\ 1 \end{pmatrix}. \tag{6.2}$$

So matrix $\mathbf{P}$ has $(1,1,1)^T$ as a right-hand eigenvector for eigenvalue 1. Assume that initially at time $t = 0$, the probability f_j for each driver to be in state $j = 1,2,3$ is given by the row-vector $l(0) = (f_1, f_2, f_3)$ with $f_j \geq 0$ and $f_1 + f_2 + f_3 = 1$. Often, the initial state is known to be i, and then f_i will be equal to one. The probability to start in state i and to enter state j after one year is equal to $f_i p_{ij}$, so the total probability of being in state j after one year, starting from an initial class i with probability f_i, equals $\sum_i f_i p_{ij}$. In matrix notation, the following vector $l(1)$ gives the probability distribution of drivers over the states after one year:

$$l(1) = l(0)\mathbf{P} = (f_1, f_2, f_3) \begin{pmatrix} p & q & 0 \\ p & 0 & q \\ p & 0 & q \end{pmatrix} = (p, q f_1, q(f_2 + f_3)). \tag{6.3}$$

After a claim, drivers fall back to state 1. The probability of entering that state equals $p = p(f_1 + f_2 + f_3)$. Non-claimers go to a higher state, if possible. The distribution $l(2)$ over the states after two years is independent of $l(0)$, since

$$l(2) = l(1)\mathbf{P} = (p, q f_1, q(f_2 + f_3)) \begin{pmatrix} p & q & 0 \\ p & 0 & q \\ p & 0 & q \end{pmatrix} = (p, pq, q^2). \tag{6.4}$$

The state two years from now does not depend on the current state, but only on the claims filed in the coming two years. Proceeding like this, one sees that $l(3) = l(4) = l(5) = \cdots = l(2)$. So we also have $l(\infty) := \lim_{t \to \infty} l(t) = l(2)$. The vector $l(\infty)$ is called the steady state distribution. Convergence will not always happen this quickly and thoroughly. Taking the square of a matrix, however, can be done very fast, and doing it ten times starting from $\mathbf{P}$ already gives $\mathbf{P}^{1024}$. Each element r_{ij} of this matrix can be interpreted as the probability of going from initial state i to state j in 1024 years. For regular bonus-malus systems, this probability will not depend heavily on the initial state i, nor will it differ much from the probability of reaching j from i in an infinite number of years. Hence all rows of $\mathbf{P}^{1024}$ will be virtually equal to the steady state distribution.

There is also a more formal way to determine the steady state distribution. This goes as follows. First, notice that

$$\lim_{t \to \infty} l(t+1) = \lim_{t \to \infty} l(t)\mathbf{P}, \quad \text{hence} \quad l(\infty) = l(\infty)\mathbf{P}. \tag{6.5}$$

But this means that the steady state distribution $l(\infty)$ is a left-hand eigenvector of $\mathbf{P}$ with eigenvalue 1. To determine $l(\infty)$ we only have to find a non-trivial solution for the linear system of equations (6.5), which is equivalent to the homogeneous system $(\mathbf{P}^T - \mathbf{I})l^T(\infty) = (0,0,\ldots,0)^T$, and to divide it by the sum of its components to make $l(\infty)$ a probability distribution. Note that all components of $l(\infty)$ are necessarily non-negative, because of the fact that $l_j(\infty) = \lim_{t \to \infty} l_j(t)$. Later on in this section we show how to find the steady state distribution using R.

Remark 6.3.1 (Interpretation of initial distribution over states and transition probability matrix)

It is not necessary to take $l(0)$ to be a probability distribution. It also makes sense to take for example $l(0) = (1000,0,0)$. In this way, one considers a thousand drivers with initial state 1. Contrary to $l(0)$, the vectors $l(1), l(2), \ldots$ as well as $l(\infty)$ do not represent the exact number of drivers in a particular state, but just the expected values of these numbers. The actual numbers are binomial random variables with as probability of success in a trial, the probability of being in that particular state at the given time.

American mathematical texts prefer to have the operand to the right of the operator, and in those texts the initial state as well as the steady state are column vectors. Therefore, the matrix $\mathbf{P}$ of transition probabilities must be transposed, so its generic element P_{ij} is the probability of a transition from state j to state i instead of the other way around such as in this text and many others. ∇

Remark 6.3.2 (Hunger for bonus)

Suppose a driver with claim probability p, in state 3 in the above system, causes a damage of size t in an accident. If he is not obliged to file this claim with his insurance company, for which t is it profitable for him to do so?

Assume that, as some policies allow, he only has to decide on December 31st whether to file this claim, so it is certain that he has no claims after this one concerning the same policy year. Since after two years the effect of this particular claim on his position on the bonus-malus scale will have vanished, we use a planning horizon of two years. His costs in the coming two years (premiums plus claim), depending on whether or not he files the claim and whether he is claim-free next year, are as follows:

	No claim next year	Claims next year
Claim not filed	$a+a+t$	$a+c+t$
Claim filed	$c+c$	$c+c$

Of course he should only file the claim if it makes his expected loss lower, which is the case if

$$(1-p)(2a+t) + p(a+c+t) \geq 2c \quad \Longleftrightarrow \quad t \geq (2-p)(c-a). \qquad (6.6)$$

From (6.6) we see that it is unwise to file claims smaller than the expected loss of bonus in the near future. This phenomenon, which is not unimportant in practice, is called *hunger for bonus*. The insurer is deprived of premiums that are his due because the insured in fact conceals that he is a bad driver. But this is compensated by the fact that small claims also involve handling costs.

Many articles have appeared in the literature, both on actuarial science and on stochastic operational research, about this phenomenon. The model used can be much refined, involving for example a longer or infinite time-horizon, with discounting. Also the time in the year that a claim occurs is important.   ∇

6.3.1 Loimaranta efficiency

The ultimate goal of a bonus-malus system is to make everyone pay a premium that reflects as closely as possible the expected value of his annual claims. If we want to investigate how efficient a bonus-malus system is in performing this task, we have to look at how the premium depends on the claim frequency λ. To this end, assume that the random variation around this theoretical claim frequency can be described as a Poisson process, see Chapter 4. Hence, the number of claims in each year is a Poisson(λ) variate, and the probability of a year with one or more claims equals $p = 1 - e^{-\lambda}$. The expected value of the asymptotic premium to be paid is called the steady state premium. It depends on λ, and in our example where $l(\infty) = (p, pq, q^2)$ and the premiums are (c, c, a), it equals

$$b(\lambda) = cp + cpq + aq^2 = c(1 - e^{-2\lambda}) + ae^{-2\lambda}. \tag{6.7}$$

This is the premium one pays on the average after the effects of in which state one initially started have vanished. In principle, this premium should be proportional to λ, since the average of the total annual claims for a driver with claim frequency intensity parameter λ is equal to λ times the average size of a single claim, which in all our considerations we have taken to be independent of the claim frequency. Define the following function for a bonus-malus system:

$$e(\lambda) := \frac{\lambda}{b(\lambda)} \frac{db(\lambda)}{d\lambda} = \frac{d\log b(\lambda)}{d\log \lambda}. \tag{6.8}$$

This is the so-called *Loimaranta efficiency*; the final equality follows from the chain rule. It represents the '(point) elasticity' of the steady state premium $b(\lambda)$ with respect to λ. As such, it is also the slope of the curve $b(\lambda)$ in a log-log graph. The number $e(\lambda) = db(\lambda)/b(\lambda)/d\lambda/\lambda$ represents the ratio of the incremental percentage change of the function $b(\lambda)$ with respect to an incremental percentage change of λ. For 'small' h, if λ increases by a factor $1 + h$, $b(\lambda)$ increases by a factor that is approximately $1 + e(\lambda)h$, so we have

$$b(\lambda(1 + h)) \approx b(\lambda)(1 + e(\lambda)h). \tag{6.9}$$

Ideally, the efficiency should satisfy $e(\lambda) \approx 1$, but no bonus-malus system achieves that for all λ. In view of the explicit expression (6.7) for $b(\lambda)$, for our particular three-state example the efficiency amounts to

$$e(\lambda) = \frac{2\lambda e^{-2\lambda}(c - a)}{c(1 - e^{-2\lambda}) + ae^{-2\lambda}}. \tag{6.10}$$

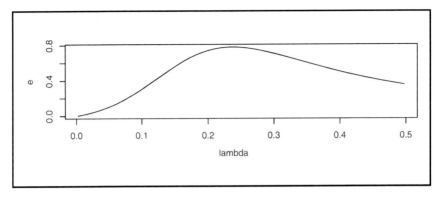

Fig. 6.1 Efficiency of the Dutch bonus-malus system

As the steady state premium does not depend on the initial state, the same holds for the efficiency, though both of course depend on the claim frequency λ.

Remark 6.3.3 (Efficiency less than one means subsidizing bad drivers)
The premium percentages in all classes are positive and finite, hence $b(0) > 0$ and $b(\infty) < \infty$ hold. In many bonus-malus systems, for example with (6.10) when $a < c$, we have $0 < e(\lambda) < 1$ over the whole range of λ. See Exercise 6.3.4. Then we get

$$\frac{d}{d\lambda} \log \frac{b(\lambda)}{\lambda} = \frac{b'(\lambda)}{b(\lambda)} - \frac{1}{\lambda} = \frac{1}{\lambda}(e(\lambda) - 1) < 0. \tag{6.11}$$

As $\log \frac{b(\lambda)}{\lambda}$ decreases with λ, so does $\frac{b(\lambda)}{\lambda}$, from ∞ as $\lambda \downarrow 0$ to 0 as $\lambda \to \infty$. So there is a claim frequency λ_0 such that the steady state premium for $\lambda = \lambda_0$ exactly equals the net premium. Drivers with $\lambda > \lambda_0$ pay less than they should, drivers with $\lambda < \lambda_0$ pay more. This means that there is a capital transfer from the good risks to the bad risks. The rules of the bonus-malus system penalize the claimers insufficiently. See again Exercise 6.3.4.

For the Dutch bonus-malus system, the efficiency never is larger than 0.8, and it is maximal near $\lambda = 0.23$. See Figure 6.1. ∇

6.4 Finding steady state premiums and Loimaranta efficiency

In this section we describe how to determine the steady state premium as well as the Loimaranta efficiency for a general bonus-malus system. Let n denote the number of states. For notational convenience, introduce the functions $t_{ij}(k)$ with $i, j = 1, 2, \ldots, n$ to describe the transition rules, as follows:

$$
\begin{aligned}
t_{ij}(k) &= 1 && \text{if by } k \text{ claims in a year, one goes from state } i \text{ to } j; \\
t_{ij}(k) &= 0 && \text{otherwise.}
\end{aligned}
\tag{6.12}
$$

When the parameter equals λ, the probability to go from state i to state j is

$$P_{ij}(\lambda) = \sum_{k=0}^{\infty} p_k(\lambda) t_{ij}(k). \tag{6.13}$$

Now consider the initial distribution $l(0) = (l_1(0), \ldots, l_n(0))$, where $l_j(0)$ is the probability of finding a contract initially, at time $t = 0$, in state j, for $j = 1, 2, \ldots, n$. Then the vector of probabilities to find a driver in class j at time $t + 1$ can be expressed in the state vector $l(t)$ at time t as follows:

$$l_j(t+1) = \sum_{i=1}^{n} l_i(t) P_{ij}(\lambda), \quad t = 0, 1, 2, \ldots \tag{6.14}$$

The sum of the $l_j(t)$ is unity for each t. In the steady state we find, taking limits for $t \to \infty$:

$$l_j(\infty) = \sum_{i=1}^{n} l_i(\infty) P_{ij}(\lambda) \quad \text{with} \quad \sum_{j=1}^{n} l_j(\infty) = 1. \tag{6.15}$$

As noted before, the steady state vector $l(\infty) = (l_1(\infty), \ldots, l_n(\infty))$ is a left-hand eigenvector of the matrix $\mathbf{P}$ corresponding to the eigenvalue 1. In the steady state, we get for the asymptotic average premium (steady state premium) with claim frequency λ:

$$b(\lambda) = \sum_{j=1}^{n} l_j(\infty) b_j, \tag{6.16}$$

with b_j the premium for state j. Note that $l_j(\infty)$ depends on λ, but not on the initial distribution over the states.

Having an algorithm to compute $b(\lambda)$ as in (6.16), we can easily approximate the Loimaranta efficiency $e(\lambda)$. All it takes is to apply (6.9). But it is also possible to compute the efficiency $e(\lambda)$ exactly. Write $l_j(\lambda) = l_j(\infty)$ and $g_j(\lambda) = \frac{dl_j(\lambda)}{d\lambda}$, then

$$\frac{db(\lambda)}{d\lambda} = \sum_{j=1}^{n} b_j \frac{dl_j(\lambda)}{d\lambda} = \sum_{j=1}^{n} b_j g_j(\lambda), \tag{6.17}$$

The $g_j(\lambda)$ can be found by taking derivatives in the system (6.15). One finds the following equations:

$$g_j(\lambda) = \sum_{i=1}^{n} g_i(\lambda) P_{ij}(\lambda) + \sum_{i=1}^{n} l_i(\lambda) P'_{ij}(\lambda), \tag{6.18}$$

where the derivatives of $P_{ij}(\lambda)$ can be found as

$$P'_{ij}(\lambda) = \frac{d}{d\lambda} \sum_{k=0}^{\infty} e^{-\lambda} \frac{\lambda^k}{k!} t_{ij}(k) = \sum_{k=0}^{\infty} e^{-\lambda} \frac{\lambda^k}{k!} \left[t_{ij}(k+1) - t_{ij}(k) \right]. \tag{6.19}$$

Using the fact that $\sum_j g_j(\lambda) = 0$, the efficiency $e(\lambda)$ can be computed for every λ by solving the resulting system of linear equations. In this way, one can compare various bonus-malus systems as regards efficiency, for example by comparing the graphs of $e(\lambda)$ for the plausible values of λ ranging from 0.05 to 0.2, or by looking at some weighted average of $e(\lambda)$ values.

To compute Loimaranta's efficiency $e(\lambda)$ for the system of Table 6.1 with R, there are three possible approaches:

Approximation: Compute the steady state distribution using any row of $\mathbf{P}^{(2^{10})}$ for $\lambda(1-\varepsilon)$ and $\lambda(1+\varepsilon)$, and use $e(\lambda) \approx \frac{\Delta \log b(\lambda)}{\Delta \log(\lambda)}$.

Exact: Implement the method described in (6.17)–(6.19).
The steady state distribution arises as a left-hand eigenvector of $\mathbf{P}$, and $b'(\lambda)$ is computed by solving a system of linear equations as well.

Simulation: Use the simulated BM positions after T years for M drivers to estimate the steady state distribution, and from the average premiums paid in year T, approximate $e(\lambda)$ like in the first method.

The following R-statements fill a vector of the BM percentages paid and an array with, in row k+1, the next BM-step Next[k+1,b] after b in case of k claims in the BM-system of Table 6.1. Using this matrix we can fill the transition matrix $\mathbf{P}$ by $P_{ij} = \sum \Pr[k \text{ claims}]$; the sum is over those k for which $t_{ij}(k) = 1$ in (6.12), so k claims lead to a transition from i to j, see (6.13). Here, p[1:4] denote the probabilities of $0, 1, 2, 3^+$ claims. Later on, the same function is used to fill $\mathbf{P}'$, with the derivatives $P'_{ij}(\lambda)$ in (6.19) replacing the $P_{ij}(\lambda)$.

```
BM.frac <- c(1.2,1,.9,.8,.7,.6,.55,.5,.45,.4,.375,.35,.325,.3)
Next <- rbind(   ##  see Table 6.1
    c( 2,  3,  4,  5,  6,  7,  8,  9,10,11,12,13,14,14),
    c( 1,  1,  1,  1,  2,  3,  4,  5,  6,  7,  7,  8,  8,  9),
    c( 1,  1,  1,  1,  1,  1,  1,  1,  2,  3,  3,  4,  4,  5),
    c( 1,  1,  1,  1,  1,  1,  1,  1,  1,  1,  1,  1,  1,  1))
FillP <- function (p)
{ PP <- matrix(0,nrow=14,ncol=14)
  for (k1 in 1:4) for (b in 1:14)
  {j <- Next[k1,b]; PP[b,j] <- PP[b,j] + p[k1]}
  return(PP)}
```

Approximate method For $\lambda \times (1-\varepsilon)$ and $\lambda \times (1+\varepsilon)$, compute $\mathbf{P}^k$ with $k = 2^{10} = 1024$. This matrix consists of the probabilities P_{ij} of going from initial state i to state j in k steps. It is evident that after k steps, it is irrelevant in which state we started. So we find the steady state distribution by using the last of the virtually equal rows of $\mathbf{P}^k$. Next, compute the corresponding steady state premiums. Finally, use

$$e(\lambda) = \frac{d \log b(\lambda)}{d \log(\lambda)} \approx \frac{\Delta \log b(\lambda)}{\Delta \log(\lambda)}. \qquad (6.20)$$

To implement this method, do:

```
b <- c(0,0); lbs <- c(0.05*(1-0.0001), 0.05*(1+0.0001))
for (i in 1:2)
{ pp <- dpois(0:2, lbs[i])
  P <- FillP(c(pp[1],pp[2],pp[3],1-sum(pp)))
  for (k in 1:10) P <- P %*% P ## i.e., P <- P^(2^10)
  stst <- P[14,] ## bottom row is near steady state db
  b[i] <- sum(stst*BM.frac)} ## b(lambda)
(log(b[2])-log(b[1])) / (log(lbs[2])-log(lbs[1]))
## = 0.1030403315
```

Exact method For $\lambda = 0.05$, fill **P** as above. First solve $l = l\mathbf{P}$ to find the steady state distribution l as the (real) left-hand eigenvector with eigenvalue 1 having total 1, as follows:

```
pp <- dpois(0:2, 0.05)
P <- FillP(c(pp[1],pp[2],pp[3],1-sum(pp)))
stst <- eigen(t(P))$vectors[,1]
stst <- stst/sum(stst); stst <- Re(stst)
```

Note that we need the *largest* eigenvalue with the *left-hand* eigenvector of **P**, hence the *first* (right-hand) eigenvector produced by `eigen(t(P))`. See `?eigen`.

Being an irreducible stochastic matrix, **P** has one eigenvalue 1; all 13 others are less than one in absolute value. This follows from Perron-Frobenius' theorem, or by reasoning as follows:

- In $t \geq 13$ steps, any state can be reached from any other. So $P_{ij}^t > 0 \; \forall i, j$.
- P_{ij}^∞ is the limiting probability to reach j from i in a very large number of steps t.
- Since the initial state i clearly becomes irrelevant as $t \to \infty$, $\mathbf{P}^\infty$ has constant rows.
- So $\mathbf{P}^\infty$, having rank 1, has one eigenvalue 1 and 13 eigenvalues 0.
- If a satisfies $a\mathbf{P} = \alpha a$, then $a\mathbf{P}^t = \alpha^t a$. Then $\alpha \neq 1 \Longrightarrow \alpha^t \to 0 \Longrightarrow |\alpha| < 1$.

Clearly, l depends on λ. Let $g_j(\lambda)$ be as in (6.17). Differentiating $l(\lambda) = l(\lambda)\mathbf{P}(\lambda)$ leads to

$$g(\lambda)(\mathbf{I} - \mathbf{P}(\lambda)) \overset{(6.18)}{=} l(\lambda)\mathbf{P}'(\lambda). \qquad (6.21)$$

The system $g(\mathbf{I} - \mathbf{P}) = l\mathbf{P}'$ is underdetermined. This is because all row sums in $\mathbf{I} - \mathbf{P}$ and $\mathbf{P}'$ are zero, so adding up all equations gives $0 = 0$. This means that the last equation is the negative of the sum of the other ones. To resolve this, replace one of the equations with $\sum g_j(\lambda) = 0$, which must hold since

$$\sum g_j(\lambda) = \frac{d}{d\lambda} \sum_j b_j(\lambda) = \frac{d}{d\lambda} 1 = 0. \qquad (6.22)$$

The sum of the coefficients of this last equation is positive, so it cannot be a linear combination of the other equations of which the coefficients all sum to zero.

To compute $\mathbf{P}'$, just replace every entry $P_{ij}(\lambda)$ in **P** by $P_{ij}'(\lambda)$ [see (6.19)]. So we get our exact efficiency by the following R-function calls:

```
P.prime <- FillP(c(-pp[1],pp[1]-pp[2],pp[2]-pp[3],pp[3]))
IP <- diag(14)-P
```

```
lP <- stst %*% P.prime            ## system (6.21) is g.IP = lP'
IP[,14] <- 1; lP[14] <- 0         ## replace last eqn by sum(g)=0
g <- lP %*% solve(IP)             ## g = lP (IP)^{-1}
0.05*sum(g*BM.frac)/sum(stst*BM.frac)      ## = 0.1030403317
```

Simulation method Generate 10000 BM-histories spanning 50 year as follows:

```
TMax <- 50; NSim <- 10000; FinalBM <- numeric(NSim)
lbs <- c(0.05*(1-0.1), 0.05*(1+0.1)); b <- c(0,0);
for (ii in 1:2) ## just as in Method 1
{ for (n in 1:NSim)
  { cn1 <- rpois(TMax,lbs[ii]); cn1 <- pmin(cn1, 3) + 1
    BM <- 14; for (i in 1:TMax) BM <- Next[cn1[i],BM]
    FinalBM[n] <- BM
  }
  print(table(FinalBM)/NSim); b[ii] <- mean(BM.frac[FinalBM])
}
(log(b[2])-log(b[1])) / (log(lbs[2])-log(lbs[1]))
```

This last method is certainly the slowest by far. Some test results reveal that to reach two decimals precision, about a million simulations are needed, which takes ± 15 min. This is definitely an issue, since generally, $e(\lambda)$ is needed for many λ-values. For example we might need to produce a graph of $e(\lambda)$, see Figure 6.1, or to compute $\max e(\lambda)$ or $\int e(\lambda) dU(\lambda)$. Also, we may wonder if after 50 years, the effect of the choice of the initial state has indeed vanished. On the other hand, simulation is undeniably more flexible in case of changes in the model.

6.5 Exercises

Section 6.2

1. Determine the percentage of the basic premium to be paid by a Dutch driver, who originally entered the bonus-malus scale at level 100%, drove without claim for 7 years, then filed one claim during the eighth policy year, and has been driving claim-free for the three years since then. Would the total of the premiums he paid have been different if his one claim occurred in the second policy year?

Section 6.3

1. Prove (6.9).

2. Determine P^2 with P as in (6.1). What is the meaning of its elements? Can you see directly from this that $l(2) = l(\infty)$ must hold?

3. Determine $e(\lambda)$ in the example with three steps in this section if in state 2, instead of c the premium is a. Argue that the system can now be described by only two states, and determine P and $l(\infty)$.

4. Show that $e(\lambda) < 1$ in (6.10) for every a and c with $a < c$. When is $e(\lambda)$ close to 1?

5. Recalculate (6.6) for a claim at the end of the policy year when the interest is i.

6. Determine the value of α such that the transition probability matrix $\mathbf{P}$ has vector $(\frac{5}{12}, \frac{7}{12})$ as its steady state vector, if $\mathbf{P} = \left(\begin{smallmatrix} \alpha & 1-\alpha \\ 1/2 & 1/2 \end{smallmatrix}\right)$.

7. If for the steady state premium we have $b(\lambda) = 100$ if $\lambda = 0.050$ and $b(\lambda) = 101$ for $\lambda = 0.051$, estimate the Loimaranta efficiency $e(\lambda) = \frac{d\log b(\lambda)}{d\log \lambda}$ at $\lambda = 0.05$.

8. For the transition probability matrix $\mathbf{P} = \left(\begin{smallmatrix} t & 1-t \\ s & 1-s \end{smallmatrix}\right)$, find the relation between s and t that holds if the steady state vector equals $(p, 1-p)$.

Section 6.4

1. In the Dutch BM-system, assume that there is a so-called 'no-claim protection' in operation: the first claim does not influence next year's position on the BM-scale. Plot the resulting efficiency of the system. Explain why it is so much lower. Also give the resulting matrix of transition probabilities. Use the same `FillP` function and the same array `Next`; change only the argument `p` of `FillP`.

2. In the Dutch BM-system, determine when it is not best to file a claim of size t caused on December, 31. Consider a time-horizon of 15 years without inflation and discounting, and take the claim frequency to be $\lambda = 0.05$ in that entire period. Compute the boundary value for each current class $k \in \{1, \ldots, 14\}$. Do the same using an annual discount factor of 2%.

Chapter 7
Ordering of risks

Through their knowledge of statistics, finance, and business, actuaries assess the risk of events occurring and help create policies that minimize risk and its financial impact on companies and clients. One of the main functions of actuaries is to help businesses assess the risk of certain events occurring and formulate policies that minimize the cost of that risk. For this reason, actuaries are essential to the insurance industry —
U.S. Department of Labor, www.bls.gov/oco

7.1 Introduction

Comparing risks is the very essence of the actuarial profession. This chapter offers mathematical concepts and tools to do this, and derives some important results of non-life actuarial science. There are two reasons why a *risk*, representing a non-negative random financial loss, would be universally preferred to another. One is that the other risk is *larger*, see Section 7.2, the second is that it is *thicker-tailed (riskier)*, see Section 7.3. Thicker-tailed means that the probability of extreme values is larger, making a risk with equal mean less attractive because it is more spread and therefore less predictable. We show that having thicker tails means having larger stop-loss premiums. We also show that preferring the risk with uniformly lower stop-loss premiums describes the *common* preferences between risks of *all* risk averse decision makers. From the fact that a risk is smaller or less risky than another, one may deduce that it is also preferable in the mean-variance order that is used quite generally. In this ordering, one prefers the risk with the smaller mean, and the variance serves as a tie-breaker. This ordering concept, however, is inadequate for actuarial purposes, since it leads to decisions about the attractiveness of risks about which there is no consensus in a group of decision makers all considered sensible.

We give several invariance properties of the stop-loss order. The most important one for actuarial applications is that it is preserved under compounding, when either the number of claims or the claim size distribution is replaced by a riskier one.

In Section 7.4 we give a number of actuarial applications of the theory of ordering risks. One is that the individual model is less risky than the collective model. In Chapter 3, we saw that the canonical collective model has the same mean but a larger variance than the corresponding individual model, while the open collective model has a larger mean (and variance). We will prove some stronger assertions, for example that any risk averse decision maker would prefer a loss with the distributional properties of the individual model to a loss distributed according to the usual collective model, and also that all stop-loss premiums for it are smaller.

From Chapter 4 we know that the non-ruin probability can be written as the cdf of a compound geometric random variable L, which represents the maximal aggregate loss. We will show that if we replace the individual claims distribution in a ruin model by a distribution that is preferred by all risk averse decision makers, this is reflected in the ruin probability getting lower, for all initial capitals u. Under somewhat more general conditions, the same holds for Lundberg's exponential upper bound for the ruin probability.

Many parametric families are monotonic in their parameters, in the sense that the risk increases (or decreases) with each of the parameters. We will show that if we look at the subfamily of the gamma(α, β) distributions with a fixed mean $\alpha/\beta = \mu$, the stop-loss premiums at each d grow with the variance $\alpha/\beta^2 = \mu^2/\alpha$, hence with decreasing α. In this way, it is possible to compare all gamma distributions with the gamma(α_0, β_0) distribution. Some will be preferred by all decision makers with increasing utility, some only by those who are also risk averse, while for others, the opinions of risk averse decision makers will differ.

In Chapter 1, we showed that stop-loss reinsurance is optimal in the sense that it gives the lowest variance for the retained risk when the mean is fixed. In this chapter we are able to prove the stronger assertion that stop-loss reinsurance leads to a retained loss that is preferable for any risk averse decision maker.

We also will show that quite often, but not always, the common good opinion of all risk averse decision makers about some risk is reflected in a premium to be asked for it. If every risk averse decision maker prefers X to Y as a loss, X has lower zero utility premiums, including for example exponential premiums.

Another field of application is given in Section 7.5. Sometimes one has to compute a stop-loss premium for a single risk of which only incomplete information is known, to be precise, the mean value μ, a maximal loss b and possibly the variance σ^2. We will determine risks with these characteristics that produce upper and lower bounds for such premiums.

It is quite conceivable that the constraints of non-negativity and independence of the terms of a sum imposed above are too restrictive. Many invariance properties depend crucially on non-negativity. But in financial actuarial applications, we must be able to incorporate both gains and losses in our models. The independence assumption is often not even approximately fulfilled, for example if the terms of a sum are consecutive payments under a random interest force, or in case of earthquake and flooding risks. Also, the mortality patterns of husband and wife are obviously related, both because of the 'broken heart syndrome' and the fact that their environments and personalities will be alike ('birds of a feather flock together'). Nevertheless, most traditional insurance models assume independence. One can force a portfolio of risks to satisfy this requirement as much as possible by diversifying, therefore not including too many related risks like the fire risks of different floors of a building, or the risks concerning several layers of the same large reinsured risk.

The assumption of independence plays a very crucial role in insurance. In fact, the basis of insurance is that by undertaking many small independent risks, an insurer's random position gets more and more predictable because of the two fundamental laws of statistics, the Law of Large Numbers and the Central Limit Theorem.

One risk is hedged by other risks, since a loss on one policy might be compensated by more favorable results on others. Moreover, assuming independence is very convenient, because mostly, the statistics gathered only give information about the marginal distributions of the risks, not about their joint distribution, that is, the way these risks are interrelated. Also, independence is mathematically much easier to handle than most other structures for the joint cdf. Note by the way that the Law of Large Numbers does not entail that the variance of an insurer's random capital goes to zero when his business expands, but only that the coefficient of variation, that is, the standard deviation expressed as a multiple of the mean, does so.

In Section 7.6 we will try to determine how to make safe decisions in case we have a portfolio of insurance policies that produce gains and losses of which the stochastic dependency structure is unknown. It is obvious that the sum of random variables is risky if these random variables exhibit a positive dependence, which means that large values of one term tend to go hand in hand with large values of the other terms. If the dependence is absent such as is the case for stochastic independence, or if it is negative, the losses will be hedged. Their total becomes more predictable and hence more attractive in the eyes of risk averse decision makers. In case of positive dependence, the independence assumption would probably underestimate the risk associated with the portfolio. A negative dependence means that the larger the claim for one risk, the smaller the other ones. The central result here is that sums of random variables are the riskiest if these random variables are maximally dependent (comonotonic).

In Section 7.7, we study the actuarial theory of dependent risks. When comparing a pair of risks with another having the same marginals, to determine which pair is more related we can of course look at the usual dependence measures, such as Pearson's correlation coefficient r, Spearman's ρ, Kendall's τ and Blomqvist's β. We noticed that stop-loss order is a more reliable ordering concept than mean-variance order, since the latter often leads to a conclusion that is contrary to the opinion of some decision makers we consider sensible. The same holds for the conclusions based on these dependence measures. We will only call a pair (X,Y) *more related* than (X',Y') if it has the same marginal cdfs but a larger joint cdf. This cdf equals $\Pr[X \leq x, Y \leq y]$, so it represents the probability of X and Y being 'small' simultaneously. Note that at the same time, the probability of being 'large' simultaneously is bigger, because $\Pr[X > x, Y > y] = 1 - \Pr[X \leq x] - \Pr[Y \leq y] + \Pr[X \leq x, Y \leq y]$. If (X,Y) is more related than an independent pair, it is positively related. If the cdf of (X,Y) is maximal, they are comonotonic. If (X,Y) is more related than (X',Y'), it can be shown that all dependence measures mentioned are larger, but not the other way around. We also introduce the concept of the *copula* of a pair (X,Y), which is just the cdf of the so-called ranks of X and Y. These ranks are the random variables $F(X)$ and $G(Y)$, respectively, if F and G are the marginal cdfs. The copula and the marginals determine a joint cdf. It will be shown that using a specific copula, a joint cdf can be found with prescribed marginals and any value of Spearman's ρ, which is nothing but the correlation of these ranks $F(X)$ and $G(Y)$.

7.2 Larger risks

In Sections 7.2–7.5, we compare risks, that is, non-negative random variables. It is easy to establish a condition under which we might call one risk Y larger than (more correctly, larger than or equal to) another risk X: without any doubt a decision maker with increasing utility will consider loss X to be preferable to Y if it is smaller with certainty, hence if $\Pr[X \leq Y] = 1$. This leads to the following definition:

Definition 7.2.1 ('Larger' risk)
For two risks, Y is 'larger' than X if a pair (X', Y) exists with $X' \sim X$ and $\Pr[X' \leq Y] = 1$. ∇

Note that in this definition, we do not just look at the marginal cdfs F_X and F_Y, but at the joint distribution of X' and Y. See the following example.

Example 7.2.2 (Binomial random variables)
Let X denote the number of times heads occur in seven tosses with a fair coin, and Y the same in ten tosses with a biased coin having probability $p > \frac{1}{2}$ of heads. If X and Y are independent, the event $X > Y$ has a positive probability. Can we set up the experiment in such a way that we can define random variables X' and Y on it, such that X' has the same cdf as X, and such that Y is always at least equal to X'?

To construct an $X' \sim X$ such that $\Pr[X' \leq Y] = 1$, we proceed as follows. Toss a biased coin with probability p of falling heads ten times, and denote the number of heads by Y. Every time heads occurs in the first seven tosses, toss another coin that falls heads with probability $\frac{1}{2p}$. Let X' be the number of heads shown by the second coin. Then $X' \sim \text{binomial}(7, \frac{1}{2})$, just as X, because the probability of a success with each potential toss of the second coin is $p \times \frac{1}{2p}$. As required, $\Pr[X' \leq Y] = 1$. The random variables X' and Y are not independent. ∇

The condition in Definition 7.2.1 for Y to be 'larger' than X proves to be equivalent to a simple requirement on the marginal cdfs:

Theorem 7.2.3 (A larger random variable has a smaller cdf)
A pair (X', Y) with $X' \sim X$ and $\Pr[X' \leq Y] = 1$ exists if, and only if, $F_X(x) \geq F_Y(x)$ for all $x \geq 0$.

Proof. The 'only if'-part of the theorem is evident, since $\Pr[X' \leq x] \geq \Pr[Y \leq x]$ must hold for all x. For the 'if'-part, we only give a proof for two important special cases. If both $F_X(\cdot)$ and $F_Y(\cdot)$ are continuous and strictly increasing, we can simply take $X' = F_X^{-1}(F_Y(Y))$. Then $F_Y(Y)$ can be shown to be uniform$(0,1)$, and therefore $F_X^{-1}(F_Y(Y)) \sim X$. Also, $X' \leq Y$ holds.

For X and Y discrete, look at the following functions, which are actually the inverse cdfs with $F_X(\cdot)$ and $F_Y(\cdot)$, and are defined for all u with $0 < u < 1$:

$$\begin{aligned} f(u) &= x \quad \text{if } F_X(x-0) < u \leq F_X(x); \\ g(u) &= y \quad \text{if } F_Y(y-0) < u \leq F_Y(y). \end{aligned} \tag{7.1}$$

Next, take $U \sim \text{uniform}(0,1)$. Then $g(U) \sim Y$ and $f(U) \sim X$, while $F_X(x) \geq F_Y(x)$ for all x implies that $f(u) \leq g(u)$ for all u, so $\Pr[f(U) \leq g(U)] = 1$. ∇

Remark 7.2.4 ('Larger' vs. larger risks)
To compare risks X and Y, we look only at their marginal cdfs F_X and F_Y. Since the joint distribution does not matter, we can, without loss of generality, look at any copy of X. But this means we can assume that, if Y is 'larger' than X in the sense of Definition 7.2.1, actually the stronger assertion $\Pr[X \leq Y] = 1$ holds. So instead of just stochastically larger, we may assume the risk to be larger with probability one. All we do then is replace X by an equivalent risk. This approach can be very helpful to prove statements about 'larger' risks. ∇

In many situations, we consider a model involving several random variables as input. Quite often, the output of the model increases if we replace any of the input random variables by a larger one. This is for example the case when comparing $X + Z$ with $Y + Z$, for a risk Z that is independent of X and Y (convolution). A less trivial example is compounding, where both the number of claims and the claim size distributions may be replaced. We have:

Theorem 7.2.5 (Compounding)
If the individual claims X_i are 'smaller' than Y_i for all i, the counting variable M is 'smaller' than N, and all these random variables are independent, then $X_1 + X_2 + \cdots + X_M$ is 'smaller' than $Y_1 + Y_2 + \cdots + Y_N$.

Proof. In view of Remark 7.2.4 we can assume without loss of generality that $X_i \leq Y_i$ as well as $M \leq N$ hold with probability one. Then the second expression has at least as many non-negative terms, all of them being at least as large. ∇

The order concept 'larger than' used above is called *stochastic order*, and the notation is as follows:

Definition 7.2.6 (Stochastic order)
Risk X precedes risk Y in stochastic order, written $X \leq_{st} Y$, if Y is 'larger' than X. ∇

In the literature, often the term 'stochastic order' is used for any ordering concept between random variables or their distributions. In this book, it is reserved for the specific order of Definition 7.2.6.

Remark 7.2.7 (Stochastically larger risks have a larger mean)
A consequence of stochastic order $X \leq_{st} Y$, that is, a *necessary condition* for it, is obviously that $E[X] \leq E[Y]$, and even $E[X] < E[Y]$ unless $X \sim Y$. See for example formula (1.33) at $d = 0$. The opposite does not hold: $E[X] \leq E[Y]$ is not sufficient to conclude that $X \leq_{st} Y$. A counterexample is $X \sim \text{Bernoulli}(p)$ with $p = \frac{1}{2}$ and $\Pr[Y = c] = 1$ for a c with $\frac{1}{2} < c < 1$. ∇

Remark 7.2.8 (Once-crossing densities are stochastically ordered)
An important *sufficient condition* for stochastic order is that the densities have the property that $f_X(x) \geq f_Y(x)$ for small x, and the opposite for large x. A proof of this statement is asked in Exercise 7.2.1. ∇

It can be shown that the order $\leq_{\text{st}}$ has a natural interpretation in terms of utility theory. We have

Theorem 7.2.9 (Stochastic order and increasing utility functions)
$X \leq_{\text{st}} Y$ holds if and only if $E[u(-X)] \geq E[u(-Y)]$ for every non-decreasing utility function $u(\cdot)$.

Proof. If $E[u(-X)] \geq E[u(-Y)]$ holds for every non-decreasing $u(\cdot)$, then it holds especially for the utility functions $u_d(y) = 1 - I_{(-\infty,-d]}(y)$. But $E[u_d(-X)]$ is just $\Pr[X \leq d]$. For the 'only if' part, if $X \leq_{\text{st}} Y$, then $\Pr[X' \leq Y] = 1$ for some $X' \sim X$, and therefore $E[u(-X)] \geq E[u(-Y)]$. ∇

So the pairs of risks X and Y with $X \leq_{\text{st}} Y$ are exactly those pairs of losses about which *all* profit seeking decision makers agree on their order.

7.3 More dangerous risks

In economics, when choosing between two potential losses, one usually prefers the loss with the smaller mean. If two risks have the same mean, the tie is broken by the variance. This mean-variance order concept is the basis of the CAPM-models in economic theory. It is inadequate for the actuary, because it might declare a risk to be preferable to another if not all risk-averse individuals feel the same way.

7.3.1 Thicker-tailed risks

It is evident that all risk averse actuaries would agree that one risk is riskier than another if its extreme values have larger probability. This is formalized as follows:

Definition 7.3.1 (Thicker-tailed)
Risk Y is *thicker-tailed* than X, written $Y \geq_{\text{tt}} X$, if they have the following properties:

Equal means $E[X] = E[Y]$
Once-crossing cdfs   A real number x_0 exists with $\Pr[X \leq x] \leq \Pr[Y \leq x]$ for small
$\quad$ x with $x < x_0$, but $\Pr[X \leq x] \geq \Pr[Y \leq x]$ when $x > x_0$. ∇

Example 7.3.2 (Dispersion and concentration)
The class of risks Y with a known upper bound b and mean μ contains an element Z with the thickest possible tails. It is the random variable with

$$\Pr[Z = b] = 1 - \Pr[Z = 0] = \frac{\mu}{b}. \tag{7.2}$$

It is clear that if Y also has mean μ and upper bound b, then $Y \leq_{\text{tt}} Z$ must hold because their cdfs cross exactly once. See Figure 7.1. The distribution of Z arises

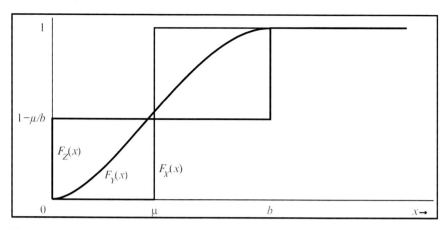

Fig. 7.1 Cdfs with thickest and thinnest tails for random variables with mean μ and support $[0,b]$, generated by *dispersion* and *concentration*

from the one of Y by *dispersion* of the probability mass to the boundaries 0 and b. The random variable Z is the most dispersed one with this given mean and upper bound. The variance $\mathrm{Var}[Z]$ is maximal, since $E[Y^2] \leq E[bY] = b\mu = E[Z^2]$.

This same class of risks also contains a least dispersed element. It arises by *concentration* of the probability mass on μ. If $X \equiv \mu$, then $X \leq_{tt} Y$, see again Figure 7.1, and its variance is minimal. The problem of finding a minimal variance ('best-case') is less interesting for practical purposes.

Dispersal and concentration can also be restricted to only the probability mass in some interval, still resulting in risks with thicker and thinner tails respectively. ∇

A sufficient condition for two crossing cdfs to cross exactly once is that the difference of these cdfs increases first, then decreases, and increases again after that. This leads to:

Theorem 7.3.3 (Densities crossing twice means cdfs crossing once)
Let X and Y be two risks with equal finite means but different densities. If intervals I_1, I_2 and I_3 exist with $I_1 \cup I_2 \cup I_3 = [0,\infty)$ and I_2 between I_1 and I_3, such that the densities of X and Y satisfy $f_X(x) \leq f_Y(x)$ both on I_1 and I_3, while $f_X(x) \geq f_Y(x)$ on I_2, then the cdfs of X and Y cross only once. Here $I_1 = [0,0]$ or $I_2 = [b,b]$ may occur if the densities are discrete.

Proof. Because of $E[X] = E[Y] < \infty$, the cdfs F_X and F_Y must cross at least once, since we assumed that $f_X \not\equiv f_Y$. This is because if they would not cross, one of the two would be larger in stochastic order by Theorem 7.2.3, and the means would then be different by Remark 7.2.7. Both to the left of 0 and in the limit at ∞, the difference of the cdfs equals zero. Both if the densities represent the derivatives of the cdfs or their jumps, it is seen that the difference of the cdfs increases first to a maximum, then decreases to a minimum, and next increases to zero again. So there

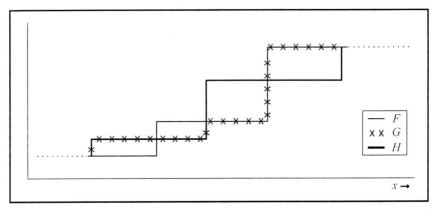

Fig. 7.2 Counterexample of the transitivity of being thicker-tailed: H is thicker-tailed than G, G is thicker-tailed than F, but H is not thicker-tailed than F.

is just one point, somewhere in I_2, where the difference in the cdfs crosses the x-axis, hence the cdfs cross exactly once. The property also holds if the densities are differentials. Note that $a > b\,dx > 0$ holds for all $a > 0$ and $b > 0$. $\qquad\qquad\nabla$

Example 7.3.4 (Binomial has thinner tails than Poisson)

If we compare a binomial(n,p) distribution with a Poisson(np) distribution, we know that they have the same mean, while the latter has a greater variance. Is it also thicker-tailed than the binomial distribution?

We will show that the discrete densities, say $f(x)$ and $g(x)$ respectively, have the crossing properties of the previous theorem. Since $f(x) > g(x) > 0$ is the same as $r(x) := f(x)/g(x) > 1$, we can do this by showing that $r(x)$ increases up to a certain value of x, and decreases after that. Writing $q = 1 - p$ as usual, we get for this ratio

$$r(x) = \frac{f(x)}{g(x)} = \frac{\binom{n}{x}p^x q^{n-x}}{(np)^x e^{-np}/x!} = \frac{n(n-1)\cdots(n-x+1)}{n^x q^x}q^n e^{np}. \qquad (7.3)$$

Now consider the ratio of successive values of $r(x)$. For $x = 1, 2, \ldots$ we have

$$\frac{r(x)}{r(x-1)} = \frac{n-x+1}{nq} \le 1 \quad \text{if and only if} \quad x \ge np + 1. \qquad (7.4)$$

As $f(\cdot)$ and $g(\cdot)$ have the same mean, they must cross at least twice. But then $r(x)$ must cross the horizontal level 1 twice, so $r(x) < 1$ must hold for small as well as for large values of x, while $r(x) > 1$ must hold for intermediate values x in an interval around $np + 1$. Now apply the previous theorem to see that the Poisson distribution indeed has thicker tails than a binomial distribution with the same mean. $\qquad\nabla$

Remark 7.3.5 (Thicker-tailed is not a transitive ordering)

It is easy to construct examples of random variables X, Y and Z where Y is thicker-tailed than X, Z is thicker-tailed than Y, but Z is not thicker-tailed than X. In Figure 7.2, the cdfs $F = F_X$ and $G = F_Y$ cross once, as do G and $H = F_Z$, but F and H cross three times. So being thicker-tailed is not a well-behaved ordering concept: order relations should be *transitive*. Transitivity can be enforced by extending the relation $\leq_{tt}$ to pairs X and Z such that a sequence of random variables $Y_1, Y_2, \ldots, Y_n$ exists with $X \leq_{tt} Y_1$, $Y_j \leq_{tt} Y_{j+1}, j = 1, 2, \ldots, n-1$, as well as $Y_n \leq_{tt} Z$. Extending the relation in this way, we get the finite transitive closure of the relation $\leq_{tt}$. This order will be called *indirectly thicker-tailed* from now on.   ▽

If X precedes Y in stochastic order, their cdfs do not cross. If Y is (indirectly) thicker-tailed than X, it can be shown that their stop-loss transforms $\pi_X(d) = \mathrm{E}[(X-d)_+]$ and $\pi_Y(d)$ do not cross. By proceeding inductively, it suffices to prove this for the case where Y is directly thicker-tailed than X. But in that case, the difference $\pi_Y(d) - \pi_X(d)$ can be seen to be zero at $d = 0$ because the means of X and Y are equal, zero at $d \to \infty$, increasing as long as the derivative of the difference $\pi_Y'(d) - \pi_X'(d) = F_Y(d) - F_X(d)$ is positive, and decreasing thereafter. Hence, Y thicker-tailed than X means that Y has higher stop-loss premiums.

Remark 7.3.6 (Odd number of sign changes not sufficient for indirectly thicker-tailed)

It is easy to see that in case of equal means, two cdfs with an even number of sign changes cannot be stop-loss ordered. One sign change is sufficient for stop-loss order, but to show that other odd numbers are not, compare the stop-loss premiums of random variables $U \sim \mathrm{uniform}(0, b)$ and $M \sim \mathrm{binomial}(b, \frac{1}{2})$, for $b = 2, 3$. As one can see from Figure 7.3, for $b = 2$ the stop-loss premiums of M are all larger. To see that the cdf is indirectly thicker-tailed, compare F_U and F_M with a cdf $G(x) = F_U(x)$ on $(-\infty, 1)$, $G(x) = F_M(x)$ on $[1, \infty)$. But for $b = 3$, the slps are not uniformly ordered, see the slp at $d = 3 - \varepsilon$ and the one at $d = 2$. There are five sign changes in the difference of the cdfs.

Note that when comparing a discrete and a continuous distribution, one should not just look at the densities, but at the corresponding differentials. They are $\mathrm{d}F_U(x) = f_U(x)\,\mathrm{d}x$, and $\mathrm{d}F_M(x) > 0$ only when $x \in \{0, \ldots, b\}$, $\mathrm{d}F_M(x) = 0$ otherwise. The former is positive on $[0, b]$, though infinitesimally small, the other one is mostly zero, but positive on some points.   ▽

We can prove that higher stop-loss premiums must imply (indirectly) thicker tails.

Theorem 7.3.7 (Thicker-tailed vs. higher stop-loss premiums)
If $\mathrm{E}[X] = \mathrm{E}[Y]$ and $\pi_X(d) \leq \pi_Y(d)$ for all $d > 0$, then there exists a sequence $F_1, F_2, \ldots$ of increasingly thicker-tailed cdfs with $X \sim F_1$ and $Y \sim \lim_{n \to \infty} F_n$.

Proof. First suppose that Y is a random variable with finitely many possible values. Then the cdf of Y is a step-function, so the stop-loss transform is a piecewise linear continuous convex function. Hence, for certain linear functions $A_1, \ldots, A_n$ it can be written in the form

$$\pi_Y(d) = \max\{\pi_X(d), A_1(d), A_2(d), \ldots, A_n(d)\}, \quad -\infty < d < \infty. \tag{7.5}$$

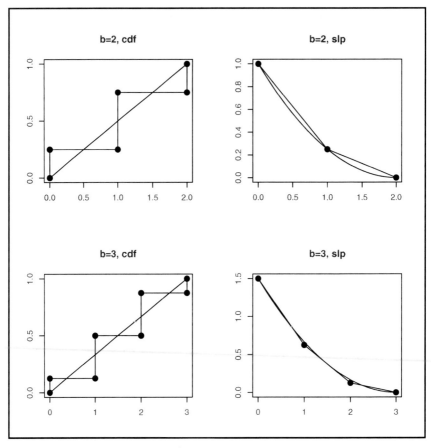

Fig. 7.3 Uniform$(0,b)$ and binomial$(b, \frac{1}{2})$ cdfs and slps, for $b = 2,3$. Note that the slps are ordered for $b = 2$, but not for $b = 3$.

Now define the following functions $\pi_i(\cdot), i = 1, 2, \ldots, n$:

$$\pi_i(d) = \max\left\{\pi_X(d), A_1(d), A_2(d), \ldots, A_{i-1}(d)\right\}, \quad -\infty < d < \infty. \qquad (7.6)$$

These functions are stop-loss transforms, say with the cdfs $F_i, i = 1, 2, \ldots, n$. As the reader may check, $X \sim F_1$, $Y \sim F_n$, and F_i has thicker tails than F_{i-1}, $i = 2, 3, \ldots, n$. See also Exercise 7.3.25.

If the support of Y is infinite, we must take the limit of the cdfs F_n in the sense of convergence in distribution. ∇

7.3.2 Stop-loss order

For pairs of random variables with ordered stop-loss premiums we have the following definition.

Definition 7.3.8 (Stop-loss order)
If X has smaller stop-loss premiums than Z, we say that X is smaller than Z in stop-loss order, and write $X \leq_{SL} Z$.   ▽

A random variable that is stop-loss larger than another risk with the same mean will be referred to as 'more dangerous' in the sequel. Note that for stop-loss order, equality of the means $E[X] = E[Z]$ is not required. We may show that any risk Z that is stop-loss larger than X is more dangerous than an intermediate risk Y that in turn is larger than X:

Theorem 7.3.9 (Separation theorem for stop-loss order)
If $X \leq_{SL} Z$ and $E[X] < E[Z]$, there exists a random variable Y for which

1. $X \leq_{st} Y$;
2. $Y \leq_{SL} Z$ and $E[Y] = E[Z]$.

Proof. The random variable $Y = \max\{X, b\}$, with $b > 0$ chosen such that $E[Y] = E[Z]$, satisfies both these requirements, as the reader is asked to verify in Exercise 7.3.12.   ▽

So Y separates X and Z, in a sense stronger than merely $\leq_{SL}$. For another separator with the stochastic inequality signs swapped, see Exercise 7.3.13.

Just like stochastic order, stop-loss order can be expressed in a utility context as the common risk preferences of a group of sensible decision makers:

Theorem 7.3.10 (Stop-loss order and concave increasing utility functions)
$X \leq_{SL} Y$ holds if and only if $E[u(-X)] \geq E[u(-Y)]$ for every *concave increasing* utility function $u(\cdot)$.

Proof. In view of Theorem 7.3.9, it suffices to give the proof for the case that $E[X] = E[Y]$. Then, it follows as a special case of Theorem 7.6.2 later on. See also Exercise 7.3.17.   ▽

So stop-loss order represents the common preferences of all risk averse decision makers. Stop-loss order applies to losses, that is, non-negative risks. Two general random variables with the same mean and ordered stop-loss premiums for all d, including $d < 0$, are called convex ordered, see Section 7.6. As a consequence of Theorem 7.3.10, expected values of convex functions are ordered. Since all functions x^α with $\alpha \geq 1$ are convex, for the moments of X and Y we have $E[X^k] \leq E[Y^k]$, $k = 1, 2, \ldots$ In particular, a more dangerous risk (with the same mean) has a higher variance. But if the means of X and Y are not equal, this is not always the case. A trivial counterexample is $X \sim$ Bernoulli$(\frac{1}{2})$ and $Y \equiv 1$.

7.3.3 Exponential order

Next to stochastic order and stop-loss order, there is another useful ordering concept to be derived from the expected utility model.

Definition 7.3.11 (Exponential order)
If for all $\alpha > 0$, decision makers with an exponential utility function with risk aversion α prefer loss X to Y, we say that X precedes Y in *exponential order*, written $X \leq_e Y$. ∇

The inequality $X \leq_e Y$ is equivalent to X having a smaller mgf than Y on the interval $(0, \infty)$. A sufficient condition for exponential order between risks is stop-loss order, since the function e^{tx} is a convex function on $[0, \infty)$ for $t > 0$, hence $\mathrm{E}[e^{tX}] \leq \mathrm{E}[e^{tY}]$ holds for all $t > 0$. But this can be seen from utility considerations as well, because the exponential order represents the preferences common to the subset of the risk-averse decision makers for which the risk attitude is independent of current wealth.

Since exponential order represents the common preferences of a smaller group of decision makers, it is to be expected that there exist pairs of random variables that are exponentially ordered, but not stop-loss ordered. See Exercise 7.4.10.

Remark 7.3.12 (Exponential order and current wealth)
The capital of an insurer with exponential utility equals $w - Z$ for some risk Z (possibly $Z = c$) and for some w. If $X \leq_e Y$ and X, Y, Z independent, will the insurer still prefer loss X to Y?

Comparing $\mathrm{E}[u(w - Z - X)]$ with $\mathrm{E}[u(w - Z - Y)]$ one sees that the factors involving w and Z cancel out. So for exponential utility, the current wealth, random or fixed, known or not, has no influence on the decisions made. This is very convenient, but on the other hand, such risk behavior is not very plausible, so this is both a strong point and a weak point of exponential utility. ∇

If $X \leq_e Y$ holds, then the following can be said:

- For the corresponding adjustment coefficients in a ruin process, $R_X \geq R_Y$ holds for each premium income c, and vice versa (Exercise 7.4.4).
- If additionally $\mathrm{E}[X] = \mathrm{E}[Y]$, then $\mathrm{Var}[X] \leq \mathrm{Var}[Y]$ (Exercise 7.3.14).
- If both $\mathrm{E}[X] = \mathrm{E}[Y]$ and $\mathrm{Var}[X] = \mathrm{Var}[Y]$, then $\gamma_X \leq \gamma_Y$ (Exercise 7.3.14).
- If $M \leq_e N$, then $X_1 + \cdots + X_M \leq_e Y_1 + \cdots + Y_N$ for compound random variables (Exercise 7.4.3). So exponential order is invariant for the operation of compounding risks.

7.3.4 Properties of stop-loss order

For stop-loss order, many invariance properties we derived for stochastic order still hold. So if we replace a particular component of a model by a more dangerous input, we often obtain a stop-loss larger result. For actuarial purposes, it is important whether the order is retained in case of compounding. First we prove that adding independent random variables, as well as taking mixtures, does not disturb the stop-loss order.

Theorem 7.3.13 (Convolution preserves stop-loss order)
If for risks X and Y we have $X \leq_{SL} Y$, and risk Z is independent of X and Y, then $X + Z \leq_{SL} Y + Z$. If further S_n is the sum of n independent copies of X and T_n is the same for Y, then $S_n \leq_{SL} T_n$.

Proof. The first stochastic inequality can be proved by using the relation:

$$E[(X + Z - d)_+] = \int_0^\infty E[(X + z - d)_+] dF_Z(z). \qquad (7.7)$$

The second follows by iterating the first inequality. ∇

Example 7.3.14 (Order in Poisson multiples)
Let $X \equiv 1$ and $1 - \Pr[Y = 0] = \Pr[Y = \alpha] = 1/\alpha$; then $X \leq_{SL} Y$. Further, let $N \sim \text{Poisson}(\lambda)$, then $X_1 + \cdots + X_N \sim \text{Poisson}(\lambda)$ and $Y_1 + \cdots + Y_N \sim \alpha M$ with $M \sim \text{Poisson}(\lambda/\alpha)$. By the previous theorem, $N \leq_{SL} \alpha M$. So letting N_μ denote a Poisson(μ) random variable, we have $N_\lambda \leq_{SL} \alpha N_{\lambda/\alpha}$ for $\alpha > 1$. It is easy to see that also $N_\lambda \geq_{SL} \alpha N_{\lambda/\alpha}$ for $\alpha < 1$.

As a consequence, there is stop-loss order in the family of multiples of Poisson random variables; see also Chapter 9. ∇

Theorem 7.3.15 (Mixing preserves stop-loss order)
Let cdfs F_y and G_y satisfy $F_y \leq_{SL} G_y$ for all real y, let $U(y)$ be any cdf, and let $F(x) = \int_{\mathbb{R}} F_y(x) dU(y)$, $G(x) = \int_{\mathbb{R}} G_y(x) dU(y)$. Then $F \leq_{SL} G$.

Proof. Using Fubini, one sees that the stop-loss premiums with F are equal to

$$\int_d^\infty [1 - F(x)] dx = \int_d^\infty \left[1 - \int_{\mathbb{R}} F_y(x) dU(y)\right] dx$$
$$= \int_d^\infty \int_{\mathbb{R}} [1 - F_y(x)] dU(y) dx = \int_{\mathbb{R}} \int_d^\infty [1 - F_y(x)] dx dU(y). \qquad (7.8)$$

Hence $F \leq_{SL} G$ follows immediately by (1.33). ∇

Corollary 7.3.16 (Mixing ordered random variables)
The following conclusions are immediate from Theorem 7.3.15:

1. If $F_n(x) = \Pr[X \leq x | N = n]$, $G_n(x) = \Pr[Y \leq x | N = n]$, and $F_n \leq_{SL} G_n$ for all n, then we obtain $X \leq_{SL} Y$ by taking the cdf of N to be $U(\cdot)$. The event $N = n$ might for example indicate the nature of a particular claim (small or large, liability or comprehensive, bonus-malus class n, and so on).
2. Taking especially $F_n(x) = F^{*n}$ and $G_n(x) = G^{*n}$, where F and G are the cdfs of individual claims X_i and Y_i, respectively, produces $X_1 + \cdots + X_N \leq_{SL} Y_1 + \cdots + Y_N$ if $F \leq_{SL} G$. Hence stop-loss order is preserved under compounding, if the individual claim size distribution is replaced by a stop-loss larger one.

3. If Λ is a structure variable with cdf U, and conditionally on the event $\Lambda = \lambda$, $X \sim F_\lambda$ and $Y \sim G_\lambda$, then $F_\lambda \leq_{SL} G_\lambda$ for all λ implies $X \leq_{SL} Y$.

4. Let G_λ denote the conditional cdf of X, given the event $\Lambda = \lambda$, and F_λ the cdf of the degenerate random variable on $\mu(\lambda) \stackrel{\text{def}}{=} E[X \mid \Lambda = \lambda]$. Then by concentration, $F_\lambda \leq_{tt} G_\lambda$. The function $\int_\mathbb{R} F_\lambda(x)\,dU(\lambda)$ is the cdf of the random variable $\mu(\Lambda) = E[X \mid \Lambda]$, while $\int_\mathbb{R} G_\lambda(x)\,dU(\lambda)$ is the cdf of X. So $E[X \mid \Lambda] \leq_{SL} X$ for all X and Λ. We have proved that *conditional means* are always less dangerous than the original random variable. $\triangledown$

Remark 7.3.17 (Rao-Blackwell theorem)

The fact that the conditional mean $E[Y \mid X]$ is less dangerous than Y itself is the basis of the Rao-Blackwell theorem, to be found in many texts on mathematical statistics. It states that if Y is an unbiased estimator for a certain parameter, then $E[Y \mid X]$ is a better unbiased estimator, provided it is a *statistic*, that is, it contains no unknown parameters. For every event $X = x$, the conditional distribution of Y is concentrated on its mean $E[Y \mid X = x]$, leading to a less dispersed and, for that reason, better estimator. $\triangledown$

We saw that if the terms of a compound sum are replaced by stop-loss larger ones, the result is also stop-loss larger. It is tougher to prove that the same happens when we replace the claim number M by the stop-loss larger random variable N. The general proof, though short, is not easy, so we start by giving the important special case with $M \sim$ Bernoulli(q) and $E[N] \geq q$. Recall that $\sum_{i-1}^n x_i = 0$ if $n = 0$.

Theorem 7.3.18 (Compounding with a riskier claim number, 1)

If $M \sim$ Bernoulli(q), N is a counting random variable with $E[N] \geq q$, $X_1, X_2, \ldots$ are copies of a risk X, and all these risks are independent, then we have

$$MX \leq_{SL} X_1 + X_2 + \cdots + X_N. \tag{7.9}$$

Proof. First we prove that if $d \geq 0$ and $x_i \geq 0$ $\forall i$, the following always holds:

$$(x_1 + \cdots + x_n - d)_+ \geq (x_1 - d)_+ + \cdots + (x_n - d)_+. \tag{7.10}$$

There only is something to prove if the right hand side is non-zero. If, say, the first term is positive, the first two $(\cdot)_+$-operators can be dropped, leaving

$$x_2 + \cdots + x_n \geq (x_2 - d)_+ + \cdots + (x_n - d)_+. \tag{7.11}$$

Since $d \geq 0$ as well as $x_i \geq 0$ for all i, inequality (7.10) is proved.

If $q_n = \Pr[N = n]$, replacing realizations in (7.10) by random variables and taking the expectation we have:

$$E[(X_1 + X_2 + \cdots + X_N - d)_+]$$

$$= \sum_{n=1}^{\infty} q_n E[(X_1 + X_2 + \cdots + X_n - d)_+]$$

$$\geq \sum_{n=1}^{\infty} q_n E[(X_1 - d)_+ + (X_2 - d)_+ + \cdots + (X_n - d)_+] \qquad (7.12)$$

$$= \sum_{n=1}^{\infty} n q_n E[(X - d)_+] \geq q E[(X - d)_+] = E[(MX - d)_+].$$

The last inequality is valid since we assumed $\sum_n n q_n \geq q$. $\qquad \triangledown$

Theorem 7.3.19 (Compounding with a riskier claim number, 2)
If for two counting random variables M and N we have $M \leq_{SL} N$, and $X_1, X_2, \ldots$ are independent copies of a risk X, then $X_1 + X_2 + \cdots + X_M \leq_{SL} X_1 + X_2 + \cdots + X_N$.

Proof. It is sufficient to prove that $f_d(n) = E[(X_1 + \cdots + X_n - d)_+]$ is a convex and increasing function of n, since by Theorem 7.3.10 this implies $E[f_d(M)] \leq E[f_d(N)]$ for all d, which is the same as $X_1 + \cdots + X_M \leq_{SL} X_1 + \cdots + X_N$. Because $X_{n+1} \geq 0$, it is obvious that $f_d(n + 1) \geq f_d(n)$. To prove convexity, we need to prove that $f_d(n + 2) - f_d(n + 1) \geq f_d(n + 1) - f_d(n)$ holds for all n. By taking the expectation over the random variables X_{n+1}, X_{n+2} and $S = X_1 + X_2 + \cdots + X_n$, one sees that for this it is sufficient to prove that for all $d \geq 0$ and all $x_i \geq 0, i = 1, 2, \ldots, n+2$, we have

$$(s + x_{n+1} + x_{n+2} - d)_+ - (s + x_{n+1} - d)_+ \geq (s + x_{n+2} - d)_+ - (s - d)_+, \quad (7.13)$$

where $s = x_1 + x_2 + \cdots + x_n$. If both middle terms of this inequality are zero, so is the last one, and the inequality is valid. If at least one of them is positive, say the one with x_{n+1}, on the left hand side of (7.13), x_{n+2} remains, and the right hand side is equal to this if $s \geq d$, and smaller otherwise, as can be verified easily. $\qquad \triangledown$

Combining Theorems 7.3.15 and 7.3.19, we see that a compound sum is riskier if the number of claims, the claim size distribution, or both are replaced by stop-loss larger ones.

Remark 7.3.20 (Functional invariance)
Just like stochastic order (see Exercise 7.2.8), stop-loss order has the property of functional invariance. Indeed, if $f(\cdot)$ and $v(\cdot)$ are non-decreasing convex functions, the composition $v \circ f$ is convex and non-decreasing as well, and hence we see immediately that $f(X) \leq_{SL} f(Y)$ holds if $X \leq_{SL} Y$. This holds in particular for the claims paid in case of excess of loss reinsurance, where $f(x) = (x - d)_+$, and proportional reinsurance, where $f(x) = \alpha x$ for $\alpha > 0$. $\qquad \triangledown$

7.4 Applications

In this section, we give some important actuarial applications of the theory of ordering of risks.

7.4.1 Individual versus collective model

In Section 3.7 we described how *the* collective model resulted from replacing every policy by a Poisson(1) distributed number of independent copies of it. But from Theorem 7.3.18 with $q = 1$ we see directly that doing this, we in fact replace the claims of every policy by a more dangerous random variable. If subsequently we add up all these policies, which we have assumed to be stochastically independent, then for the portfolio as a whole, a more dangerous total claims distribution results. This is because stop-loss order is preserved under convolution, see Theorem 7.3.13.

As an alternative for the canonical collective model, in Remark 3.8.2 we introduced an *open* collective model. If the claims of policy i are $I_i b_i$ for some fixed amount at risk b_i and a Bernoulli(q_i) distributed random variable I_i, the term in *the* collective model corresponding to this policy is $M_i b_i$, with $M_i \sim \text{Poisson}(q_i)$. In the open collective model, it is $N_i b_i$, with $N_i \sim \text{Poisson}(t_i)$ for $t_i = -\log(1 - q_i)$, and hence $I_i \leq_{\text{st}} N_i$. So in the open model each policy is replaced by a compound Poisson distribution with a *stochastically* larger claim number distribution than with the individual model. Hence the open model will not only be less attractive than the individual model for all risk averse decision makers, but even for the larger group of all decision makers with increasing utility functions. Also, the canonical collective model is preferable to the open model for this same large group of decision makers. Having a choice between the individual and *the* collective model, some decision makers might prefer the latter. Decision makers preferring the individual model over the collective model are not consistently risk averse.

7.4.2 Ruin probabilities and adjustment coefficients

In Section 4.7, we showed that the non-ruin probability $1 - \psi(u)$ is the cdf of a compound geometric random variable $L = L_1 + L_2 + \cdots + L_M$, where $M \sim \text{geometric}(p)$ is the number of record lows in the surplus process, L_i is the amount by which a previous record low in the surplus was broken, and L represents the maximal aggregate loss. Recall (4.59) and (4.60):

$$p = 1 - \psi(0) = \frac{\theta}{1 + \theta} \quad \text{and} \quad f_{L_1}(y) = \frac{1 - P(y)}{\mu_1}, \, y > 0. \qquad (7.14)$$

Here θ is the safety loading, and $P(y)$ is the cdf of the claim sizes in the ruin process. Now suppose that we replace cdf P by Q, where $P \leq_{SL} Q$ and Q has the same mean as P. From (7.14) it is obvious that since the stop-loss premiums with Q are larger than those with P, the probability $\Pr[L_1 > y]$ is increased when P is replaced by Q. This means that we get a new compound geometric distribution with the same geometric parameter p because μ_1 and hence θ are unchanged, but a *stochastically* larger distribution of the individual terms L_i. This leads to a smaller cdf for L, and hence a larger ruin probability. Note that the equality of the means μ_1 of P and Q is essential here, to ensure that p remains the same and that the L_1 random variables increase stochastically.

If P is a degenerate cdf, R has support $\{0, b\}$ with $b = \max[Q]$, and $\mu_P = \mu_Q = \mu_R$, then $P \leq_{SL} Q \leq_{SL} R$ holds, see Example 7.3.2. So the ruin probability $\psi_R(u)$ with claims $\sim R$ is maximal for every initial capital u, while $\psi_P(u)$ is minimal in the class of risks with mean μ_Q and maximum $\max[Q]$. Notice that claims $\sim R$ lead to a ruin process with claims zero or b, hence in fact to a process with only one possible claim size, just as with claims $\sim P$. This means that algorithm (4.79) with $m = 1$ can be used to find bounds for the ruin probabilities; see also Section 4.9.

Now suppose that we replace the claim size cdf Q by R with $Q \leq_{st} R$, but leave the premium level c unchanged. This means that we replace the ruin process by a process with the same premium per unit time and the same claim number process, but 'larger' claims. By Remark 7.2.4, without loss of generality we can take each claim to be larger with probability one, instead of just stochastically larger. This means that also with probability one, the new surplus $U_R(t)$ will be lower than or equal to $U_Q(t)$, at each instant $t > 0$. This in turn implies that for the ruin probabilities, we have $\psi_R(u) \geq \psi_Q(u)$. It may happen that one gets ruined in the R-process, but not in the Q-process; the other way around is impossible. Because in view of the Separation Theorem 7.3.9, when P is replaced by R with $P \leq_{SL} R$ we can always find a separating Q with the same expectation as P and with $P \leq_{SL} Q \leq_{st} R$, we see that whenever we replace the claims distribution by any stop-loss larger distribution, the ruin probabilities are increased for every value of the initial capital u, assuming that the annual premium remains the same.

From Figure 4.2 we see directly that when the mgf with the claims is replaced by one that is larger on $(0, \infty)$, the resulting adjustment coefficient is smaller. This is the case when we replace the claims distribution by an exponentially larger one, see Remark 7.3.12. So we get larger ruin probabilities by replacing the claims by stop-loss larger ones, but for the Lundberg exponential upper bound to increase, exponential order suffices.

We saw that stop-loss larger claims lead to uniformly larger ruin probabilities. The weaker exponential order is not powerful enough to enforce this. To give a counterexample, first observe that pairs of exponentially ordered random variables exist that have the same mean and variance. Take for example $\Pr[X = 0, 1, 2, 3] = \frac{1}{3}, 0, \frac{1}{2}, \frac{1}{6}$ and $Y \sim 3 - X$; see also Exercise 7.4.10. Now if $\psi_X(u) \leq \psi_Y(u)$ for all u would hold, the cdfs of the maximal aggregate losses L_X and L_Y would not cross, so $L_X \leq_{st} L_Y$. Therefore either $E[L_X] < E[L_Y]$, or $\psi_X \equiv \psi_Y$. But the former is not possible since

$$E[L_X] = E[M]E[L_i^{(X)}] = E[M] \int_0^\infty E[(X-x)_+] \, dx/E[X]$$

$$= E[M]\frac{1}{2}E[X^2]/E[X] = E[M]\frac{1}{2}E[Y^2]/E[Y] = \cdots = E[L_Y]. \tag{7.15}$$

The latter is also not possible. If the two ruin probability functions are equal, the mgfs of L_X and L_Y are equal, and therefore also the mgfs of $L_i^{(X)}$ and $L_i^{(Y)}$, see (4.64), hence in view of (4.60), the claim size distribution must be the same.

7.4.3 Order in two-parameter families of distributions

In many two-parameter families of densities, given any parameter pair, the parameter space can be divided into areas containing stocastically smaller/larger distributions, stop-loss larger/smaller distributions and distributions not comparable with our reference pair in stop-loss order. Apart from the gamma distributions studied in detail below, we mention the binomial, beta, inverse Gaussian, lognormal, overdispersed Poisson and Pareto families, as well as families of random variables of type bIX with $I \sim$ Bernoulli(q) and X, for example, constant, exponential(1) or uniform$(0,1)$.

The gamma distribution is important as a model for the individual claim size, for example for damage to the own vehicle, see also Chapters 3 and 9. In general when one thinks of a gamma distribution, one pictures a density that is unimodal with a positive mode, like a tilted normal density. But if the shape parameter $\alpha = 1$, we get the exponential distribution, which is unimodal with mode 0. In general, the gamma(α,β) density has mode $(\alpha-1)_+/\beta$. The skewness of a gamma distribution is $2/\sqrt{\alpha}$, so distributions with $\alpha < 1$ are more skewed than an exponential distribution with the same mean, and ultimately have larger tail probabilities.

Suppose we want to compare two gamma distributions, say with parameters α_0, β_0 (our reference point) and α_1, β_1. It is easy to compare means and variances, and hence to find which α_1, β_1 combinations lead to a larger distribution in mean-variance ordering. Is there perhaps more to be said about order between such distributions, for example about certain tail probabilities or stop-loss premiums?

From the form of the mgf $m(t;\alpha,\beta) = (1-t/\beta)^{-\alpha}$, one sees that gamma random variables are additive in α. Indeed we have $m(t;\alpha_1,\beta)m(t;\alpha_2,\beta) = m(t;\alpha_1 + \alpha_2,\beta)$, so if X and Y are independent gamma random variables with the same β, their sum is also gamma. From $E[e^{t(\beta X)}] = (1-t)^{-\alpha}$ one sees that $\beta X \sim$ gamma$(\alpha,1)$ if $X \sim$ gamma(α,β), and in this sense, the gamma distributions are multiplicative in the scale parameter β. But from these two properties we have immediately that a gamma(α,β) random variable gets 'larger' if α is replaced by $\alpha + \varepsilon$, and 'smaller' if β is replaced by $\beta + \varepsilon$ for $\varepsilon > 0$. There is monotonicity in stochastic order in both parameters, see also Exercise 7.2.2.

Now let us compare the gamma(α_1,β_1) with the gamma(α_0,β_0) distribution when it is known that they have the same mean, so $\alpha_1/\beta_1 = \alpha_0/\beta_0$. Suppose that

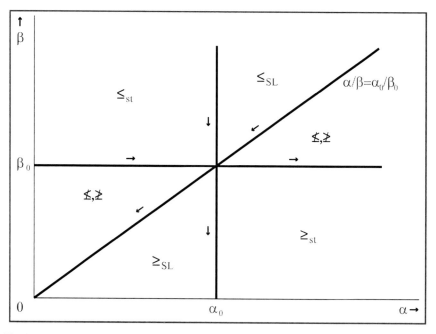

Fig. 7.4 Ordering of gamma(α, β) distributions. The arrows indicate directions of increase in $\leq_{SL}$.

$\alpha_1 < \alpha_0$, therefore also $\beta_1 < \beta_0$. We will show by investigating the densities that the gamma(α_1, β_1) distribution not just has a larger variance $\alpha_1/\beta_1^2 > \alpha_0/\beta_0^2$, but is in fact more dangerous. A sufficient condition for this is that the densities cross exactly twice. Consider the ratio of these two densities (where the symbol $\propto$ denotes equality up to a constant not depending on x):

$$\frac{\frac{1}{\Gamma(\alpha_1)}\beta_1^{\alpha_1}x^{\alpha_1-1}e^{-\beta_1 x}}{\frac{1}{\Gamma(\alpha_0)}\beta_0^{\alpha_0}x^{\alpha_0-1}e^{-\beta_0 x}} \propto x^{\alpha_1-\alpha_0}e^{-(\beta_1-\beta_0)x} = (x^\mu e^{-x})^{\beta_1-\beta_0}. \tag{7.16}$$

The derivative of $x^\mu e^{-x}$ is positive if $0 < x < \mu$, negative if $x > \mu$, so the ratio (7.16) crosses each horizontal level at most twice. But because both densities have the same mean, there is no stochastic order, which means that they must intersect more than once. So apparently, they cross exactly twice, therefore one of the two random variables is more dangerous than the other. One can find out which by looking more closely at where each density is larger than the other. But we already know which one is the more dangerous, since it must necessarily be the one having the larger variance, that is, the one with parameters α_1, β_1.

We may conclude that going along any diagonal in the (α, β) plane from (α_0, β_0) toward the origin, one finds increasingly more dangerous parameter combinations. Also we see in Figure 7.4 that if a point (α, β) can be reached from (α_0, β_0) by first going along the diagonal in the direction of the origin, and next either to the right or straight down, this points corresponds to a stop-loss larger gamma distribution, because it is stochastically larger than a separating more dangerous distribution (found on the diagonal). In Figure 7.4, one sees the distributions stochastically larger than (α_0, β_0) in the quarter-plane to the right and below this point. In the opposite quarter-plane are the stochastically smaller ones. The quarter-plane to the left and below (α_0, β_0) has stop-loss larger distributions below the diagonal, while for the distributions above the diagonal one may show that the means are lower, but the stop-loss premiums for $d \to \infty$ are higher than for (α_0, β_0). The latter can be proved by applying l'Hôpital's rule twice. Hence, there is a difference of opinion about such risks between the risk averse decision makers. See also Exercise 7.4.8.

This can be summarized as follows. Suppose we compare a gamma(α_0, β_0) and a gamma(α, β) random variable with for the means, $\alpha_0/\beta_0 \le \alpha/\beta$. Then for *all* stop-loss premiums with the parameter pair (α, β) to be larger, it is necessary but also sufficient that the stop-loss premiums are larger for very large retentions d.

A similar statement holds for the two-parameter families of densities mentioned at the beginning of this subsection. In each case, the proof of stop-loss order is given by showing stochastic increase with the mean if one of the parameters is kept constant, and stop-loss increase with the variance if the mean is kept constant. In the gamma case, the set of parameter pairs with equal mean μ is a diagonal $\alpha/\beta = \mu$, but it often is hyperbolic, for example for the binomial distributions with mean np or the negative binomial distributions with mean $r(1-p)/p$. The stop-loss premiums for high retentions being lower in these cases is equivalent to one of the parameters being lower (or higher, in case of monotonic decrease). Stop-loss order can only hold if the conditions on the mean and the asymptotic stop-loss premiums hold. If so, stop-loss order is proved by a separation argument.

7.4.4 Optimal reinsurance

Theorem 1.4.3 states that among the reinsurance contracts with the same expected value of the retained risk, stop-loss reinsurance gives the lowest possible variance. Suppose the random loss equals X, and compare the cdf of the retained loss $Z = X - (X-d)_+$ under stop-loss reinsurance with another retained loss $Y = X - I(X)$, where $E[Y] = E[Z]$. Assume that the function $I(\cdot)$ is non-negative, then it follows that $Y \le X$ holds, and hence $F_Y(x) \ge F_X(x)$ for all $x > 0$. Further, $Z = \min\{X, d\}$, so $F_Z(x) = F_X(x)$ for all $x < d$, and $F_Z(x) = 1$ for $x \ge d$. Clearly, the cdfs of Z and Y cross exactly once, at d, and Y is the more dangerous risk. So $Z \le_{\mathrm{SL}} Y$.

This has many consequences. First, we have $E[u(-Z)] \ge E[u(-Y)]$ for every concave increasing utility function $u(\cdot)$. Also, we see confirmed that Theorem 1.4.3 holds, because obviously $\mathrm{Var}[Y] \ge \mathrm{Var}[Z]$. We can also conclude that excess of loss coverage is more effective than any other reinsurance with the same mean that operates on individual claims. Note that these conclusions depend crucially on the fact that the premiums asked for different form of reinsurance depend only on the expected values of the reimbursements.

7.4.5 Premiums principles respecting order

If a loss X is stop-loss smaller than Y, all risk averse decision makers prefer losing X. Perhaps surprisingly, this does not always show in the premiums that are needed to compensate for this loss. Consider for example the standard deviation premium $\pi[X] = \mathrm{E}[X] + \alpha\sqrt{\mathrm{Var}[X]}$, see (5.7). If $X \sim \mathrm{Bernoulli}(\frac{1}{2})$ and $Y \equiv 1$, while $\alpha > 1$, the premium for X is larger than the one for Y even though $\Pr[X \leq Y] = 1$.

The zero utility premiums, including the exponential premiums, do respect stop-loss order. For these, the premium $\pi[X]$ for a risk X is calculated by solving the utility equilibrium equation (1.11), in this case:

$$\mathrm{E}[u(w + \pi[X] - X)] = u(w). \tag{7.17}$$

The utility function $u(\cdot)$ is assumed to be risk averse, that is, concave increasing, and w is the current wealth. If $X \leq_{\mathrm{SL}} Y$ holds, we also have $\mathrm{E}[u(w + \pi[Y] - X)] \geq \mathrm{E}[u(w + \pi[Y] - Y)]$. The right hand side equals $u(w)$. Since $\mathrm{E}[u(w + P - X)]$ increases in P and because $\mathrm{E}[u(w + \pi[X] - X)] = u(w)$ must hold, it follows that $\pi[X] \leq \pi[Y]$.

7.4.6 Mixtures of Poisson distributions

In Chapter 8, we study, among other things, mixtures of Poisson distributions as a model for the number of claims on an automobile policy, assuming heterogeneity of the risk parameters. Quite often, the estimated structure distribution has the sample mean as its mean. If we assume that there is a gamma structure distribution and a risk model such as in Example 3.3.1, the resulting number of claims is negative binomial. If we estimate the negative binomial parameters by maximum likelihood or by the method of moments, in view of (3.76) and (3.78) we get parameters leading to the same mean, but different variances. Therefore also the corresponding estimated gamma structure distributions have the same mean. If we replace the structure distribution by a more dangerous one, we increase the uncertainty present in the model. Does it follow from this that the resulting marginal claim number distribution is also stop-loss larger?

A partial answer to this question can be given by combining a few facts that we have seen before. In Exercise 7.3.9, we saw that a negative binomial distribution is stop-loss larger than a Poisson distribution with the same mean. Hence, a gamma(α, β) mixture of Poisson distributions such as in Example 3.3.1 is stop-loss larger than a pure Poisson distribution with the same mean $\mu = \alpha/\beta$.

To give a more general answer, we first introduce some more notation. Suppose that the structure variables are $\Lambda_j, j = 1, 2$, and assume that given $\Lambda_j = \lambda$, the random variables N_j have a Poisson(λ) distribution. Let W_j be the cdf of Λ_j. We want to prove that $W_1 \leq_{\mathrm{SL}} W_2$ implies $N_1 \leq_{\mathrm{SL}} N_2$. To this end, we introduce the functions $\pi_d(\lambda) = \mathrm{E}[(M_\lambda - d)_+]$, with $M_\lambda \sim \mathrm{Poisson}(\lambda)$. Then $N_1 \leq_{\mathrm{SL}} N_2$ holds if and only if $E[\pi_d(\Lambda_1)] \leq E[\pi_d(\Lambda_2)]$ for all d. So all we have to do is to prove that the function $\pi_d(\lambda)$ is convex increasing, hence to prove that $\pi'_d(\lambda)$ is positive and increasing in λ. This proof is straightforward:

$$\pi'_d(\lambda) = \sum_{n>d}(n-d)\frac{d}{d\lambda}\frac{\lambda^n e^{-\lambda}}{n!} = \sum_{n>d}(n-d)\left(\frac{\lambda^{n-1}e^{-\lambda}}{(n-1)!} - \frac{\lambda^n e^{-\lambda}}{n!}\right)$$

$$= \pi_{d-1}(\lambda) - \pi_d(\lambda) = \int_{d-1}^{d}[1 - F_{M_\lambda}(t)]\,dt. \tag{7.18}$$

The last expression is positive, and increasing in λ because $M_\lambda \leq_{st} M_\mu$ for all $\lambda < \mu$.

7.4.7 Spreading of risks

Suppose one can invest a total amount 1 in n possible funds. These funds produce iid yields G_i per share. How should one choose the fraction p_i of a share to buy from fund i if the objective is to maximize the expected utility?

Assume that the utility of wealth is measured by the risk averse function $u(\cdot)$. We must solve the following constrained optimization problem:

$$\max_{p_1,\ldots,p_n} E\left[u\left(\sum_i p_i G_i\right)\right] \quad \text{subject to} \quad \sum_i p_i = 1. \tag{7.19}$$

We will prove that the solution with $p_i = \frac{1}{n}, i = 1,2,\ldots,n$ is optimal. Write $A = \frac{1}{n}\sum_i G_i$ for the average yield. Note that as $\sum_i E[G_i|A] \equiv E[\sum_i G_i|A] \equiv nA$, we have $E[G_i|A] \equiv A$, because for symmetry reasons the outcome should be the same for every i. This implies

$$E\left[\sum_i p_i G_i \,\Big|\, A\right] \equiv \sum_i p_i E[G_i|A] \equiv \sum_i p_i A \equiv A. \tag{7.20}$$

By part 4 of Corollary 7.3.16, we have $E[\sum_i p_i G_i|A] \leq_{SL} \sum_i p_i G_i$, hence because $u(\cdot)$ is concave, the maximum in (7.19) is found when $p_i = \frac{1}{n}, i = 1,2,\ldots,n$.

7.4.8 Transforming several identical risks

Consider a sequence of iid risks $X_1,\ldots,X_n$ and non-negative functions ρ_i, $i = 1,\ldots,n$. Then we can prove that

$$\sum_{i=1}^{n}\overline{\rho}(X_i) \leq_{SL} \sum_{i=1}^{n}\rho_i(X_i), \quad \text{where} \quad \overline{\rho}(x) = \frac{1}{n}\sum_{i=1}^{n}\rho_i(x). \tag{7.21}$$

This inequality expresses the fact that given identical risks, to get the least variable result the same treatment should be applied to all of them. To prove this, we show that if V is the random variable on the right and W the one on the left in (7.21), we have $W \equiv E[V \mid W]$. Next, we use that $E[V \mid W] \leq_{SL} V$, see part 4 of Corollary 7.3.16. We have

$$E\left[\sum_{i=1}^{n} \rho_i(X_i) \mid \sum_{i=1}^{n} \overline{\rho}(X_i)\right] \equiv \sum_{i=1}^{n} E\left[\rho_i(X_i) \mid \sum_{k=1}^{n} \frac{1}{n} \sum_{l=1}^{n} \rho_k(X_l)\right]. \tag{7.22}$$

For symmetry reasons, the result is the same if we replace the X_i by X_j, for each $j = 1, \ldots, n$. But this means that we also have

$$\sum_{i=1}^{n} E\left[\rho_i(X_i) \mid \sum_{k=1}^{n} \frac{1}{n} \sum_{l=1}^{n} \rho_k(X_l)\right] \equiv \sum_{i=1}^{n} E\left[\frac{1}{n} \sum_{j=1}^{n} \rho_i(X_j) \mid \sum_{k=1}^{n} \frac{1}{n} \sum_{l=1}^{n} \rho_k(X_l)\right]. \tag{7.23}$$

This last expression can be rewritten as

$$E\left[\sum_{i=1}^{n} \frac{1}{n} \sum_{j=1}^{n} \rho_i(X_j) \mid \sum_{k=1}^{n} \frac{1}{n} \sum_{l=1}^{n} \rho_k(X_l)\right] \equiv \sum_{k=1}^{n} \frac{1}{n} \sum_{l=1}^{n} \rho_k(X_l) \equiv \sum_{l=1}^{n} \overline{\rho}(X_l). \tag{7.24}$$

So we have proved that indeed $W \equiv E[V \mid W]$, and the required stop-loss inequality in (7.21) follows immediately from Corollary 7.3.16.

Remark 7.4.1 (Law of large numbers and stop-loss order)
The weak law of large numbers states that for sequences of iid observations $X_1, X_2, \ldots \sim X$ with finite mean μ and variance σ^2, the average $\overline{X}_n = \frac{1}{n} \sum_{i=1}^{n} X_i$ converges to μ, in the sense that when $\varepsilon > 0$ and $\delta > 0$, we have

$$\Pr[|\overline{X}_n - \mu| < \varepsilon] \geq 1 - \delta \quad \text{for all } n > \sigma^2/\varepsilon^2 \delta. \tag{7.25}$$

In terms of stop-loss order, we may prove the following assertion:

$$\overline{X}_1 \geq_{SL} \overline{X}_2 \geq_{SL} \cdots \geq_{SL} \mu. \tag{7.26}$$

Hence the sample averages $\overline{X}_n$, all having the same mean μ, decrease in dangerousness. As $n \to \infty$, the stop-loss premiums $E[(\overline{X}_n - d)_+]$ at each d converge to $(\mu - d)_+$, which is the stop-loss premium of the degenerate random variable on μ. The proof of (7.26) can be given by taking, in the previous remark, $\rho_i(x) = \frac{x}{n-1}, i = 1, \ldots, n-1$ and $\rho_n(x) \equiv 0$, resulting in $\overline{\rho}(x) \equiv \frac{x}{n}$. $\qquad \nabla$

7.5 Incomplete information

In this section we study the situation that we only have limited information about the distribution $F_Y(\cdot)$ of a certain risk Y, and try to determine a safe stop-loss premium at retention d for it. From past experience, from the policy conditions, or from the particular reinsurance that is operative, it is often possible to fix a practical upper bound for the risk. Hence in this section we will assume that we know an upper bound b for the payment Y. We will also assume that we have a good estimate for the mean risk μ as well as for its variance σ^2. In reinsurance proposals, sometimes these values are prescribed. Also it is conceivable that we have deduced mean and variance from scenario analyses, where for example the mean payments and the variance about this mean are calculated from models involving return times of catastrophic spring tides or hurricanes. With these data the actuary, much more than the statistician, will tend to base himself on the worst case situation where under the given conditions on μ, σ^2 and the upper bound b, the distribution is chosen that leads to the maximal possible stop-loss premium. Best case scenarios, that is, risks with prescribed characteristics leading to minimal stop-loss premiums, are also mathematically interesting, but less relevant in practice.

First notice that the following conditions are necessary for feasible distributions with mean μ, variance σ^2 and support $[0,b]$ to exist at all:

$$0 \le \mu \le b, \quad 0 \le \sigma^2 \le \mu(b-\mu). \tag{7.27}$$

The need for the first three inequalities is obvious. The last one can be rewritten as $E[Y^2] = \sigma^2 + \mu^2 \le \mu b$, which must obviously hold as well. We will assume the inequalities in (7.27) to be strict, so as to have more than one feasible distribution.

Note that if the risks X and Y have the same mean and variance, stop-loss order is impossible, because their stop-loss transforms must cross at least once. This is because in view of (3.116), if $\pi_Y(d) \ge \pi_X(d)$ for all d, either $\mathrm{Var}[X] < \mathrm{Var}[Y]$ or $X \sim Y$ must hold. Later on we will prove that the random variable Z_d with the largest stop-loss premium at d necessarily has a support consisting of two points only. Which support this is depends on the value of d. In case the variance is not specified, by concentration and dispersion we find the *same* smallest and largest distributions for every d, see Example 7.3.2. So then we also get attainable upper bounds for compound stop-loss premiums and ruin probabilities. If the variance is prescribed, that is not possible.

First we study the class of two-point distributions with mean μ and variance σ^2.

Lemma 7.5.1 (Two-point distributions with given mean and variance)
Suppose a random variable T with $E[T] = \mu$, $\mathrm{Var}[T] = \sigma^2$, but not necessarily $\Pr[0 \le T \le b] = 1$, has a two-point support $\{r, \bar{r}\}$. Then r and $\bar{r}$ are related by

$$\bar{r} = \mu + \frac{\sigma^2}{\mu - r}. \tag{7.28}$$

Proof. We know that $E[(T - \bar{r})(T - r)] = 0$ must hold. This implies

$$0 = E[T^2 - (r + \bar{r})T + r\bar{r}] = \mu^2 + \sigma^2 - (r + \bar{r})\mu + r\bar{r}. \tag{7.29}$$

For a given r, we can solve for $\bar{r}$, obtaining (7.28). ∇

So for any given r, the number $\bar{r}$ denotes the unique point that, together with r, can form a two-point support with known μ and σ^2. Note the special points $\bar{0}$ and $\bar{b}$. The probability $p_r = \Pr[T = r]$ is uniquely determined by

$$p_r = \frac{\mu - \bar{r}}{r - \bar{r}} = \frac{\sigma^2}{\sigma^2 + (\mu - r)^2}. \tag{7.30}$$

This means that there is exactly one two-point distribution containing $r \neq \mu$. The bar operator assigning $\bar{r}$ to r has the following properties:

$$\overline{(\bar{r})} = r;$$
$$\text{for } r \neq \mu, \ \bar{r} \text{ is increasing in } r; \tag{7.31}$$
$$\text{if } r < \mu, \text{ then } \bar{r} > \mu.$$

So if $\{r,s\}$ and $\{u,v\}$ are two possible two-point supports with $r < s$, $u < v$ and $r < u$, then $r < u < \mu < s < v$ must hold, in line with the fact that because the distributions have equal mean and variance, their stop-loss transforms must cross at least once, their cdfs at least twice, and their densities three or more times.

In our search for the maximal stop-loss premiums, we prove next that the maximal stop-loss premium in any retention d cannot be attained by a distribution with a support contained in $[0,b]$ that consists of more than two points. For this purpose, assume that we have a support $\{a,c,e\}$ of a feasible distribution with $0 \leq a < c < e \leq b$. It can be verified that $c \leq \bar{a} \leq e$, as well as $a \leq \bar{e} \leq c$. From a sketch of the stop-loss transforms, see Figure 7.5, it is easy to see that on $(-\infty, c]$, the two-point distribution on $\{a, \bar{a}\}$ has a stop-loss premium at least equal to the one corresponding to $\{a,c,e\}$, while on $[c, \infty)$, the same holds for $\{e, \bar{e}\}$. In the same fashion, a distribution with n mass points is dominated by one with $n-1$ mass points. To see why, just let a, c and e above be the last three points in the n-point support. The conclusion is that the distribution with a maximal stop-loss premium at retention d is to be found among the distributions with a two-point support.

So to find the random variable X that maximizes $\mathrm{E}[(X-d)_+]$ for a particular value of d and for risks with the properties $\Pr[0 \leq X \leq b] = 1$, $\mathrm{E}[X] = \mu$ and $\mathrm{Var}[X] = \sigma^2$, we only have to look at random variables X with two-point support $\{c, \bar{c}\}$. Note that in case either $d < \bar{c} < c$ or $\bar{c} < c < d$, we have $\mathrm{E}[(X-d)_+] = (\mu - d)_+$, which is in fact the minimal possible stop-loss premium, so we look only at the case $c \geq d \geq \bar{c}$. First we ignore the support constraint $0 \leq \bar{c} < c \leq b$, and solve the following maximization problem:

$$\max_{c > \mu} \ \mathrm{E}[(X-d)_+] \quad \text{for } \Pr[X = c] = \Pr[X \neq \bar{c}] = \frac{\sigma^2}{\sigma^2 + (c - \mu)^2}. \tag{7.32}$$

This is equivalent to

$$\max_{c > \mu} \ \frac{\sigma^2(c-d)}{\sigma^2 + (c-\mu)^2}. \tag{7.33}$$

Dividing by σ^2 and taking the derivative with respect to c leads to

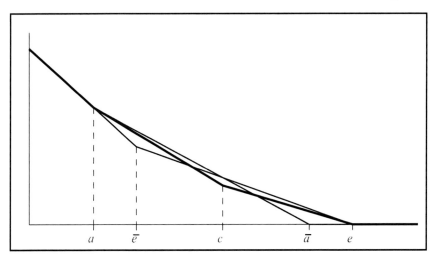

Fig. 7.5 Proof that the stop-loss premium of a 3-point distribution cannot be maximal. For each retention, one of the two-point distributions $\{a,\bar{a}\}$ or $\{\bar{e},e\}$ has a larger stop-loss premium than the three-point distribution.

$$\frac{d}{dc}\frac{c-d}{\sigma^2+(c-\mu)^2} = \frac{(c-\mu)^2+\sigma^2-2(c-d)(c-\mu)}{[\sigma^2+(c-\mu)^2]^2}. \tag{7.34}$$

Setting the numerator equal to zero gives a quadratic equation in c:

$$-c^2+2dc+\mu^2+\sigma^2-2d\mu = 0. \tag{7.35}$$

The solution with $c > \mu > \bar{c}$ is given by

$$c^* = d+\sqrt{(d-\mu)^2+\sigma^2}, \quad \bar{c}^* = d-\sqrt{(d-\mu)^2+\sigma^2}. \tag{7.36}$$

Notice that we have $d = \frac{1}{2}(c^*+\bar{c}^*)$. The numbers c^* and $\bar{c}^*$ of (7.36) constitute the optimal two-point support if one ignores the requirement that $\Pr[0 \le X \le b] = 1$. Imposing this restriction additionally, we may get boundary extrema. Since $0 \le \bar{c}$ implies $\bar{0} \le c$, we no longer maximize over $c > \mu$, but only over the values $\bar{0} \le c \le b$. If $c^* > b$, which is equivalent to $d > \frac{1}{2}(b+\bar{b})$, the optimum is $\{b,\bar{b}\}$. If $\bar{c}^* < 0$, hence $d < \frac{1}{2}\bar{0}$, the optimum is $\{0,\bar{0}\}$.

From this discussion we can establish the following theorem about the supports leading to the maximal stop-loss premiums; we leave it to the reader to actually compute the optimal values.

Theorem 7.5.2 (Maximal stop-loss premiums)
For values $\frac{1}{2}\bar{0} \le d \le \frac{1}{2}(b+\bar{b})$, the maximal stop-loss premium for a risk with given mean μ, variance σ^2 and upper bound b is the one with the two-point support $\{c^*,\bar{c}^*\}$ with c^* and $\bar{c}^*$ as in (7.36). For $d > \frac{1}{2}(b+\bar{b})$, the distribution with support

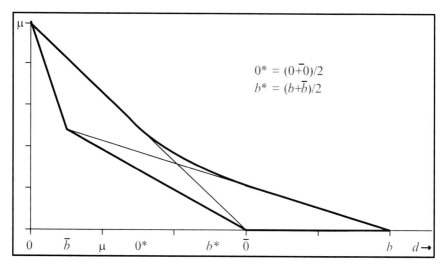

Fig. 7.6 Extremal stop-loss premiums at retention d for $\mu = 1$, $\sigma^2 = 2$ and $b = 5$.

$\{b,\bar{b}\}$ has the maximal stop-loss premium, and for $d < \frac{1}{2}\bar{0}$, the optimal support is $\{0,\bar{0}\}$.   ∇

Example 7.5.3 (Minimal and maximal stop-loss premiums)
In Figure 7.6, the thick lines are the minimal and maximal stop-loss premiums for all $d \in [0,b]$ for the case $\mu = 1$, $\sigma^2 = 2$, and $b = 5$. It can be seen that both the minimal possible stop-loss premiums and the maximal stop-loss premiums constitute a convex decreasing function, hence both are the stop-loss transform with certain risks. It is evident from the diagram that both these risks have the correct mean μ and upper bound b, but not the right variance σ^2. There are no cdfs that lead to large stop-loss premiums uniformly. For example the risk with support $\{0,\bar{0}\}$ has a stop-loss transform that coincides with the upper bounds on $d \in (0,0^*)$, is the thin line on $d \in (0^*,\bar{0})$, and is minimal when $d > \bar{0}$.

For reinsurance as occurring in practice, it is the large retentions with $d > \mu + \sigma$, say, that are of interest. One may show that if b is small, for all these d-values the stop-loss premium is maximal for the support $\{b,\bar{b}\}$; this support is optimal as long as $d > \frac{1}{2}(b+\bar{b})$, and $\mu + \sigma > \frac{1}{2}(b+\bar{b})$ holds if $0 < \frac{b-\mu}{\sigma} < 1 + \sqrt{2}$, as the reader may check. See Exercise 7.5.8.

The distributions that produce the maximum stop-loss premium have a two-point support, and their stop-loss transforms are tangent lines at d to the graph with the upper bounds. Minima are attained at $(\mu - d)_+$ when $d \le \bar{b}$ or $d \ge \bar{0}$. In those cases, the support is $\{d,\bar{d}\}$. For intermediate values of d we will argue that the minimal stop-loss premium $l(d)$ is attained by a distribution with support $\{0,d,b\}$. In a sense, these distributions have a two-point support as well, if one counts the boundary points 0 and b, of which the location is fixed but the associated probability can be chosen freely, for one half. In Figure 7.6 one sees that connecting the points $(0,\mu)$,

$(d, l(d))$ and $(b, 0)$ gives a stop-loss transformation $\pi(\cdot)$ with not only the right mean $\pi(0) = \mu$, but also with an upper bound b since $\pi(b) = 0$. Moreover, the variance is equal to σ^2. This is because the area below the stop-loss transform, which equals one half the second moment of the risk, is equal to the corresponding area for the risks with support $\{0, \bar{0}\}$ as well as with $\{b, \bar{b}\}$. To see this, use the areas of triangles with base line $(b, 0)$ to $(0, \mu)$. In fact, on the interval $\bar{b} < d < \bar{0}$ the function $l(d)$ runs parallel to the line connecting $(b, 0)$ to $(0, \mu)$. Note that $l(d)$ is the minimal value of a stop-loss premium at d, because any stop-loss transform through a point (d, h) with $h < \pi(d)$ leads to a second moment strictly less than $\mu^2 + \sigma^2$. ∇

Remark 7.5.4 (Related problems)
Other problems of this type have been solved as well. There are analytical results available for the extremal stop-loss premiums given up to four moments, and algorithms for when the number of known moments is larger than four. The practical relevance of these methods is questionable, since the only way to have reliable estimates of the moments of a distribution is to have many observations, and from these one may estimate a stop-loss premium directly. There are also results for the case that Y is unimodal with a known mode M. As well as extremal stop-loss premiums, also extremal tail probabilities can be computed. ∇

Example 7.5.5 (Verbeek's inequality; mode zero)
Let Y be a *unimodal* risk with mean μ, upper bound b and mode 0. As F_Y is concave on $[0, b]$, $2\mu \leq b$ must hold. Further, let X and Z be risks with $\Pr[Z = 0] = 1 - 2\frac{\mu}{b}$, and

$$F_Z'(y) = \frac{2\mu}{b^2}, \quad 0 < y < b;$$

$$F_X'(y) = \frac{1}{2\mu}, \quad 0 < y < 2\mu, \tag{7.37}$$

and zero otherwise. Then X and Z are also unimodal with mode zero, and $E[X] = E[Y] = E[Z]$, as well as $X \leq_{SL} Y \leq_{SL} Z$. See Exercise 7.5.2. So this class of risks also has elements that have *uniformly* minimal and maximal stop-loss premiums, respectively, allowing results extending to compound distributions and ruin probabilities. ∇

7.6 Comonotonic random variables

Up to now, we only considered random variables representing non-negative losses. In order to be able to handle both gains and losses, we start by extending the concept of stop-loss order somewhat to account for more general random variables, with possibly negative values as well as positive ones, instead of the non-negative risks that we studied so far. We state and prove the central result in this theory, which is that the least attractive portfolios are those for which the claims are maximally dependent. Next, we give some examples of how to apply the theory.

With stop-loss order, we are concerned with large values of a random loss, and call random variable Y less attractive than X if the expected values of all top parts $(Y - d)_+$ are larger than those of X. Negative values for these random variables are actually gains. But with stability in mind, excessive gains are also unattractive for the decision maker. Hence X will be more attractive than Y only if both the top parts $(X - d)_+$ and the bottom parts $(d - X)_+$ have a lower mean value than the corresponding quantities for Y. This leads to the following definition:

Definition 7.6.1 (Convex order)
If both the following conditions hold for every $d \in (-\infty, \infty)$:

$$E[(X - d)_+] \le E[(Y - d)_+], \quad \text{and}$$
$$E[(d - X)_+] \le E[(d - Y)_+], \tag{7.38}$$

then the random variable X is less than Y in convex order, written $X \le_{cx} Y$. ∇

Note that adding d to the first set of inequalities and letting $d \to -\infty$ leads to $E[X] \le E[Y]$. Subtracting d in the second set of inequalities and letting $d \to +\infty$, on the other hand, produces $E[X] \ge E[Y]$. Hence $E[X] = E[Y]$ must hold for two random variables to be convex ordered. Also note that the first set of inequalities combined with equal means implies the second set of (7.38), since $E[(X - d)_+] - E[(d - X)_+] = E[X] - d$. So two random variables with equal means and ordered stop-loss premiums are convex ordered, while random variables with unequal means are never convex ordered.

Stop-loss order is the same as having ordered expected values $E[f(X)]$ for all *non-decreasing* convex functions $f(\cdot)$, see Theorem 7.3.10. Hence it represents the common preferences of all risk averse decision makers. Convex order is the same as ordered expectations for *all* convex functions. This is where the name convex order derives from. In a utility theory context, it represents the common preferences of all risk averse decision makers between random variables with equal means. One way to prove that convex order implies ordered expectations of convex functions is to use the fact that any convex function can be obtained as the uniform limit of a sequence of piecewise linear functions, each of them expressible as a linear combination of functions $(x - t)_+$ and $(t - x)_+$. This is the proof that one usually finds in the literature. A simpler proof, involving partial integrations, is given below.

Theorem 7.6.2 (Convex order means ordered convex expectations)
If $X \le_{cx} Y$ and $f(\cdot)$ is convex, then $E[f(X)] \le E[f(Y)]$.
If $E[f(X)] \le E[f(Y)]$ for every convex function $f(\cdot)$, then $X \le_{cx} Y$.

Proof. To prove the second assertion, consider the convex functions $f(x) = x$, $f(x) = -x$ and $f(x) = (x - d)_+$ for arbitrary d. The first two functions lead to $E[X] = E[Y]$, the last one gives $E[(X - d)_+] \le E[(Y - d)_+]$.

To prove the first assertion, consider $g(x) = f(x) - f(a) - (x - a)f'(a)$, where a is some point where the function f is differentiable. Since $E[X] = E[Y]$, the inequality $E[f(X)] \le E[f(Y)]$, assuming these expectations exist, is equivalent to $E[g(X)] \le E[g(Y)]$. Write $F(x) = \Pr[X \le x]$ and $\overline{F}(x) = 1 - F(x)$. Since $g(a) = g'(a) = 0$, the integrated terms below vanish, so by four partial integrations we get

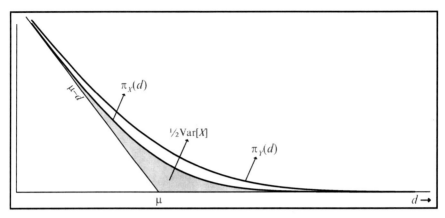

Fig. 7.7 Two stop-loss transforms $\pi_X(d) = \mathrm{E}[(X-d)_+]$ and $\pi_Y(d)$ when $X \leq_{cx} Y$. Note that the asymptotes are equal.

$$
\begin{aligned}
\mathrm{E}[g(X)] &= \int_{-\infty}^{a} g(x)\,\mathrm{d}F(x) - \int_{a}^{\infty} g(x)\,\mathrm{d}\overline{F}(x) \\
&= -\int_{-\infty}^{a} g'(x)F(x)\,\mathrm{d}x + \int_{a}^{\infty} g'(x)\overline{F}(x)\,\mathrm{d}x \qquad (7.39) \\
&= \int_{-\infty}^{a} \mathrm{E}[(x-X)_+]\,\mathrm{d}g'(x) + \int_{a}^{\infty} \mathrm{E}[(X-x)_+]\,\mathrm{d}g'(x),
\end{aligned}
$$

from which the result immediately follows because since $f(\cdot)$ is convex, so is $g(\cdot)$, and therefore $\mathrm{d}g'(x) \geq 0$ for all x. $\qquad \nabla$

The stop-loss transforms $\mathrm{E}[(X-d)_+]$ of two random variables with equal mean μ have common asymptotes. One is the x-axis, the other the line $y = \mu - x$. Generalizing (3.115), it can be shown that $\int_{-\infty}^{\infty}\{\mathrm{E}[(X-t)_+] - (\mu-t)_+\}\,\mathrm{d}t = \frac{1}{2}\mathrm{Var}[X]$. Hence, just as for risks, the integrated difference between the stop-loss transforms of two arbitrary random variables with the same mean is half the difference in their variances. See Figure 7.7.

Consider some univariate cumulative distribution function F. It is well-known that if $U \sim \mathrm{uniform}(0,1)$, the random variable $F^{-1}(U)$ is distributed according to F (probability integral transform). Note that it is irrelevant how we define $y = F^{-1}(u)$ for arguments u where there is an ambiguity, that is, where $F(y) = u$ holds for an interval of y-values; see Section 5.6. Just as the cdf of a random variable can have only countably many jumps, it can be shown that there can only be countably many such horizontal segments. To see this, observe that in the interval $[-2^n, 2^n]$ there are only finitely many intervals with a length over 2^{-n} where $F(y)$ is constant, and let $n \to \infty$. Hence, if $g(\cdot)$ and $h(\cdot)$ are two different choices for the inverse cdf, $g(U)$ and $h(U)$ will be equal with probability one. The customary choice, as in Definition 5.6.1 of the VaR, is to take $F^{-1}(u)$ to be the left-hand endpoint of the interval of y-values (generally containing one point only) with $F(y) = u$. Then, $F^{-1}(\cdot)$ is nondecreasing and continuous from the left.

Now consider any random n-vector $(X_1,\ldots,X_n)$. Define a set in $\mathbb{R}^n$ to be *comonotonic* if each two vectors in it are ordered componentwise, that is, all components of the one are at least the corresponding components of the other. We will call a distribution comonotonic if its support is comonotonic. Also, any random vector having such a distribution is comonotonic. We have:

Theorem 7.6.3 (Comonotonic joint distribution)
For a random vector $\vec{X} := (X_1,\ldots,X_n)$, define a *comonotone equivalent* as follows:

$$\vec{Y} := (Y_1,\ldots,Y_n) = (F_{X_1}^{-1}(U),\ldots,F_{X_n}^{-1}(U)), \qquad (7.40)$$

where $U \sim \text{uniform}(0,1)$. It has the following properties:

1. $\vec{Y}$ has the same marginals as $\vec{X}$, so $Y_j \sim X_j$ for all j.
2. It has a comonotonic support.
3. Its joint cdf equals the so-called *Fréchet/Höffding* upper bound:

$$\Pr[Y_1 \le y_1,\ldots,Y_n \le y_n] = \min_{j=1,\ldots,n} \Pr[X_j \le y_j]. \qquad (7.41)$$

Proof. First, we have for all $j = 1,\ldots,n$:

$$\Pr[Y_j \le y_j] = \Pr[F_{X_j}^{-1}(U) \le y_j] = \Pr[U \le F_{X_j}(y_j)] = F_{X_j}(y_j). \qquad (7.42)$$

Next, the support of $(Y_1,\ldots,Y_n)$ is a curve $\{(F_{X_1}^{-1}(u),\ldots,F_{X_n}^{-1}(u)) \,|\, 0 < u < 1\}$ that increases in all its components. If $(y_1,\ldots,y_n)$ and $(z_1,\ldots,z_n)$ are two elements of it with $F_{X_i}^{-1}(u) = y_i < z_i = F_{X_i}^{-1}(v)$ for some i, then $u < v$ must hold, and hence $y_j \le z_j$ for *all* $j = 1,2,\ldots,n$.
Finally, we have

$$\begin{aligned}
\Pr[Y_1 \le y_1,\ldots,Y_n \le y_n] &= \Pr[F_{X_1}^{-1}(U) \le y_1,\ldots,F_{X_n}^{-1}(U) \le y_n] \\
&= \Pr[U \le F_{X_1}(y_1),\ldots,U \le F_{X_n}(y_n)] \\
&= \min_{j=1,\ldots,n} \Pr[X_j \le y_j],
\end{aligned} \qquad (7.43)$$

which proves the final assertion of the theorem. $\qquad\qquad\qquad\qquad \nabla$

The support S of $(Y_1,\ldots,Y_n)$ consists of a series of closed curves, see Figures 7.8 and 7.9, possibly containing just one point. Together they form a comonotonic set. By connecting the endpoints of consecutive curves by straight lines, we get the connected closure $\overline{S}$ of S; it is a continuous curve that is also comonotonic. Note that this has to be done only at discontinuities of one of the inverse cdfs in the components. The set $\overline{S}$ is a continuously increasing curve in $\mathbb{R}^n$.
Note that by (7.41), the joint cdf of $Y_1,\ldots,Y_n$, that is, the probability that all components have small values simultaneously, is as large as it can be without violating the marginal distributions; trivially, the right hand side of this equality is an upper bound for any joint cdf with the prescribed marginals. Also note that comonotonicity entails that no Y_j is in any way a hedge for another component Y_k, since in that

case, large values of one can never be compensated by small values of the other. In view of the remarks made in the introduction of this chapter, it is not surprising that the following theorem holds.

Theorem 7.6.4 (Comonotonic equivalent has the convex largest sum)
The random vector $(Y_1, \ldots, Y_n)$ in Theorem 7.6.3 has the following property:

$$Y_1 + \cdots + Y_n \geq_{cx} X_1 + \cdots + X_n. \tag{7.44}$$

Proof. Since the means of these two random variables are equal, it suffices to prove that the stop-loss premiums are ordered. The following holds for all $x_1, \ldots, x_n$ when $d_1 + \cdots + d_n = d$:

$$
\begin{aligned}
(x_1 + \cdots + x_n - d)_+ &= \{(x_1 - d_1) + \cdots + (x_n - d_n)\}_+ \\
&\leq \{(x_1 - d_1)_+ + \cdots + (x_n - d_n)_+\}_+ \\
&= (x_1 - d_1)_+ + \cdots + (x_n - d_n)_+.
\end{aligned}
\tag{7.45}
$$

Assume that d is such that $0 < \Pr[Y_1 + \cdots + Y_n \leq d] < 1$ holds; if not, the stop-loss premiums of $Y_1 + \cdots + Y_n$ and $X_1 + \cdots + X_n$ can be seen to be equal. The connected curve $\overline{S}$ containing the support S of the comonotonic random vector $(Y_1, \ldots, Y_n)$ points upwards in all coordinates, so it is obvious that $\overline{S}$ has exactly one point of intersection with the hyperplane $\{(x_1, \ldots, x_n) \mid x_1 + \cdots + x_n = d\}$. From now on, let $(d_1, \ldots, d_n)$ denote this point of intersection. In specific examples, it is easy to determine this point, but for now, we only need the fact that such a point exists. For all points $(y_1, \ldots, y_n)$ in the support S of $(Y_1, \ldots, Y_n)$, we have the following *equality:*

$$(y_1 + \cdots + y_n - d)_+ \equiv (y_1 - d_1)_+ + \cdots + (y_n - d_n)_+. \tag{7.46}$$

This is because for this particular choice of $(d_1, \ldots, d_n)$, by the comonotonicity, when $y_j > d_j$ for any j, we also have $y_k \geq d_k$ for all k; when all $y_j \leq d_j$, both sides of (7.46) are zero. Now replacing constants by the corresponding random variables in the two relations above and taking expectations, we get

$$
\begin{aligned}
E[(Y_1 + \cdots + Y_n - d)_+] &= E[(Y_1 - d_1)_+] + \cdots + E[(Y_n - d_n)_+] \\
&= E[(X_1 - d_1)_+] + \cdots + E[(X_n - d_n)_+] \\
&\geq E[(X_1 + \cdots + X_n - d)_+],
\end{aligned}
\tag{7.47}
$$

since X_j and Y_j have the same marginal distribution for all j. ∇

Example 7.6.5 (A three-dimensional continuous random vector)
Let $X \sim$ uniform on the set $(0, \frac{1}{2}) \cup (1, \frac{3}{2})$, $Y \sim \text{Beta}(2,2)$, $Z \sim \text{N}(0,1)$. The support of the comonotonic distribution is the set

$$\{(F_X^{-1}(u), F_Y^{-1}(u), F_Z^{-1}(u)) \mid 0 < u < 1\}. \tag{7.48}$$

See Figure 7.8. Actually, not all of the support is depicted. The part left out corresponds to $u \notin (\Phi(-2), \Phi(2))$ and extends along the asymptotes, that is, the vertical

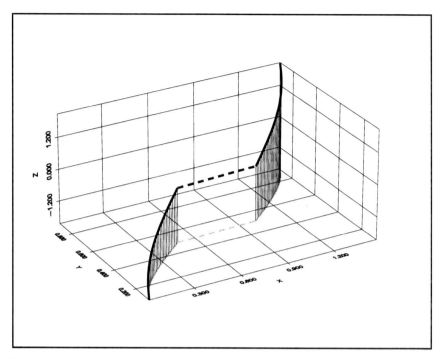

Fig. 7.8 Comonotonic support with (X,Y,Z) as in Example 7.6.5, and the marginal comonotonic support of (X,Y). The dotted lines serve to make the comonotonic support connected.

lines $(0,0,z)$ and $(\frac{3}{2},1,z)$. The thick continuous line is the support S, while the dotted line is the straight line needed to transform S into the connected curve $\overline{S}$. Note that $F_X(x)$ has a horizontal segment between $x = \frac{1}{2}$ and $x = 1$. The projection of $\overline{S}$ along the z-axis can also be seen to constitute an increasing curve, as do projections along the other axes. $\qquad\nabla$

Example 7.6.6 (A two-dimensional discrete example)
For a discrete example, take X uniform on $\{0,1,2,3\}$ and $Y \sim \text{binomial}(3,\frac{1}{2})$, so

$$\left(F_X^{-1}(u), F_Y^{-1}(u)\right) = \begin{cases} (0,0) & \text{for } 0 < u < \frac{1}{8}, \\ (0,1) & \text{for } \frac{1}{8} < u < \frac{2}{8}, \\ (1,1) & \text{for } \frac{2}{8} < u < \frac{4}{8}, \\ (2,2) & \text{for } \frac{4}{8} < u < \frac{6}{8}, \\ (3,2) & \text{for } \frac{6}{8} < u < \frac{7}{8}, \\ (3,3) & \text{for } \frac{7}{8} < u < 1. \end{cases} \qquad (7.49)$$

At the boundaries of the intervals for u, by convention we take the limit from the left. The points $(1,1)$ and $(2,2)$ have probability $\frac{1}{4}$, the other points of the support S of the comonotonic distribution have probability $\frac{1}{8}$. The curve $\overline{S}$ arises by simply

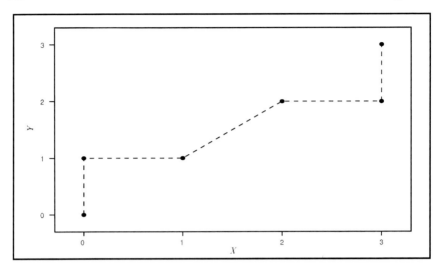

Fig. 7.9 Comonotonic support of (X,Y) as in Example 7.6.6; the dots represent the points with positive probability, the dotted lines connect the support.

connecting these points consecutively with straight lines, the dotted lines in Figure 7.9. The straight line connecting $(1,1)$ and $(2,2)$ is not along one of the axes. This is because at level $u = \frac{1}{2}$, both $F_X(y)$ and $F_Y(y)$ have horizontal segments. Any non-decreasing curve connecting $(1,1)$ and $(2,2)$ leads to a feasible $\overline{S}$.   ∇

Example 7.6.7 (Mortality risks of husband and wife)
Let $n = 2$, $X \sim$ Bernoulli(q_x), and $\frac{1}{2}Y \sim$ Bernoulli(q_y). Assume the risks to be dependent, and write $z = \Pr[X = 1, Y = 2]$. We can represent the joint distribution of (X,Y) as follows:

	$X = 0$	$X = 1$	Total
$Y = 0$	$1 - q_x - q_y + z$	$q_x - z$	$1 - q_y$
$Y = 2$	$q_y - z$	z	q_y
Total	$1 - q_x$	q_x	1

For each convex function $f(\cdot)$, we have $f(3) - f(2) \geq f(1) - f(0)$, so the following is non-decreasing in z:

$$E[f(X+Y)] = f(0)(1 - q_x - q_y) + f(1)q_x + f(2)q_y$$
$$+ [f(0) - f(1) - f(2) + f(3)]z. \tag{7.50}$$

Hence, one gets the maximal $X + Y$ in convex order by taking the z as large as possible, so $z = \min\{q_x, q_y\}$. Assume that $q_x < q_y$ holds, then we get:

	$X = 0$	$X = 1$	Total
$Y = 0$	$1 - q_y$	0	$1 - q_y$
$Y = 2$	$q_y - q_x$	q_x	q_y
Total	$1 - q_x$	q_x	1

The joint distribution can only be comonotonic if one or both of the events $X = 1$, $Y = 0$ and $X = 0$, $Y = 2$ have probability zero. In the comonotonic distribution for $q_x < q_y$, if $X = 1$ occurs, necessarily event $Y = 2$ occurs as well. If $q_x > q_y$, the situation is reversed.

This example describes the situation of life insurances on two lives, one male of age x and with amount at risk 1, and one female of age y with amount at risk 2. The comonotonic joint mortality pattern is such that if the person with the smaller mortality probability dies, so does the other. For $q_x = q_y$, we have $Y = 2X$ with probability one in case of comonotonic mortality.   ▽

7.7 Stochastic bounds on sums of dependent risks

Assume that we have to make payments equal to 1 at the end of each year for the coming n years. The interest is not fixed, but it varies randomly. We assume that the discount factor for a payment to be made at time k is equal to

$$X_k = e^{-(Y_1 + \cdots + Y_k)}, \tag{7.51}$$

where the annual logreturns Y_j are assumed to have, for example, a multinormal distribution such as in case the returns follow a Brownian motion. Then X_k is lognormal, and the total present value $S = \sum X_k$ of all payments is the sum of dependent lognormal random variables. It is not easy to handle such random variables analytically. Though the dependency structure is fully known, it is too cumbersome to use fruitfully. In view of Theorem 7.6.4, however, it is easy to find an upper bound in convex order for S that is easy to handle. It suffices to replace the random vector $(X_1, X_2, \dots)$ by its comonotonic equivalent with components $X_k^U = F_{X_k}^{-1}(U)$. For such an approximation by a comonotonic sum to be good, the dependence between the terms should be strong to begin with. To see why this might well be the case here, note that for $k < l$ we have $X_l = X_k \exp\left(-(Y_{k+1} + \cdots + Y_l)\right)$. So for $k < l$ large, X_k and X_l have a large correlation in many random interest term structures.

7.7.1 Sharper upper and lower bounds derived from a surrogate

Sometimes the information known about the dependencies allows us to derive sharper upper and lower bounds (in convex order) for a sum of random variables.

In view of Corollary 7.3.16, a convex lower bound for $S = X_1 + \cdots + X_n$ is $\mathrm{E}[S \mid V]$, for any random variable V. If $\mathrm{E}[S \mid V]$ is easy to compute and V is a good surrogate for S in the sense that it explains much of the variation in S, we have found a useful improved lower bound $\mathrm{E}[S] \leq_{\mathrm{cx}} \mathrm{E}[S \mid V] \leq_{\mathrm{cx}} S$. By a similar technique, we can also improve on the comonotonic upper bound.

Theorem 7.7.1 (Sharper lower and upper bounds)
Let $(X_1, \ldots, X_n)$ be a random n-vector and S the sum of its components. For some $U \sim \mathrm{uniform}(0,1)$, write the sum of the components of the comonotone equivalent of $(X_1, \ldots, X_n)$ as $S^U = F_{X_1}^{-1}(U) + \cdots + F_{X_n}^{-1}(U)$. For any V, write $S^{U \mid V} = h(U,V)$ with

$$h(u,v) \stackrel{\text{def}}{=} F_{X_1 \mid V=v}^{-1}(u) + \cdots + F_{X_n \mid V=v}^{-1}(u), \quad 0 < u < 1. \tag{7.52}$$

Then the following stochastic inequalities hold:

$$\mathrm{E}[S] \leq_{\mathrm{cx}} \mathrm{E}[S \mid V] \leq_{\mathrm{cx}} S \leq_{\mathrm{cx}} S^{U \mid V} \leq_{\mathrm{cx}} S^U. \tag{7.53}$$

Proof. For all v we have

$$\mathrm{E}[S \mid V = v] \leq_{\mathrm{cx}} S \mid V = v \leq_{\mathrm{cx}} h(U,v), \tag{7.54}$$

where $S \mid V = v$ denotes a random variable that has as its distribution the conditional distribution of S given $V = v$. This random variable has mean $\mathrm{E}[S \mid V = v]$. The random variable $h(U,v)$ is the sum of the comonotonic versions of $X_k \mid V = v$.

Multiplying the cdfs of the random variables in (7.54) by the differential $\mathrm{d}F_V(v)$ and integrating gives the following lower and upper bounds (see Theorem 7.3.15):

$$F_{\mathrm{E}[S]} \leq_{\mathrm{cx}} F_{\mathrm{E}[S \mid V]} \leq_{\mathrm{cx}} F_S \leq_{\mathrm{cx}} F_{S^{U \mid V}} \leq_{\mathrm{cx}} F_{S^U}. \tag{7.55}$$

The first cdf is the degenerate cdf of the constant $\mathrm{E}[S]$, the second is the one of $\mathrm{E}[S \mid V]$. The improved upper bound is the cdf of $S^{U \mid V} = h(U,V)$. This is a sum of random variables with the prescribed marginals, hence it is lower than the comonotonic upper bound, which is the convex largest sum with these marginal distributions. See Exercise 7.7.4.   ▽

For these bounds to be useful, the distribution of the surrogate V must be known, as well as all the conditional distributions of X_k, given $V = v$. A structure variable such as one encounters in credibility contexts is a good example for V. How to choose V best will be discussed below.

Example 7.7.2 (Stochastic bounds for a sum of two lognormal risks)
We illustrate the technique of conditioning on the value of a structure random variable by looking at the sum of two lognormal random variables. The multinormal distribution is very useful in this context, because conditional and marginal distributions are easily derived. Let $n = 2$, and take Y_1, Y_2 to be independent $N(0,1)$ random variables. Look at the sum $S = X_1 + X_2$ where $X_1 = e^{Y_1} \sim \mathrm{lognormal}(0,1)$, and $X_2 = e^{Y_1 + Y_2} \sim \mathrm{lognormal}(0,2)$. For the surrogate V, we take a linear combination of Y_1, Y_2, in this case $V = Y_1 + Y_2$. For the lower bound $\mathrm{E}[S \mid V]$, note that $\mathrm{E}[X_2 \mid V] = e^V$,

while $Y_1 | Y_1 + Y_2 = v \sim N(\frac{1}{2}v, \frac{1}{2})$, and hence

$$E[e^{Y_1} | Y_1 + Y_2 = v] = m(1; \frac{1}{2}v, \frac{1}{2}), \tag{7.56}$$

where $m(t; \mu, \sigma^2) = e^{\mu t + \frac{1}{2}\sigma^2 t^2}$ is the $N(\mu, \sigma^2)$ moment generating function. This leads to

$$E[e^{Y_1} | V] = e^{\frac{1}{2}V + \frac{1}{4}}. \tag{7.57}$$

So the sharper lower bound in (7.53) is

$$E[X_1 + X_2 | V] = e^{\frac{1}{2}V + \frac{1}{4}} + e^V. \tag{7.58}$$

When $U \sim$ uniform$(0, 1)$, the random variable $S^U = e^W + e^{\sqrt{2}W}$ with $W = \Phi^{-1}(U)$, so $W \sim N(0, 1)$, is a comonotonic upper bound for S. The improved upper bound $S^{U|V}$ as in (7.53) has as its second term again e^V. The first term equals $e^{\frac{1}{2}V + \frac{1}{2}\sqrt{2}W}$, with V and W mutually independent, $V \sim N(0, 2)$ and $W \sim N(0, 1)$. All terms occurring in these bounds are lognormal random variables, so the variances of the bounds are easy to compute.

Note that to compare variances is meaningful when comparing stop-loss premiums of convex ordered random variables. This is because half the variance difference between two convex ordered random variables equals the integrated difference of their stop-loss premiums, see for example Figure 7.7. This implies that if $X \leq_{cx} Y$ and in addition $Var[X] = Var[Y]$, then X and Y must necessary be equal in distribution. Moreover, the ratio of the variances for random variables with the same mean is roughly equal to the ratio of the stop-loss premiums, less their minimal possible value, see Rule of thumb 3.10.6. We have, as the reader may verify,

$$
\begin{aligned}
(E[S])^2 &= e^1 + 2e^{\frac{3}{2}} + e^2, \\
E[(E[S|V])^2] &= e^{\frac{3}{2}} + 2e^{\frac{5}{2}} + e^4, \\
E[S^2] = E[(S^{U|V})^2] &= e^2 + 2e^{\frac{5}{2}} + e^4, \\
E[(S^U)^2] &= e^2 + 2e^{\frac{3}{2}+\sqrt{2}} + e^4.
\end{aligned} \tag{7.59}
$$

Hence,

$$
\begin{aligned}
Var[E[S]] &= 0, \\
Var[E[S|V]] &= 64.374, \\
Var[S] = Var[S^{U|V}] &= 67.281, \\
Var[S^U] &= 79.785.
\end{aligned} \tag{7.60}
$$

So a stochastic lower bound $E[S|V]$ for S, much better than just $E[S]$, is obtained by conditioning on $V = Y_1 + Y_2$, and the improved upper bound $S^{U|V}$ has in fact the same distribution as S. It can be proved that for all pairs of random variables, the distributions of $S^{U|V}$ and S coincide when one uses one of these variables as a surrogate. See Exercise 7.6.22.

Since $\text{Var}[S] = \text{E}\big[\text{Var}[S\,|\,V]\big] + \text{Var}\big[\text{E}[S\,|\,V]\big]$, the variance of the lower bound is just the component of the variance of S left 'unexplained' by V. To maximize the second term is to minimize the first, so we look for a V that represents S as faithfully as possible. Approximating e^{Y_1} and $e^{Y_1+Y_2}$ by $1+Y_1$ and $1+Y_1+Y_2$ respectively, we see that $S \approx 2+2Y_1+Y_2$, hence taking $2Y_1+Y_2$ instead of Y_1+Y_2 as our conditioning random variable might lead to a better lower bound. It is left as Exercise 7.7.11 to check whether this is indeed the case. In Exercise 7.7.12, an even better choice is discussed.   ▽

7.7.2 Simulating stochastic bounds for sums of lognormal risks

The random variable X_k in (7.51) has a lognormal(v, τ^2) distribution with $v = -\text{E}[Y_1 + \cdots + Y_k]$, $\tau^2 = \text{Var}[Y_1 + \cdots + Y_k]$. Assuming payments $\alpha_1, \ldots, \alpha_n$ must be made at the end of the year, the total payments discounted to time 0 amounts to

$$S = \alpha_1 X_1 + \cdots + \alpha_n X_n. \tag{7.61}$$

Many interest models lead to $\vec{Y} = (Y_1, \ldots, Y_n)'$ having a multivariate normal distribution, say with mean vector $\vec{\mu}_Y$ and variance matrix Σ_Y. A simple model is for example to assume that the annual log-returns Y_k are iid $N(\mu, \sigma^2)$ random variables.

In case all $\alpha_k \geq 0$, the random variable S can be seen to be a sum of dependent lognormal random variables, since $\alpha_k X_k$ is then again lognormal. In case of mixed signs, we have the difference of two such sums of lognormals. Random variables of this type occur quite frequently in financial contexts. Even for the simplest covariance structures, it is in general impossible to compute their distribution function analytically. To handle this situation, we can do Monte Carlo simulations, simulating outcomes of S to draw a histogram, or to estimate, for example, the 99.75% percentile. As only one in 400 simulations will exceed the 99.75% percentile, to get a reasonably precise estimate of it requires many thousands of simulations. Another approach is to approximate the distribution of S by some suitable parametric distribution, like the normal, the lognormal, the gamma or the inverse gamma, which is simply the distribution of $1/X$ when $X \sim$ gamma. By adding a shift parameter such as is done in Section 2.5 for the gamma distribution, we get a much better fitting cdf, because three moments can be fitted. See the exercises in this section. A third approach is to look at upper or lower bounds, in convex order, for S. Since these bounds turn out to be comonotonic, exact expressions for the inverse cdfs (VaRs) and other risk measures for these bounds can be given, thus avoiding the need for time-consuming simulations. But here we will look at sample cdfs of these bounds, and compare them to a sample cdf of S. Since the terms in S are dependent lognormal random variables, we start by showing how to generate multivariate normal random vectors, generalizing the technique used in Appendix A.2.

Generating random drawings from normal distributions It is easy to generate a stream of pseudo-random numbers from a univariate normal distribution, simply invoking the R-function `rnorm`. There also exists a function to sample from the multivariate normal distribution with mean vector $\vec{\mu}$ and arbitrary variance matrix Σ. It is the function mvrnorm, to be found in the library MASS of R that consists of functions and datasets from the main package of Venables & Ripley (2002). To generate a multinormal random vector $\vec{Y}$, mvrnorm uses the fact that if $V_1, \ldots, V_n$ are iid $N(0,1)$ random variables and $\mathbf{A}$ is an $n \times n$ matrix, then $\vec{W} = \vec{\mu} + \mathbf{A}\vec{V}$ is multivariate normal with mean $E[\vec{W}] = \vec{\mu}$ and covariances $Cov[W_i, W_j] = E[(W_i - \mu_i)(W_j - \mu_j)]$ being the (i,j) element of the variance matrix. By the linearity property of expectations, this covariance matrix is equal to

$$E[(\vec{W} - \vec{\mu})(\vec{W} - \vec{\mu})'] = E[\mathbf{A}\vec{V}(\mathbf{A}\vec{V})'] = E[\mathbf{A}\vec{V}\vec{V}'\mathbf{A}']$$
$$= \mathbf{A}E[\vec{V}\vec{V}']\mathbf{A}' = \mathbf{A}\mathbf{I}\mathbf{A}' = \mathbf{A}\mathbf{A}'. \tag{7.62}$$

So to generate $N(\vec{\mu}, \Sigma)$ random vectors $\vec{W}$ all we have to do is find any 'square root' $\mathbf{A}$ of matrix Σ, with $\Sigma = \mathbf{A}\mathbf{A}'$. Such square roots can easily be found. One is the Cholesky decomposition. It has the particular property that it produces a lower triangular matrix, making computations by hand easier. The function mvrnorm in fact uses the eigenvalue decomposition, by decomposing $\Sigma = \mathbf{R}\Lambda\mathbf{R}'$ with $\mathbf{R}$ an orthogonal matrix of eigenvectors and Λ the diagonal matrix of eigenvalues of Σ. All eigenvalues are positive when Σ is positive definite. Then $\mathbf{R}\Lambda^{1/2}$ is also a 'square root' of Σ.

Drawing from S To simulate 4000 outcomes of the sum of lognormals S above, with $n = 20$ payments $\alpha_k \equiv +1$ and iid $N(20\%, (13.5\%)^2)$ annual log-returns Y_k, do:

```
nsim <- 4000; n <- 20; alpha <- rep(1, n)
mu.Y <- rep(20/100,n); Sigma.Y <- diag((13.5/100)^2, n)
L <- lower.tri(Sigma.Y, diag=TRUE)
    ## TRUE (=1) below, and on, diagonal
mu.Z <- as.vector(L %*% mu.Y)        ## means of (Z1,..Zn)
Sigma.Z <- L %*% Sigma.Y %*% t(L) ## and variance matrix
library(MASS)
S <- exp(-mvrnorm(nsim, mu.Z, Sigma.Z)) %*% alpha
```

Note that the function `rnorm` uses the *standard deviation* as a parameter, while mvrnorm uses the variance matrix, which has marginal *variances* on its diagonal. Matrix L is lower triangular with ones (actually, TRUE) on and below the diagonal, so $\vec{Z} = \mathbf{L}\vec{Y} = (Y_1, Y_1 + Y_2, \ldots, Y_1 + \cdots + Y_n)'$ has variance matrix Sigma.Z and mean vector mu.Z. To compute the total discounted value of the payments, we draw $\vec{Z}$ from a multinormal distribution, take exponents to obtain lognormal random variables X_k, multiply these with α_k and take the total for all $k = 1, \ldots, n$.

Drawing from S^U A comonotonic upper bound (in the sense of convex order) S^U for S arises by replacing the terms $\alpha_k X_k$ of S by their comonotone equivalents, hence by assuming $Z_k^U = a_k + b_k W \text{sign}(\alpha_k)$ for a_k and b_k such that $E[Z_k^U] = E[Z_k]$ and $\text{Var}[Z_k^U] = \text{Var}[Z_k]$, and with $W = \Phi^{-1}(U)$ a standard normal variate (the same one for all k). So for the iid annual log-returns case, $a_k = k$ and $b_k = \sqrt{k}$ must be chosen. We will further assume that all $\alpha_k \geq 0$ (it is left to the reader to make the necessary adjustments in case of mixed signs, using R's `sign` function). The following lines produce a sample of the comonotonic upper bound:

```
w <- rnorm(nsim) %o% sqrt(diag(Sigma.Z)) + rep(1,nsim) %o% mu.Z
S.upp <- exp(-w) %*% alpha
```

The operation `%o%` for two vectors denotes the *outer product* $\mathbf{A} = \vec{a} \otimes \vec{b} = \vec{a}\vec{b}'$, so $A_{ij} = a_i b_j$ for all i and j. See Appendix A.

Replacing random variables by variables with correlation 1 makes sense for the discounted payments $\alpha_k X_k$ since they are positively dependent, and increasingly so as k grows. In case of iid annual log-returns for example, we have for $i \leq j$, see Exercise 7.7.8:

$$r(\log(\alpha_i X_i), \log(\alpha_j X_j)) = r(Z_i, Z_j) = \frac{\sqrt{i}}{\sqrt{j}}. \tag{7.63}$$

Drawing from $E[S|V]$ For a comonotonic lower bound (again in the sense of convex order) for the cdf of S, we use the conditional mean $E[S|V]$. As argued at the end of Example 7.7.2, our conditioning variable V must have a simple dependence structure with the components of S. Also, it must behave like S in the sense that $E\big[\text{Var}[S|V]\big]$ is small, or equivalently, $\text{Var}\big[E[S|V]\big]$ is large. We take $\alpha_k \equiv 1$, with n arbitrary, and $V = \vec{b}'\vec{Y} = b_1 Y_1 + \cdots + b_n Y_n$ for some vector $\vec{b}$. Initially we let $b_k \equiv 1$ like in Example 7.7.2, so $V = Z_n$, but we study better choices in the exercises.

To compute the terms of $E[S|V]$, we need the conditional distribution of one component of a bivariate normal random pair, given the value of the other. It is well-known, see for example Theorem 5.4.8 of Bain & Engelhardt (1992), that if (U,V) has a bivariate normal distribution with parameters $\mu_U, \mu_V, \sigma_U^2, \sigma_V^2, \rho$, then for any v, the following holds for the conditional distribution of U, given $V = v$:

$$U|V = v \quad \sim \quad N\Big(\mu_U + \rho \frac{\sigma_U}{\sigma_V}(v - \mu_V), \sigma_U^2(1 - \rho^2)\Big). \tag{7.64}$$

For some fixed $k \in \{1,\ldots,n\}$, let $\vec{a} = (1,\ldots,1,0,\ldots,0)'$, with k ones and $n - k$ zeros. Since $\vec{Y}$ has a multivariate normal distribution, the pair $(\vec{a}'\vec{Y}, \vec{b}'\vec{Y}) = (Y_1 + \cdots + Y_k, b_1 Y_1 + \cdots + b_n Y_n)$ is bivariate normal. The means are $\vec{a}'\vec{\mu} = \mu_1 + \cdots + \mu_k$ and $b'\vec{\mu}$. To determine the covariance matrix we write

$$\mathbf{B} = \begin{pmatrix} a_1 & a_2 & \cdots & a_n \\ b_1 & b_2 & \cdots & b_n \end{pmatrix} \quad \text{so} \quad \mathbf{B}\vec{Y} = \begin{pmatrix} \vec{a}'\vec{Y} \\ \vec{b}'\vec{Y} \end{pmatrix}. \tag{7.65}$$

Then, see (7.62), we have $\Omega = \mathbf{B}\Sigma\mathbf{B}'$ for the variance matrix of $\mathbf{B}\vec{Y}$. For the kth discount factor X_k, conditionally given $V = v$, we have, writing $m(t; \mu, \sigma^2) = \exp(\mu t + \sigma^2 t^2 / 2)$ for the $N(\mu, \sigma^2)$ mgf:

$$E[X_k | V = v] = E[e^{-Z_k} | V = v]$$
$$= m(-1; E[Z_k | V = v], \text{Var}[Z_k | V = v]). \tag{7.66}$$

It is often possible to compute the quantiles and the cdf of $f(V) = E[S | V]$ explicitly, using $\Pr[f(V) \leq v] = \Pr[V \leq f^{-1}(v)]$. This is the case if, such as in this example with all $\alpha_k > 0$ and $V = Z_n$, the function $f(v) = E[S | V = v]$ is monotone increasing in v; see Exercise 7.7.15. We confine ourselves here to approximating the cdf through simulation. First, we compute the means and the covariance matrix Ω of $C\vec{Y} = (Z_1, \ldots, Z_n, V)$. From this, the marginal variances and the correlations follow.

```
b <- rep(1,n) ## for the choice V = b'Y = Y1+..Yn
C <- rbind(L,b)
mu.Zk <- (C %*% mu.Y)[1:n]; mu.V <- (C %*% mu.Y)[n+1]
Omega <- C %*% Sigma.Y %*% t(C)
sd.Zk <- sqrt(diag(Omega))[1:n]; sd.V <- sqrt(diag(Omega)[n+1])
rho.ZkV <- pmin(cov2cor(Omega)[n+1,1:n],1)
```

The call of `pmin` in the last line proved necessary because sometimes correlations numerically larger than 1 resulted.

Next, we fill `EZkV` with `nsim` conditional means $E[Z_k | V]$, $k = 1, 2, \ldots, n$, using (7.64). The vector `VarZkV` holds the conditional variances. Finally each element of `S.low` is computed as $\sum_k \alpha_k E[X_k | V = v]$, using (7.66). Note that `alpha` as well as `mu.Zk`, `sd.Zk` and `rho.ZkV` are vectors of length n.

```
V <- rnorm(nsim, mu.V, sd.V)
EZkV <- rep(1,nsim) %o% mu.Zk + ## nsim x n matrix
        (V-mu.V) %o% (rho.ZkV * sd.Zk / sd.V)
VarZkV <- sd.Zk^2*(1-rho.ZkV^2) ## n vector
S.low <- exp(-EZkV + 0.5 * rep(1,nsim) %o% VarZkV) %*% alpha
```

Drawing from $S^{U|V}$ In $S^{U|V=v}$, the terms are comonotone equivalents, given $V = v$, of $\alpha_k X_k$ with $X_k \sim$ lognormal(μ_k, σ_k^2). The parameters are $\mu_k = E[Z_k | V = v]$ and $\sigma_k^2 = \text{Var}[Z_k | V = v]$. So drawings from this random variable $S^{U|V=v}$ can be generated by adding σ_k times a standard normal random variable to $E[Z_k | V = v]$, taking the exponent and then the sum, weighted by α_k.

```
ZkV <- EZkV + rnorm(nsim) %o% sqrt(VarZkV)
S.imp <- exp(-ZkV) %*% alpha
```

Plotting the results To plot the empirical cdfs of S, the comonotonic upper bound S^U, the improved upper bound $S^{U|V}$ and the lower bound $E[S | V]$ for the choice $V = Z_n = Y_1 + \cdots + Y_n$, we do:

```
y <- (-0.5+1:nsim)/nsim
plot(sort(S), y, type="l", yaxp=c(0,1,1), xaxp=c(2,10,4),
     ylab="", xlab="", lwd=1.5)
lines(sort(S.upp), y, lty="solid")
lines(sort(S.imp), y, lty="dashed")
lines(sort(S.low), y, lty="dotted")
mu.S <- sum(alpha*exp(-(mu.Z-diag(Sigma.Z)/2)))
lines(c(min(S.upp),mu.S,mu.S,max(S.upp)), c(0,0,1,1))
```

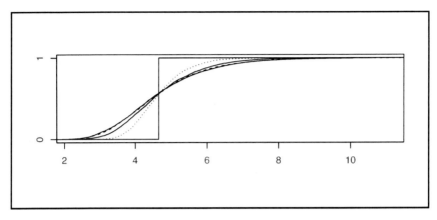

Fig. 7.10 Empirical cdfs of S (thick) and stochastic bounds S.low (dotted), S.imp (dashed) and S.upp (solid)

For the result, see Figure 7.10. The final call of `lines` adds the plot of the cdf of the random variable that is degenerate on $E[S]$, which is a trivial lower bound for S in the convex order sense. It has jump of 1 at the mean $E[S]$ of S. Actually, it is a special case of the class of lower bounds $E[S|V]$ for different V, since $E[S|V] \equiv E[S]$ if V is independent of S.

As one sees, the improved upper bound $S^{U|V}$ is not much better in this case than the upper bound S^U. We chose here $V = Z_n$, the accumulated log-returns at the end. Some better choices for surrogates can be found in the exercises.

7.8 More related joint distributions; copulas

Throughout this section, we assume that all random variables $X, X', X^{\bullet}, X^{\perp}$ and X^U have the same continuous marginal cdf F, and all corresponding random variables Y have marginal cdf G, also continuous. The joint cdfs will be denoted by H, adorned correspondingly with an upper index.

7.8.1 More related distributions; association measures

We have seen that two random variables are maximally related if their joint distribution is comonotonic, hence, by Theorem 7.6.3, if their joint cdf is as large as possible. Following this idea, we define a partial order between pairs of random variables having the same marginals; a pair with a uniformly larger joint cdf will be called 'more related'.

Definition 7.8.1 (More related joint distributions)

The pair of random variables (X, Y) is *more related* than (X', Y') if $X \sim X'$ and $Y \sim Y'$, as well as

$$H(x, y) := F_{X,Y}(x, y) \geq F_{X',Y'}(x, y) =: H'(x, y) \text{ for all } x \text{ and } y. \tag{7.67}$$

So the probability that X and Y are both small is larger than this probability for X' and Y'.   ▽

From (7.67) we see that $\Pr[X > x, Y > y] = 1 - F(x) - G(y) + H(x, y)$ (both 'large') is also larger than the corresponding probability for (X', Y').

With independence acting as 'zero dependence', we can introduce positive dependence as follows.

Definition 7.8.2 (Positive quadrant dependence)

X and Y are PQD, which is short for *positive quadrant dependent*, if

$$\Pr[X \leq x, Y \leq y] \geq \Pr[X \leq x] \Pr[Y \leq y] \text{ for all } x \text{ and } y. \tag{7.68}$$

So any pair that is more related than an independent pair $(X^{\perp}, Y^{\perp})$ with the same marginal distributions is PQD.   ▽

Since by Theorem 7.6.3, $H(x, y) \leq \min\{F(x), G(y)\}$ holds, a comonotonic pair (X^U, Y^U) is 'most related'. There is also a joint distribution with the right marginals that is 'least related', or 'most antithetic'. It is derived from the following lower bound for the joint cdf, also studied by Fréchet/Höffding:

$$\Pr[X \leq x, Y \leq y] \geq \max\{0, F(x) + G(y) - 1\}. \tag{7.69}$$

This inequality follows directly from Bonferroni's inequality, see Exercise 7.6.8. The pair $(X, Y) = (F^{-1}(U), G^{-1}(1 - U))$ has this cdf; Y is small when X is large and vice versa. In fact, in this case X and $-Y$ are most related; X and Y are not comonotonic, but countermonotonic.

A different approach to comparing pairs of random variables as regards degree of relatedness is to simply compare their values of well-known association measures.

Example 7.8.3 (Association measures)

The most widely used association measures for pairs of random variables are:

- Pearson's r: the correlation coefficient $r(X, Y) = \frac{E[XY] - E[X]E[Y]}{\sigma_X \sigma_Y}$;
- Spearman's ρ: the rank correlation coefficient $\rho(X, Y) = r(F(X), G(Y))$;
- Kendall's τ: the (normed) probability of concordance with another sample pair, so $\tau(X, Y) = 2 \times \Pr[(X - X^\bullet)(Y - Y^\bullet) > 0] - 1$ if $(X, Y) \sim (X^\bullet, Y^\bullet)$ independent;
- Blomqvist's β: looks at concordance with the medians $\tilde{x}$ and $\tilde{y}$ having $F(\tilde{x}) = G(\tilde{y}) = 0.5$, so $\beta(X, Y) = 2 \times \Pr[(X - \tilde{x})(Y - \tilde{y}) > 0] - 1$.

The *rank* of an element in a random sample is simply its index in the sorted sample. The population equivalent is the rank of a random variable $X \sim F$, which is simply the random variable $F(X)$. This r.v. is uniform$(0, 1)$ if X is continuous. If for some x

we have $F(x) = 0.2$, it means that in large samples, the rank of an observation equal to x will be about 20% of the sample size. Two pairs of real numbers are concordant if one lies to the right and above the other.

By using $2\Pr[\text{"concordance"}] - 1$ in both last measures, we ensure that all these association measures have the value zero in case of independence, and are in the interval $[-1, +1]$. Note that (X,Y) being comonotonic implies $\rho(X,Y) = \tau(X,Y) = \beta(X,Y) = 1$ (and v.v. for ρ and τ), but $r(X,Y) < 1$ is possible, see Exercises 7.8.4–5. The last three association measures depend only on the joint cdf of the ranks $F(X)$ and $G(Y)$, and they are the same for the original random variables and their ranks. But the Pearson correlation r does not have this property; see Exercises 7.8.5–6. Just like mean-variance order compared with stop-loss order, comparing the values of association measures has the advantage of leading to a total order between pairs of risks with the same marginals. $\quad\triangledown$

An important property of the concept of being 'more related' is that the sum of the components of the more related pair is larger in convex order.

Property 7.8.4 (More related means having a convex larger sum)
If $(X,Y) \sim H$ is more related than $(X',Y') \sim H'$, then $X + Y \geq_{cx} X' + Y'$.

Proof. This can be inferred from combining the equality $E[(X + Y - d)_+] = E[(d - X - Y)_+] + E[X] + E[Y] - d$ with the following one, derived by swapping the order of integration (Fubini):

$$E[(d - X - Y)_+] = \iint_{x+y\leq d} \int_{t=y}^{d-x} dt \, dH(x,y)$$
$$= \int_{t=-\infty}^{\infty} \iint_{y\leq t,\, x\leq d-t} dH(x,y)\, dt \qquad (7.70)$$
$$= \int_{t=-\infty}^{\infty} H(t, d-t)\, dt.$$

Now use $H \geq H'$ to finish the proof. $\quad\triangledown$

The property of being more related is reflected in the association measures, in the sense that the more related pair has a larger value for all association measures mentioned in Example 7.8.3.

Property 7.8.5 (More related means higher association)
If $(X,Y) \sim H$ is more related than $(X',Y') \sim H'$, then $r(X,Y) \geq r(X',Y')$. The same holds for $\rho(\cdot,\cdot)$, $\tau(\cdot,\cdot)$ and $\beta(\cdot,\cdot)$.

Proof. By Property 7.8.4, $X + Y \geq_{cx} X' + Y'$, so $\text{Var}[X+Y] \geq \text{Var}[X'+Y']$. This means that $\text{Cov}[X,Y] \geq \text{Cov}[X',Y']$, hence $r(X,Y) \geq r(X',Y')$.

An alternative proof of this uses a relation that parallels $E[X] = \int_0^\infty \Pr[X > x]\,dx$, see (1.33). For convenience assume that X and Y are both bounded from below, say, $\geq -M$. By proceeding with $X + M$ and $Y + M$ we can assume $M = 0$ without further loss of generality. Then applying Fubini (for example $0 < t < x < \infty$):

$$E[XY] = \iint xy \, dH(x,y) = \iint \left(\int_0^x dt \int_0^y du \right) dH(x,y)$$

$$= \int_0^\infty \int_0^\infty \left(\int_t^\infty \int_u^\infty dH(x,y) \right) dt \, du \qquad (7.71)$$

$$= \int_0^\infty \int_0^\infty \{ 1 - F(t) - G(u) + H(t,u) \} dt \, du,$$

so $E[XY] \geq E[X'Y']$ follows. Since the marginal distributions of (X,Y) and (X',Y') are the same, so are their marginal moments, hence $r(X,Y) \geq r(X',Y')$.

Write $U = F(X)$, $V = G(Y)$, $S = F(X')$ and $T = G(Y')$ for the ranks. Then (U,V) is more related than (S,T) as $\Pr[U \leq u, V \leq v] = H(F^{-1}(u), G^{-1}(v))$. So $\rho(X,Y) \geq \rho(X',Y')$ by the above.

For Kendall's τ, let $C(\cdot,\cdot)$ be the joint cdf of (U,V). Let $(X^\bullet, Y^\bullet) \sim (X,Y)$ independent, with ranks $(U^\bullet, V^\bullet)$. Then

$$\Pr[(X - X^\bullet)(Y - Y^\bullet) > 0] = \Pr[(U - U^\bullet)(V - V^\bullet) > 0]$$

$$= 2 \iint \Pr[U < u, V < v \mid U^\bullet = u, V^\bullet = v] \, dC(u,v) \qquad (7.72)$$

$$= 2 \iint \Pr[U < u, V < v] \, dC(u,v) = 2 \iint C(u,v) \, dC(u,v),$$

so $\tau(X,Y) = 4E[C(U,V)] - 1$.

Now let again (S,T) denote the ranks with (X',Y'), and denote their joint cdf by D. Then also $C \geq D$ holds, so $E[C(S,T)] \geq E[D(S,T)]$. So it suffices to prove that $E[C(S,T)] \leq E[C(U,V)]$. We have

$$E[C(S,T)] - E[C(U,V)] = \int_0^1 \int_0^1 C(u,v) \, dD(u,v) - \int_0^1 \int_0^1 C(u,v) \, dC(u,v). \quad (7.73)$$

Using $C(u,v) = \int_0^u \int_0^v dC(x,y)$ and Fubini, for example with $0 < x < u < 1$, we can transform (7.73) into:

$$\int_0^1 \int_0^1 \int_0^u \int_0^v dC(x,y) \, dD(u,v) - \int_0^1 \int_0^1 \int_0^u \int_0^v dC(x,y) \, dC(u,v)$$

$$= \int_0^1 \int_0^1 \int_x^1 \int_y^1 dD(u,v) \, dC(x,y) - \int_0^1 \int_0^1 \int_x^1 \int_y^1 dC(u,v) \, dC(x,y) \qquad (7.74)$$

$$= \iint \{ \Pr[S > x, T > y] - \Pr[U > x, V > y] \} \, dC(x,y)$$

$$= \iint \{ D(x,y) - C(x,y) \} \, dC(x,y) \leq 0.$$

For the final equality, use that, for example, $\Pr[S > x, T > y] = 1 - x - y + D(x,y)$. From (7.74) we conclude that $\tau(X,Y) \geq \tau(X',Y')$.

Finally, Blomqvist's β is just $4C(\frac{1}{2}, \frac{1}{2}) - 1$, so $\beta(X,Y) \geq \beta(X',Y')$ is trivial. $\quad \nabla$

See also the exercises for some characteristics of the PQD property. In particular, as one would expect, the pair (X,X) is PQD, as well as, when X, Y and Z are independent, the pairs $(X,X+Z)$ and $(X+Y,X+Z)$. The concept can also be generalized to dimension $n > 2$.

7.8.2 Copulas

Consider a two-dimensional random vector (X,Y) with joint distribution $H(x,y)$. The marginals are assumed given and continuous, and again written as $H(x,\infty) = F(x)$ and $H(\infty,y) = G(y)$.

Definition 7.8.6 (Copula)
The *copula* of a pair (X,Y) is the joint cdf of the ranks $F(X)$ and $G(Y)$.   ∇

The copula was already used in the proof of Property 7.8.5. It couples marginal and joint cdfs in the following way: $H(x,y) = C(F(x),G(y))$. The well-known Sklar's theorem states that the copula function satisfying this relation is unique also in case the marginals are not continuous. To compute the last three association measures in Example 7.8.3, it suffices to know the copula rather than the whole joint cdf.

The cdf $C(u,v)$ is a two-dimensional cdf with uniform$(0,1)$ marginals, hence it has support $0 < u < 1, 0 < v < 1$, and $C(u,1) = u$ as well as $C(1,v) = v$. If a pair is more related than another, the same holds for the corresponding ranks. Among the families of copulas in common use are the Student, Gaussian, Archimedean, Clayton, Frank, Gumbel, Farlie-Gumbel-Morgenstern (see Exercise 7.8.3) and many others. One of the uses of copulas is that they provide a means to draw a random pair with prescribed marginals and such that the distribution from which we sample resembles the one that generated our observations. An important criterion, especially when the objective is to model extremal events, for that is that the joint tail behavior fits the observations. So the conditional probability that one risk is large, given that the other is, should be fitted. We will not pursue this further, and only present a method to construct a copula in such a way that a random pair with this copula, apart from the right marginals, also has a prescribed rank correlation. For this purpose, it suffices to look at the following important special cases of copula functions:

1. Comonotonic or Maximum copula: $C_1(u,v) = \min\{u,v\}$;
2. Independence or Product copula: $C_2(u,v) = uv, \quad 0 < u < 1, 0 < v < 1$;
3. Countermonotonic or Minimum copula: $C_3(u,v) = \max\{0, u+v-1\}$.

As one sees, $C_1(u,v)$ is the Fréchet/Höffding upper bound for any copula function; it is the copula of the most related (comonotonic) pair. The lower bound $C_3(u,v)$ corresponds to the most antithetic pair. The product copula $C_2(u,v)$ simply represents the case that X and Y are independent. So if (U,V) is a random vector of which the marginals are both uniform$(0,1)$, then if its joint cdf (copula function) $C = C_1$, we have $U \equiv V$, if $C = C_3$, we have $U \equiv 1 - V$, and if $C = C_2$, U and V are independent.

We will show how by taking a convex combination (mixture) of the three copulas used above, we can get a random vector with uniform marginals that has any (rank) correlation between -1 and $+1$. Indeed if for $p_1, p_2, p_3 \geq 0$ with $p_1 + p_2 + p_3 = 1$, we write

$$C(u,v) = p_1 C_1(u,v) + p_2 C_2(u,v) + p_3 C_3(u,v), \qquad (7.75)$$

then the random vector (U,V) has this distribution if V is given by:

$$V = I_1 U + I_2 U^\perp + I_3 (1-U), \qquad (7.76)$$

where I_i, $i = 1,2,3$ are dependent Bernoulli(p_i) random variables with $I_1 + I_2 + I_3 \equiv 1$, and $U^\perp \sim$ uniform$(0,1)$, independent of U. To determine the correlation $r(U,V)$, or the rank correlation $\rho(U,V)$, which is the same for this case since $V \sim$ uniform$(0,1)$ as well, note that by conditioning on which $I_i = 1$, we have from (7.76)

$$E[UV] = p_1 E[U^2] + p_2 E[UU^\perp] + p_3 E[U(1-U)] = \frac{1}{3}p_1 + \frac{1}{4}p_2 + \frac{1}{6}p_3, \quad (7.77)$$

so with $E[U] = E[V] = \frac{1}{2}$, $\text{Var}[U] = \text{Var}[V] = \frac{1}{12}$ and $p_1 + p_2 + p_3 = 1$, we get

$$r(U,V) = \frac{E[UV] - E[U]E[V]}{\sqrt{\text{Var}[U]}\sqrt{\text{Var}[V]}} = p_1 - p_3. \qquad (7.78)$$

Hence $r(U,V) = 1$ if $p_1 = 1$ (the comonotonic upper bound), $r(U,V) = -1$ if $p_3 = 1$ (the countermonotonic lower bound), and $r(U,V) = 0$ holds if $p_1 = p_3$. Independence holds only if $p_1 = p_3 = 0$. This mechanism makes it possible to construct $(X,Y) = (F^{-1}(U), G^{-1}(V))$ with given marginal cdfs F and G and any arbitrary (rank) correlation $\rho(X,Y) = r(U,V) = p_1 - p_3 \in [-1,1]$.

The first and third copula used here have the diagonals $u = v$ and $u = 1 - v$, respectively, as their support. To get a more realistic distribution, it is best to take p_2 as large as possible. But other copulas exist that are flexible enough to produce lookalikes of many realistic joint distributions, allowing us to simulate drawings from more and less dangerous sums of random variables.

Remark 7.8.7 (Generating random drawings from a copula)

To simulate a random drawing from a joint cdf if this cdf is generated by a copula, first generate outcome u of $U \sim$ uniform$(0,1)$, simply taking a computer generated random number, for example using `runif`. Then, draw an outcome v for V from the conditional distribution of V, given $U = u$. This is easy in the three cases considered above: for C_1, take $v = u$; for C_2, draw another independent uniform number; for C_3, take $v = 1 - u$. In general, draw from $\frac{\partial C(u,v)}{\partial u}$. Next, to produce an outcome of (X,Y), simply take $x = F^{-1}(u)$ and $y = G^{-1}(v)$. Note that in case of non-uniform marginals, the above calculation (7.78) does not produce the ordinary Pearson's correlation coefficien $r(X,Y)$, but rather the Spearman rank correlation $\rho(X,Y) = r(F(X), G(Y))$. $\triangledown$

7.9 Exercises

Section 7.2

1. Let $f_X(\cdot)$ and $f_Y(\cdot)$ be two continuous densities (or two discrete densities) that cross exactly once, in the sense that for a certain c, we have $f_X(x) \geq f_Y(x)$ if $x < c$, and $f_X(x) \leq f_Y(x)$ if $x > c$. Show that $X \leq_{st} Y$. Why do the densities $f_X(\cdot)$ and $f_Y(\cdot)$ cross *at least* once?

2. Show that if $X \sim \text{gamma}(\alpha, \beta)$ and $Y \sim \text{gamma}(\alpha, \beta')$ with $\beta > \beta'$, then $X \leq_{st} Y$. The same if $Y \sim \text{gamma}(\alpha', \beta)$ with $\alpha < \alpha'$.

3. Prove that the binomial(n, p) distributions increase in p with respect to stochastic order, by constructing a pair (X, Y) just as in Example 7.2.2 with $X \sim \text{binomial}(n, p_1)$ and $Y \sim \text{binomial}(n, p_2)$ for $p_1 < p_2$, with additionally $\Pr[X \leq Y] = 1$.

4. Prove the assertion in the previous exercise with the help of Exercise 7.2.1.

5. As Exercise 7.2.3, but now for the case that $X \sim \text{binomial}(n_1, p)$ and $Y \sim \text{binomial}(n_2, p)$ for $n_1 < n_2$. Then, give the proof with the help of Exercise 7.2.1.

6. If $N \sim \text{binomial}(2, 0.5)$ and $M \sim \text{binomial}(3, p)$, show that $(1 - p)^3 \leq \frac{1}{4}$ is necessary and sufficient for $N \leq_{st} M$.

7. For two risks X and Y having marginal distributions $\Pr[X = j] = \frac{1}{4}, j = 0, 1, 2, 3$ and $\Pr[Y = j] = \frac{1}{4}, j = 0, 4$, $\Pr[Y = 2] = \frac{1}{2}$, construct a simultaneous distribution with the property that $\Pr[X \leq Y] = 1$.

8. Prove that $\leq_{st}$ is functionally invariant, in the sense that for every non-decreasing function f, we have $X \leq_{st} Y$ implies $f(X) \leq_{st} f(Y)$. Apply this property especially to the excess of loss part $f(x) = (x - d)_+$ of a claim x, and to proportional (re-)insurance $f(x) = \alpha x$ for some $\alpha > 0$.

Section 7.3

1. Prove that $M \leq_{SL} N$ if $M \sim \text{binomial}(n, p)$ and $N \sim \text{binomial}(n + 1, \frac{np}{n+1})$. Show that in the limit for $n \to \infty$, the Poisson stop-loss premium is found for any retention d.

2. If $N \sim \text{binomial}(2, 0.5)$ and $M \sim \text{binomial}(3, p)$, show that $p \geq \frac{1}{3}$ is necessary and sufficient for $N \leq_{SL} M$.

3. Show that if $X \sim Y$, then $\frac{1}{2}(X + Y) \leq_{SL} X$. Is it necessary that X and Y are independent?

4. Let X and Y be two risks with the same mean and with the same support $\{a, b, c\}$ with $0 \leq a < b < c$. Show that either $X \leq_{SL} Y$, or $Y \leq_{SL} X$ must hold. Also give an example of two random variables, with the same mean and both with support $\{0, 1, 2, 3\}$, that are not stop-loss ordered.

5. Compare the cdf F of a risk with another cdf G with the same mean and with $F(x) = G(x)$ on $(-\infty, a)$ and $[b, \infty)$, but $G(x)$ is constant on $[a, b)$. Note that G results from F by dispersion of the probability mass on (a, b) to the endpoints of this interval. Show that $F \leq_{SL} G$ holds. Sketch the stop-loss transforms with F and G.

6. As the previous exercise, but now for the case that the probability mass of F on (a, b) has been concentrated on an appropriate m, that is, such that $G(x)$ is constant both on $[a, m)$ and $[m, b)$. Also consider that case that all mass on the closed interval $[a, b]$ is concentrated.

7. Consider the following differential of mixed cdf F:

$$dF(x) = \begin{cases} \frac{dx}{3} & \text{for } x \in (0, 1) \cup (2, 3) \\ \frac{1}{12} & \text{for } x \in \{1, 2\} \\ \frac{1}{6} & \text{for } x = \frac{3}{2} \end{cases}$$

Show that this cdf is indirectly more dangerous than the uniform$(0,3)$ cdf.

8. Let A_1, A_2, B_1 and B_2 be independent Bernoulli random variables with parameters p_1, p_2, q_1 and q_2. If $p_1 + p_2 = q_1 + q_2$, when is $A_1 + A_2 \leq_{\text{SL}} B_1 + B_2$, when is $A_1 + A_2 \geq_{\text{SL}} B_1 + B_2$, and when does neither of these stochastic inequalities hold?

9. Show that a negative binomial random variable N is stop-loss larger than any Poisson random variable M having $\text{E}[M] \leq \text{E}[N]$. The same for $M \sim$ binomial.

10. Suppose it is known that for every value of the risk aversion α, the exponential premium for the risk X is less than for Y. Which order relation holds between X and Y?

11. Show that the stop-loss transforms $\pi_i(d)$ in (7.6) correspond to cdfs F_i that increase in dangerousness.

12. Complete the proof of Theorem 7.3.9 by proving that the random variable Y satisfies the requirements, using sketches of the stop-loss transform and the cdf.

13. Let $X \leq_{\text{SL}} Z$ and $\text{E}[X] < \text{E}[Z]$. Consider the function $\pi(\cdot)$ that has $\pi(t) = \text{E}[(X-t)_+]$ for $t \leq 0$, $\pi(t) = \text{E}[(Z-t)_+]$ for $t \geq c$, and $\pi(t) = A(t)$ for $0 \leq t \leq c$. Here c and $A(\cdot)$ are chosen in such a way that $A(0) = \mu$ and $A(t)$ is the tangent line to $\text{E}[(Z-t)_+]$ at $t = c$. Show that $\pi(\cdot)$ is convex, and hence the stop-loss transform of a certain risk Y. Sketch the cdfs of X, Y and Z. Show that $X \leq_{\text{SL}} Y \leq_{\text{SL}} Z$, as well as $\text{E}[X] = \text{E}[Y]$ and $Y \leq_{\text{st}} Z$. [In this way, Y is another separator between X and Z in a sense analogous to Theorem 7.3.9.]

14. Show that if $X \leq_{\text{e}} Y$ and $\text{E}[X^k] = \text{E}[Y^k]$ for $k = 1, 2, \ldots, n-1$, then $\text{E}[X^n] \leq \text{E}[Y^n]$. [This means especially that if $X \leq_{\text{e}} Y$ and $\text{E}[X] = \text{E}[Y]$, then $\text{Var}[X] \leq \text{Var}[Y]$. The moments of X and Y are called *lexicographically ordered*.]

15. For risks X and Y and for a certain $d > 0$ we have $\Pr[Y > d] > 0$ while $\Pr[X > d] = 0$. Can $X \leq_{\text{SL}} Y$, $X \leq_{\text{st}} Y$ or $X \leq_{\text{e}} Y$ hold?

16. Let $X \sim$ uniform$(0,1)$, $V = \frac{1}{2}X$, and $W = \min\{X, d\}$ for a certain $d > 0$. Sketch the cdfs of V, W and X. Investigate for which d we have $V \leq_{\text{SL}} W$ and for which we have $W \leq_{\text{SL}} V$.

17. Prove Theorem 7.3.10 for the case $\text{E}[X] = \text{E}[Y]$ by using partial integrations. Use the fact that the stop-loss transform is an antiderivative of $F_X(x) - 1$, and consider again $v(x) = -u(-x)$. To make things easier, look at $\text{E}[v(X) - v(0) - Xv'(0)]$ and assume that $v(\cdot)$ is differentiable at 0.

18. The following risks $X_1, \ldots, X_5$ are given.

    1. $X_1 \sim$ binomial$(10, \frac{1}{2})$;
    2. $X_2 \sim$ binomial$(15, \frac{1}{3})$;
    3. $X_3 \sim$ Poisson(5);
    4. $X_4 \sim$ negative binomial$(2, \frac{2}{7})$;
    5. $X_5 \sim 15I$, where $I \sim$ Bernoulli$(\frac{1}{3})$.

    Do any two decision makers with increasing utility function agree about preferring X_1 to X_2? For each pair (i, j) with $i, j = 1, 2, 3, 4$, determine if $X_i \leq_{\text{SL}} X_j$ holds. Determine if $X_j \leq_{\text{SL}} X_5$ or its reverse holds, $j = 2, 3$. Does $X_3 \leq_{\text{e}} X_5$?

19. Consider the following class of risks $X_p = pY + (1-p)Z$, with Y and Z independent exponential(1) random variables, and p a number in $[0, \frac{1}{2}]$. Note that $X_0 \sim$ exponential(1), while $X_{0.5} \sim$ gamma$(2,2)$. Are the risks in this class stochastically ordered? Show that decision makers with an exponential utility function prefer losing X_p to X_q if and only if $p \geq q$. Prove that $X_{1/2} \leq_{\text{SL}} X_p \leq_{\text{SL}} X_0$.

20. The cdfs $G(\cdot)$ and $V(\cdot)$ are given by

$$G(x) = \frac{1}{4}\left\{F^{*0}(x) + F^{*1}(x) + F^{*2}(x) + F^{*3}(x)\right\}$$

$$V(x) = q_0 F^{*0}(x) + q_1 F^{*1}(x) + q_2 F^{*2}(x)$$

Here F is the cdf of an arbitrary risk, and F^{*n} denotes the nth convolution power of cdf F. For $n = 0$, F^{*0} is the cdf of the constant 0. Determine $q_0, q_1, q_2 \geq 0$ with $q_0 + q_1 + q_2 = 1$ such that $V \leq_{SL} G$ and moreover V and G have equal mean.

21. Compare two compound Poisson random variables S_1 and S_2 in the three stochastic orders $\leq_e$, $\leq_{st}$, and $\leq_{SL}$, if the parameters of S_1 and S_2 are given by

    1. $\lambda_1 = 5$, $p_1(x) = \frac{1}{5}$ for $x = 1,2,3,4,5$;
    2. $\lambda_2 = 4$, $p_1(x) = \frac{1}{4}$ for $x = 2,3,4,5$.

22. Investigate the order relations $\leq_e$, $\leq_{st}$, and $\leq_{SL}$ for risks X and Y with $Y \equiv CX$, where C and X are independent and $\Pr[C = 0.5] = \Pr[C = 1.5] = 0.5$.

23. Let $N \sim \text{binomial}(2, \frac{1}{2})$ and $M \sim \text{Poisson}(\lambda)$. For which λ do $N \geq_{st} M$, $N \leq_{st} M$ and $N \leq_{SL} M$ hold?

24. In the proof of Theorem 7.3.7, sketch the functions $\pi_i(d)$ for the case that $Y \sim \text{uniform}(0,3)$ and Z integer-valued with $\pi_Z(0) = 2$ and $\pi_Z(k) = \pi_Y(k)$ for $k = 1,2,3$. Describe the transitions $\pi_Y(d) \to \pi_2(d) \to \cdots \to \pi_Z(d)$ in terms of dispersion.

25. Let $A_j \sim \text{Bernoulli}(p_j)$, $j = 1,2,\ldots,n$ be independent random variables, and let $\bar{p} = \frac{1}{n}\sum_j p_j$. Show that $\sum_j A_j \leq_{SL} \text{binomial}(n, \bar{p})$.
    [This exercise proves the following statement: Among all sums of n independent Bernoulli random variables with equal mean total μ, the binomial$(n, \frac{\mu}{n})$ is the stop-loss largest. Note that in this case by replacing all probabilities of success by their average, thus eliminating variation from the underlying model, we get a more spread result.]

26. Let $\Pr[X = i] = \frac{1}{6}$, $i = 0,1,\ldots,5$, and $Y \sim \text{binomial}(5, p)$. For which p do $X \leq_{st} Y$, $Y \leq_{st} X$, $X \leq_{SL} Y$ and $Y \leq_{SL} X$ hold?

Section 7.4

1. Consider the family of distributions $F(\cdot; p, \mu)$, defined as $F(x; p, \mu) = 1 - pe^{-x/\mu}$ for some $p \in (0,1)$ and $\mu > 0$. Investigate for which parameter values p and μ the cdf $F(\cdot; p, \mu)$ is stochastically or stop-loss larger or smaller than $F(\cdot; p_0, \mu_0)$, and when it is neither stop-loss larger, nor stop-loss smaller.

2. Investigate the order relations $\leq_{SL}$ and $\leq_{st}$ in the class of binomial(n, p) distributions, $n = 0,1,\ldots$, $0 \leq p \leq 1$.

3. Show that exponential order is preserved under compounding: if $X \leq_e Y$ and $M \leq_e N$, then $X_1 + X_2 + \cdots + X_M \leq_e Y_1 + Y_2 + \cdots + Y_N$.

4. What can be said about two individual claim amount random variables X and Y if for two risk processes with the same claim number process and the same premium c per unit of time, and individual claims such as X and Y respectively, it proves that for each c, the adjustment coefficient with the second ruin process is at most the one with the first?

5. Let S have a compound Poisson distribution with individual claim sizes $\sim X$, and let t_1, t_2 and α be such that $E[(S - t_1)_+] = \lambda E[(X - t_2)_+] = \alpha E[S]$. For an arbitrary $d > 0$, compare $E[(\min\{S, t_1\} - d)_+]$, $E[(S - \sum_1^N (X_i - t_2)_+ - d)_+]$ and $E[((1 - \alpha)S - d)_+]$.

6. If two risks have the same mean μ and variance σ^2, but the skewness of the first risk is larger, what can be said about the stop-loss premiums?

7. Compare the risks S and T in Exercise 3.8.6 as regards exponential, stochastic and stop-loss order.

8. In Section 7.4.3, show that, in areas where the separation argument does not lead to the con-
    clusion that one risk is stop-loss larger than the other, the stop-loss premiums are sometimes
    larger, sometimes smaller.

9. Prove that indeed $\overline{X}_{n+1} \leq_{\text{SL}} \overline{X}_n$ in Remark 7.4.1.

10. Show that the random variables X and Y at the end of Section 7.4.2 are exponentially ordered,
    but not stop-loss ordered.

Section 7.5

1. Let $0 < d < b$ hold. Risk X has $\Pr[X \in \{0, d, b\}] = 1$, risk Y has $\Pr[0 \leq Y \leq b] = 1$. If the means
    and variances of X and Y are equal, show that $E[(X - d)_+] \leq E[(Y - d)_+]$.

2. Show that $X \leq_{\text{SL}} Y \leq_{\text{SL}} Z$ holds in Example 7.5.5. Use the fact that a unimodal continuous
    density with mode 0 is the same as a concave cdf on $[0, \infty)$. Consider the case that Y is not
    continuous separately.

3. Compute the minimal and the maximal stop-loss premium at retention $d = 0.5$ and $d = 3$ for
    risks with $\mu = \sigma^2 = 1$ and a support contained in $[0, 4]$.

4. Give expressions for the minimal and the maximal possible values of the stop-loss premium
    in case of mean μ, variance σ^2 and a support contained in $[0, b]$, see Figure 7.6. In this figure,
    sketch the stop-loss transform of the feasible risk that has the minimal stop-loss premium at
    retention $d = 2$.

5. Which two-point risk with mean μ, variance σ^2 and support contained in $[0, b]$ has the largest
    skewness? Which one has the smallest?

6. Show that the solutions of the previous exercise also have the extremal skewnesses in the class
    of arbitrary risks with mean μ, variance σ^2 and support contained in $[0, b]$.

7. Let $T = Y_1 + \cdots + Y_N$ with $N \sim \text{Poisson}(\lambda)$, $\Pr[0 \leq Y \leq b] = 1$ and $E[Y] = \mu$. Show that
    $\mu N \leq_{\text{SL}} T \leq_{\text{SL}} bM$, if $M \sim \text{Poisson}(\lambda \mu / b)$. What are the means and variances of these three
    random variables?

8. Verify the assertions in the middle paragraph of Example 7.5.3.

Section 7.6

1. Prove that the first set of inequalities of (7.38), together with equal means, implies the second
    set. Use that $E[(X - d)_+] - E[(d - X)_+] = E[X] - d$.

2. Show that equality (3.115) can be generalized from risks to arbitrary random variables X with
    mean μ, leading to $\int_{-\infty}^{\infty} \{E[(X - t)_+] - (\mu - t)_+\} \, dt = \frac{1}{2} \text{Var}[X]$.

3. The function $f(x) = (d - x)_+$ is convex decreasing. Give an example with $X \leq_{\text{SL}} Y$ but not
    $E[(d - X)_+] \leq E[(d - Y)_+]$.

4. Consider n married couples with one-year life insurances, all having probability of death 1%
    for her and 2% for him. The amounts insured are unity for both sexes. Assume that the mor-
    tality between different couples is independent. Determine the distribution of the individual
    model for the total claims, as well as for the collective model approximating this, assuming
    a) that the mortality risks are also independent within each couple, and b) that they follow a
    comonotonic distribution. Compare the stop-loss premiums for the collective model in case of
    a retention of at least $0.03n$.

5. In Example 7.6.7, sketch the stop-loss transform of $X_1 + X_2$ for various values of x. In this way, show that $X_1 + X_2$ increases with x in stop-loss order.

6. Show that $X \leq_{\text{st}} Y$ holds if and only if their comonotonic joint density has the property $h(x,y) = 0$ for $x > y$.

7. If X has support $\{x_1, \ldots, x_n\}$ and Y has support $\{y_1, \ldots, y_m\}$, describe their comonotonic joint density if $x_1 < \cdots < x_n$ and $y_1 < \cdots < y_m$.

8. Prove Bonferroni's inequality: $\Pr[A \cap B] \geq \Pr[A] + \Pr[B] - 1$. Use it to derive the lower bound (7.69) for joint cdfs. Check that the right hand side of (7.69) has the right marginal distributions. Prove that $(F^{-1}(U), G^{-1}(1-U))$ has this lower bound as its cdf.

9. Prove that the following properties hold for TVaR, see Section 5.7:

    - TVaR is comonotonic additive: $\text{TVaR}[S+T;p] = \text{TVaR}[S;p] + \text{TVaR}[T;p]$ for all p if S and T are comonotonic;
    - TVaR respects stochastic order: $F_S \geq F_T$ implies that $\text{TVaR}[S;\cdot] \leq \text{TVaR}[T;\cdot]$;
    - TVaR respects stop-loss order: $\pi_S \leq \pi_T$ implies that $\text{TVaR}[S;\cdot] \leq \text{TVaR}[T;\cdot]$.
    - Hint: use Theorem 7.3.7 and 7.3.9, or use (5.43).

10. Prove that if random variables X and Y are comonotonic, then $\text{Cov}[X,Y] \geq 0$. Can X and Y be at the same time comonotonic and independent?

11. Let $(X,Y) \sim$ bivariate normal, and let (X^U, Y^U) be comonotonic with the same marginals. Show that the cdfs of $X + Y$ and $X^U + Y^U$ cross only once, and determine where.

Section 7.7

1. Let X_1 and X_2 be the length of two random persons. Suppose that these lengths are iid random variables with $\Pr[X_i = 160, 170, 180] = \frac{1}{4}, \frac{1}{2}, \frac{1}{4}$. What is the distribution of the comonotonic upper bound S^U? Determine the distribution of the lower bound if we take as a surrogate V for S the gender of person 1, of which we know it is independent of the length of person 2, while $\Pr[V = 0] = \Pr[V = 1] = \frac{1}{2}$ as well as $\Pr[X_1 = 160, 170 \mid V = 0] = \Pr[X_1 = 170, 180 \mid V = 1] = \frac{1}{2}$. What is the distribution of the improved upper bond $S^{U|V}$? Compare the variances of the various convex upper and lower bounds derived.

2. Let X and Y be independent $N(0,1)$ random variables, and let $S = X + Y$. Assume $V = X + aY$ for some real a. What is the conditional distribution of X, given $V = v$? Determine the distribution of the convex lower bound $\text{E}[S \mid V]$. Also determine the distribution of the comonotonic upper bound and the improved convex upper bound. Compare the variances of these bounds for various values of a. Consider especially the cases $V \equiv S$, $V \perp S$, $V \equiv X$ and $V \propto Y$, that is, $|a| \to \infty$.

3. In Example 7.7.2, compute the variance of the lower bound in case we take $V = aY_1 + Y_2$ instead of $V = Y_1 + Y_2$ as the surrogate random variable. For which a is this variance maximal?

4. In case the event $V = v$ occurs, the improved upper bound of Example 7.7.2 can be written as $F_{X_1 \mid V=v}^{-1}(U) + \cdots + F_{X_n \mid V=v}^{-1}(U)$. Write the terms of this sum as $g_i(U,v)$, then $g_i(U,v)$ is the unconditional contribution of component i to the improved upper bound $S^{U|V} = \sum_i g_i(U,V)$. In general, these random variables will not be comonotonic. Show that $g_i(U,V)$ has the same marginal distribution as X_i. Conclude that the improved upper bound is indeed an improvement over the comonotonic upper bound.

5. Prove that for a pair of random variables (X,Y), if one conditions on $V \equiv X$, the distributions of $S^{U|V} = F_{X|V}^{-1}(U) + F_{Y|V}^{-1}(U)$ and $S = X + Y$ coincide.

6. For the payments (7.51), assume that the logreturns $Y_j \sim N(\mu, \sigma^2)$ are iid. If $X_1^U, X_2^U, \ldots$ denote comonotonic versions of $X_1, X_2, \ldots$, determine the distribution of the random variables $\log X_{k+1}^U - \log X_k^U$. Find $r\left(\log(\alpha_k X_k), \log(\alpha_l X_l)\right)$ for $k < l$ and positive α_k, α_l.

7. Even if n is large, there is no reason to expect the Central Limit Theorem to hold for the cdf of S in (7.61). This is because the terms in S are very much dependent. Still, we want to compare our results to the normal cdf with the same mean and variance as S. Verify that the theoretical mean and variance of S can be computed by:

```
mu.S <- sum(alpha*exp(-(mu.Z-diag(Sigma.Z)/2)))
s <- 0 ## see (2.11) in Vanduffel et al. (2008)
for (i in 1:n) for (j in 1:n)
{s <- s + alpha[i] * alpha[j] *
            exp(- mu.Z[i] - mu.Z[j] + Sigma.Z[i,i]/2
            + Sigma.Z[j,j]/2 + Sigma.Z[i,j])}
sd.S <- sqrt(s-mu.S^2)
```

To the plot in Figure 7.10, add the approximating normal cdf, as follows:

```
lines(qnorm(y, mu.S, sd.S), y, lty="longdash")
```

Compare theoretical mean and variance with `mean(S)` and `sd(S)`.
Also compute the mean and variance of $E[S \,|\, Y_1 + \cdots + Y_n]$.
Discuss the use of the normal approximation to the cdf of S for determining, for example, the 99.75% quantile.
Plot the *kernel density estimate* for the empirical cdf of S, see `?density`.
Also, plot the cdf of a lognormal random variable with the same two first moments as S. The same for a gamma cdf. Which of these three cdfs is the best approximation?

8. Verify the variances found in Example 7.7.2. To do this, run the examples in R as above, and compute the sample variances.

9. At the end of Example 7.7.2, it is suggested that instead of using $V = Z_n = Y_1 + \cdots + Y_n$ as a conditioning variable, it might be better to use $V = Y_1 + (Y_1 + Y_2) + \cdots + (Y_1 + \cdots + Y_n)$ (in case $\alpha_k \equiv +1$). This is based on the observation that the terms in S are lognormal random variables $\alpha_k e^{-Z_k}$ with 'small' Z_k, therefore 'close to' $\alpha_k(1 + Z_k)$, so this choice of V will closely resemble S, apart from a constant. Verify that this is indeed a better choice by inspecting the fits of the empirical cdfs of $E[S \,|\, V]$ and $E[S \,|\, Z_n]$ with the one of S, and by comparing the resulting sample variances.
Try also `alpha <- c(rep(1,5),rep(-1,5))` as a payment pattern for $n = 10$.

10. In the paper Vanduffel *et al.* (2008, formula (2.16)) it is argued that

$$V = \sum_{k=1}^{n} \alpha_k \exp\left(-E[Z_k] + \text{Var}[Z_k]/2\right) Z_k$$

might lead to an even better bound. It does not actually maximize $\text{Var}\left[E[S \,|\, V]\right]$ itself, but a first order Taylor approximation for it. Verify if this is indeed a better choice in the example of this section.
Again, also try `alpha <- c(rep(1,5),rep(-1,5))` as a payment pattern for $n = 10$.

11. An estimator for the skewness based on the sample S is `cum3(S)/sd(S)^1.5`. Fit a translated gamma cdf to mean, variance and skewness of S.
To get an unbiased estimate of the variance, `var` and `sd` divide `sum((S-mean(S))^2)` by n-1. Verify that by using a multiplier `n/((n-1)*(n-2))`, see source text and help-file, we get an unbiased estimate of the third cumulant. Verify if `cum3(S)` uses this multiplier. But are the resulting estimates of σ_S and γ_S unbiased?
In Vanduffel *et al.* (2008), also the inverse gamma distribution (the one of $1/X$ when $X \sim$ gamma) is suggested as an approximation, and its performance is compared with the comonotonic lower bound. Compare it with the translated gamma approximation as well.

12. If in the method to generate random drawings for the upper bound and the lower bound for S, instead of comonotonic random normal drawings we substitute $\Phi^{-1}(u)$, we get an expression for the inverse cdfs of these random variables. In this way, simulations can be avoided altogether, leading to very fast and accurate approximate algorithms. Using these, give exact expressions for the quantiles, and compare with the sample quantiles. See also Vanduffel *et al.* (2008).

13. Show that in case $\alpha_k > 0$ for all k, the function $f(v) = E[S|V = v]$ with S as in (7.61) and $V = Z_n$ is monotone increasing in v.

Section 7.8

1. If (X,Y) are PQD, what can be said of $\Pr[X \leq x | Y \leq y]$?

2. Show that the random pairs (X,X), $(X,X+Z)$ and $(X+Y,X+Z)$ are all PQD if X, Y and Z are independent random variables.

3. The Farlie-Gumbel-Morgenstern copula has $C(u,v) = uv[1 + \alpha(1-u)(1-v)]$ on its support $0 < u < 1, 0 < v < 1$, for $|\alpha| \leq 1$. Show that the corresponding joint density $\partial^2 C(u,v)/\partial u \partial v$ is non-negative. Show that C has uniform$(0,1)$ marginals. Find Spearman's rank correlation ρ, Kendall's τ and Blomqvist's β for random variables with copula C.

4. For continuous random variables, compute ρ, τ and β for the comonotonic random variables. Prove that $\rho = 1$ or $\tau = 1$ imply comonotonicity, but $\beta = 1$ does not.

5. Determine the correlation $r(X,Y)$, as a function of σ, if $X \sim \text{lognormal}(0,1)$ and $Y \sim \text{lognormal}(0,\sigma^2)$, and $r(\log X, \log Y) = 1$. Verify that it equals 1 for $\sigma = 1$, and tends to zero for $\sigma \to \infty$. Also compute ρ and τ.

6. For $X \sim N(0,1)$, determine $r(X^2, (X+b)^2)$.

Chapter 8
Credibility theory

Credibility, as developed by American actuaries, has provided us with a very powerful and intuitive formula. The European Continental School has contributed to its interpretation. The concepts are fundamental to insurance and will continue to be most important in the future. I find it deplorable that the world of finance has not yet realized the importance of collateral knowledge far beyond insurance and the power of credibility-type formulae — Hans Bühlmann, 1999

8.1 Introduction

In insurance practice it often occurs that one has to set a premium for a group of insurance contracts for which there is some claim experience regarding the group itself, but a lot more on a larger group of contracts that are more or less related. The problem is then to set up an experience rating system to determine next year's premium, taking into account not only the individual experience with the group, but also the collective experience. Two extreme positions can be taken. One is to charge the same premium to everyone, estimated by the overall mean $\overline{X}$ of the data. This makes sense if the portfolio is homogeneous, which means that all risk cells have identical mean claims. But if this is not the case, the 'good' risks will take their business elsewhere, leaving the insurer with only 'bad' risks. The other extreme is to charge to group j its own average claims $\overline{X}_j$ as a premium. Such premiums are justified if the portfolio is heterogeneous, but they can only be applied if the claims experience with each group is large enough. As a compromise, one may ask a premium that is a weighted average of these two extremes:

$$z_j\overline{X}_j + (1 - z_j)\overline{X}. \tag{8.1}$$

The factor z_j that expresses how 'credible' the individual experience of cell j is, is called the *credibility factor*; a premium such as (8.1) is called a *credibility premium*. Charging a premium based on collective as well as individual experience is justified because the portfolio is in general neither completely homogeneous, nor completely heterogeneous. The risks in group j have characteristics in common with the risks in other groups, but they also possess unique group properties.

One would choose z_j close to one under the following circumstances: the risk experience with cell j is vast, it exhibits only little variation, or the variation between groups is substantial. There are two methods to find a meaningful value for z_j. In *limited fluctuation credibility theory*, a cell is given full credibility $z_j = 1$ if the experience with it is large enough. This means that the probability of having at least a certain relative error in the individual mean does not exceed a given threshold. If the experience falls short of full credibility, the credibility factor is taken as the ratio of the experience actually present and the experience needed for full credibility. More interesting is the *greatest accuracy credibility theory*, where the credibility factors are derived as optimal coefficients in a Bayesian model with variance components. This model was developed in the 1960s by Bühlmann.

Note that apart from claim amounts, the data can also concern *loss ratios*, that is claims divided by premiums, or claims as a percentage of the sum insured, and so on. Quite often, the claims experience in a cell relates to just one contract, observed in a number of periods, but it is also possible that a cell contains various 'identical' contracts.

In practice, one should use credibility premiums only if one only has very few data. If one has additional information in the form of collateral variables, for example, probably using a generalized linear model (GLM) such as described in the following chapter is indicated, or a mixed model. The main problem is to determine how much virtual experience, see Remark 8.2.7 and Exercise 8.4.7, one should incorporate.

In Section 8.2 we present a basic model to illustrate the ideas behind credibility theory. In this model the claims total X_{jt} for contract j in period t is decomposed into three separate components. The first component is the overall mean m. The second a deviation from this mean that is specific for this contract. The third is a deviation for the specific time period. By taking these deviations to be independent random variables, we see that there is a covariance structure between the claim amounts, and under this structure we can derive estimators of the components that minimize a certain sum of squares. In Section 8.3 we show that exactly these covariance structures, and hence the same optimal estimators, also arise in more general models. Furthermore, we give a short review of possible generalizations of the basic model. In Section 8.4, we investigate the Bühlmann-Straub model, in which the observations are measured in different precision. In Section 8.5 we give an application from motor insurance, where the numbers of claims are Poisson random variables with as a parameter the outcome of a structure parameter that is assumed to follow a gamma distribution.

8.2 The balanced Bühlmann model

To clarify the ideas behind credibility theory, we study in this section a stylized credibility model. Consider the random variable X_{jt}, representing the claim figure of cell j, $j = 1, 2, \ldots, J$, in year t. For simplicity, we assume that the cell contains a single contract only, and that every cell has been observed during T observation periods. So for each j, the index t has the values $t = 1, 2, \ldots, T$. Assume that this claim statistic is the sum of a cell mean m_j plus 'white noise', that is, that all X_{jt} are independent and $N(m_j, s^2)$ distributed, with possibly unequal mean m_j for each cell, but with the same variance $s^2 > 0$. We can test for equality of all group means using the familiar statistical technique of *analysis of variance* (ANOVA). If the null-hypothesis that all m_j are equal fails to hold, this means that there will be more variation between the cell averages $\overline{X}_j$ around the overall average $\overline{X}$ than can be expected in view of the observed variation within the cells. For this reason we look at the following random variable, called the *sum-of-squares-between*:

$$SSB = \sum_{j=1}^{J} T(\overline{X}_j - \overline{X})^2. \tag{8.2}$$

One may show that, under the null-hypothesis that all group means m_j are equal, the random variable SSB has mean $(J-1)s^2$. Since s^2 is unknown, we must estimate this parameter separately. This estimate is derived from the *sum-of-squares-within*, defined as

$$SSW = \sum_{j=1}^{J}\sum_{t=1}^{T}(X_{jt} - \overline{X}_j)^2. \tag{8.3}$$

It is easy to show that the random variable SSW has mean $J(T-1)s^2$. Dividing SSB by $J-1$ and SSW by $J(T-1)$ we get two random variables, each with mean s^2, called the *mean-square-between* (MSB) and the *mean-square-within* (MSW) respectively. We can perform an F-test now, where large values of the MSB compared to the MSW indicate that the null-hypothesis that all group means are equal should be rejected. The test statistic to be used is the so-called *variance ratio* or F-ratio:

$$F = \frac{MSB}{MSW} = \frac{\frac{1}{J-1}\sum_j T(\overline{X}_j - \overline{X})^2}{\frac{1}{J(T-1)}\sum_j \sum_t (X_{jt} - \overline{X}_j)^2}. \tag{8.4}$$

Under the null-hypothesis, SSB divided by s^2 has a $\chi^2(J-1)$ distribution, while SSW divided by s^2 has a $\chi^2(J(T-1))$ distribution. Furthermore, it is possible to show that these random variables are independent. Therefore, the ratio F has an $F(J-1, J(T-1))$ distribution. Proofs of these statements can be found in many texts on mathematical statistics, under the heading 'one-way analysis of variance'. The critical values of F can be found in an F-table (Fisher distribution).

Example 8.2.1 (A heterogeneous portfolio)
Suppose that we have the following observations for 3 groups and 5 years:

	$t=1$	$t=2$	$t=3$	$t=4$	$t=5$	$\overline{X}_j$
$j=1$	99.3	93.7	103.9	92.5	110.6	100.0
$j=2$	112.5	108.3	118.0	99.4	111.8	110.0
$j=3$	129.2	140.9	108.3	105.0	116.6	120.0

As the reader may verify, the MSB equals 500 with 2 degrees of freedom, while the MSW is 109 with 12 degrees of freedom. This gives a value $F = 4.588$, which is significant at the 95% level, the critical value being 3.885. The conclusion is that the data show that the mean claims per group are not all equal.

To get R to do the necessary calculations, do the following:

```
J <- 3; K <- 5; X <- scan(n=J*K)
 99.3   93.7 103.9   92.5 110.6
112.5 108.3 118.0   99.4 111.8
129.2 140.9 108.3 105.0 116.6
j <- rep(1:J, each=K); j <- as.factor(j)
X.bar <- mean(X); Xj.bar <- tapply(X, j, mean)
MSB <- sum((Xj.bar-X.bar)^2) * K / (J-1)
MSW <- sum((X-rep(Xj.bar,each=K))^2)/J/(K-1)
MSB/MSW; qf(0.95, J-1, J*(K-1))   ## 4.588 and 3.885
```

The use of K instead of T to denote time avoids problems with the special identifiers T and t in R. The vector Xj.bar is constructed by applying the mean function to all groups of elements of X with the same value of j.

It is also possible to let R do the analysis of variance. Use a linear model, explaining the responses X from the group number j (as a factor). This results in:

```
> anova(lm(X~j))
Analysis of Variance Table

Response: X
          Df  Sum Sq Mean Sq F value  Pr(>F)
j          2 1000.00  500.00  4.5884 0.03311 *
Residuals 12 1307.64  108.97
```

The probability of obtaining a larger F-value than the one we observed here is 0.03311, so the null-hypothesis that the group means are all equal is rejected at the 5% level.   ∇

If the null-hypothesis fails to be rejected, there is apparently no convincing statistical evidence that the portfolio is heterogeneous. So there is no reason not to ask the same premium for each contract. In case of rejection, apparently there is variation between the cell means m_j. In this case one may treat these numbers as fixed unknown numbers, and try to find a system behind these numbers, for example by doing a regression on collateral data. Another approach is to assume that the numbers m_j have been produced by a chance mechanism, hence by 'white noise' similar to the one responsible for the deviations from the mean within each cell. This means that we can decompose the claim statistics as follows:

$$X_{jt} = m + \Xi_j + \Xi_{jt}, \quad j = 1, \ldots, J, \, t = 1, \ldots, T, \tag{8.5}$$

with Ξ_j and Ξ_{jt} independent random variables for which

$$\mathrm{E}[\Xi_j] = \mathrm{E}[\Xi_{jt}] = 0, \quad \mathrm{Var}[\Xi_j] = a, \quad \mathrm{Var}[\Xi_{jt}] = s^2. \tag{8.6}$$

Because the variance of X_{jt} in (8.5) equals the sum of the variances of its components, models such as (8.5) are called *variance components models*. Model (8.5) is a simplified form of the so-called classical Bühlmann model, because we assumed independence of the components where Bühlmann only assumes the correlation to be zero. We call our model that has equal variance for all observations, as well as equal numbers of policies in all cells, the *balanced Bühlmann model*.

The interpretation of the separate components in (8.5) is the following.

1. m is the overall mean; it is the expected value of the claim amount for an arbitrary policyholder in the portfolio.
2. Ξ_j denotes a random deviation from this mean, specific for contract j. The conditional mean, given $\Xi_j = \xi$, of the random variables X_{jt} equals $m + \xi$. It represents the long-term average of the claims each year if the length of the observation period T goes to infinity. The component Ξ_j describes the risk quality of this particular contract; the mean $E[\Xi_j]$ equals zero, its variation describes differences between contracts. The distribution of Ξ_j depicts the risk structure of the portfolio, hence it is known as the *structure distribution*. The parameters m, a and s^2 characterizing the risk structure are called the *structural parameters*.
3. The components Ξ_{jt} denote the deviation for year t from the long-term average. They describe the within-variation of a contract. It is the variation of the claim experience in time through good and bad luck of the policyholder.

Note that in the model described above, the random variables X_{jt} are *dependent* for fixed j, since they share a common risk quality component Ξ_j. One might say that stochastically independent random variables with the same probability distribution involving unknown parameters in a sense are dependent anyway, since their values all depend on these same unknown parameters.

In the next theorem, we are looking for a predictor of the as yet unobserved random variable $X_{j,T+1}$. We require this predictor to be a linear combination of the observable data $X_{11}, \ldots, X_{JT}$ with the same mean as $X_{j,T+1}$. Furthermore, its mean squared error must be minimal. We prove that under model (8.5), this predictor has the credibility form (8.1), so it is a weighted average of the individual claims experience and the overall mean claim. The theorem also provides us with the optimal value of the credibility factor z_j. We want to know the optimal predictor of the amount to be paid out in the next period $T + 1$, since that is the premium we should ask for this contract. The distributional assumptions are assumed to hold for all periods $t = 1, \ldots, T + 1$. Note that in the theorem below, normality is not required.

Theorem 8.2.2 (Balanced Bühlmann model; homogeneous estimator)
Assume that the claim figures X_{jt} for contract j in period t can be written as the sum of stochastically independent components, as follows:

$$X_{jt} = m + \Xi_j + \Xi_{jt}, \quad j = 1, \ldots, J, \, t = 1, \ldots, T + 1, \tag{8.7}$$

where the random variables Ξ_j are iid with mean $E[\Xi_j] = 0$ and $Var[\Xi_j] = a$, and also the random variables Ξ_{jt} are iid with mean $E[\Xi_{jt}] = 0$ and $Var[\Xi_{jt}] = s^2$ for all j and t. Furthermore, assume the random variables Ξ_j to be independent of the Ξ_{jt}. Under these conditions, the homogeneous linear combination $g_{11}X_{11} + \cdots + g_{JT}X_{JT}$ that is the best unbiased predictor of $X_{j,T+1}$ in the sense of minimal mean squared error (MSE)

$$E[\{X_{j,T+1} - g_{11}X_{11} - \cdots - g_{JT}X_{JT}\}^2] \tag{8.8}$$

equals the credibility premium

$$z\overline{X}_j + (1-z)\overline{\overline{X}}, \tag{8.9}$$

where

$$z = \frac{aT}{aT + s^2} \tag{8.10}$$

is the resulting best credibility factor (which in this case is equal for all j),

$$\overline{\overline{X}} = \frac{1}{JT} \sum_{j=1}^{J} \sum_{t=1}^{T} X_{jt} \tag{8.11}$$

is the collective estimator of m, and

$$\overline{X}_j = \frac{1}{T} \sum_{t=1}^{T} X_{jt} \tag{8.12}$$

is the individual estimator of m.

Proof. Because of the independence assumptions and the equal distributions, the random variables X_{it} with $i \neq j$ are interchangeable. By convexity, (8.8) has a unique minimum. In the optimum, all values of $g_{it}, i \neq j$ must be identical, for reasons of symmetry. If not, by interchanging coefficients we can show that more than one extremum exists. The same goes for all values $g_{jt}, t = 1, \ldots, T$. Combining this with the unbiasedness restriction, we see that the homogeneous linear estimator with minimal MSE must be of the form (8.9) for some z. We only have to find its optimal value.

Since X_{jt}, $\overline{X}_j$ and $\overline{\overline{X}}$ all have mean m, we can rewrite the MSE (8.8) as:

$$\begin{aligned}
\mathrm{E}[\{X_{j,T+1} - (1-z)\overline{\overline{X}} - z\overline{X}_j\}^2] &= \mathrm{E}[\{X_{j,T+1} - \overline{\overline{X}} - z(\overline{X}_j - \overline{\overline{X}})\}^2] \\
&= \mathrm{E}[\{X_{j,T+1} - \overline{\overline{X}}\}^2] - 2z\,\mathrm{E}[\{X_{j,T+1} - \overline{\overline{X}}\}\{\overline{X}_j - \overline{\overline{X}}\}] + z^2\,\mathrm{E}[\{\overline{X}_j - \overline{\overline{X}}\}^2] \quad (8.13) \\
&= \mathrm{Var}[X_{j,T+1} - \overline{\overline{X}}] - 2z\,\mathrm{Cov}[X_{j,T+1} - \overline{\overline{X}}, \overline{X}_j - \overline{\overline{X}}] + z^2\,\mathrm{Var}[\overline{X}_j - \overline{\overline{X}}].
\end{aligned}$$

This quadratic form in z is minimal for the following choice of z:

$$z = \frac{\mathrm{Cov}[X_{j,T+1} - \overline{\overline{X}}, \overline{X}_j - \overline{\overline{X}}]}{\mathrm{Var}[\overline{X}_j - \overline{\overline{X}}]} = \frac{aT}{aT + s^2}, \tag{8.14}$$

where it is left to the reader (Exercise 8.2.1) to verify the final equality in (8.13) by proving and filling in the necessary covariances:

$$\begin{aligned}
&\mathrm{Cov}[X_{jt}, X_{ju}] = a \text{ for } t \neq u; \\
&\mathrm{Var}[X_{jt}] = a + s^2; \\
&\mathrm{Cov}[X_{jt}, \overline{X}_j] = \mathrm{Var}[\overline{X}_j] = a + \frac{s^2}{T}; \\
&\mathrm{Cov}[\overline{X}_j, \overline{\overline{X}}] = \mathrm{Var}[\overline{\overline{X}}] = \frac{1}{J}\left(a + \frac{s^2}{T}\right).
\end{aligned} \tag{8.15}$$

So indeed predictor (8.9) leads to the minimal MSE (8.8) for the value of z given in (8.10). ∇

Remark 8.2.3 (Asymptotic properties of the optimal credibility factor)
The credibility factor z in (8.10) has plausible asymptotic properties:

1. If $T \to \infty$, then $z \to 1$. The more claims experience there is, the more faith we can have in the individual risk premium. This asymptotic case is not very relevant in practice, as it assumes that the risk does not change over time.
2. If $a \downarrow 0$, then $z \downarrow 0$. If the expected individual claim amounts are identically distributed, there is no heterogeneity in the portfolio. But then the collective mean m, when known, or its best homogeneous estimator $\overline{X}$ are optimal linear estimators of the risk premium. See (8.16) and (8.9).
3. If $a \to \infty$, then $z \to 1$. This is also intuitively clear. In this case, the result on the other contracts does not provide information about risk j.
4. If $s^2 \to \infty$, then $z \to 0$. If for a fixed risk parameter, the claims experience is extremely variable, the individual experience is not especially useful for estimating the real risk premium. ∇

Note that (8.9) is only a *statistic* if the ratio s^2/a is known; otherwise its distribution will contain unknown parameters. In Example 8.2.5 below we show how this ratio can be estimated as a by-product of the ANOVA. The fact that the credibility factor (8.14) does not depend on j is due to the simplifying assumption we have made that the number of observation periods is the same for each j, as well as that all observations have the same variance.

If we allow our linear estimator to contain a constant term, looking in fact at the best *inhomogeneous* linear predictor $g_0 + g_{11}X_{11} + \cdots + g_{JT}X_{JT}$, we get the next theorem. Two things should be noted. One is that it will prove that the unbiasedness restriction is now superfluous. The other is that (8.16) below looks just like (8.9), except that the quantity $\overline{X}$ is replaced by m. But this means that the inhomogeneous credibility premium for group j does not depend on the data from other groups $i \neq j$. The homogeneous credibility premium assumes the ratio s^2/a to be known; the inhomogeneous credibility premium additionally assumes that m is known.

Theorem 8.2.4 (Balanced Bühlmann model; inhomogeneous estimator)
Under the same distributional assumptions about X_{jt} as in the previous theorem, the inhomogeneous linear combination $g_0 + g_{11}X_{11} + \cdots + g_{JT}X_{JT}$ to predict next year's claim total $X_{j,T+1}$ that is optimal in the sense of mean squared error is the credibility premium

$$z\overline{X}_j + (1-z)m, \tag{8.16}$$

where z and $\overline{X}_j$ are as in (8.10) and (8.12).

Proof. The same symmetry considerations as in the previous proof tell us that the values of $g_{it}, i \neq j$ are identical in the optimal solution, just as those of $g_{jt}, t = 1, \ldots, T$. So for certain g_0, g_1 and g_2, the inhomogeneous linear predictor of $X_{j,T+1}$ with minimal MSE is of the following form:

$$g_0 + g_1\overline{X} + g_2\overline{X}_j. \tag{8.17}$$

The MSE can be written as variance plus squared bias, as follows:

$$
\begin{aligned}
&\mathrm{E}[\{X_{j,T+1} - g_0 - g_1\overline{X} - g_2\overline{X}_j\}^2] \\
&= \mathrm{Var}[X_{j,T+1} - g_1\overline{X} - g_2\overline{X}_j] + \{\mathrm{E}[X_{j,T+1} - g_0 - g_1\overline{X} - g_2\overline{X}_j]\}^2.
\end{aligned} \tag{8.18}
$$

The second term on the right hand side is zero, and hence minimal, if we choose $g_0 = m(1 - g_1 - g_2)$. This entails that the estimator we are looking for is necessarily unbiased. The first term on the right hand side of (8.18) can be rewritten as

$$
\begin{aligned}
&\mathrm{Var}[X_{j,T+1} - (g_2 + g_1/J)\overline{X}_j - g_1(\overline{X} - \overline{X}_j/J)] \\
&= \mathrm{Var}[X_{j,T+1} - (g_2 + g_1/J)\overline{X}_j] + \mathrm{Var}[g_1(\overline{X} - \overline{X}_j/J)] + 0,
\end{aligned} \tag{8.19}
$$

because the covariance term vanishes since $g_1(\overline{X} - \overline{X}_j/J)$ depends only of X_{it} with $i \neq j$. Hence any solution (g_1, g_2) with $g_1 \neq 0$ can be improved, since a lower value of (8.19) is obtained by taking $(0, g_2 + g_1/J)$. Therefore choosing $g_1 = 0$ is optimal. So all that remains to be done is to minimize the following expression for g_2:

$$\mathrm{Var}[X_{j,T+1} - g_2\overline{X}_j] = \mathrm{Var}[X_{j,T+1}] - 2g_2\mathrm{Cov}[X_{j,T+1}, \overline{X}_j] + g_2^2\mathrm{Var}[\overline{X}_j], \tag{8.20}$$

which has as an optimum

$$g_2 = \frac{\mathrm{Cov}[X_{j,T+1}, \overline{X}_j]}{\mathrm{Var}[\overline{X}_j]} = \frac{aT}{aT + s^2}, \tag{8.21}$$

so the optimal g_2 is just z as in (8.10). The final equality can be verified by filling in the relevant covariances (8.15). This means that the predictor (8.16) for $X_{j,T+1}$ has minimal MSE. ∇

Example 8.2.5 (Credibility estimation in Example 8.2.1)

Consider again the portfolio of Example 8.2.1. It can be shown (see Exercise 8.2.8), that in model (8.5) the numerator of F in (8.4) (the MSB) has mean $aT + s^2$, while the denominator MSW has mean s^2. Hence $1/F$ will be close to $s^2/\{aT + s^2\}$, which means that we can use $1 - 1/F$ to estimate z. Note that this is not an unbiased estimator, since $\mathrm{E}[1/MSB] \neq 1/\mathrm{E}[MSB]$. The resulting credibility factor is $z = 0.782$ for each group. So the optimal forecasts for the claims next year in the three groups are $0.782\overline{X}_j + (1 - 0.782)\overline{X}, j = 1, 2, 3$, resulting in 102.18, 110 and 118.82. Notice the 'shrinkage effect': the credibility estimated premiums are closer together than the original group means 100, 110 and 120. ∇

Remark 8.2.6 (Estimating the risk premium)

One may argue that instead of aiming to predict next year's claim figure $X_{j,T+1}$, including the fluctuation $\Xi_{j,T+1}$, we actually should estimate the *risk premium* $m + \Xi_j$ of group j. But whether we allow a constant term in our estimator or not, in each case we get the same optimum. Indeed for every random variable Y:

$$E[\{m + \Xi_j + \Xi_{j,T+1} - Y\}^2]$$
$$= E[\{m + \Xi_j - Y\}^2] + \text{Var}[\Xi_{j,T+1}] + 2\text{Cov}[m + \Xi_j - Y, \Xi_{j,T+1}]. \quad (8.22)$$

If Y depends only on the X_{jt} that are already observed, hence with $t \leq T$, the covariance term must be equal to zero. Since it follows from (8.22) that the MSEs for Y as an estimator of $m + \Xi_j$ and of $X_{j,T+1} = m + \Xi_j + \Xi_{j,T+1}$ differ only by a constant $\text{Var}[\Xi_{j,T+1}] = s^2$, we conclude that both MSEs are minimized by the same estimator Y. $\quad \triangledown$

The credibility premium (8.16) is a weighted average of the estimated individual mean claim, with as a weight the credibility factor z, and the estimated mean claim for the whole portfolio. Because we assumed that the number of observation years T for each contract is the same, by asking premium (8.16) on the lowest level we receive the same premium income as when we would ask $\overline{X}$ as a premium from everyone. For $z = 0$ the individual premium equals the collective premium. This is acceptable in a homogeneous portfolio, but in general not in a heterogeneous one. For $z = 1$, a premium is charged that is fully based on individual experience. In general, this individual information is scarce, making this estimator unusable in practice. Sometimes it even fails completely, like when a prediction is needed for a contract that up to now has not produced any claim.

The quantity $a > 0$ represents the heterogeneity of the portfolio as depicted in the risk quality component Ξ_j, and s^2 is a global measure for the variability within the homogeneous groups.

Remark 8.2.7 (Virtual experience)
Write $X_{j\Sigma} = X_{j1} + \cdots + X_{jT}$, then an equivalent expression for the credibility premium (8.16) is the following:

$$\frac{s^2 m + aT\overline{X}_j}{s^2 + aT} = \frac{ms^2/a + X_{j\Sigma}}{s^2/a + T}. \quad (8.23)$$

So if we extend the number of observation periods T by an extra s^2/a periods and also add ms^2/a as virtual claims to the actually observed claims $X_{j\Sigma}$, the credibility premium is nothing but the average claims, adjusted for virtual experience. $\quad \triangledown$

8.3 More general credibility models

In model (8.5) of the previous section, we assumed the components Ξ_j and Ξ_{jt} to be independent random variables. But from (8.14) and (8.15) one sees that actually only the covariances of the random variables X_{jt} are essential. We get the same results if we impose a model with weaker requirements, as long as the covariance structure remains the same. An example is to only require independence and identical distributions of the Ξ_{jt}, conditionally given Ξ_j, with $E[\Xi_{jt}|\Xi_j = \xi] = 0$ for all ξ. If the joint distribution of Ξ_j and Ξ_{jt} is like that, the Ξ_{jt} are not necessarily independent, but they are uncorrelated, as can be seen from the following lemma:

Lemma 8.3.1 (Conditionally iid random variables are uncorrelated)
Suppose that given Ξ_j, the random variables $\Xi_{j1}, \Xi_{j2}, \ldots$ are iid with mean zero. Then we have

$$\text{Cov}[\Xi_{jt}, \Xi_{ju}] = 0, \, t \neq u; \quad \text{Cov}[\Xi_j, \Xi_{jt}] = 0. \tag{8.24}$$

Proof. Because of the decomposition rule for conditional covariances, see Exercise 8.3.1, we can write for $t \neq u$:

$$\text{Cov}[\Xi_{ju}, \Xi_{jt}] = \text{E}\big[\text{Cov}[\Xi_{ju}, \Xi_{jt}|\Xi_j]\big] + \text{Cov}\big[\text{E}[\Xi_{ju}|\Xi_j], \text{E}[\Xi_{jt}|\Xi_j]\big]. \tag{8.25}$$

This equals zero since, by our assumptions, $\text{Cov}[\Xi_{ju}, \Xi_{jt}|\Xi_j] \equiv 0$ and $\text{E}[\Xi_{ju}|\Xi_j] \equiv 0$. Clearly, $\text{Cov}[\Xi_j, \Xi_{jt}|\Xi_j] \equiv 0$ as well. Because

$$\text{Cov}[\Xi_j, \Xi_{jt}] = \text{E}\big[\text{Cov}[\Xi_j, \Xi_{jt}|\Xi_j]\big] + \text{Cov}\big[\text{E}[\Xi_j|\Xi_j], \text{E}[\Xi_{jt}|\Xi_j]\big], \tag{8.26}$$

the random variables Ξ_j and Ξ_{jt} are uncorrelated as well. ∇

Note that in the model of this lemma, the random variables X_{jt} are not marginally uncorrelated, let alone independent.

Example 8.3.2 (Mixed Poisson distribution)
Assume that the X_{jt} random variables represent the numbers of claims in a year on a particular motor insurance policy. The driver in question has a number of claims in that year that has a Poisson(Λ_j) distribution, where the parameter Λ_j is a drawing from a certain non-degenerate structure distribution. Then the first component of (8.5) represents the expected number of claims $m = \text{E}[X_{jt}] = \text{E}[\Lambda_j]$ of an arbitrary driver. The second is $\Xi_j = \Lambda_j - m$; it represents the difference in average numbers of claims between this particular driver and an arbitrary driver. The third term $\Xi_{jt} = X_{jt} - \Lambda_j$ equals the annual fluctuation around the mean number of claims of this particular driver. In this case, the second and third component, though uncorrelated in view of Lemma 8.3.1, are not independent, for example because $\text{Var}[X_{jt} - \Lambda_j|\Lambda_j - m] \equiv \text{Var}[X_{jt}|\Lambda_j] \equiv \Lambda_j$. See also Section 8.5. ∇

Remark 8.3.3 (Parameterization through risk parameters)
The variance components model (8.5), even with relaxed independence assumptions, sometimes is too restricted for practical applications. Suppose that X_{jt} as in (8.5) now represents the annual claims total of the driver from Example 8.3.2, and also suppose that this has a compound Poisson distribution. Then apart from the Poisson parameter, there are also the parameters of the claim size distribution. The conditional variance of the noise term, given the second term (mean annual total claim costs), is now no longer a function of the second term. To remedy this, Bühlmann studied slightly more general models, having a latent random variable Θ_j, that might be vector-valued, as a structure parameter. The risk premium is the conditional mean $\mu(\Theta_j) := \text{E}[X_{jt}|\Theta_j]$ instead of simply $m + \Xi_j$. If $\text{E}[X_{jt}|\Theta_j]$ is not a one-to-one function of Θ_j, it might occur that contracts having the same Ξ_j in the

basic model above, have a different pattern of variation $\text{Var}[\Xi_{jt}|\Theta_j]$ in Bühlmann's model. Therefore the basic model is insufficient here. But it can be shown that in this case the same covariances, and hence the same optimal estimators, are found.

An advantage of using a variance components model over Bühlmann's Bayesian way of describing the risk structure is that the resulting models are technically as well as conceptually easier, at only a slight cost of generality and flexibility.   ▽

It is possible to extend credibility theory to models that are more complicated than (8.5). Results resembling the ones from Theorems 8.2.2 and 8.2.4 can be derived for such models. In essence, to find an optimal predictor in the sense of least squares one minimizes the quadratic MSE over its coefficients, if needed with an additional unbiasedness restriction. Because of the symmetry assumptions in the balanced Bühlmann model, only a one-dimensional optimization was needed there. But in general we must solve a system of linear equations that arises by differentiating either the MSE or a Lagrange function. The latter situation occurs when there is an unbiasedness restriction. One should not expect to obtain analytical solutions such as above.

Some possible generalizations of the basic model are the following.

Example 8.3.4 (Bühlmann-Straub model; varying precision)
Credibility models such as (8.5) can be generalized by looking at X_{jt} that are averages over a number of policies. It is also conceivable that there are other reasons to assume that not all X_{jt} have been measured equally precisely, therefore have the same variance. For this reason, it may be expedient to introduce weights in the model. By doing this, we get the Bühlmann-Straub model. In principle, these weights should represent the total number of observation periods of which the figure X_{jt} is the mean (*natural weights*). Sometimes this number is unknown. In that case, one has to make do with approximate relative weights, like for example the total premium paid. If the actuary deems it appropriate, he can adjust these numbers to express the degree of confidence he has in the individual claims experience of particular contracts. In Section 8.4 we prove a result, analogous to Theorem 8.2.2, for the homogeneous premium in the Bühlmann-Straub model.   ▽

Example 8.3.5 (Jewell's hierarchical model)
A further generalization is to subdivide the portfolio into sectors, and to assume that each sector p has its own deviation from the overall mean. The claims experience for contract j in sector p in year t can then be decomposed as follows:

$$X_{pjt} = m + \Xi_p + \Xi_{pj} + \Xi_{pjt}. \tag{8.27}$$

This model is called Jewell's hierarchical model. Splitting up each sector p into subsectors q, each with its own deviation $\Xi_p + \Xi_{pq}$, and so on, leads to a hierarchical chain of models with a tree structure.   ▽

Example 8.3.6 (Cross classification models)
It is conceivable that X_{pjt} is the risk in sector p, and that index j corresponds to some other general factor to split up the policies, for example if p is the region and j the gender of the driver. For such two-way cross classifications it does not make

sense to use a hierarchical structure for the risk determinants. Instead, one could add to (8.27) a term Ξ'_j, to describe the risk characteristics of group j. In this way, one gets

$$X_{pjt} = m + \Xi_p + \Xi'_j + \Xi_{pj} + \Xi_{pjt}. \tag{8.28}$$

This is a cross-classification model. In Chapter 9, we study similar models, where the row and column effects are fixed but unknown, instead of being modeled as random variables such as here.   ▽

Example 8.3.7 (De Vijlder's credibility model for IBNR)
Credibility models are also useful to tackle the problem of estimating IBNR reserves to be held, see also Chapter 10. These are provisions for claims that are not, or not fully, known to the insurer. In a certain calendar year T, realizations are known for random variables X_{jt} representing the claim figure for policies written in year j, in their tth year of development, $t = 0, 1, \ldots, T - j$. A credibility model for this situation is

$$X_{jt} = (m + \Xi_j)d_t + \Xi_{jt}, \tag{8.29}$$

where the numbers d_t are development factors, for example with a sum equal to 1, that represent the fraction of the claims paid on average in the tth development period, and where $m + \Xi_j$ represents the claims, aggregated over all development periods, on policies written in year j.   ▽

Example 8.3.8 (Regression models; Hachemeister)
We can also generalize (8.5) by introducing collateral data. If for example y_{jt} represents a certain risk characteristic of contract j, like for example the age of the policy holder in year t, Ξ_j might be written as a linear, stochastic, function of y_{jt}. Then the claims in year t are equal to

$$\{m^{(1)} + \Xi_j^{(1)}\} + \{m^{(2)} + \Xi_j^{(2)}\}y_{jt} + \Xi_{jt}, \tag{8.30}$$

which is a credibility-regression model. Classical one-dimensional regression arises when $\Xi_j^{(k)} \equiv 0, k = 1, 2$. This means that there are no latent risk characteristics. Credibility models such as (8.30) were first studied by Hachemeister.   ▽

8.4 The Bühlmann-Straub model

Just as in (8.7), in the Bühlmann-Straub model the observations can be decomposed as follows:

$$X_{jt} = m + \Xi_j + \Xi_{jt}, \quad j = 1, \ldots, J, \; t = 1, \ldots, T + 1, \tag{8.31}$$

where the unobservable risk components $\Xi_j, j = 1, 2, \ldots, J$ are iid with mean zero and variance a; the Ξ_{jt} are also independent with mean zero. The components Ξ_j and Ξ_{jt} are assumed to be independent, too. The difference between the Bühlmann and the Bühlmann-Straub models is that in the latter the variance of the Ξ_{jt} components is s^2/w_{jt}, where w_{jt} is the *weight* attached to observation X_{jt}. This weight represents the relative precision of the various observations. Observations with variances like this arise when X_{jt} is an average of w_{jt} replications, hence $X_{jt} = \sum_k X_{jtk}/w_{jt}$ where $X_{jtk} = m + \Xi_j + \Xi_{jtk}$ with Ξ_{jtk} iid with zero mean and variance s^2. The random variables Ξ_{jtk} then denote deviations from the risk premium $m + \Xi_j$ for the kth individual contract in time period t and group j. In this case, the weights are called *natural weights*. Sometimes these weights are unknown, or there is another mechanism that leads to differing variances. In that case we can, for example, approximate the volume by the total premium for a cell.

To find the best homogeneous unbiased linear predictor $\sum h_{it} X_{it}$ of the *risk premium* $m + \Xi_j$ (see Remark 8.2.6), we minimize its MSE. In Theorem 8.4.1 below, we derive the optimal values in (8.33) for the coefficients h_{it}, under the unbiasedness restriction. The following notation will be used, see (8.10)–(8.12):

$$
\begin{aligned}
&w_{j\Sigma} = \sum_{t=1}^{T} w_{jt}; && w_{\Sigma\Sigma} = \sum_{j=1}^{J} w_{j\Sigma}; \\[2mm]
&z_j = \frac{aw_{j\Sigma}}{s^2 + aw_{j\Sigma}}; && z_{\Sigma} = \sum_{j=1}^{J} z_j; \\[2mm]
&X_{jw} = \sum_{t=1}^{T} \frac{w_{jt}}{w_{j\Sigma}} X_{jt}; && X_{ww} = \sum_{j=1}^{J} \frac{w_{j\Sigma}}{w_{\Sigma\Sigma}} X_{jw}; && X_{zw} = \sum_{j=1}^{J} \frac{z_j}{z_{\Sigma}} X_{jw}.
\end{aligned}
\tag{8.32}
$$

Notice the difference between for example X_{jw} and X_{ju}. If a w appears as an index, this indicates that there has been a weighted summation over this index, using the (natural or other) weights of the observations. An index z denotes a weighted summation with credibility weights, while a Σ is used for an unweighted summation. The simplest way to allow for different numbers of observation periods T_j is to include some observations with weight zero when necessary.

Theorem 8.4.1 (Bühlmann-Straub model; homogeneous estimator)
The MSE-best homogeneous unbiased predictor $\sum_{i,t} h_{it} X_{it}$ of the risk premium $m + \Xi_j$ in model (8.31), that is, the solution to the following restricted minimization problem

$$
\begin{aligned}
&\min_{h_{it}} \mathrm{E}[\{m + \Xi_j - \sum_{i,t} h_{it} X_{it}\}^2] \\
&\text{subject to } \mathrm{E}[m + \Xi_j] = \sum_{i,t} h_{it} \mathrm{E}[X_{it}],
\end{aligned}
\tag{8.33}
$$

is the following credibility estimator (see also (8.9)):

$$
z_j X_{jw} + (1 - z_j) X_{zw}.
\tag{8.34}
$$

Here X_{jw} as in (8.32) is the individual estimator of the risk premium, X_{zw} is the credibility weighted collective estimator and z_j the credibility factor for contract j.

Proof. To prove that of all the linear combinations of the observations to estimate $m + \Xi_j$ with the same mean, (8.34) has the smallest MSE, we could do a Lagrange optimization, solving the first order conditions to find an extremum. But it is simpler to prove the result by making use of the result that linear combinations of uncorrelated random variables with a given mean have minimal variance if the coefficients are inversely proportional to the variances; see Exercise 8.4.1. First we derive the optimal 'mix' $h_{it}/h_{i\Sigma}$ of the contracts in group i. The best choice proves to be $h_{it}/h_{i\Sigma} = w_{it}/w_{i\Sigma}$; from this we see that the observations X_{it} have to appear in (8.33) in the form X_{iw}. Then we derive that the totals $h_{i\Sigma}$ of the coefficients with group $i \neq j$ are best taken proportional to z_j. Finally, the optimal value of $h_{j\Sigma}$ is derived.

From (8.33) we see that the following problem must be solved to find the best predictor of $m + \Xi_j$:

$$\min_{h_{it}:h_{\Sigma\Sigma}=1} \mathrm{E}\left[\{m + \Xi_j - \sum_{i,t} h_{it} X_{it}\}^2\right]. \tag{8.35}$$

The restriction $h_{\Sigma\Sigma} = 1$ is the unbiasedness constraint in (8.33). By this constraint, the expectation in (8.35) is also the variance. Substituting decomposition (8.31) for X_{it}, we get from (8.35):

$$\min_{h_{it}:h_{\Sigma\Sigma}=1} \mathrm{Var}\left[(1 - h_{j\Sigma})\Xi_j - \sum_{i \neq j} h_{i\Sigma}\Xi_i - \sum_{i,t} h_{it}\Xi_{it}\right], \tag{8.36}$$

or, what is the same because of the variances of the components Ξ_j and Ξ_{jt} and the independence of these components:

$$\min_{h_{it}:h_{\Sigma\Sigma}=1} (1 - h_{j\Sigma})^2 a + \sum_{i \neq j} h_{i\Sigma}^2 a + \sum_i h_{i\Sigma}^2 \sum_t \frac{h_{it}^2}{h_{i\Sigma}^2} \frac{s^2}{w_{it}}. \tag{8.37}$$

First we optimize the inner sum, extending over t. Because of Exercise 8.4.1 the optimal values of $h_{it}/h_{i\Sigma}$ are $w_{it}/w_{i\Sigma}$. So we can replace the observations $X_{it}, t = 1, 2, \ldots, T$ by their weighted averages X_{iw}. We see that the credibility estimator has the form $\sum_i h_{i\Sigma} X_{iw}$, where the values of $h_{i\Sigma}$ are still to be found.

The minimal value for the inner sum equals $s^2/w_{i\Sigma}$. From (8.32) we see that $a + s^2/w_{i\Sigma} = a/z_i$. So we can rewrite (8.37) in the form

$$\min_{h_{i\Sigma}:h_{\Sigma\Sigma}=1} (1 - h_{j\Sigma})^2 a + h_{j\Sigma}^2 \frac{s^2}{w_{j\Sigma}} + (1 - h_{j\Sigma})^2 \sum_{i \neq j} \frac{h_{i\Sigma}^2}{(1 - h_{j\Sigma})^2} \frac{a}{z_i}. \tag{8.38}$$

As $h_{\Sigma\Sigma} = 1$, we have $\sum_{i \neq j} h_{i\Sigma}/(1 - h_{j\Sigma}) = 1$. So again because of Exercise 8.4.1, the optimal choice in (8.38) for the factors $h_{i\Sigma}$, $i \neq j$ is

$$\frac{h_{i\Sigma}}{1 - h_{j\Sigma}} = \frac{z_i}{z_\Sigma - z_j}. \tag{8.39}$$

The minimal value for the sum in (8.38) is $a/(z_\Sigma - z_j)$, so (8.38) leads to

$$\min_{h_{j\Sigma}}(1 - h_{j\Sigma})^2 (a + \frac{a}{z_\Sigma - z_j}) + h_{j\Sigma}^2 \frac{s^2}{w_{j\Sigma}}. \tag{8.40}$$

The optimal value for $h_{j\Sigma}$, finally, can be found by once again applying Exercise 8.4.1. This optimal value is, as the reader may verify,

$$h_{j\Sigma} = \frac{w_{j\Sigma}}{\frac{s^2}{a + a/(z_\Sigma - z_j)} + w_{j\Sigma}} = \frac{1}{\frac{1/z_j - 1}{1 + 1/(z_\Sigma - z_j)} + 1}$$

$$= \frac{z_j(z_\Sigma - z_j + 1)}{(1 - z_j)(z_\Sigma - z_j) + z_j(z_\Sigma - z_j + 1)} = z_j + (1 - z_j)\frac{z_j}{z_\Sigma}. \tag{8.41}$$

Because of (8.39) we see that $h_{i\Sigma} = (1 - z_j)z_i/z_\Sigma$, which implies that (8.34) is indeed the MSE-optimal homogeneous unbiased linear predictor of the risk premium $m + \Xi_j$.   ∇

Notice that if we replace Ξ_j in (8.31) by the constant ξ_j, that is, we take $a = 0$, we get the classical weighted mean X_{ww}. This is because in that case the relative weight $w_{j\Sigma}$ for X_{jw} is equal to the credibility weight z_j.

The *inhomogeneous* estimator of $m + \Xi_j$ contains a constant h, next to the homogeneous linear combination of the X_{jt} in (8.33). One may show, just as in Theorem 8.2.4, that the unbiasedness restriction is superfluous in this situation. The inhomogeneous estimator is equal to the homogeneous one, except that X_{zw} in (8.34) is replaced by m. The observations outside group j do not occur in the estimator. For the inhomogeneous estimator, both the ratio s^2/a and the value of m must be known. By replacing m by its best estimator X_{zw} under model (8.31), we get the homogeneous estimator again. Just as in Remark 8.2.6, the optimal predictor of $m + \Xi_j$ is also the optimal predictor of $X_{j,T+1}$. The asymptotic properties of (8.34) are analogous to those given in Remark 8.2.3. Also, the credibility premium can be found by combining the actual experience with virtual experience, just as in Remark 8.2.7. See the exercises.

8.4.1 Parameter estimation in the Bühlmann-Straub model

The credibility estimators of this chapter depend on the generally unknown structure parameters m, a and s^2. To be able to apply them in practice, one has to estimate these portfolio characteristics. Some unbiased estimators (not depending on the structure parameters that are generally unknown) are derived in the theorem below. We can replace the unknown structure parameters in the credibility estimators by these estimates, hoping that the quality of the resulting estimates is still good. The estimators of s^2 and a are based on the weighted sum-of-squares-within:

$$SSW = \sum_{j,t} w_{jt}(X_{jt} - X_{jw})^2, \tag{8.42}$$

and the weighted sum-of-squares-between

$$SSB = \sum_j w_{j\Sigma}(X_{jw} - X_{ww})^2. \tag{8.43}$$

Note that if all weights w_{jt} are taken equal to one, these expressions reduce to (8.2) and (8.3), defined in the balanced Bühlmann model.

Theorem 8.4.2 (Unbiased parameter estimates)
In the Bühlmann-Straub model, the following statistics are unbiased estimators of the corresponding structure parameters:

$$
\begin{aligned}
\widetilde{m} &= X_{ww}, \\
\widetilde{s^2} &= \frac{1}{J(T-1)} \sum_{j,t} w_{jt}(X_{jt} - X_{jw})^2, \\
\widetilde{a} &= \frac{\sum_j w_{j\Sigma}(X_{jw} - X_{ww})^2 - (J-1)\widetilde{s^2}}{w_{\Sigma\Sigma} - \sum_j w_{j\Sigma}^2 / w_{\Sigma\Sigma}}.
\end{aligned}
\tag{8.44}
$$

Proof. The proof of $E[X_{ww}] = m$ is easy. Using a weighted version of the covariance relations (8.15), we get for $\widetilde{s^2}$:

$$
\begin{aligned}
J(T-1)E[\widetilde{s^2}] &= \sum_{j,t} w_{jt} \left[\text{Var}[X_{jt}] + \text{Var}[X_{jw}] - 2\text{Cov}[X_{jt}, X_{jw}] \right\} \\
&= \sum_{j,t} w_{jt} \left\{ a + \frac{s^2}{w_{jt}} + a + \frac{s^2}{w_{j\Sigma}} - 2(a + \frac{s^2}{w_{j\Sigma}}) \right\} \\
&= J(T-1)s^2.
\end{aligned}
\tag{8.45}
$$

For $\widetilde{a}$ we have

$$
\begin{aligned}
&E\left[\sum_j w_{j\Sigma}(X_{jw} - X_{ww})^2 \right] \\
&= \sum_j w_{j\Sigma} \left\{ \text{Var}[X_{jw}] + \text{Var}[X_{ww}] - 2\text{Cov}[X_{jw}, X_{ww}] \right\} \\
&= \sum_j w_{j\Sigma} \left\{ a + \frac{s^2}{w_{j\Sigma}} + a \sum_k \frac{w_{k\Sigma}^2}{w_{\Sigma\Sigma}^2} + \frac{s^2}{w_{\Sigma\Sigma}} - 2\left(\frac{s^2}{w_{\Sigma\Sigma}} + \frac{a w_{j\Sigma}}{w_{\Sigma\Sigma}} \right) \right\} \\
&= a \sum_j w_{j\Sigma} \left(1 + \sum_k \frac{w_{k\Sigma}^2}{w_{\Sigma\Sigma}^2} - 2\frac{w_{j\Sigma}}{w_{\Sigma\Sigma}} \right) + s^2 \sum_j w_{j\Sigma} \left(\frac{1}{w_{j\Sigma}} - \frac{1}{w_{\Sigma\Sigma}} \right) \\
&= a\left(w_{\Sigma\Sigma} - \sum_j \frac{w_{j\Sigma}^2}{w_{\Sigma\Sigma}} \right) + (J-1)s^2.
\end{aligned}
\tag{8.46}
$$

Taking $E[\widetilde{a}]$ in (8.44), using (8.45) and (8.46) we see that $\widetilde{a}$ is unbiased as well. ∇

Remark 8.4.3 (Negativity of estimators)
The estimator $\tilde{s}^2$ is non-negative, but $\tilde{a}$ might well be negative. Although this may be an indication that $a = 0$ holds, it can also happen if $a > 0$. Let us elaborate on Example 8.2.1, returning to the balanced Bühlmann model where all weights w_{jt} are equal to one. In that case, defining MSW and MSB as in (8.4), the estimators of s^2 and a in Theorem 8.4.2 reduce to

$$\tilde{s}^2 = MSW; \qquad \tilde{a} = \frac{MSB - MSW}{T}. \tag{8.47}$$

To estimate z, we substitute these estimators into $z = \frac{aT}{aT + s^2}$, and we get the following statistic:

$$\tilde{z} = 1 - \frac{MSW}{MSB}. \tag{8.48}$$

Using $X_{jt} = m + \Xi_j + \Xi_{jt}$ and defining $\overline{\Xi}_j = \frac{1}{T}\sum_t \Xi_{jt}$, we see that the SSW can be written as

$$SSW = \sum_{j=1}^{J}\sum_{t=1}^{T}(X_{jt} - \overline{X}_j)^2 = \sum_{j=1}^{J}\sum_{t=1}^{T}(\Xi_{jt} - \overline{\Xi}_j)^2. \tag{8.49}$$

Under the assumption that the Ξ_{jt} are iid $N(0, s^2)$, the right hand side divided by s^2 has a $\chi^2(J(T-1))$ distribution. It is independent of the averages $\overline{\Xi}_j$, and hence also of the averages $\overline{X}_j = m + \Xi_j + \overline{\Xi}_j$. So MSW is independent of the $\overline{X}_j$, hence also of MSB.

Assuming that the components Ξ_j are iid $N(0, a)$, we find in similar fashion that

$$\frac{SSB}{a + s^2/T} = \frac{J-1}{aT + s^2}MSB \tag{8.50}$$

is $\chi^2(J-1)$ distributed. So under the normality assumptions made, if it is multiplied by the constant $s^2/(aT + s^2) = 1 - z$, the variance ratio MSB/MSW of Section 8.2 is still $F(J-1, J(T-1))$ distributed. Thus,

$$(1-z)\frac{MSB}{MSW} = \frac{1-z}{1-\tilde{z}} \sim F(J-1, J(T-1)). \tag{8.51}$$

In this way, $\Pr[\tilde{a} < 0]$ can be computed for different values of J, T and s^2/a, see for example Exercise 8.4.9.

Note that by (8.47), the event $\tilde{a} < 0$ is the same as $MSB/MSW < 1$. In Section 8.2 we established that the data indicate rejection of equal means, which boils down to $a = 0$ here, only if MSB/MSW exceeds the right-hand $F(J-1, J(T-1))$ critical value, which is larger than one for all J, T. Thus we conclude that, although $\Pr[\tilde{a} < 0] > 0$ for every $a > 0$, obtaining such a value means that a Fisher test for $a = 0$ based on these data would not have led to rejection. This in turn means that there is in fact no statistical reason not to charge every contract the same premium.

In order to estimate $a = \mathrm{Var}[\Xi_j]$, one would use $\max\{0, \tilde{a}\}$ in practice, but, though still consistent, this is not an unbiased estimator.   ∇

Remark 8.4.4 (Credibility weighted mean and ordinary weighted mean)
The best unbiased estimator of m in model (8.31) is not X_{ww}, but X_{zw}. This does not contradict Exercise 8.4.1, since both X_{ww} and X_{zw} are linear combinations of the random variables X_{jw}, the variances of which are not proportional to the original weights $w_{j\Sigma}$, but rather to the credibility adjusted weights z_j. So a lower variance is obtained if we estimate m by the credibility weighted mean X_{zw} instead of by the ordinary weighted mean X_{ww}. A problem is that we do not know the credibility factors z_j to be used, as they depend on the unknown parameters that we are actually estimating. One way to achieve better estimators is to use iterative *pseudo-estimators*, which find estimates of the structure parameters by determining a fixed point of certain equations. See Example 8.4.6, as well as the more advanced literature on credibility theory.   ▽

Example 8.4.5 (Computing the estimates in the Bühlmann-Straub model)
First, we generate a dataset consisting of J=10 contracts, with K=5 years of exposure each, satisfying the distributional assumptions (8.31) of the Bühlmann-Straub model. For that, we execute the following R-statements.

```
J <- 10; K <- 5; j <- rep(1:J, each=K); j <- as.factor(j)
m <- 100; a <- 100; s2 <- 64;
set.seed(6345789)
w <- 0.50 + runif(J*K)
X <- m + rep(rnorm(J, 0, sqrt(a)), each=K) +
         rnorm(J*K, 0, sqrt(s2/w))
```

Note that we attach a random weight in the interval $(0.5, 1.5)$ to each observation. In the last line, the second term is a vector of J independent $N(0, a)$ random drawings Ξ_j, replicated K times each, the last a vector of independent $N(0, s^2/w_{jk})$ random drawings Ξ_{jk}, $j = 1, \ldots, J, k = 1, \ldots, K$.

Just as in Example 8.2.1, we apply ANOVA to determine if there is any significant variation in the group means; if not, there is no heterogeneity between contracts in the portfolio, therefore no reason to apply credibility theory.

```
> anova(lm(X~j,weight=w))
Analysis of Variance Table

Response: X
          Df Sum Sq Mean Sq F value    Pr(>F)
j          9 5935.0   659.4  14.836 3.360e-10 ***
Residuals 40 1778.0    44.5
```

In this example the data clearly exhibit an effect of the factor group number. The Sum Sq values 5935 and 1778 are the *SSB* and the *SSW*, respectively; see (8.42) and (8.43), and see also below. The *MSB* and *MSW*, see (8.4), arise by dividing by Df.

In these laboratory conditions, the parameter values a, m and s^2 are known. Therefore we can directly compute the credibility premiums (8.34) for this case. First we compute the quantities in (8.32).

```
w.js <- tapply(w, j, sum); w.ss <- sum(w.js)
z.j <- 1 / (1 + s2/(a*w.js)); z.s <- sum(z.j)
X.jw <- tapply(X*w, j, sum)/w.js
X.ww <- sum(X.jw * w.js) / w.ss
X.zw <- sum(X.jw * z.j) / z.s
pr.j <- z.j * X.jw + (1-z.j) * X.zw #(8.34)
```

In the real world, these parameters m, s^2 and a are of course unknown and have to be estimated from the data. In (8.42)–(8.44), we find formulas for unbiased estimators $\tilde{m}$, $\tilde{s}^2$ and $\tilde{a}$. Using R, they can be found as follows:

```
m.tilde <- X.ww
SSW <- sum(w*(X-X.jw[j])^2)
s2.tilde <- SSW/J/(K-1)
SSB <- sum(w.js*(X.jw-X.ww)^2)
a.tilde <- (SSB - (J-1)*s2.tilde) / (w.ss - sum(w.js^2)/w.ss)
```

Using the statements:

```
z.j.tilde <- 1 / (1 + s2.tilde / (a.tilde * w.js))
z.s.tilde <- sum(z.j.tilde)
X.zw.tilde <- sum(X.jw * z.j.tilde)/ z.s.tilde
pr.j.tilde <- z.j.tilde * X.jw + (1-z.j.tilde) * X.zw.tilde
```

we can recompute the credibility premiums (8.34) using the unbiased parameter estimates and (8.32). ∇

Example 8.4.6 (A pseudo-estimator for the heterogeneity parameter)
The estimator $\tilde{a}$ of the heterogeneity parameter a given in (8.44) is unbiased, but it is also awkward looking and unintuitive. Consider the unbiased estimate of s^2, the heterogeneity in time of the results of the contract, in (8.44):

$$\tilde{s}^2 = \frac{1}{J(T-1)} \sum_{j,t} w_{jt}(X_{jt} - X_{jw})^2. \tag{8.52}$$

It adds up the squared differences of the observations with the contract mean, weighted by the natural weight w_{jt} that was used to construct X_{jw}. To get an unbiased estimate of s^2, we divide by the total experience JT, corrected for the fact that J means have been estimated so only $J(T-1)$ independent terms remain.

To get an analogous estimate of the between-groups heterogeneity a, consider

$$A = \frac{1}{J-1} \sum_j z_j(X_{jw} - X_{zw})^2. \tag{8.53}$$

In this case, there are J groups, and one mean is estimated. The squared differences between group mean and the best estimate X_{zw} for the overall mean m have weight proportional to the credibility weight z_j, the same set of weights that produces the minimal variance estimator X_{zw}. The reader is invited to show that $E[A] = a$, see the exercises.

There is, however, a problem with the random variable A. If we fill in X_{zw} in the previous equation we get:

$$A = \frac{1}{J-1} \sum_j z_j \left(X_{jw} - \sum_i \frac{z_i}{z_\Sigma} X_{iw} \right)^2 \quad \text{with} \quad z_j = \left(1 + \frac{s^2}{aw_{j\Sigma}} \right)^{-1}. \qquad (8.54)$$

So the right hand side depends on the unknown structure parameters a and s^2 (actually only on the ratio s^2/a), let us say as $A = f(a, s^2)$ with f the appropriate function. As a result, the random variable A is not a statistic, hence not an estimator. In such cases, we speak of *pseudo-estimators*. So we look at the following estimate of A:

$$A_1 = f(\widetilde{a}, \widetilde{s^2}). \qquad (8.55)$$

But A_1 does not use the 'best' credibility weights available, therefore we look at $A_2 = f(A_1, \widetilde{s^2})$. Optimizing, we then look iteratively at $A_{n+1} = f(A_n, \widetilde{s^2})$, $n = 2, 3, \ldots$. Taking the limit for $n \to \infty$, and calling the limiting random variable $\lim_{n \to \infty} A_n =: \widehat{a}$, we see that the random variable $\widehat{a}$ is the solution to the following implicit equation:

$$\widehat{a} = \frac{1}{J-1} \sum_j \widehat{z}_j \left(X_{jw} - \sum_i \frac{\widehat{z}_i}{\widehat{z}_\Sigma} X_{iw} \right)^2 \quad \text{with} \quad \widehat{z}_j = \left(1 + \frac{\widetilde{s^2}}{\widehat{a} w_{j\Sigma}} \right)^{-1}. \qquad (8.56)$$

The exact statistical properties of this estimator $\widehat{a}$ are hard to determine; even proving existence and uniqueness of the solution is a problem. But it is very easy to solve this equation by successive substitution, using the following R statements:

```
a.hat <- a.tilde
repeat {
    a.hat.old <- a.hat
    z.j.hat <- 1/(1+s2.tilde/(w.js*a.hat))
    X.zw.hat <- sum(z.j.hat * X.jw) / sum(z.j.hat)
    a.hat <- sum(z.j.hat*(X.jw-X.zw.hat)^2)/(J-1)
    if (abs((a.hat-a.hat.old)/a.hat.old) < 1e-6) break}
```

Here `a.tilde`, assumed positive, and `s2.tilde` are the unbiased estimates, `a.hat.old` is the current best guess A_n of $\widehat{a}$, and `a.hat` is the next one A_{n+1}. ∇

8.5 Negative binomial model for the number of car insurance claims

In this section we expand on Example 8.3.2 by considering a driver with a random accident proneness drawn from a non-degenerate distribution, and, given that his accident proneness equals λ, a Poisson(λ) distributed number of claims in a year. Charging a credibility premium in this situation leads to an experience rating system that resembles the bonus-malus systems we described in Chapter 6.

If for a motor insurance policy, all relevant variables for the claim behavior of the policyholder can be observed as well as used, the number of claims still is generated by a stochastic process. Assuming that this process is a Poisson process, the rating factors cannot do more than provide us with the exact Poisson intensity, that is, the Poisson parameter of the number of claims each year. Of the claim size, we know the probability distribution. The cell with policies sharing common values for all the risk factors would be homogeneous, in the sense that all policy holders have the same Poisson parameter and the same claims distribution. In reality, however, some uncertainty about the parameters remains, because it is impossible to obtain all relevant information on these parameters. So the cells are heterogeneous. This heterogeneity is the actual justification of using a bonus-malus system. In case of homogeneity, each policy represents the same risk, and there is no ground for asking different premiums within a cell.

The heterogeneity of the claim frequency can be modeled by assuming that the Poisson parameter λ has arisen from a structure variable Λ, with distribution $U(\lambda) = \Pr[\Lambda \leq \lambda]$. Just as in (8.5), we decompose the number of claims X_{jt} for driver $j = 1,\ldots,J$ in time period $t = 1,\ldots,T_j$ as follows:

$$X_{jt} = \mathrm{E}[\Lambda] + \{\Lambda_j - \mathrm{E}[\Lambda]\} + \{X_{jt} - \Lambda_j\}. \tag{8.57}$$

Here $\Lambda_j \sim \Lambda$ iid. The last two components are uncorrelated, but not independent; see Exercise 8.5.6. Component $\Lambda_j - \mathrm{E}[\Lambda]$ has variance $a = \mathrm{Var}[\Lambda]$; for $X_{jt} - \Lambda_j$, just as in Example 3.3.1, $\mathrm{Var}[X_{jt}] - \mathrm{Var}[\Lambda_j] = \mathrm{E}[\Lambda]$ remains. The structural parameters m and s^2 coincide because of the Poisson distributions involved.

Up to now, except for its first few moments, we basically ignored the structure distribution. Several models for it are possible. Because of its mathematical properties and good fit (see later on for a convincing example), we will prefer the gamma distribution. Another possibility is the structure distribution that produces a 'good' driver, having claim frequency λ_1, with probability p, or a 'bad' driver with claim frequency $\lambda_2 > \lambda_1$. The number of claims of an arbitrary driver then has a mixed Poisson distribution with a two-point mixing distribution. Though one would expect more than two types of drivers to be present, this 'good driver/bad driver' model quite often fits rather closely to data found in practice.

For convenience, we drop the index j, except when we refer back to earlier sections. It is known, see again Example 3.3.1, that if the structure distribution of the Poisson parameter is gamma(α, τ), the marginal distribution of the number of claims X_t of driver j in time period t is negative binomial$(\alpha, p = \tau/(\tau + 1))$. In Lemaire (1985), we find data from a Belgian portfolio with $J = 106\,974$ policies, see Table 8.1. The numbers n_k, $k = 0,1,\ldots$, denote the number of policies with k accidents. If $X_t \sim \mathrm{Poisson}(\lambda)$ for all j, the maximum likelihood estimate $\widehat{\lambda}$ for λ equals the average number of claims over all j. In Section 3.9 we showed how to find the negative binomial parameter estimates $\widehat{\alpha}$ and $\widehat{p}$ by maximum likelihood, solving (3.76) and (3.77). Then $\widehat{\tau} = \widehat{p}/(1 - \widehat{p})$ in the Poisson-gamma mixture model follows from the invariance property of ML-estimators in case of reparameterization. Equation (3.76) ensures that the first moment of the estimated structure distribution, hence also of the marginal distribution of the number of claims, coincides with the first sample moment. The parameters p, λ_1 and λ_2 of the good driver/bad driver model have been estimated by the method of moments. Note that this method might fail to

Table 8.1 Observed numbers of accidents in some portfolio, and fitted values for a pure Poisson model and a negative binomial model fitted with ML, and a mixed Poisson model fitted by the method of moments.

k	n_k	$\widehat{n}_k$ (Poisson)	$\widehat{n}_k$ (Neg.Bin.)	$\widehat{n}_k$ (good/bad)
0	96 978	96 689.5	96 980.8	96 975.1
1	9 240	9 773.4	9 230.9	9 252.0
2	704	494.0	708.6	685.0
3	43	16.6	50.0	56.9
4	9	0.4	3.4	4.6
5+	0	0.0	0.2	0.3
χ^2		191.	0.1	2.1

produce admissible estimates $\widehat{\lambda}_i \geq 0$ and $0 \leq \widehat{p} \leq 1$. The resulting estimates for the three models considered were

$$\widehat{\lambda} = 0.1010806;$$
$$\widehat{\alpha} = 1.631292, \quad \widehat{\tau} = 16.13852; \qquad\qquad (8.58)$$
$$\widehat{\lambda}_1 = 0.07616114, \quad \widehat{\lambda}_2 = 0.3565502, \quad \widehat{p} = 0.8887472.$$

Observed and estimated frequencies are in Table 8.1. The bottom row contains $\chi^2 = \sum_k (n_k - \widehat{n}_k)^2 / \widehat{n}_k$. When computing such χ^2-statistics, one usually combines cells with estimated numbers less than 5 with neighboring cells. So the last three rows are joined together into one row representing 3 or more claims. The two mixed models provide an excellent fit; in fact, the fit of the negative binomial model is almost too good to be true. Note that we fit 4 numbers using 2 or 3 parameters. But homogeneity for this portfolio is rejected without any doubt whatsoever.

Though the null-hypothesis that the numbers of claims for each policy holder are independent Poisson random variables with the same parameter is rejected, while the mixed Poisson models are not, we cannot just infer that policy holders have a fixed unobservable risk parameter, drawn from a structure distribution. It might well be that the numbers of claims are just independent negative binomial random variables, for example because the number of claims follows a Poisson process in which each year a new intensity parameter is drawn independently from a gamma structure distribution.

With the model of this section, we want to predict as accurately as possible the number of claims that a policy holder produces in the next time period $T + 1$. This number is a Poisson(λ) random variable, with λ an observation of Λ, of which the prior distribution is known to be, say, gamma(α, τ). Furthermore, observations $X_1, \ldots, X_T$ from the past are known. The posterior distribution of Λ, given $X_1 = x_1, \ldots, X_T = x_T$, is also a gamma distribution, with adjusted parameters $\tau' = \tau + T$ and $\alpha' = \alpha + x_\Sigma$ with $x_\Sigma = x_1 + \cdots + x_T$; see Exercise 8.5.2. Assuming a quadratic loss function, in view of Exercise 8.2.9 the best predictor of the number of claims next year is the posterior expectation of Λ:

$$\lambda_{T+1}(x_1,\ldots,x_T) = \frac{\alpha + x_\Sigma}{\tau + T}. \tag{8.59}$$

This is just the observed average number of claims per time unit, provided we include a virtual prior experience of α claims in a time period of length τ. See also Remark 8.2.7. The forecasted premium (8.59) is also a credibility forecast, being a linear combination of a priori premium and policy average, because, see (8.10):

$$\frac{\alpha + x_\Sigma}{\tau + T} = z\frac{x_\Sigma}{T} + (1-z)\frac{\alpha}{\tau} \quad \text{for} \quad z = \frac{T}{\tau + T}. \tag{8.60}$$

Remark 8.5.1 (Non-linear estimators; exact credibility)
In Theorems 8.2.2 and 8.2.4 it was required that the predictors of $X_{j,T+1}$ were linear in the observations. Though such linear observations are in general the easiest to deal with, one may also look at more general functions of the data. Without linearity restriction, the best predictor in the sense of MSE for $X_{j,T+1}$ is the so-called *posterior Bayes estimator*, which is just the conditional mean $E[X_{j,T+1}|X_{11},\ldots,X_{JT}]$. See also (8.59). If the Ξ_j and the Ξ_{jt} are independent *normal* random variables, the optimal linear estimator coincides with the Bayes estimator. In the literature, this is expressed as 'the credible mean is exact Bayesian'. Also combining a gamma prior and a Poisson posterior distribution gives such 'exact credibility', because the posterior Bayes estimator happens to be linear in the observations. See Exercise 8.5.2. The posterior mean of the claim figure is equal to the credibility premium (8.60).   ▽

If we split up the premium necessary for the whole portfolio according to the mean value principle, we get an experience rating system based on credibility, which is a solid system for the following reasons:

1. The system is fair. Upon renewal of the policy, every insured pays a premium that is proportional to his estimated claim frequency (8.59), taking into account all information from the past.
2. The system is balanced financially. Write $X_\Sigma = X_1 + \cdots + X_T$ for the total number of claims generated, then $E[X_\Sigma] = E[E[X_\Sigma|\Lambda]] = TE[\Lambda]$, so

$$E\left[\frac{\alpha + X_\Sigma}{\tau + T}\right] = \frac{\alpha + T\frac{\alpha}{\tau}}{\tau + T} = \frac{\alpha}{\tau}. \tag{8.61}$$

This means that for every policy, the mean of the proportionality factor (8.59) is equal to its overall mean α/τ. So the expected value of the premium to be paid by an arbitrary driver remains constant over the years.
3. The premium only depends on the number of claims filed in the previous T years, and not on how these are distributed over this period. So for the premium next year, it makes no difference if the claim in the last five years was in the first or in the last year of this period. The bonus-malus system in Section 6.2 does not have this property. But it is questionable if this property is even desirable. If one assumes, like here, the intensity parameter λ to remain constant, K is a sufficient statistic. In practice, however, the value of λ is not constant. People get past their

Table 8.2 Optimal estimates (8.62) of the claim frequency next year compared with a new driver

Nr. of claims x_Σ	0	1	2	3	4	5	6	7	8	9	10
					Number of years T						
0	100	94	89	84	80	76	73	70	67	64	62
1		153	144	137	130	124	118	113	108	104	100
2		212	200	189	180	171	164	157	150	144	138
3		271	256	242	230	219	209	200	192	184	177
4		329	311	295	280	267	255	243	233	224	215

youth or past their prime, or the offspring gets old enough to drive the family car. Following this reasoning, later observations should count more heavily than old ones.

4. Initially, at time $t = 0$, everyone pays the same premium, proportional to α/τ. If T tends to ∞, the difference between the premium $(\alpha + x_\Sigma)/(\tau + T)$ asked and the actual average payments on the policy x_Σ/T vanishes. The variance $(\alpha + x_\Sigma)/(\tau + T)^2$ of the posterior distribution converges to zero. So in the long run, everyone pays the premium corresponding to his own risk; the influence of the virtual experience vanishes.

Using the values $\alpha = 1.6$ and $\tau = 16$, see (8.58), we have constructed Table 8.2 giving the optimal estimates of the claim frequencies in case of various lengths T of the observation period and numbers $k = x_\Sigma$ of claims observed. The initial premium is set to 100%, the a posteriori premiums are computed as:

$$100 \frac{\lambda_{T+1}(x_1, \ldots, x_T)}{\lambda_1} = \frac{100 \frac{\alpha + x_\Sigma}{\tau + T}}{\alpha/\tau} = 100 \frac{\tau(\alpha + x_\Sigma)}{\alpha(\tau + T)} \qquad (8.62)$$

One sees that in Table 8.2, a driver who caused exactly one claim in the past ten years represents the same risk as a new driver, who is assumed to carry with him a virtual experience of 1.6 claims in 16 years. A person who drives claim-free for ten years gets a discount of $1 - \tau/(\tau + 10) = 38\%$. After a claims experience of 16 years, actual and virtual experience count just as heavily in the premium.

Example 8.5.2 (Comparison with the bonus-malus system of Chapter 6)
As an example, look at the premiums to be paid in the 6th year of insurance by a driver who has had one claim in the first year of observation. In Table 8.2, his premium next year equals 124%. In the system of Table 6.1, his path on the ladder was $2 \to 1 \to 2 \to 3 \to 4$, so now he pays the premium of step 5, that is, 70%. The total of the premiums paid (see Table 8.2) is $100 + 153 + 144 + 137 + 130 + 124 = 788\%$ of the premium for a new entrant. In the system of Table 6.1, he has paid only $100 + 120 + 100 + 90 + 80 + 70 = 560\%$. Note that for the premium next year in Table 8.2, it makes no difference if the claim occurred in the first or the last year of observation, though this affects the total claims paid. $\qquad \triangledown$

Remark 8.5.3 (Overlapping claim frequencies)
Consider a policyholder with T years of claims experience. The posterior distribution of the expected number of claims Λ is gamma$(\alpha + x_\Sigma, \tau + T)$ if x_Σ claims were filed. If $T = 3$, in case $x_\Sigma = 0$ and $x_\Sigma = 2$, the premium to be paid next year differs by a factor $189/84 = 2.25$. But the posterior distributions of both claim frequencies overlap to a large extent. Indeed, in the first case, the probability is 60.5% to have a claim frequency lower than the average $\alpha/(\tau + T) = 0.0842$ for drivers with a similar claims experience, since $G(0.0842; \alpha, \tau + T) = 0.605$. But in the second case, there also is a substantial probability to have a better Poisson parameter than the average of drivers as above, since $G(0.0842; \alpha + x_\Sigma, \tau + T) = 0.121$ for $x_\Sigma = 2$ and $T = 3$. Experience rating by any bonus-malus system may be quite unfair for 'good' drivers who are unlucky enough to produce claims. ∇

8.6 Exercises

Section 8.2

1. Finish the proofs of Theorems 8.2.2 and 8.2.4 by filling in and deriving the relevant covariance relations (8.15). Use and verify the linearity properties of covariances: for all random variables X, Y and Z, we have $\mathrm{Cov}[X, Y + Z] = \mathrm{Cov}[X, Y] + \mathrm{Cov}[X, Z]$, while for all real α, $\mathrm{Cov}[X, \alpha Y] = \alpha \mathrm{Cov}[X, Y]$.

2. Let $X_1, \ldots, X_T$ be uncorrelated random variables with mean m and variance s^2. Consider the weighted average $X_w = \sum_t w_t X_t$, where the weights $w_t \geq 0, t = 1, \ldots, T$ satisfy $\sum_t w_t = 1$. Show that $\mathrm{E}[X_w] = m$, $\mathrm{Cov}[X_t, X_w] = w_t s^2$ and $\mathrm{Var}[X_w] = \sum_t w_t^2 s^2$.
    [If especially $w_t \equiv \frac{1}{T}$, we get $X_w = \overline{X}$ and $\mathrm{E}[\overline{X}] = m; \mathrm{Cov}[X_t, \overline{X}] = \mathrm{Var}[\overline{X}] = \frac{s^2}{T}$.]

3. Show that the sample variance $S^2 = \frac{1}{T-1} \sum_1^T \{X_t - \overline{X}\}^2$ is an unbiased estimator of s^2.

4. Show that the best predictor of $X_{j,T+1}$ is also the best estimator of the risk premium $m + \Xi_j$ in the situation of Theorem 8.2.2. What is the best linear unbiased estimator (BLUE) of Ξ_j?

5. Determine the variance of the credibility premium (8.9). What is the MSE? Also determine the MSE of (8.9) as an estimator of $m + \Xi_j$.

6. Determine the credibility estimator if the unbiasedness restriction is not imposed in Theorem 8.2.2. Also investigate the resulting bias.

7. Show that if each contract pays the homogeneous premium, the sum of the credibility premiums equals the average annual outgo in the observation period.

8. Show that in model (8.5), the *MSB* has mean $aT + s^2$, while the *MSW* has mean s^2.

9. Prove that for each random variable Y, the real number p that is the best predictor of it in the sense of *MSE* is $p = \mathrm{E}[Y]$.

10. Let $\vec{X} = (X_{11}, \ldots, X_{1T}, X_{21}, \ldots, X_{2T}, \ldots, X_{J1}, \ldots, X_{JT})^T$ be the vector containing the observable random variables in (8.7). Describe the covariance matrix $\mathrm{Cov}[\vec{X}, \vec{X}]$.

Section 8.3

1. Derive the formula $\mathrm{Cov}[X,Y] = \mathrm{E}[\mathrm{Cov}[X,Y|Z]] + \mathrm{Cov}[\mathrm{E}[X|Z],\mathrm{E}[Y|Z]]$ for the decomposition of covariances into conditional covariances.

Section 8.4

1. Let $X_1,\ldots,X_T$ be independent random variables with variances $\mathrm{Var}[X_t] = s^2/w_t$ for certain positive numbers (weights) w_t, $t = 1,\ldots,T$. Show that the variance $\sum_t \alpha_t^2 s^2/w_t$ of the linear combination $\sum_t \alpha_t X_t$ with $\alpha_\Sigma = 1$ is minimal when we take $\alpha_t \propto w_t$, where the symbol $\propto$ means 'proportional to'. Hence the optimal solution has $\alpha_t = w_t/w_\Sigma$. Prove also that the minimal value of the variance in this case is s^2/w_Σ.

2. Prove that in model (8.31), we have $\mathrm{Var}[X_{zw}] \le \mathrm{Var}[X_{ww}]$. See Remark 8.4.4.

3. Determine the best homogeneous linear estimator of m.

4. Show that in determining the best inhomogeneous linear estimator of $m + \Xi_j$, the unbiasedness restriction is superfluous.

5. Show that, just as in Remark 8.2.6, the optimal predictors of $X_{j,T+1}$ and $m + \Xi_j$ coincide in the Bühlmann-Straub model.

6. Describe the asymptotic properties of z_j in (8.32); see Remark 8.2.3.

7. In the same way as in Remark 8.2.7, describe the credibility premium (8.34) as a mix of actual and virtual experience.

8. Show that (8.9) follows from (8.34) in the special case (8.5)–(8.6) of the Bühlmann-Straub model given in (8.31).

9. In the situation of Remark 8.4.3, for $s^2/a = 0.823$, $J = 5$ and $T = 4$, show that the probability of the event $\tilde{a} < 0$ equals 0.05.

10. Estimate the credibility premiums in the Bühlmann-Straub setting when the claims experience for three years is given for three contracts, each with weight $w_{jt} \equiv 1$. Find the estimates both by hand and by using R, if the claims on the contracts are as follows:

	$t = 1$	$t = 2$	$t = 3$
$j = 1$	10	12	14
$j = 2$	13	17	15
$j = 3$	14	10	6

11. Show that the pseudo-estimator A in (8.53) has indeed mean a.

12. Compare the quality of the iterative estimator $\hat{a}$ in (8.56) and the unbiased one $\tilde{a}$, by generating a large sample (say, 100 or 1000 replications of the laboratory portfolio as above). Look at sample means and variances, and plot a histogram. Count how often the iterative estimate is closer to the real value a.

Section 8.5

1. Verify that the parameters estimates given in (8.58) are as they should be.

2. [♠] Suppose that Λ has a gamma(α, τ) prior distribution, and that given $\Lambda = \lambda$, the annual numbers of claims $X_1, \ldots, X_T$ are independent Poisson(λ) random variables. Prove that the posterior distribution of Λ, given $X_1 = x_1, \ldots, X_T = x_T$, is gamma$(\alpha + x_\Sigma, \tau + T)$, where $x_\Sigma = x_1 + \cdots + x_T$.

3. By comparing $\Pr[X_2 = 0]$ with $\Pr[X_2 = 0 | X_1 = 0]$ in the previous exercise, show that the numbers of claims X_t are not marginally independent. Also show that they are not uncorrelated.

4. Show that the mode of a gamma(α, τ) distribution, that is, the argument where the density is maximal, is $(\alpha - 1)_+ / \tau$.

5. [♠] Determine the estimated values for n_k and the χ^2-test statistic if α and τ are estimated by the method of moments.

6. Show that in the model (8.57) of this section, Λ_j and $X_{jt} - \Lambda_j$ are uncorrelated. Taking $\alpha = 1.6$ and $\tau = 16$, determine the ratio $\mathrm{Var}[\Lambda_j]/\mathrm{Var}[X_{jt}]$. [Since no model for X_{jt} can do more than determine the value of Λ_j as precisely as possible, this ratio provides an upper bound for the attainable 'percentage of explained variation' on an individual level.]

7. [♠] What is the Loimaranta efficiency of the system in Table 8.2? What is the steady state distribution?

8. Verify the estimated values for n_k and the χ^2-test statistic if the estimates $\hat{\lambda}_1$, $\hat{\lambda}_2$, $\hat{p}$ in (8.58) are determined by maximum likelihood.

Chapter 9
Generalized linear models

R, an open-source programming environment for data analysis and graphics, has in only a decade grown to become a de-facto standard for statistical analysis against which many popular commercial programs may be measured. The use of R for the teaching of econometric methods is appealing. It provides cutting-edge statistical methods which are, by R's open-source nature, available immediately. The software is stable, available at no cost, and exists for a number of platforms —
Jeff Racine & Rob Hyndman, 2002

9.1 Introduction

Multiple linear regression is the most widely used statistical technique in practical econometrics. In actuarial statistics, situations occur that do not fit comfortably in that setting. Regression assumes normally distributed disturbances with a constant variance around a mean that is linear in the collateral data. In many actuarial applications, a symmetric normally distributed random variable with a variance that is the same whatever the mean does not adequately describe the situation. For counts, a Poisson distribution is generally a good model, if the assumptions of a Poisson process described in Chapter 4 are valid. For these random variables, the mean and variance are the same, but the datasets encountered in practice generally exhibit a variance greater than the mean. A distribution to describe the claim size should have a thick right-hand tail. The distribution of claims expressed as a multiple of their mean would always be much the same, so rather than a variance not depending of the mean, one would expect the coefficient of variation to be constant. Furthermore, the phenomena to be modeled are rarely additive in the collateral data. A multiplicative model is much more plausible. If other policy characteristics remain the same, moving from downtown to the country would result in a reduction in the average total claims by some fixed percentage of it, not by a fixed amount independent of the original risk. The same holds if the car is replaced by a lighter one.

These problems can be solved in an elegant way by working with Generalized Linear Models (GLM) instead of ordinary linear models. The generalization is in two directions. First, it is allowed that the random deviations from the mean have a distribution different from the normal. In fact, one can take any distribution from the exponential dispersion family, which includes apart from the normal distribution also the Poisson, the (negative) binomial, the gamma and the inverse Gaussian distributions. Second, in ordinary Linear Models the mean of the random variable is a linear function of the explanatory variables, but in GLMs it may be linear on some other scale. If this scale for example is logarithmic, we have in fact a multiplicative model instead of an additive model.

Often, one does not look at the observations themselves, but at transformed values that are better suited for the ordinary multiple regression model, with normality, hence symmetry, with a constant variance and with additive systematic effects. But this is not always possible. Look at the following example.

Example 9.1.1 (Transformations of a Poisson random variable)
Let $X \sim \text{Poisson}(\mu)$. Then the transformation $X^{2/3}$ makes X more or less symmetric, taking the square root stabilizes the variance and taking a log-transform reduces multiplicative systematic effects to additive effects.

We will in fact demonstrate that

a) if $Y = X^{2/3}$, then its skewness $\gamma_Y \approx 0$;
b) if $Y = X^{1/2}$, then its variance $\sigma_Y^2 \approx \frac{1}{4}$ is stable;
c) if $Y = \log(X + \frac{1}{2})$, then $E[Y] \approx \log\mu$ (log-linearity).

See Exercise 9.1.1 for how well these approximations work.

Proof. These approximations are based on the so-called *delta method*. It says that for a function $g(\cdot)$ and a random variable X with mean $E[X] = \mu$ and variance $\text{Var}[X] = \sigma^2$, the following μ-asymptotic approximations hold:

$$E[g(X)] \approx g(\mu) + \frac{1}{2}g''(\mu)\sigma^2; \quad \text{Var}[g(X)] \approx (g'(\mu))^2\sigma^2. \tag{9.1}$$

They can be justified by looking at Taylor expansions around μ:

$$g(X) \approx g(\mu) + (X - \mu)g'(\mu) + \frac{1}{2}(X - \mu)^2 g''(\mu). \tag{9.2}$$

The first approximation in (9.1) follows from this by simply taking expectations. Leaving out the last term and taking variances, we get the second one.

We successively have

a) If $Y = X^{2/3}$, by (9.1) with $g(x) = x^{2/3}$ we have

$$E[Y^3] = E[(X^{2/3})^3] = E[X^2] = \mu^2 + \mu;$$
$$E[Y] \approx g(\mu) + \frac{1}{2}\sigma^2 g''(\mu) = \mu^{2/3} - \frac{1}{9}\mu^{-1/3}; \tag{9.3}$$
$$\text{Var}[Y] \approx \sigma^2 g'(\mu)^2 = \frac{4}{9}\mu^{1/3}.$$

So for the third central moment we get

$$E[(Y - E[Y])^3] = E[Y^3] - 3E[Y]\text{Var}[Y] - (E[Y])^3$$
$$\approx \mu^2 + \mu - \frac{4}{3}\mu + \frac{4}{27} - \left(\mu^2 - \frac{1}{3}\mu + \frac{1}{27} - \frac{1}{729\mu}\right) \tag{9.4}$$
$$= \frac{1}{9} + \frac{1}{729\mu}.$$

Since $\sigma_Y^3 \approx \frac{8}{27}\sqrt{\mu}$, for the skewness we have $\gamma_Y \approx 0$ for large μ.

b) If $Y = X^{1/2}$, then by (9.1) with $g(x) = \sqrt{x}$,

$$\text{Var}[Y] \approx \text{Var}[X]\left(\frac{1}{2\sqrt{E[X]}}\right)^2 = \frac{1}{4}. \tag{9.5}$$

c) If $Y = \log(X + \frac{1}{2})$, then using (9.1) and $\log(1 + u) = u + O(u^2)$,

$$E[Y] \approx \log\left(\mu + \frac{1}{2}\right) + \frac{1}{2}\mu\frac{-1}{(\mu + \frac{1}{2})^2}$$

$$= \log\mu + \log\frac{\mu + \frac{1}{2}}{\mu} - \frac{\mu}{2(\mu + \frac{1}{2})^2} \tag{9.6}$$

$$= \log\mu + \frac{\frac{1}{2}}{\mu} + O(\mu^{-2}) - \frac{\mu}{2(\mu + \frac{1}{2})^2} = \log\mu + O(\mu^{-2}).$$

Note that looking at $Y = \log X$ gives problems since $\Pr[X = 0] > 0$. ∇

So using transformations does not always achieve 'normality'; none of the above transformations at the same time leads to skewness zero, homoskedasticity and additive systematic effects. Note also that an unbiased estimator in the new scale is no longer unbiased when returning to the original scale, by Jensen's inequality (1.7).

In this chapter, we will not deal with Generalized Linear Models in their full generality. Mostly, we restrict to cross-classified observations, which can be put into a two-dimensional table in a natural way. The relevant collateral data with random variable X_{ij} are the row number i and the column number j. In the next chapter, we will also include the 'diagonal number' $i + j - 1$ as an explanatory variable. For more general models, we refer to Chapter 11. In general, the observations are arranged in a vector of n outcomes of independent but not identically distributed random variables, and there is a design matrix containing the explanatory variables in a suitable form.

Many actuarial problems can be tackled using specific Generalized Linear Models such as ANOVA, Poisson regression and logit and probit models. GLMs can also be applied to IBNR problems, see the next chapter, to survival data, and to compound Poisson distributions. Furthermore, it proves that some venerable actuarial techniques are in fact instances of GLMs. In the investigation that led to the bonus-malus system of Chapter 6, estimation techniques were chosen with a simple heuristic foundation that also turn out to produce maximum likelihood estimates in specific GLMs. The same holds for some widely used IBNR techniques. As opposed to credibility models, there is a lot of software that is able to handle GLMs. The first was the specialized program GLIM (Generalized Linear Interactive Modeling), originally developed by the Numerical Algorithms Group (NAG) under the auspices of Nelder. Apart from the open-source program R that we use, we mention as commercial alternatives S-Plus also implementing the programming language S, the module GenMod included in SAS, and the program Stata.

The study of Generalized Linear Models was initiated by Nelder and Wedderburn in the early 1970s. They gave a unified description, in the form of a GLM, of a multitude of statistical methods, including ANOVA, probit-analysis and many others. Also, they gave an algorithm to estimate all these models optimally and efficiently; this is described in Chapter 11. In later versions of GLIM, other algorithms were implemented to improve stability in some situations.

In Section 9.2, we briefly present generalized linear models. In Section 9.3, we show how some rating systems used in actuarial practice in fact are instances of GLMs. In Section 9.4, we study the deviance (and the scaled deviance) as a measure for the goodness of fit. For normal distributions, these quantities are sums of squared residuals, hence χ^2-like statistics, but in general they derive from the loglikelihood. In Section 9.5, we do an analysis of deviance on a portfolio such as the one generated in Appendix A.3. In Section 9.6, as an example we analyze a portfolio of motor insurance data, reminiscent of the study that led to the bonus-malus system of Chapter 6. For the application of GLMs to IBNR problems, see Chapter 10. More on GLMs can be found also in Chapter 11.

9.2 Generalized Linear Models

In a standard linear model, the observations are assumed to be normally distributed around a mean that is a linear function of parameters and covariates. See also Section 11.2. Generalized Linear Models generalize this in two directions. The random variables involved need not be normal with a variance independent of the mean, and also the scale in which the means are linear in the covariates may vary. For example, it may be loglinear.

Generalized Linear Models have three characteristics:

1. The *stochastic component* of the model states that the observations are *independent* random variables Y_i, $i = 1, \ldots, n$ with a density in the exponential dispersion family. The most important examples for our goal are:

   - $N(\mu_i, \psi_i)$ random variables;
   - $\text{Poisson}(\mu_i)$ random variables;
   - multiples ψ_i times $\text{Poisson}(\mu_i/\psi_i)$ distributed random variables; also known as quasi-Poisson or overdispersed Poisson random variables (ODP); usually $\psi_i > 1$ holds, for example in case a Poisson count is to be modeled with parameter uncertainty (Example 3.3.1);
   - $\psi_i \times \text{binomial}(n = \frac{1}{\psi_i}, p = \mu_i)$ random variables (hence, the proportions of successes in $1/\psi_i$ trials);
   - $\text{gamma}(\alpha = \frac{1}{\psi_i}, \beta = \frac{1}{\psi_i \mu_i})$ random variables;
   - inverse $\text{Gaussian}(\alpha = \frac{1}{\psi_i \mu_i}, \beta = \frac{1}{\psi_i \mu_i^2})$ random variables.

It can be shown that in all these examples, the mean is μ_i for each ψ_i. The variance depends on μ_i and ψ_i as $\text{Var}[Y_i] = \psi_i V(\mu_i)$ for some function V called the *variance function*.

2. The *systematic component* of the model attributes to every observation a *linear predictor* $\eta_i = \sum_j x_{ij}\beta_j$, linear in the parameters $\beta_1, \ldots, \beta_p$. The x_{ij} are called covariates or regressors.

3. The *link function* links the expected value μ_i of Y_i to the linear predictor as $\eta_i = g(\mu_i)$.

Remark 9.2.1 (Parameterization of the exponential dispersion family)
Note that the parameterizations used in the stochastic component above are not always the usual ones. The customary parameterization for gamma and inverse Gaussian random variables, for example, involves a scale parameter β and a shape parameter α. The μ_i parameter is the mean; the variance is $V(\mu_i)\psi_i$ with $V(\cdot)$ the variance function. We take ψ_i to be equal to ϕ/w_i, where ϕ is a common *dispersion parameter*, known or unknown, and w_i the known (natural) weight of observation i. Just as in the Bühlmann-Straub setting of Chapter 8, in principle a natural weight represents the number of iid observations of which our observation Y_i is the arithmetic average. Note that, for example, halving ϕ has the same effect on the variance as doubling the weight (sample size) has.

Weights w_i are needed for example to model the average claim frequency of a driver in a cell with w_i policies in it. By not taking the weights into account, one disregards the fact that the observations in cells with many policies in them have been measured with much more precision than the ones in practically empty cells. See also Appendix A.3 as well as Example 3.3.1.

In statistics, some other interpretations of weights can be found. The weight might represent the number of duplicated observations (frequency weight). Storing the data this way may lead to a substantial reduction in file size. Or, the researcher might want to attach a sampling weight to observations that is inversely proportional to the probability that this observation is included in the sample due to the sampling design. A weight of 1000 means that this observation is representative of 1000 subjects in the population. Our natural weights, as stated before, are denominators in averages taken, and are also known as exposure weights.

In GLM theory one generally considers the so-called *natural parameterization*, in which $\theta = \theta(\mu)$ replaces μ as a parameter. See Chapter 11.   ▽

Remark 9.2.2 (Variance functions)
Using mean μ and dispersion ϕ as parameters, the distributions listed above have a variety of variance functions $V(\cdot)$, making it possible to model many actuarial statistical problems with different heteroskedasticity patterns adequately. Assume that, for every observation i, we have weight $w_i = 1$, hence $\psi_i = \phi$. In increasing order of the exponent of μ in the variance function, we have in the list above:

1. The normal distribution with a constant variance $\sigma^2 = \mu^0 \phi$ (homoskedasticity).
2. The Poisson distribution with a variance equal to the mean, hence $\sigma^2 = \mu^1$, and the class of Poisson multiples having a variance proportional to the mean, hence $\sigma^2 = \mu^1 \phi$.

3. The gamma($\alpha = \frac{1}{\phi}, \beta = \frac{1}{\phi\mu}$) distributions, with a fixed shape parameter, and hence a constant coefficient of variation σ/μ, so $\sigma^2 = \mu^2 \phi$.

4. The inverse Gaussian($\alpha = \frac{1}{\phi\mu}, \beta = \frac{1}{\phi\mu^2}$) distributions, with a variance equal to $\sigma^2 = \frac{\alpha}{\beta^2} = \mu^3 \phi$.

The variance of Y_i describes the precision of the ith observation. Apart from weight, this precision is constant for the normally distributed random variables. Poisson measurements are less precise for large parameter values than for small ones, so the residuals in a fit should be smaller for small observations than for large ones. This is even more strongly the case for gamma distributions, as well as for the inverse Gaussian distributions in the parameterization as listed.

When estimating the mean, the mean-variance relationship is very important, because this determines how 'credible' we should consider observations to be. If we have reason to believe that the variance grows proportionally to the mean, we may choose to use a GLM in which the random variables are a fixed multiple times Poisson variates, even though this is obviously not the case.   ▽

Remark 9.2.3 (Canonical link)
Each of the distributions has a natural link function associated with it, called the *canonical link function*. Using these link functions has some technical advantages, see Chapter 11. For the normal distribution, the canonical link is the identity, leading to additive models. For the Poisson it is the logarithmic function, leading to loglinear, multiplicative models. For the gamma, it is the reciprocal,   ▽

Remark 9.2.4 ('Null' and 'full' models)
The least refined linear model that we study uses as a systematic component only the constant term, hence it ascribes all variation to chance and does not attach any influence to the collateral data. In the GLM-literature, this model is called the *null model*. Every observation is assumed to have the same distribution, and the weighted average Y_w is the best estimator of every μ_i. At the other extreme, one finds the so-called *full model* or *saturated model*, where every unit of observation i has its own parameter. Maximizing the total likelihood then produces the observation Y_i as an estimator of $E[Y_i]$. The model merely repeats the data, without condensing it at all, and without imposing any structure. In this model, all variation between the observations is due to the systematic effects.

The null model in general is too crude, the full model has too many parameters for practical use. Somewhere between these two extremes, one has to find an 'optimal' model. This model has to fit well, in the sense that the predicted outcomes should be close to the actually observed values. On the other hand, the more parsimonious it is in the sense that it has fewer parameters, the easier the model is to 'sell', not as much to potential policy holders, but especially to the manager, who favors thin tariff books and a workable and understandable model. There is a trade-off between the *predictive power* of a model and its *manageability*.   ▽

In GLM analyses, the criterion to determine the quality of a model is the loglikelihood of the model. It is known that under the null-hypothesis that a certain refinement of the model (adding extra parameters or relaxing constraints on parameters) is not an actual improvement, the gain in loglikelihood ($\times 2$, and divided by the dispersion parameter ϕ), approximately has a $\chi^2(k)$ distribution with k the number of parameters that have to be estimated additionally. Based on this, one can look at a chain of ever refined models and judge which of the refinements lead to a significantly improved fit, expressed in the maximal likelihood. A bound for the loglikelihood is the one of the full model, which can serve as a yardstick. Not only should the models to be compared be nested, with subsets of parameter sets, possibly after reparameterization by linear combinations, but also should the link function and the error distribution be the same.

Remark 9.2.5 (Residuals)
To judge if the fit a model is good enough and where it can be improved, we look at the *residuals*. These are the differences between actual observations and the values predicted for them by the model, standardized by taking into account the variance function as well as parameter estimates. We might look at the ordinary Pearson residuals, but in this context it is preferable to look at residuals based on the contribution of this observation to the maximized loglikelihood, the so-called *deviance residuals*. See also Section 11.4. For the normal distribution with as a link the identity function, the sum of the squares of the standardized (Pearson) residuals has a χ^2 distribution and is proportional to the difference in maximized likelihoods; for other distributions, this quantity provides an alternative for the difference in maximized likelihoods to compare the goodness of fit. ∇

9.3 Some traditional estimation procedures and GLMs

In this section, we illustrate the ideas behind GLMs using $I \times J$ contingency tables. We have a table of observed insurance losses Y_{ij}, $i = 1, \ldots, I$, $j = 1, \ldots, J$, classified by two rating factors into I and J risk classes. Hence, we have $I \cdot J$ independent observations (but some cells may be empty) indexed by i and j instead of n observations indexed by i as before. Generalization to more than two dimensions is straightforward. The collateral data with each observation consist of the row number i and the column number j in the table. The numbers in each cell represent averages over all w_{ij} observations in that cell (natural weights). With these factors, we try to construct a model for the expected values of the observations. Many situations are covered by this example. For example, the column number may indicate a certain region/usage combination such as in Table 9.6, the row number may be a weight class for a car or a step in the bonus-malus scale. The observations might then be the observed average number of accidents for all drivers with the characteristics i and j. Other examples, see also the next chapter, arise if i is the year that a certain policy was written, j is the development year, and the observations denote the total amount paid in year $i + j - 1$ regarding claims pertaining to policies of the year i. The calendar year $i + j - 1$ can be used as a third collateral variable. We will assume that the probability distribution of the observations Y_{ij} obeys a GLM, more specifically, a loglinear GLM with i and j as explanatory classifying variables. This means that for the expected values of the Y_{ij} we have

$$E[Y_{ij}] = \mu \, \alpha_i \beta_j, \quad i = 1, \ldots, I, \; j = 1, \ldots, J. \tag{9.7}$$

The parameters of the model are μ, α_i and β_j. Two parameters are superfluous; without loss of generality we will first assume that $\mu = \beta_1 = 1$ holds. Later on, we will find it more convenient to fix $\alpha_1 = 1$ instead of $\mu = 1$, so μ can be interpreted as the expected value of the reference cell $(i, j) = (1, 1)$. One gets an additive model in (9.7) by adding the parameters instead of multiplying them. As stated earlier, such models are often not relevant for actuarial practice.

Remark 9.3.1 (Connection with loglinear models)
One may wonder how our model (9.7) can be reconciled with the second and third characteristic of a GLM as listed above. A loglinear model in i and j arises, obviously, when $E[Y_{ij}] = \exp(i \log \alpha + j \log \beta + \log \mu)$ for some α, β and μ. In that case we call the regressors i and j *variates*. They must be measured on an interval scale; the contribution of i to the linear predictor has the form $i \log \alpha$ and the parameter α_i associated with observations from row i has the special form $\alpha_i = \alpha^i$ (the first i is an index, the second an exponent). If, as in (9.7), variable i classifies the data, and the numerical values of i act only as labels, we call i a *factor*. The parameters with a factor are arbitrary numbers α_i, $i = 1, \ldots, I$. To achieve this within the GLM model as stated, that is, to express $E[Y_{ij}]$ as a loglinear form of the collateral data, for each observation we recode the row number by a series of I *dummy variables* $d_1, \ldots, d_I$, of which $d_i = 1$ if the row number for this observation is i, the d_j with $j \neq i$ are zero. The contribution to (9.7) of a cell in row i can then be written in the loglinear form $\alpha_i = \exp(\sum_{t=1}^{I} d_t \log \alpha_t)$. ∇

Remark 9.3.2 (Aliasing)
To avoid the identification problems arising from redundant parameters in the model such as occur when a constant term as well as a factor are present in the model or when more than one factor is replaced by a set of dummies, we leave out the redundant dummies. In GLIM parlance, these parameters are *aliased*. This phenomenon is also known as 'multicollinearity' and as the 'dummy trap'. ∇

Remark 9.3.3 (Interaction between variables)
Sometimes two factors, or a factor and a variate, 'interact', for example when gender and age (class) are regressors, but the age effect for males and females is different. Then these two variables can be combined into one that describes the combined effect of these variables and is called their *interaction*. If two factors have I and J levels, their interaction has $I \cdot J$ levels. See further Section 9.5–6 and Appendix A.3. ∇

Remark 9.3.4 (Weights of observations)
For every cell (i, j), next to an observed claim figure Y_{ij} there is a weight w_{ij}, see also Remark 9.2.1. In actuarial applications, several interpretations are possible for these quantities:

1. Y_{ij} is the average claim frequency if $S_{ij} = Y_{ij} w_{ij}$ is the number of claims and w_{ij} is the *exposure* of cell (i, j), which is the total number of years that policies in it have been insured;
2. Y_{ij} is the average claim size if S_{ij} is the total claim amount for the cell and w_{ij} is the number of claims;
3. Y_{ij} is the observed pure premium if S_{ij} is the total claim amount for the cell and w_{ij} is the exposure.

Any of these interpretations may apply in the examples below. The weights w_{ij} are assumed to be constants, measured with full precision, while the S_{ij} and hence the Y_{ij} are random variables with outcomes denoted as s_{ij} and y_{ij}. $\qquad\nabla$

In the sequel, we give some methods to produce estimates $\widehat{\alpha}_i$ and $\widehat{\beta}_j$ of the parameters α_i and β_j in such a way that the values $\widehat{\alpha}_i \widehat{\beta}_j$ are close to y_{ij}; we fix the parameter μ to be equal to 1 for identifiability. These methods have been used in actuarial practice without some users being aware that they were actually statistically quite well founded methods. For each method we give a short description, and indicate also for which GLM this method computes the maximum likelihood estimates, or which other estimates are computed.

Method 9.3.5 (Bailey-Simon = Minimal chi-square with Poisson)
In the Bailey-Simon method, the parameter estimates $\widehat{\alpha}_i$ and $\widehat{\beta}_j$ in the multiplicative model are determined as the solution of

$$\min_{\alpha_i, \beta_j} BS \quad \text{with} \quad BS = \sum_{i,j} \frac{w_{ij}(y_{ij} - \alpha_i \beta_j)^2}{\alpha_i \beta_j}. \tag{9.8}$$

A justification of this method is that if the S_{ij} denote Poisson distributed numbers of claims, BS in (9.8) is just the χ^2-statistic, since (9.8) can be rewritten as

$$BS = \sum_{i,j} \frac{(s_{ij} - w_{ij}\alpha_i\beta_j)^2}{w_{ij}\alpha_i\beta_j} = \sum_{i,j} \frac{(s_{ij} - \mathrm{E}[S_{ij}])^2}{\mathrm{Var}[S_{ij}]}. \tag{9.9}$$

So minimizing BS is nothing but determining the minimal-χ^2 estimator. The model hypotheses can be easily tested.

Solving the normal equations arising from differentiating BS in (9.8) with respect to each parameter, we get a system of equations that can be written as follows:

$$\begin{aligned}
\alpha_i &= \left(\sum_j \frac{w_{ij} y_{ij}^2}{\beta_j} \Big/ \sum_j w_{ij}\beta_j \right)^{1/2}, \quad i = 1, \dots, I; \\
\beta_j &= \left(\sum_i \frac{w_{ij} y_{ij}^2}{\alpha_i} \Big/ \sum_i w_{ij}\alpha_i \right)^{1/2}, \quad j = 1, \dots, J.
\end{aligned} \tag{9.10}$$

The procedure to determine the optimal values is the method of *successive substitution*. In this method, equations are derived that describe the optimal choice, because these equations being satisfied is considered desirable in itself, like for the marginal totals equations, and/or because they are the normal equations for some optimization problem, like the maximization of a loglikelihood or the minimization of some distance between fitted and observed values such as above. These equations are then written in a form $\vec{\alpha} = f(\vec{\beta})$, $\vec{\beta} = g(\vec{\alpha})$ for some vector-valued functions f and g. For an arbitrary initial choice of $\vec{\beta}$, for example $\vec{\beta} = (1,\ldots,1)'$, one computes the corresponding $\vec{\alpha} = f(\vec{\beta})$, and then executes $\vec{\beta} := g(\vec{\alpha})$; $\vec{\alpha} := f(\vec{\beta})$ successively until convergence is reached. Actually, an old $\vec{\beta}$ is transformed into a new one $\vec{\beta}^*$ through $\vec{\beta}^* = g(f(\vec{\beta}))$ (using $\vec{\alpha}$ only as an auxiliary variable), so successive substitution is a *fixed point* algorithm.

Successive substitution can be implemented in R quite easily. We can use the functions `rowSums` and `colSums` for two-dimensional data stored in a matrix. In general, having stored the data in linear arrays, we use the function `tapply` to compute the sums in (9.10) extending over only subsets of all four values. Its first argument is a vector of values to be processed. Its second argument gives a factor or a list of factors, splitting up the elements of the vector into different groups with different levels of the factor(s). Its third argument is the function to be applied, in this case `sum`. Simply doing 20 iterations whatever the circumstances (to see a description of the control-flow constructs of the R language, use for example ?break), the following R code solves our optimization problem for the Bailey-Simon case:

```
y <- c(10,15,20,35); w <- c(300,500,700,100)
i <- c(1,1,2,2); j <- c(1,2,1,2); beta <- c(1,1)
for (iter in 1:20){
  alpha <- sqrt(tapply(w*y^2/beta[j],i,sum)/
               tapply(w*beta[j],i,sum))
  beta <- sqrt(tapply(w*y^2/alpha[i],j,sum)/
               tapply(w*alpha[i],j,sum))}
```

The vector `w*y^2/beta[j]` expands into the values

```
w[1]*y[1]^2/beta[j[1]],  ..., w[4]*y[4]^2/beta[j[4]]
```

which is just what we need.

The method of successive substitution is simple to implement, once the system of equations has been written in a suitable form. Of course other algorithms may be used to handle the likelihood maximization.   ▽

Remark 9.3.6 (Compound Poisson distributions)
In the case of compound Poisson distributed total claims we can apply χ^2-tests under some circumstances. Let S_{ij} denote the total claim amount and w_{ij} the total exposure of cell (i,j). Assume that the number of claims caused by each insured is Poisson(λ_{ij}) distributed. The individual claim amounts are iid random variables, distributed as X. Hence the mean claim frequency varies, but the claim size distribution is the same for each cell. Then we have

$$E[S_{ij}] = w_{ij}\,\lambda_{ij}\,E[X]; \quad \text{Var}[S_{ij}] = w_{ij}\,\lambda_{ij}\,E[X^2], \tag{9.11}$$

hence with $E[Y_{ij}] = \alpha_i\beta_j$ we get

$$\text{Var}[Y_{ij}] = \frac{\alpha_i \beta_j}{w_{ij}} \frac{\text{E}[X^2]}{\text{E}[X]}. \tag{9.12}$$

So the random variable BS in (9.8) is the sum of the squares of random variables with zero mean and constant variance. This is already the case when the ratio $\text{E}[X^2]/\text{E}[X]$ is the same for all cells. If we correct BS for this factor and if moreover our estimation procedure produces best asymptotic normal estimators (BAN), such as maximum likelihood estimation does, asymptotically we get a χ^2-distribution, with $(I-1)(J-1)$ degrees of freedom. $\qquad \triangledown$

Property 9.3.7 (Bailey-Simon leads to a 'safe' premium)

The Bailey-Simon method in the multiplicative model has a property that will certainly appeal to actuaries. It proves that with this method, the resulting fitted loss, that is, the total premium to be asked next year, is larger than the observed loss. In fact, this already holds for each group (a row or a column). In other words, we can prove that, assuming that $\widehat{\alpha}_i$ and $\widehat{\beta}_j$ solve (9.10), we have

$$\sum_{i(j)} w_{ij} \widehat{\alpha}_i \widehat{\beta}_j \geq \sum_{i(j)} w_{ij} y_{ij} \quad \text{for all } j(i). \tag{9.13}$$

A summation over $i(j)$ for all $j(i)$ means that the sum has to be taken either over i for all j, or over j for all i, so (9.13) is just shorthand for a system like (9.10). To prove (9.13), we rewrite the first set of equations in (9.10) as

$$\widehat{\alpha}_i^2 = \sum_j \frac{w_{ij} \widehat{\beta}_j}{\sum_h w_{ih} \widehat{\beta}_h} \frac{y_{ij}^2}{\widehat{\beta}_j^2}, \quad i = 1, 2, \ldots, I. \tag{9.14}$$

But this is just $\text{E}[U^2]$ if U is a random variable with $\Pr[U = d_j] = p_j$, where

$$p_j = \frac{w_{ij} \widehat{\beta}_j}{\sum_h w_{ih} \widehat{\beta}_h} \quad \text{and} \quad d_j = \frac{y_{ij}}{\widehat{\beta}_j}. \tag{9.15}$$

Since $\text{E}[U^2] \geq (\text{E}[U])^2$ for any random variable U, we have immediately

$$\widehat{\alpha}_i \geq \sum_j \frac{w_{ij}}{\sum_h w_{ih} \widehat{\beta}_h} y_{ij}, \quad \text{hence} \quad \sum_j w_{ij} \widehat{\alpha}_i \widehat{\beta}_j \geq \sum_j w_{ij} y_{ij}. \tag{9.16}$$

In the same way one proves that the fitted column totals are at least the observed totals. $\qquad \triangledown$

Method 9.3.8 (Marginal Totals)

The basic idea behind the method of marginal totals is the same as the one behind the actuarial equivalence principle: in a 'good' tariff system, for large groups of insureds, the total premium equals the observed loss (apart from loadings). We determine the values $\widehat{\alpha}_i$ and $\widehat{\beta}_j$ in such a way that this condition is met for all groups of risks for which one of the risk factors, either the row number i or the column number j, is constant. The equivalence does not hold for each cell, but it does on the next-higher aggregation level of rows and columns.

In the multiplicative model, to estimate the parameters we have to solve the following system of equations consisting of $I + J$ equations in as many unknowns:

$$\sum_{i(j)} w_{ij} \alpha_i \beta_j = \sum_{i(j)} w_{ij} y_{ij} \quad \text{for all } j(i). \tag{9.17}$$

If all estimated and observed row totals are the same, the same holds for the sum of all these row totals. So the total of all observations equals the sum of all estimates. Since adding up the columns leads to the same equation as adding up the rows, the last equation in (9.17) can be written as a linear combination of all the others. The fact that one of the equations in this system is superfluous is in line with the fact that the α_i and the β_j in (9.17) are only identified up to a multiplicative constant.

One way to solve (9.17) is again by successive substitution, starting from any positive initial value for the β_j. For this, rewrite the system in the form:

$$\alpha_i = \sum_j w_{ij} y_{ij} \Big/ \sum_j w_{ij} \beta_j, \quad i = 1, \ldots, I;$$

$$\beta_j = \sum_i w_{ij} y_{ij} \Big/ \sum_i w_{ij} \alpha_i, \quad j = 1, \ldots, J. \tag{9.18}$$

A few iterations generally suffice to produce the optimal estimates.   ▽

The heuristic justification of the method of marginal totals applies for every interpretation of the Y_{ij}. But if the Y_{ij} denote claim numbers, there is another explanation, as follows.

Property 9.3.9 (Loglinear Poisson GLM = Marginal totals method)
Suppose the number of claims caused by each of the w_{ij} insureds in cell (i, j) has a Poisson(λ_{ij}) distribution with $\lambda_{ij} = \alpha_i \beta_j$. Then estimating α_i and β_j by maximum likelihood gives the same results as the marginal totals method.

Proof. The total number of claims in cell (i, j) has a Poisson$(w_{ij} \lambda_{ij})$ distribution. The likelihood of the parameters λ_{ij} with the observed numbers of claims s_{ij} then equals

$$L = \prod_{i,j} e^{-w_{ij} \lambda_{ij}} \frac{(w_{ij} \lambda_{ij})^{s_{ij}}}{s_{ij}!}. \tag{9.19}$$

By substituting into (9.19) the relation

$$E[Y_{ij}] = E[S_{ij}]/w_{ij} = \lambda_{ij} = \alpha_i \beta_j \tag{9.20}$$

and maximizing (9.19) or its logarithm for α_i and β_j we get exactly the equations (9.17). See Exercise 9.3.1.   ▽

Method 9.3.10 (Least squares = ML with normality)
In the method of least squares, estimators are determined that minimize the total of the squared differences of observed loss and estimated premium, weighted by the exposure in a cell. If the variance of Y_{ij} is proportional to $1/w_{ij}$, which is for example the case when Y_{ij} is the mean of w_{ij} iid random variables with the same variance, all terms in (9.21) below (with y_{ij} replaced by Y_{ij}) have the same mean. So using these weights ensures that the numbers added have the same order of magnitude. The parameters α_i and β_j are estimated by solving:

$$\min_{\alpha_i, \beta_j} SS \quad \text{with} \quad SS = \sum_{i,j} w_{ij}(y_{ij} - \alpha_i \beta_j)^2. \tag{9.21}$$

The normal equations produce the following system, which is written in a form that is directly suitable to be tackled by successive substitution:

$$\alpha_i = \sum_j w_{ij} y_{ij} \beta_j \Big/ \sum_j w_{ij} \beta_j^2, \quad i = 1, \dots, I;$$

$$\beta_j = \sum_i w_{ij} y_{ij} \alpha_i \Big/ \sum_i w_{ij} \alpha_i^2, \quad j = 1, \dots, J. \tag{9.22}$$

Because of the form of the likelihood of the normal distribution, one may show that minimizing SS is tantamount to maximizing the normal loglikelihood. See also Exercise 9.3.7. In an additive model where $\alpha_i \beta_j$ is replaced by $\alpha_i + \beta_j$, the normal equations with (9.21) are a linear system, which can be solved directly. $\quad \nabla$

Method 9.3.11 (Direct method = ML with gamma distribution)
The direct method determines estimates for the parameters α_i and β_j by solving, for example by successive substitution, the following system:

$$\alpha_i = \sum_j w_{ij} \frac{y_{ij}}{\beta_j} \Big/ \sum_j w_{ij}, \quad i = 1, \dots, I;$$

$$\beta_j = \sum_i w_{ij} \frac{y_{ij}}{\alpha_i} \Big/ \sum_i w_{ij}, \quad j = 1, \dots, J. \tag{9.23}$$

The justification for this method is as follows. Assume that we know the correct values of the parameters β_j, $j = 1, \dots, J$. Then all random variables Y_{ij}/β_j have mean α_i. Estimating α_i by a weighted average, we get the equations (9.23) of the direct method. The same reasoning applied to Y_{ij}/α_i gives estimates for β_j. See also Exercise 9.3.4.

The direct method also amounts to determining the maximum likelihood in a certain GLM. In fact, it produces ML-estimators when $S_{ij} \sim \text{gamma}(\gamma w_{ij}, \frac{\gamma}{\alpha_i \beta_j})$. This means that S_{ij} is the sum of w_{ij} gamma$(\gamma, \frac{\gamma}{\alpha_i \beta_j})$ random variables, with a *fixed* coefficient of variation $\gamma^{-1/2}$, and a mean $\alpha_i \beta_j$. The likelihood of the observation in cell (i, j) can be written as

$$f_{S_{ij}}(s_{ij}; \alpha_i, \beta_j) = \frac{1}{\Gamma(\gamma w_{ij})} \left(\frac{\gamma}{\alpha_i \beta_j} \right)^{\gamma w_{ij}} s_{ij}^{\gamma w_{ij}-1} e^{\frac{-\gamma s_{ij}}{\alpha_i \beta_j}}. \tag{9.24}$$

With $L = \prod_{i,j} f_{S_{ij}}(s_{ij}; \alpha_i, \beta_j)$, we find by differentiating with respect to α_k:

$$\frac{\partial \log L}{\partial \alpha_k} = \frac{\partial}{\partial \alpha_k} \sum_{i,j} \{ \gamma w_{ij} \log \frac{\gamma}{\alpha_i \beta_j} - \frac{\gamma s_{ij}}{\alpha_i \beta_j} \} + 0 = \sum_j \{ \frac{-\gamma w_{kj}}{\alpha_k} + \frac{\gamma s_{kj}}{\alpha_k^2 \beta_j} \}. \tag{9.25}$$

The derivatives with respect to β_h are similar. Writing down the normal equations for the ML-estimation, that is, setting these derivatives and (9.25) equal to zero, after a little algebra produces exactly the system (9.23) of the direct method. ∇

Example 9.3.12 (Numerical illustration of the above methods)
We applied the four methods given above to the data given in the following table, which gives $w_{ij} \times y_{ij}$ for $i, j = 1, 2$:

	$j = 1$	$j = 2$
$i = 1$	300×10	500×15
$i = 2$	700×20	100×35

The following fitted values $\widehat{\alpha}_i \widehat{\beta}_j$ were found for the given methods:

Bailey-Simon	marginal totals	least squares	direct method
9.40 15.38	9.39 15.37	9.04 15.34	9.69 15.29
20.27 33.18	20.26 33.17	20.18 34.24	20.27 31.97
$\Delta = 28.85$	$\Delta = 28.86$	$\Delta = 37.06$	$\Delta = 36.84$

Here $\Delta = \sum_{i,j} w_{ij}(y_{ij} - \widehat{\alpha}_i \widehat{\beta}_j)^2/(\widehat{\alpha}_i \widehat{\beta}_j)$ describes the goodness of the fit; it is the quantity minimized by the Bailey-Simon method. The systems of equations from which the $\widehat{\alpha}_i$ and the $\widehat{\beta}_j$ have to be determined are alike, but not identical. See also Exercise 9.3.2.

Observe that for small y_{ij}, least squares fits worst, the direct method best, and vice versa for large y_{ij}. This is because for the least squares criterion, all observations are equally credible, while the direct method intrinsically assumes that larger values are much more imprecise ($\sigma \propto \mu$). Bailey-Simon and the marginal totals method give very similar results, which was to be expected since they are both good estimation methods in the same Poisson model. ∇

9.4 Deviance and scaled deviance

As a measure for the difference between vectors of fitted values and of observations, one generally looks at the Euclidean distance, that is, the sum of the squared differences such as in (9.21). If the observations are from a normal distribution, minimizing this distance is the same as maximizing the likelihood of the parameter values with the given observations. In GLM-analyses, one looks at the difference of the 'optimal' likelihood of a certain model, compared with the maximally attainable likelihood if one does not impose a model on the parameters, hence for the *full model* with a parameter for every observation.

The *scaled deviance* of a model is -2 times the logarithm of the likelihood ratio. This is the likelihood maximized under our particular model, divided by the likelihood of the full model. The *deviance* equals the scaled deviance multiplied by the dispersion parameter ϕ. From the theory of mathematical statistics it is known that the scaled deviance is approximately χ^2 distributed, with as degrees of freedom the number of observations less the number of estimated parameters. Also, if one model is a submodel of another, it is known that the difference between the scaled deviances has a χ^2-distribution.

For three suitable choices of the distribution of the random variation around the mean in a GLM, we give expressions for their deviances. We saw in the preceding section that minimizing these deviances gives the same parameter estimates as those obtained by some heuristic methods. If the observations are Poisson, we get the method of marginal totals, see Example 9.3.8. With normality, we get the least squares method (9.21). If the individual claim sizes are gamma distributed, the normal equations are the equations (9.23) of the direct method of Example 9.3.11. We will always assume that the expected values μ_i of our observations Y_i, $i = 1, \ldots, n$ follow a certain model, for example a multiplicative model with rows and columns such as above. We denote by $\widehat{\mu}_i$ the optimally estimated means under this model, and by $\widetilde{\mu}_i$ the mean, optimally estimated under the full model, where every observation has its own parameter and the maximization of the total likelihood can be done term by term. We will always take the ith observation to be the mean of w_i single iid observations. All these have a common dispersion parameter ϕ. We already remarked that this dispersion parameter is proportional to the variances, which, as a function of the mean μ, are equal to $\phi V(\mu)/w$, where the function $V(\cdot)$ is the variance function.

Example 9.4.1 (Normal distribution)
Let $Y_1, \ldots, Y_n$ be independent normal random variables, where Y_i is the average of w_i random variables with an $N(\mu_i, \phi)$ distribution, hence $Y_i \sim N(\mu_i, \phi/w_i)$. Let L denote the likelihood of the parameters with the given observations. Further let $\widehat{L}$ and $\widetilde{L}$ denote the values of L when $\widehat{\mu}_i$ and $\widetilde{\mu}_i$ are substituted for μ_i. We have

$$L = \prod_{i=1}^{n} \frac{1}{\sqrt{2\pi\phi/w_i}} \exp \frac{-(y_i - \mu_i)^2}{2\phi/w_i}. \tag{9.26}$$

It is clear that in the full model, maximizing (9.26) term by term, we can simply take $\mu_i = \tilde{\mu}_i = y_i$ for each i. It turns out that this holds for every member of the exponential dispersion family; see also Examples 9.4.2 and 9.4.3, as well as Property 11.3.6.

If D denotes the deviance, we have

$$\frac{D}{\phi} = -2\log\frac{\widehat{L}}{\widetilde{L}} = \frac{1}{\phi}\sum_i w_i(\widehat{\mu}_i - y_i)^2. \tag{9.27}$$

This means that for the normal distribution, minimizing the deviance, or what is the same, maximizing the likelihood, is the same as determining the parameter estimates by least squares. ∇

Example 9.4.2 (Poisson sample means and multiples)
Let $Y_i = \phi M_i/w_i$ with $M_i \sim \text{Poisson}(w_i\mu_i/\phi)$. We write $Y_i \sim \text{Poisson}(\mu_i, \phi/w_i)$. Notice that $\text{E}[Y_i] = \mu_i$ and $\text{Var}[Y_i] = \frac{\phi}{w_i}\mu_i$, so $V(\mu) = \mu$.

In the special case that $w_i \equiv 1$ as well as $\phi = 1$, we have ordinary Poisson random variables. If w_i/ϕ is an integer, Y_i can be regarded as the sample mean of w_i/ϕ Poisson(μ_i) random variables, but without this restriction we also have a valid model. For the likelihood we have

$$L(\mu_1,\ldots,\mu_n;\ \phi,w_1,\ldots,w_n,y_1,\ldots,y_n)$$
$$= \prod_{i=1}^{n}\text{Pr}[w_iY_i/\phi = w_iy_i/\phi] = \prod_{i=1}^{n}\frac{e^{-\mu_iw_i/\phi}(\mu_iw_i/\phi)^{w_iy_i/\phi}}{(w_iy_i/\phi)!}. \tag{9.28}$$

For every ϕ, the ith term in this expression is maximal for the value of μ_i that maximizes $e^{-\mu_i}\mu_i^{y_i}$, which is for $\mu_i = y_i$, so we see that just as with the normal distribution, we get $\tilde{\mu}_i$ by simply taking the ith residual equal to zero.

It is easy to see that the scaled deviance is equal to the following expression:

$$\frac{D}{\phi} = -2\log\frac{\widehat{L}}{\widetilde{L}} = \frac{2}{\phi}\sum_i w_i\left(y_i\log\frac{y_i}{\widehat{\mu}_i} - (y_i - \widehat{\mu}_i)\right). \tag{9.29}$$

By taking $\phi \neq 1$, we get distributions of which the variance is not equal to the mean, but remains proportional to it. One speaks of *overdispersed* Poisson distributions in this case, the case $\phi > 1$ being much more common. The random variable Y_i in this example has as a support the integer multiples of ϕ/w_i, but obviously the deviance (9.29) allows minimization for other non-negative values of y_i as well. This way, one gets *quasi-likelihood* models, where only the mean-variance relation is specified. See also the second part of Section 11.4. ∇

Example 9.4.3 (Gamma distributions)
Now let $Y_i \sim \text{gamma}(w_i/\phi, w_i/\{\phi\mu_i\})$. If w_i is integer, Y_i has the distribution of an average of w_i gamma$(1/\phi, 1/\{\phi\mu_i\})$ random variables, or equivalently, of w_i/ϕ random variables with an exponential$(1/\mu_i)$ distribution. We have

$$E[Y_i] = \mu_i, \quad \text{Var}[Y_i] = \frac{\phi}{w_i} V(\mu_i) = \frac{\phi}{w_i} \mu_i^2. \tag{9.30}$$

For this case, we have $\tilde{\mu}_i = y_i$ for the full model as well, since for the density f_Y of Y_i we can write

$$
\begin{aligned}
f_Y\left(y; \frac{w}{\phi}, \frac{w}{\phi\mu}\right) &= \frac{1}{\Gamma(w/\phi)}\left(\frac{w}{\phi\mu}\right)^{w/\phi} y^{w/\phi-1} e^{-wy/(\phi\mu)} \\
&= \frac{1}{\Gamma(w/\phi)}\left(\frac{w}{\phi}\right)^{w/\phi} \frac{1}{y}\left[\frac{y}{\mu} e^{-y/\mu}\right]^{w/\phi},
\end{aligned}
\tag{9.31}
$$

which is maximal when the expression in square brackets is largest, so for $\frac{y}{\mu} = 1$. The scaled deviance in this situation is as follows:

$$\frac{D}{\phi} = -2\log\frac{\hat{L}}{\tilde{L}} = \frac{2}{\phi}\sum_i w_i\left(-\log\frac{y_i}{\hat{\mu}_i} + \frac{y_i - \hat{\mu}_i}{\hat{\mu}_i}\right). \tag{9.32}$$

The y_i of course must be strictly positive here. $\quad\triangledown$

The value of the deviance D can be computed from the data alone; it does not involve unknown parameters. This means D is a *statistic*. Notice that in each of the three classes of distributions given above, the maximization over μ_i gave results that did not depend on ϕ. The estimation of ϕ can hence be done independently from the determining of optimal values for the μ_i. Only the relative values of the parameters ϕ/w_i with each observation are relevant.

To estimate the value of ϕ, one often proceeds as follows. Under the null-hypothesis that $Y_i \sim N(\mu_i, \phi/w_i)$, the minimized sum of squares (9.27) has a $\chi^2(k)$ distribution with as its parameter k the number of observations less the number of parameter estimates needed in evaluating $\hat{\mu}_i$. Then one can estimate ϕ by the method of moments, setting (9.27) equal to its mean value k and solving for ϕ. Another possibility is to estimate ϕ by maximum likelihood. To ensure that the differences between Y_i and the fitted values are caused by chance and not by systematic deviations because one has used too crude a model, the estimation of ϕ is done in the most refined model that still can be estimated, even though there will generally be too many parameters in this model. Hence, for this model the scaled deviance equals the value of k.

The interpretation for the dispersion parameter ϕ is different for each class of distributions. For the normal distributions, it is simply the variance of the errors. For a pure Poisson distribution, we have $\phi = 1$; in case of overdispersion it is the ratio of variance and mean, as well as the factor by which all Poisson variables have been multiplied. For the gamma distributions, $\sqrt{\phi}$ denotes the coefficient of variation σ/μ for individual observations.

Remark 9.4.4 (Comparing different models)
Analysis of deviance can be used to compare two *nested* models, one of which arises from the other by relaxing constraints on the parameters. For example, if a factor or variate is added as a covariate, parameters that used to be restricted to be zero in the

linear predictor are now arbitrary. If a variate is replaced by a factor (see Remark 9.3.1), parameters with factor level i that were of prescribed form $i \cdot \alpha$ for some real α now can have an arbitrary value α_i. Another example is when interaction between factors is allowed, for example when the parameter corresponding to cell (i, j) is no longer of the specific form $\alpha_i + \beta_j$ but more generally equal to some arbitrary γ_{ij}. If the gain in deviance (scaled by dividing by an estimate for the scale parameter ϕ) exceeds the, say, 95% critical value of the $\chi^2(k)$ distribution, with k the number of extra parameters estimated, the relaxed model fits significantly better, and the restricted model is rejected. If this is not the case, it is not proper to say that the restricted model is better, or accepted. We can only say that the null-hypothesis that the extra parameters are actually equal to zero (in the linear predictor) is not rejected.

The idea behind the *Akaike information criterion* (AIC) is to examine the complexity of the model together with the goodness of its fit to the sample data, and to produce a measure which balances between the two. A model with many parameters will provide a very good fit to the data, but will have few degrees of freedom and be of limited utility. This balanced approach discourages overfitting, and encourages 'parsimony'. The preferred model is that with the lowest AIC value. AIC is the negative of twice the log-likelihood plus twice the number of linear and scale parameters. Therefore,

$$\text{AIC} = -2\ell + 2k, \tag{9.33}$$

where k is the number of parameters and ℓ is the loglikelihood.

A similar tool is the Schwarz criterion (also Schwarz information criterion (SIC) or Bayesian information criterion (BIC) or Schwarz-Bayesian information criterion). Just as AIC, it stimulates parsimony by imposing a penalty for including too many terms in a regression model. The generic formula is

$$\text{BIC} = -2\ell + k \log n, \tag{9.34}$$

with n the number of observations. ∇

9.5 Case study I: Analyzing a simple automobile portfolio

Assume we have observed drivers in some fictitious motor insurance portfolio. The structure of the portfolio is in fact the same as the aggregated pseudo-random portfolio in Appendix A.3, except that in that case, fractional exposures were taken into account. The drivers can be divided into cells on the basis of the following *risk factors*:

- sex: 1 = female, 2 = male
- region: 1 = countryside, 2 = elsewhere, 3 = big city
- type of car: 1 = small, 2 = middle, 3 = big
- job class: 1 = civil servant/actuary/..., 2 = in-between, 3 = dynamic drivers

For each of the $2 \times 3 \times 3 \times 3 = 54$ cells, the following totals are known:

- `expo`: total number of policy years with these risk factors. All policies were observed during 7 years.
- n: their observed total number of claims in a particular year.

We have a list of the claim numbers that occurred, as well as a list of the corresponding numbers of policies. These data are read as follows.

```
n <- scan(n=54) ## read 54 numbers into vector n
 1  8 10   8   5 11 14 12 11 10   5 12 13 12 15 13 12 24
12 11   6   8 16 19 28 11 14   4 12   8 18   3 17   6 11 18
12   3 10 18 10 13 12 31 16 16 13 14   8 19 20   9 23 27
expo <- scan(n=54) * 7 ## number of policies times 7
10 22 30 11 15 20 25 25 23 28 19 22 19 21 19 16 18 29
25 18 20 13 26 21 27 14 16 11 23 26 29 13 26 13 17 27
20 18 20 29 27 24 23 26 18 25 17 29 11 24 16 11 22 29
```

The variable `sex` equals 1 for the first 27 observations, 2 for the last 27 observations; the variable `region` has 3 levels that occur in blocks of 9, like this:

$$(\underbrace{1,\dots,1}_{9\times}, \underbrace{2,\dots,2}_{9\times}, \underbrace{3,\dots,3}_{9\times}, \underbrace{1,\dots,1}_{9\times}, \underbrace{2,\dots,2}_{9\times}, \underbrace{3,\dots,3}_{9\times})$$

There is a special function `gl` in R to generate factor levels, see `?gl`, but the cell characteristics can also be reconstructed as follows:

```
sex <- as.factor(rep(1:2, each=27, len=54))
region <- as.factor(rep(1:3, each=9, len=54))
type <- as.factor(rep(1:3, each=3, len=54))
job <- as.factor(rep(1:3, each=1, len=54))
AnnClFr <- round(1000 * n/expo)
data.frame(expo, n, sex, region, type, job, AnnClFr)[1:10,]
```

The last line prints the contents of the first 10 cells, with appropriate headings.

The average number of claims per contract `n/expo` in each cell is the quantity of interest in this case. We want to relate this annual claim frequency to the risk factors given, to establish or analyze a tariff for the portfolio. We will try to find a well-fitting loglinear model for the claim frequency in terms of the risk factors. First we produce a cross-tabulation of the values of `AnnClFr`, and print it in the form of a flat table (see Table 9.1), with `sex:region` combinations in the different rows, `type:job` combinations in the columns. This is how:

```
xt <- xtabs(AnnClFr ~ sex+region+type+job)
ftable(xt, row.vars=1:2, col.vars=3:4)
```

For the number of claims on a contract, it is reasonable to assume a Poisson distribution. By aggregation, we have rather high values of the Poisson means in each cell, so using a normal approximation would not be not far-fetched. But there are two reasons why it is not appropriate to fit an ordinary linear model. First, the variance depends on the mean, so there is heteroskedasticity. Second, we have a multiplicative model, not an additive one. So the linearity is not on the standard scale, but only

Table 9.1 Annual claim frequency (n/expo × 1000) for each combination of factors

sex	region	type job	1 1	2	3	2 1	2	3	3 1	2	3
1	1		14	52	48	104	48	79	80	69	68
	2		51	38	78	98	82	113	116	95	118
	3		69	87	43	88	88	129	148	112	125
2	1		52	75	44	89	33	93	66	92	95
	2		86	24	71	89	53	77	75	170	127
	3		91	109	69	104	113	179	117	149	133

on the log-scale. Therefore we are going to fit a generalized linear model, with a mean-variance relation of the poisson type, and a log-link.

Using the aggregate data in each cell, we make a model for the average number of claims per contract, that is, for n/expo. But this means that, apart from the proportionality of the variance of the response variable to the mean, it is also to be divided by expo. This is communicated to R by telling it that there is a prior weight expo attached to each observation/cell, as follows:

```
>       glm(n/expo ~ sex+region+type+job,
+       fam=poisson(link=log), wei=expo)
There were 50 or more warnings
      (use warnings() to see the first 50)
Call:   glm(formula = n/expo ~ sex + region + type + job,
            family = poisson(link = log), weights = expo)
Coefficients:
(Intercept)           sex2          region2          region3
   -3.0996          0.1030          0.2347          0.4643
      type2          type3            job2            job3
    0.3946          0.5844          -0.0362          0.0607
Degrees of Freedom: 53 Total (i.e. Null);   46 Residual
Null Deviance:        105
Residual Deviance: 41.9              AIC: Inf
```

The warnings are given because averages are used, which are not integer in most cases, therefore not Poisson distributed. This does not present any problems when estimating the coefficients, but it prohibits the glm function from computing the Akaike information criterion (AIC), see Remark 9.4.4.

An equivalent way to estimate this GLM is by using an *offset*:

```
glm(n ~ sex+region+type+job+offset(log(expo)),
        fam=poisson(link=log))
```

By this mechanism, the logarithm of expo is added to each linear predictor, with a coefficient not to be estimated from the data but fixed at 1. The effect is that because of the log-link used, each fitted value is multiplied by expo. Now the AIC can be computed; it is 288.2. Fitted values are for n, not for the average annual claim frequency.

Table 9.2 Analysis-of-deviance table

Model specification	df	dev	Δdev	Δdf	
1	53	104.7			
1 + sex	52	102.8	1.9	1	
1 + sex + region	50	81.2	21.6	2	
1 + sex + region + sex:region	48	80.2	1.0	2	
1 + type	51	68.4	36.4	2	
1 + region	51	83.1	21.6	2	
1 + region + type	49	44.9	38.2	2	♡
1 + region + type + region:type	45	42.4	2.5	4	
1 + region + type + job	47	43.8	1.1	2	
1 + region + type + sex	48	43.1	1.8	1	

From the output given above, it is easy to determine how many claims on average a driver with the worst factor levels produces, compared with someone having the best combination. The coefficients given are relative to the standard class, which are sex1, region1, type1 and job1. For these classes, the coefficients are taken to be zero. In view of the sign of the coefficients, the best class is the one with job2, and all other factor levels at 1. The corresponding average number of claims equals $\exp(-3.0996 - 0.0362) = 0.0435$, that is, one claim each 23 years on average. The worst drivers in this example happen to have the high class labels, and their number of claims equals $\exp(-3.0996 + 0.1030 + 0.4643 + 0.5844 + 0.0607) = 0.151$, which is about one claim in every 6.6 years.

In Table 9.2, we 'explain' the dependent variable $N \sim$ Poisson by the main effects sex, region, type, job, and some interaction terms, with a log-link. To test a null-hypothesis that adding factors sex, region, type or job actually has no effect, we look at the deviance, generalizing the sum-of-squares in the ordinary linear model. The difference in deviance between the null-model and the model with these factors in it has a χ^2 distribution with as degrees of freedom the number of parameters estimated additionally; see Section 9.4. Note that the deviance resulting from fitted models including region and type is close to its expected value (df) for a pure Poisson model, so there is no indication that risk factors have been ignored, nor that interactions between risk factors, like for example a different effect of region for males and females, should have been incorporated in the model.

Differences Δdf and Δdev in Table 9.2 are with respect to the model without the last term (printed in italics); the number of degrees of freedom df is the number of cells less the number of parameters estimated; for the difference in deviance, we have Δdev $\overset{\approx}{\sim} \chi^2(\Delta$df$)$. R-commands that might help produce this table are:

```
g <- glm(n/expo ~ 1+region+type+region:type, poisson, wei=expo)
anova(g, test="Chisq")
```

Adding an *interaction* term region:type to the main effects region and type (in fact looking at the model region*type; see also ?formula) one may judge if, for example, the effect of region is significantly different for the various types of car. The analysis of deviance shows this is not the case (the critical value for a

$\chi^2(4)$ distribution is about 9.5 at 95% level; for $\chi^2(2)$, it is 6, for $\chi^2(1)$, it is 3.84). Also, sex is not a meaningful rating factor in this case.

The models coarser than 1+region+type come out significantly worse, but all refinements of it give an improvement of the deviance that is less than the critical value. So this model ($\heartsuit$) comes out best.

9.6 Case study II: Analyzing a bonus-malus system using GLM

In this section, we study data from a rather large artificial motor insurance portfolio with about one million policies. The setting resembles the one described in Chapter 6. Drivers have been split up according to the following risk factors:

- R is region of residence: 1 = rural, 2 = town, 3 = big city
  There is not a lot of traffic in rural areas, so there are few claims; in big cities traffic is slow but dense: many small claims are to be expected.
- A is age class: 1 = 18–23, 2 = 24–64, 3 = 65–ω
  Young drivers cause more accidents, as do elderly drivers, but not as many.
- M is mileage: 1 = 0–7500 miles per year, 2 = 7500–12500, 3 = 12500–∞
  Drivers with low mileage tend to drive carefully, but they are not very experienced; a high mileage indicates that the driver often uses the highway, where claims are infrequent but more severe.
- U is usage: 1 = private use, 2 = business use
  Business use tends to lead to less careful driving and more claims; sometimes these drivers do not have to pay Value Added Tax (VAT), leading to lower claims.
- B is bonus-malus class: 1–14 as in Table 6.1.
  B=1 is the only malus class: more premium is due in this class than for a driver without any known history, who starts in B=2. Since claims are infrequent, it will come as no surprise that, in practice, we find about half of all drivers in the highest bonus-malus class B=14; they pay one fourth of the amount due in the malus class. In practice there is often a *bonus guarantee*: the longer one has remained in the class with the highest bonus, the more favorable the new class if a claim does occur. Therefore there is less *hunger for bonus* (see Remark 6.3.2) in this class: if a claim does not lead to more premium having to be paid, small claims will be filed as well, leading to more and somewhat smaller claims than in other classes.
- WW is the car weight class: 1 = ± 650, ..., 11 = ± 1600 kilograms
  The heavier the car causing an accident, the more damage it will cause, therefore the costlier a third-party claim. Persons who are on the road a lot, including business users, tend to drive heavier cars, so claim frequency will also increase with car weight.

We choose not to work with the individual data on each policy, ignoring the 'extra' information, if any, that can be obtained from the separate policies in a cell. See also Appendix A.3. So for the $3 \times 3 \times 3 \times 2 \times 14 \times 11 = 8316$ cells of drivers with identical values for these risk factors, the following totals have been recorded:

- Expo is the number of contracts in the cell (exposure).
  If a policy has been in force for only a fraction of the year, not a full policy year but only this fraction should be counted in Expo. See Remark 9.3.4 and Appendix A.3.
- nCl is the number of claims in a cell (we consider third party claims only).
- TotCl is the total amount of the claims paid.
- TotPrem is the total amount of premiums collected.

The annual premium for a policy in cell (r, a, m, b, u, w) is determined as follows:

$$\pi_{rambuw} = 500 \times P_b \times (W_w/W_1) \times R_r \times M_m \qquad (9.35)$$

The factors used in the tariff for region, mileage class and bonus-malus class have the following values (in %):

$$R_r = 85, 90, 100\%;$$
$$M_m = 90, 100, 110\%; \qquad (9.36)$$
$$P_b = 120, 100, 90, 80, 70, 60, 55, 50, 45, 40, 37.5, 35, 32.5, 30\%.$$

Notice that age and usage are not tariff factors. The premium is proportional to the car weight W_w. This actual weight is the coefficient in the vector $(650, \ldots, 1600)$ corresponding to the weight class $w = 1, \ldots, 11$.

This is how the portfolio is read and stored (see also Appendix A):

```
fn <- "http://www1.fee.uva.nl/ke/act/people/kaas/Cars.txt"
Cars <- read.table(fn, header=T)
Cars$A <- as.factor(Cars$A);  Cars$R <- as.factor(Cars$R)
Cars$M <- as.factor(Cars$M);  Cars$U <- as.factor(Cars$U)
Bminus1 <- Cars$B - 1;  Bis14 <- as.numeric(Cars$B==14)
Cars$B <- as.factor(Cars$B);  Cars$WW <- as.factor(Cars$WW)
ActualWt <- c(650,750,825,875,925,975,1025,1075,1175,1375,1600)
W <- log(ActualWt/650)[Cars$WW]
str(Cars); attach(Cars)
ftable(xtabs(cbind(Expo, nCl, TotCl, TotPrem) ~ R+A+M+U))
```

The first line stores the filename; note the forward slashes used. The second reads the data frame stored there, and assigns it to Cars; the first line of the file contains the names of the variables and is used as a header. The classifying variables are to be treated as factors. Bis14 is a vector with ones for observations with B==14, zeros for others. The weight class WW is converted to the variate W containing the log of the actual weight (relative to 650 kilograms) in the next two lines. In the next-to-last line, the structure of the data frame is displayed. It is then attached to the R search path, so the data frame is searched by R when evaluating a variable. Objects in it can now be accessed by simply giving their names, for example just WW denotes Cars$WW. The last line produces a condensed version of the data. In fact it prints

Table 9.3 Average claim frequencies in %, by risk group

Region R	1	2	3	Age group A	1	2	3
claims	7.6	9.6	12.7	claims	28.2	8.6	14.7

Mileage M	1	2	3	Usage U	1	2
claims	9.0	10.2	11.8	claims	9.0	13.4

a flat table of a cross table of the cell totals for all combinations of the levels of R, A, M, U, aggregated over all weight and bonus-malus classes WW, B.

We are going to analyze if the rating system works properly, since if the premiums per policy paid deviate much from the average claims caused, drivers who are overcharged might take their business elsewhere, leaving the insurer with mainly 'bad' risks and therefore future losses.

The total of the claims on a particular policy is a random variable. Policies are assumed to be statistically independent. A reasonable model is to assume that every driver in a particular cell has a Poisson distributed number of claims, the size of each of which is an independent drawing from the same distribution. So the total claims in a cell is a compound Poisson random variable. For automobile claims, especially if restrained by for example an excess of loss reinsurance where a reinsurer takes care of the large claims, statistical studies have shown that it is reasonable to take the individual claim amounts in the compound Poisson sum to be gamma(α, β) distributed random variables for certain parameters α, β.

A first step in the analysis is to tabulate the claim frequencies per policy against each of the risk factors R, A, M, U. This leads to the percentages given in Table 9.3. One sees that the denser populated the region where the policy holder lives, the more he claims. Young drivers perform worse than mature or elderly drivers. The more mileage, the more claims, but by no means in proportion to the miles driven. Business usage leads to roughly 50% more claims than private use.

Remark 9.6.1 (Danger of one-dimensional analysis)
The actuary should not be tempted to stop the analysis at this stage and use the values in Table 9.3 to construct tariff factors. One reason is that the classifications are most likely correlated. Consider the stylized example of Table 9.4, in which policies have been split up by car weight class and type of usage. There were 300 business users with a light car. They had average claims 230. A one-dimensional analysis (using the marginal totals in the table only) would lead to a surcharge for business usage of 22% ($261.05/213.33 = 1.2237$). For a heavy car rather than a light one, this would be 24% ($249.00/200.97 = 1.2397$). For a heavy car for business use, this would amount to 51.70%.

But within the table, in all weight classes there is an effect of 15% for business usage, and for both usage classes, an effect of 20% more claims for a heavy vehicle compared with a light one. This amounts to 38% more premium for a heavy business car as opposed to a light car for private use.

A more practical example is that young drivers might be charged both for being inexperienced and for the fact that they cannot be in high bonus classes yet.

Table 9.4 Claim averages and cell size by usage and weight

Weight	Private usage	Business usage	Total
Light	9000 × 200.00	300 × 230.00	9300 × 200.97
Middle	6000 × 220.00	700 × 253.00	6700 × 223.45
Heavy	3000 × 240.00	1000 × 276.00	4000 × 249.00
Total	18000 × 213.33	2000 × 261.05	20000 × 219.11

A one-dimensional analysis may also lead to undesirable results if there is *inter-action* between risk factors. Suppose drivers have light and heavy cars for business and private use in equal proportions, without correlations. Assume only business drivers with heavy cars actually claim a lot, say twice as much as all other categories. As the reader may verify, a one-dimensional analysis, looking at the marginals only, leads to heavy car business drivers paying 36%, light car private users only 16%, the other two categories each 24% of the total premium. Of course these percentages should be 40% for the first category, 20% for the other three.   ∇

To construct Tables 9.3 and 9.5, one may use `aggregate`, or one can apply the `sum` function to `nCl` in the numerator and to `Expo` in the denominator, splitting up the data by the values of one or more factors, as follows:

```
100 * tapply(nCl, R, sum) / tapply(Expo, R, sum)
## 7.649282   9.581532 12.680255
100 * tapply(nCl,list(R,A),sum) / tapply(Expo,list(R,A),sum)
```

On the two-dimensional tables in Table 9.5 the same observations can be made as on the one-dimensional tables in Table 9.3. But from the total weights given, one sees that the distribution of the contracts over the risk classes is far from orthogonal. For example, mileage and region are correlated. In region 3, there are many business users, in the other regions there are more 'Sunday drivers'. Business users drive a lot: in class M=1 they are a small minority, in class M=3, half the drivers are business users.

Table 9.6 gives more insight. It gives weighted averages of claim numbers, with their aggregated numbers of exposure (total exposure weights), for all combinations of levels of R, A, M, U. Looking closely at this table, one sees that the effect of mileage has somehow vanished. Apparently the business users with U=2 and the big city drivers with R=3 are the ones having many claims. How many miles one drives does not matter, given the other characteristics. The apparent effect of mileage seen earlier can be ascribed to the fact that mileage and usage, as well as mileage and region of residence, are very much correlated. There is also the practical problem that in registering mileage, there is ample opportunity for foul play by the policyholder or by the salespeople, who might have their own reasons to give some policyholders a discount for low mileage, even if unwarranted. In practice, the mileage entered for a policy hardly ever changes over the years.

If one plots the logarithm of the average number of claims against the bonus-malus class (see Exercise 9.6.1), one sees that this relation is close to linear, except

Table 9.5 Average numbers of claims in %, by all pairs of factors R, A, M, U

A		1	2	3
R				
1	Claims	21	6	11
	exposure	9033	239447	50803
2	Claims	27	8	14
	exposure	9004	240208	51074
3	Claims	34	11	18
	exposure	11952	323020	68206

U		1	2
R			
1	Claims	7	10
	exposure	222001	77282
2	Claims	9	12
	exposure	221815	78471
3	Claims	11	16
	exposure	266408	136770

M		1	2	3
R				
1	Claims	7	8	8
	exposure	119205	120495	59583
2	Claims	9	10	10
	exposure	119721	120236	60329
3	Claims	11	12	14

U		1	2
A			
1	Claims	25	37
	exposure	21222	8767
2	Claims	8	11
	exposure	568616	234059
3	Claims	13	19

M		1	2	3
A				
1	Claims	24	29	32
	exposure	9594	12004	8391
2	Claims	8	9	10
	exposure	255381	322335	224959
3	Claims	13	15	17
	exposure	54490	67871	47722

U		1	2
M			
1	Claims	9	12
	exposure	287573	31892
2	Claims	9	13
	exposure	281934	120276
3	Claims	10	14
	exposure	140717	140355

Table 9.6 Average numbers of claims in %, and total exposure

		R	1		2		3	
		U	1	2	1	2	1	2
A	M							
1	1	Claims	19	27	24	32	29	40
		exposure	3255	369	3219	370	2154	227
	2	Claims	20	29	25	38	29	42
		exposure	2512	1084	2526	1084	3356	1442
	3	Claims	18	26	25	32	30	44
		exposure	903	910	907	898	2390	2383
2	1	Claims	6	9	7	11	9	13
		exposure	85803	9570	86001	9484	58053	6470
	2	Claims	6	8	7	10	9	14
		exposure	68040	28337	67485	29040	90550	38883
	3	Claims	6	8	7	10	9	14
		exposure	23864	23833	23844	24354	64976	64088
3	1	Claims	10	14	13	20	15	22
		exposure	18205	2003	18609	2038	12274	1361
	2	Claims	10	13	13	18	16	21
		exposure	14374	6148	13998	6103	19093	8155
	3	Claims	10	15	13	19	16	22
		exposure	5045	5028	5226	5100	13562	13761

for the special class 14. In this class, the bonus guarantee makes the bonus hunger less compulsive, leading to more (and smaller) claims. This is why we will try if the estimation of the percentages to be asked in the various bonus-malus classes can perhaps be replaced by a system in which the premium for the next-higher step is a fixed percentage of that of the current step, with the exception of the last step 14. So later on we will try if this restricted system predicts the observed claims much worse than a system in which the percentages may be arbitrary.

The premium is proportional to the weight. Plotting average number of claims against weight, one sees an increase, somewhere between those exhibited by the functions $f(x) = x$ and $f(x) = \sqrt{x}$. Plotting the logarithm of the mean number of claims against the logarithm of the car weight, one sees that their relation is almost linear through the origin, with a slope of about 0.9. That means that the average number of claims increases as $w^{0.9}$ (see Exercise 9.6.2).

For the claim sizes, a similar exploratory data analysis can be done. Again, it proves that the factors R, A, U have an influence, while the effect of mileage M can be ascribed to the fact that it is correlated with U and R. Plotting claim sizes against bonus-malus class reveals no connection, except in class 14 where small claims tend to be filed more frequently.

9.6.1 GLM analysis for the total claims per policy

The total claims, $S = X_1 + \cdots + X_N$, say, on each policy is assumed to be compound Poisson with $E[N] = \lambda$ and $X_i \sim \mathrm{gamma}(\alpha, \beta)$. Therefore, see (3.60):

$$\mathrm{Var}[S] = \lambda E[X^2] = \lambda \left((E[X])^2 + \mathrm{Var}[X] \right) = \frac{\lambda \alpha}{\beta} \frac{\alpha + 1}{\beta} = E[S] \frac{\alpha + 1}{\beta}. \qquad (9.37)$$

Assuming that λ varies much more strongly over cells than shape α and scale β of the claim sizes do, it is reasonable to say that $\mathrm{Var}[S]$ is approximately proportional to $E[S]$. This would imply that we could fit models from the quasipoisson family (random variables with a mean-variance relationship such as the Poisson multiples in the sense of Example 9.4.2) to S. Taking the canonical log-link is indicated since our aim is to construct a tariff, so we want to find specifically the loglinear model that fits the annual claims best.

In Tables 9.7 and 9.8, we fit the *main effects* (for interactions between factors, see Remark 9.6.5), first without M, in g1, then with M, in g2. It is clear that mileage does not affect the claims total, since there is hardly any change in the deviance, nor are the coefficients with levels M=2 and M=3 very different from 1 in Table 9.8. We already saw this from the exploratory data analysis.

In both models g1 and g2 in Table 9.8, the fitted percentages in the various bonus-malus steps shrink by a fixed fraction, in this case 0.8979, for the steps 1–13, while the 14th is allowed to be chosen freely. Conceivably, a better fit is obtained by estimating a separate parameter for every bonus-malus class. This is done in

Table 9.7 Deviance analysis of various `glm` calls

```
> g1 <- glm(TotCl/Expo~R+A+U+W+Bminus1+Bis14, quasipoisson, wei=Expo)
> g2 <- glm(TotCl/Expo~R+A+U+W+Bminus1+Bis14+M, quasipoisson, wei=Expo)
> g3 <- glm(TotCl/Expo~R+A+U+W+B, quasipoisson, wei=Expo)
> anova(g1,g2)
Analysis of Deviance Table
Model g1: TotCl/Expo ~ R + A + U + W + Bminus1 + Bis14
Model g2: TotCl/Expo ~ R + A + U + W + Bminus1 + Bis14 + M
   Resid. Df Resid. Dev   Df Deviance
g1      7515   38616941
g2      7513   38614965    2     1976
> anova(g1,g3)
Analysis of Deviance Table
Model g1: TotCl/Expo ~ R + A + U + W + Bminus1 + Bis14
Model g3: TotCl/Expo ~ R + A + U + W + B
   Resid. Df Resid. Dev   Df Deviance
g1      7515   38616941
g3      7504   38544506   11    72435
```

Model g3, which contains g1 as a submodel (with restricted parameter values). By extending the model with 11 extra parameters, the deviance drops from 38 616 941 to 38 544 506, a difference of 72 435. This might seem a substantial difference. But if we assume that our last model is 'correct', actually the deviance, divided by the scale parameter, is a χ^2 random variable with df 7504, so it should be somewhere near its mean. That would mean that the scale parameter is about $38\,544\,506/7504 = 5136.528$, so the observed decrease in the deviance is only 14.1 times the scale parameter. For a random variable with a $\chi^2(11)$ distribution, the value 14.1 is not one to arouse suspicion. One sees that the coefficients for the bonus-malus steps 1–13 in Model g3 closely follow the powers of 0.8979 in Model g1. We conclude that our model g1, with a separate contribution for class 14 but otherwise a geometric decrease per bonus-malus step, fits the data virtually as well as g3 having a separate parameter for all bonus-malus classes.

We have fitted a quasi-Poisson family in Table 9.7. The distribution in the exponential family that actually has the mean-variance relationship with a variance proportional to the mean, with a proportionality factor ϕ, is the one of ϕ times a Poisson random variable, see Example 9.4.2. Essentially, we approximate the total claims likelihood by a compound Poisson likelihood with a severity distribution that is degenerate on ϕ.

Remark 9.6.2 (The loss ratios)
The ultimate test for a rating system is how it reflects the observed losses. This is because, as said before, in non-life insurance, policy holders may leave when they think they are overcharged. Also, a 'wrong' rating system may attract bad risks. Taking the ratio of total claims and total premiums for cells characterized by the relevant risk factors leads to Table 9.9. As one sees, the tariff system leads to 'subsidizing solidarity', a capital transfer to the group of young drivers. Also, business users and elderly drivers are unfavorable risks in this system and for this portfolio. But recall that the portfolio studied is completely artificial.   ∇

Table 9.8 Parameter estimates of the models in Table 9.7

Variable/level	Model g1	Model g2	Model g3
(Intercept)	524.3017	522.6628	515.5321
R2	1.0843	1.0843	1.0843
R3	1.1916	1.1914	1.1917
A2	0.4147	0.4147	0.4143
A3	0.6184	0.6185	0.6179
U2	1.3841	1.3835	1.3842
W	2.3722	2.3722	2.3722
M2	1	1.0073	1
M3	1	1.0015	1
B2	0.8979	0.8979	0.9111
B3	$0.8979^2 = 0.8062$	0.8062	0.8275
B4	$0.8979^3 = 0.7238$	0.7238	0.7404
B5	:	0.6499	0.6499	0.6843
B6	:	0.5835	0.5835	0.6089
B7	:	0.5239	0.5239	0.5416
B8	:	0.4704	0.4704	0.4489
B9	:	0.4224	0.4224	0.4152
B10	:	0.3792	0.3792	0.3889
B11	:	0.3405	0.3405	0.3459
B12	:	0.3057	0.3057	0.3143
B13	$0.8979^{12} = 0.2745$	0.2745	0.2833
B14	0.2724	0.2724	0.2773

Remark 9.6.3 (Heterogeneity in the cells)
To mimic the effect of unobserved risk factors when we generated the data, we assumed that the cells were not homogeneous. Instead, each consists of about one
third 'bad' drivers, two thirds 'good' drivers. Good drivers are three times as good
as bad ones. So for a cell containing n policies with fixed theoretical frequency
λ, we did not draw from a Poisson(λn) distribution for the number of claims. Instead, we sampled from a mixed Poisson distribution, where a bad risk with mean
claim frequency 1.8λ has probability $1/3$, while a good risk has parameter 0.6λ.
So if $N \sim$ binomial($n, 1/3$) is the number of bad risks in a cell, we drew from a
Poisson($\lambda(1.8N + 0.6(n - N))$) distribution to determine the total number of claims.
How this affects the ratio of variance and mean of the number of claims in a cell is
studied in Exercise 9.6.10.

In practice, the expected claim frequency for arbitrary drivers in a cell of course
does not have two possible values only. The assumption that the expected claim
frequencies have a gamma distribution is more natural. A cell with a bonus class
11, for example, would be a mix of very good drivers who by bad fortune had a
claim three years ago, very bad drivers with some luck, but also drivers simply too
young to have climbed higher. A driver does not assume the claim frequency of the
bonus-malus class he happens to be in, but would in normal circumstances keep his
original driving skills. ∇

Table 9.9 Loss ratios in % for different levels of the risk factors

Risk factor	Levels and loss percentages													
B	1	2	3	4	5	6	7	8	9	10	11	12	13	14
loss %	53	58	58	59	62	64	63	57	59	62	53	52	50	53
WW	1	2	3	4	5	6	7	8	9	10	11			
loss %	58	60	58	58	58	57	57	55	56	54	53			
R	1	2	3											
loss %	56	57	56											
M	1	2	3											
loss %	58	57	55											
A	1	2	3											
loss %	119	49	73											
U	1	2												
loss %	52	67												

Remark 9.6.4 (GLMs for claim frequency and claim size separately)
For each cell, we can estimate both the number of claims and their sizes using a multiplicative model. The product of mean claim frequency and mean claim size is the risk premium in a cell, and in this way, we also get a multiplicative model, just as when we estimated a model for `TotCl/Expo` directly. The mathematical analysis of this system is much more complicated, but the resulting tariff is not much different; see the exercises.   ▽

Remark 9.6.5 (Main effects and interactions)
In the models studied so far, we only included the *main effects*. That is, we assumed that there is no interaction between the tariff factors: a young driver has a fixed percentage more claims than an older one with otherwise the same characteristics. If one wants to look at the effects of all two-way interactions between the pairs of model terms in model `g1` in Table 9.7, one simply replaces the model formula by its square, as follows.

```
glm(TotCl/Expo ~ (R+A+U+W+Bminus1+Bis14)^2, quasipoisson,
    wei=Expo)
```

Specific interactions can also be introduced by adding for example R*A as a model term. The estimates of the coefficients resulting from adding all two-way interactions are in Table 9.10. The effect of such an interaction is that for someone in region R=2 of age A=2, not just the coefficients with the main effects (+0.136793 and −0.848149) appear in the linear predictor, but also the additional effect of being of age A=2 in region R=2, which equals −0.126989.

Apart from not estimating coefficients for the first factor level because a constant is present in the model, R warns us that two coefficients were not estimated because of singularities. This means that should the columns for these coefficients be included in the design matrix for the corresponding parameters, this matrix would not

Table 9.10 Coefficients for interactions between factors

```
Coefficients: (2 not defined because of singularities)
              Estimate Std. Error t value Pr(>|t|)
(Intercept)    6.281532   0.064606   97.23  < 2e-16 ***
R2             0.136793   0.061855    2.21    0.027 *
R3             0.222821   0.056979    3.91  9.3e-05 ***
A2            -0.848149   0.063505  -13.36  < 2e-16 ***
A3            -0.446048   0.072156   -6.18  6.7e-10 ***
U2             0.348047   0.050944    6.83  9.0e-12 ***
W              0.658625   0.098057    6.72  2.0e-11 ***
Bminus1       -0.109536   0.008443  -12.97  < 2e-16 ***
Bis14          0.066347   0.072282    0.92    0.359
R2:A2         -0.126989   0.053841   -2.36    0.018 *
R3:A2         -0.097544   0.049465   -1.97    0.049 *
R2:A3         -0.041894   0.060023   -0.70    0.485
R3:A3         -0.138804   0.055476   -2.50    0.012 *
R2:U2         -0.002288   0.032524   -0.07    0.944
R3:U2         -0.000214   0.029226   -0.01    0.994
R2:W          -0.002231   0.067974   -0.03    0.974
R3:W           0.063855   0.062231    1.03    0.305
R2:Bminus1     0.007076   0.005495    1.29    0.198
R3:Bminus1     0.001414   0.005003    0.28    0.777
R2:Bis14      -0.034353   0.056984   -0.60    0.547
R3:Bis14       0.026924   0.052248    0.52    0.606
A2:U2         -0.027850   0.041099   -0.68    0.498
A3:U2         -0.058451   0.046086   -1.27    0.205
A2:W           0.172445   0.088386    1.95    0.051 .
A3:W           0.124628   0.098906    1.26    0.208
A2:Bminus1    -0.005273   0.006869   -0.77    0.443
A3:Bminus1     0.001427   0.007679    0.19    0.853
A2:Bis14       0.074591   0.049825    1.50    0.134
A3:Bis14            NA         NA      NA      NA
U2:W           0.023474   0.052041    0.45    0.652
U2:Bminus1    -0.000096   0.004087   -0.02    0.981
U2:Bis14      -0.001739   0.043217   -0.04    0.968
W:Bminus1      0.005245   0.008740    0.60    0.548
W:Bis14       -0.045760   0.092255   -0.50    0.620
Bminus1:Bis14       NA         NA      NA      NA
```

be of full rank because some covariate is a linear combination of the other ones. See Exercise 9.6.5.

The proper test to judge if the interactions contribute significantly to the fit is by comparing the resulting deviances, see again Exercise 9.6.5. From the * and . symbols in Table 9.10, conveniently provided by R to reflect the degree of significance of the corresponding parameter, we see that only coefficients with the interaction of region and age level are clearly different from zero, in the sense of being more than, say, two estimated standard deviations away from it.   ▽

9.7 Exercises

Section 9.1

1. Illustrate the quality of the approximations in Example 9.1.1 by computing and plotting these quantities for $\mu = 4, 5, \ldots, 40$.

Section 9.2

1. Of the distributions mentioned in the random component of a GLM, show that the mean is always μ_i, and express the variance in μ_i, ψ_i.

2. Show that if $X_i \sim \text{gamma}(\alpha, \beta_i)$ with parameters $\alpha = 1/\phi$ and $\beta_i = 1/(\phi \mu_i)$, all X_i have the same coefficient of variation, $i = 1, \ldots, n$. What is the skewness?

Section 9.3

1. Verify (9.10), (9.22) and (9.23). Also verify if (9.17) describes the maximum of (9.19) under assumption (9.20).

2. Show that the methods of Bailey-Simon, marginal totals and least squares, as well as the direct method, can all be written as methods of *weighted* marginal totals, where the following system is to be solved:

$$\sum_{i(j)} w_{ij} z_{ij} (\alpha_i \beta_j - y_{ij}) = 0 \quad \text{for all } j(i),$$

$$\text{where} \quad z_{ij} = 1 + \frac{y_{ij}}{\alpha_i \beta_j} \qquad \text{Bailey-Simon,}$$

$$= 1 \qquad\qquad \text{marginal totals,}$$

$$= \alpha_i \beta_j \qquad\quad \text{least squares,}$$

$$= \frac{1}{\alpha_i \beta_j} \qquad\quad \text{direct method.}$$

3. Show that the additive model $E[X_{ij}] = \alpha_i + \beta_j$ of the least squares method coincides with the one of the marginal totals.

4. Which requirement should the means and variances of $Y_{ij}/(\alpha_i \beta_j)$ fulfill in order to make (9.23) produce *optimal* estimates for α_i? (See Exercise 8.4.1.)

5. Starting from $\widehat{\alpha}_1 = 1$, determine $\widehat{\alpha}_2, \widehat{\beta}_1$ and $\widehat{\beta}_2$ in Example 9.3.12. Verify if the solution found for $\widehat{\alpha}_1$ satisfies the corresponding equation in each system of equations. Determine the results for the different models after the first iteration step, with initial values $\widehat{\beta}_j \equiv 1$, and after rescaling such that $\widehat{\alpha}_1 = 1$.

6. In Example 9.3.12, compare the resulting total premium according to the different models. What happens if we divide all weights w_{ij} by 10?

7. Show that the least squares method leads to maximum likelihood estimators in case the S_{ij} have a normal distribution with variance $w_{ij}\sigma^2$.

8. What can be said about the sum of the residuals $\sum_{i,j}(s_{ij} - w_{ij}\widehat{\alpha}_i\widehat{\beta}_j)$ if the $\widehat{\alpha}_i$ and the $\widehat{\beta}_j$ are fitted by the four methods of this section?

Section 9.4

1. Verify if (9.29) is the scaled deviance for a Poisson distribution.

2. Verify if (9.32) is the scaled deviance for a gamma distribution.

3. Show that in the model of Method 9.3.9, the second term of (9.29) is always zero.

4. Also show that the second term of deviance (9.32) is zero in a multiplicative model for the expected values, if the parameters are estimated by the direct method.

5. The deviance residuals, see also Remark 9.2.5, for Poisson and gamma GLMs are the square root of the ith term of the sum in (9.29) and (9.32) with the sign of $y_i - \hat{\mu}_i$. Using the inequality $t - 1 \geq \log t$ for $t \geq 0$, prove that these ith terms are non-negative.

6. Run the following R-code:

```
set.seed(1); y <- rpois(10, 7+2*(1:10))
g <- glm(y~I(1:10), poisson(link=identity))
2-AIC(g)/2 == logLik(g)
```

Find out why TRUE results from this. How can the value of the maximized loglikelihood be reconstructed using the functions dpois and fitted? Compute the BIC for g.

7. For $X \sim N(\mu,1)$, gamma$(\mu,1)$ and Poisson(μ), compare finding the mode with maximizing the likelihood.

Section 9.5

1. Do an analysis similar to the one in this section, but now to the portfolio of Appendix A.3.

2. To the same portfolio as in the previous exercise, apply

```
g <- glm(y ~ Re*Sx, poisson); anova(g, test="Chisq")}
```

Compare with what happens if we use the aggregated data instead of the full data. Look at the five number summary of the residuals, too.

Section 9.6

1. Make a plot of the average number of claims against the bonus-malus class. Do the same, but now with the average number of claims on a log-scale.

2. Make a plot of the average number of claims against the log of the car weight. The same with both on a logarithmic scale.

3. Note that there were not 8316 cells in Table 9.7, but only 7524; this is because many cells had weight Expo = 0. Can you think of the reason why so many cells were empty? Hint: how old must one be to enjoy the maximal bonus?

4. In the model of Table 9.7, can (B==14), B or W be removed from the model without getting a significantly worse fit? Does it help to allow for separate coefficients for each weight class?

5. The deviance of the same model as in Table 9.10, but without interactions, is 38 616 941 on 7515 df, the deviance with interactions is 38 408 588 on 7491 df. Do the interaction terms improve the fit significantly?
   In Table 9.10, for Bminus1:Bis14 it is easy to explain that no coefficient can be estimated: how is it related to Bis14? Show that when Expo > 0, A3:Bis14 is always equal to Bis14 - A2:Bis14; see Exercise 9.6.3.

6. Estimate a multiplicative model explaining nCl/Expo by R, A, M, U, W, B and B==14 using a quasi-Poisson error structure. Estimate a multiplicative model explaining TotCl/nCl by the same covariates using a Gamma family. Use a log-link, and weights nCl. Combine these two models into one for the risk premium, and compare the coefficients with those obtained by estimating TotCl/Expo directly.

7. Note that loss ratios over 56% represent bad risks for the insurer, those up to 56% are good. Discuss if and how the rating system should be changed. Explain why the loss ratios evolve with weight WW as they do.

8. Determine the ratio of variance and mean of the total claims as a function of λ. Is it reasonable to treat this as a constant, in view of which values of λ, say $\lambda \in [0.05, 0.2]$, are plausible?

9. For the claim sizes, instead of gamma(α, β) random variables with $\alpha = \frac{1}{2}$, we generated lognormal(μ, σ^2) claims with the same mean and variance. Which choice of μ, σ^2 in terms of α, β ensures that the mean and variance of a lognormal(μ, σ^2) and a gamma(α, β) random variable coincide? How about the skewness?

10. For a cell containing n policies with average claim frequency $\lambda = 0.1$, determine how much overdispersion $\text{Var}[N]/\text{E}[N]$ is caused by the fact that we did not take a homogeneous Poisson model for the number of claims N but rather the mixed model of Remark 9.6.3.

11. Find out what the most and least profitable groups of policyholders are by doing

```
l <- list(Use=U,Age=A,Area=R,Mile=M)
round(ftable(100*tapply(TotCl,l,sum)/tapply(TotPrem,l,sum)))
```

Chapter 10
IBNR techniques

IBNR reserves represent an important cog in the insurance accounting machinery — Bornhuetter & Ferguson, 1978

10.1 Introduction

In the past, non-life insurance portfolios were financed through a pay-as-you-go system. All claims in a particular year were paid from the premium income of that same year, no matter in which year the claim originated. The financial balance in the portfolio was realized by ensuring that there was an equivalence between the premiums collected and the claims paid in a particular financial year. Technical gains and losses arose because of the difference between the premium income in a year and the claims paid during the year.

The claims originating in a particular year often cannot be finalized in that year. For example, long legal procedures are the rule with liability insurance claims. But there may also be other causes for delay, such as the fact that the exact size of the claim is hard to assess. Also, the claim may be filed only later, or more payments than one have to be made, such as in disability insurance. All these factors will lead to delay of the actual payment of the claims. The claims that have already occurred, but are not sufficiently known, are foreseeable in the sense that one knows that payments will have to be made, but not how much the total payment is going to be. Consider also the case that a premium is paid for the claims in a particular year, and a claim arises of which the insurer is not notified as yet. Here also, we have losses that have to be reimbursed in future years.

Such claims are now connected to the years for which the premiums were actually paid. This means that reserves have to be kept regarding claims that are known to exist, but for which the eventual size is unknown at the time the reserves have to be set. For claims like these, several acronyms are in use. One has IBNR claims (Incurred But Not Reported) for claims that have occurred but have not been filed. Hence the name IBNR methods, IBNR claims and IBNR reserves for all quantities of this type. There are also RBNS claims (Reported But Not Settled), for claims that are known but not (completely) paid. Other acronyms are IBNFR, IBNER and RB-NFS, where the F is for Fully, the E for Enough. Large claims known to the insurer are often handled on a case-by-case basis.

Table 10.1 A run-off triangle with payments by development year (horizontally) and year of origin (vertically)

Year of origin	Development year							
	1	2	3	4	5	6	7	8
2011	101	153	52	17	14	3	4	1
2012	99	121	76	32	10	3	1	
2013	110	182	80	20	21	2		
2014	160	197	82	38	19			
2015	161	254	85	46				
2016	185	201	86					
2017	178	261						
2018	168							

When modeling these situations, one generally starts from a so-called *run-off triangle*, containing loss figures, for example cumulated payments, for each combination of policy year and development year. It is compiled in the following way:

1. We start in 2011 with a portfolio of insurance contracts. Let us assume that the total claims to be paid are fully known on January 1, 2019, eight years after the end of this year of origin;
2. The claims occurring in the year 2011 have to be paid from the premiums collected in 2011;
3. These payments have been made in the year 2011 itself, but also in the years 2012–2018;
4. In the same way, for the claims pertaining to the year of origin 2012, one has the claims which are known in the years 2012–2018, and it is unknown what has to be paid in 2019;
5. For the year 2016, the known claims are the ones paid in the period 2016–2018, but there are also unknown ones that will come up in the years 2019 and after;
6. For the claims concerning the premiums paid in 2018, on December 31, 2018 only the payments made in 2018 are known, but we can expect that more payments will have to be made in and after 2009. We may expect that the claims develop in a pattern similar to the one of the claims in 2011–2018.

The development pattern can schematically be depicted as in Table 10.1. The numbers in the triangle are the known total payments, grouped by year of origin i (by row) and development year j (by column). The row corresponding to year 2013 contains the six numbers known on December 31, 2018. The third element in this row, for example, denotes the claims incurred in 2013, but paid for in the third year of development 2015. In the triangle of Table 10.1, we look at new contracts only. This situation may occur when a new type of policy was issued for the first time in 2011. The business written in this year on average has had only half a year to produce claims in 2011, which is why the numbers in the first column are somewhat lower than those in the second. The numbers on the diagonal with $i + j - 1 = k$ denote the payments that were made in calendar year k. There are many ways to group these same numbers into a triangle, but the one given in Table 10.1 is the customary one. On the basis of the claim figures in Table 10.1, which could be claim numbers but also more general losses, we want to make predictions about claims that will be paid, or filed, in future calendar years. These future years are to be found in the bottom-right part of Table 10.1. The goal of the actuarial IBNR techniques is to predict these figures, so as to complete the triangle into a square. The total of the figures found in the lower right triangle is the total of the claims that will have to be paid in the future from the premiums that were collected in the period 2011–2018. This total is precisely the reserve to be kept.

We assume that the development pattern lasts eight years. In many branches, notably in liability, claims may still be filed after a time longer than eight years. In that case, we have to make predictions about development years after the eighth, of which our run-off triangle provides no data. We not only have to extend the triangle to a square, but to a rectangle containing more development years. The usual practice is to assume that the development procedure is stopped after a number of years, and to apply a correction factor for the payments made after the development period considered.

The future payments are estimated following well-established actuarial practice. Sometimes one central estimator is given, but also sometimes a whole range of possibilities is considered, containing both the estimated values and, conceivably, the actual results. Not just estimating the mean, but also getting an idea of the variance of the results is important. Methods to determine the reserves have been developed that each meet specific requirements, have different model assumptions, and produce different estimates. In practice, sometimes the method that is the most likely to produce the 'best' estimator is used to determine the estimate of the expected claims, while the results of other methods are used as a means to judge the variation of the stochastic result, which is of course a rather unscientific approach.

To complete the triangle in Table 10.1, we can give various methods, each reflecting the influence of a number of exogenous factors. In the direction of the year of origin, variation in the size of the portfolio will have an influence on the claim figures. On the other hand, for the factor development year (horizontally), changes in the claim handling procedure as well as in the speed of finalization of the claims will produce a change. The figures on the diagonals correspond to payments in a particular calendar year. Such figures will change due to monetary inflation, but also by changing jurisprudence or increasing claim proneness. As an example, in liability insurance for the medical profession the risk increases each year, and if the amounts awarded by judges get larger and larger, this is visible along the diagonals. In other words, the separation models, which have as factors the year of development and the calendar year, would be the best choice to describe the evolution of portfolios like these.

Obviously, one should try to get as accurate a picture as possible about the stochastic mechanism that produced the claims, test this model if possible, and estimate the parameters of this model optimally to construct good predictors for the unknown observations. Very important is how the variance of claim figures is related to the mean value. This variance can be more or less constant, it can be proportional to the mean, proportional to the square of the mean (meaning that the coefficient of variation is a constant), or have some other relation with it. See the following section, as well as the chapters on Generalized Linear Models.

Just as with many rating techniques, see the previous chapter, in the actuarial literature quite often a heuristic method to complete an IBNR triangle was described first, and a sound statistical foundation was provided only later. In Section 10.2, we describe briefly the two most often used techniques to make IBNR forecasts: the chain ladder method and the Bornhuetter-Ferguson technique. There is a very basic generalized linear model (GLM) for which the ML-estimators can be computed by the well-known chain ladder method. On the other hand it is possible to give a model that involves a less rigid statistical structure and in which the calculations of the chain ladder method produce an optimal estimate in the sense of mean squared error. In Section 10.3 we give a general GLM, special cases of which can be shown to boil down to familiar methods of IBNR estimation such as the arithmetic and the geometric separation methods, as well as the chain ladder method. A numerical illustration is provided in Section 10.4, where various sets of covariates are used in GLMs to complete the triangle in Table 10.1. How to use R to do the calculations is described in Section 10.5. In Section 10.6, an analytical estimate of the prediction error of the chain ladder method is studied, as well as a bootstrap method. They were proposed by England and Verrall (1999) and England (2002). In this way, a standard error of prediction and an approximate predictive distribution for the random future losses are produced. In Section 10.7, we give another example, in which the parameters relating to the accident year are replaced by the known portfolio size, expressed in its number of policies or its premium income. A method related to Bornhuetter-Ferguson arises.

10.2 Two time-honored IBNR methods

The two methods most frequently used in practice are the chain ladder (CL) method and the Bornhuetter-Ferguson method. We give a short description here; for the R implementation, we refer to later sections.

10.2.1 Chain ladder

The idea behind the chain ladder method is that in any development year, about the same total percentage of the claims from each year of origin will be settled. In other words, in the run-off triangle, the columns are proportional. To see how in the chain ladder method predictions are computed for the unobserved part of a run-off rectangle, look at Table 10.2. Note that in most texts, the run-off figures given are cumulated by rows. This is a relic of the time when calculations had to be done by hand. In this text, we avoid this custom.

Table 10.2 Completing a run-off rectangle with CL predictions

	1	2	3	4	5
01	$\mathscr{A}$	$\mathscr{A}$	$\mathscr{A}$	$\mathscr{B}$	$\bullet$
02	$\mathscr{A}$	$\mathscr{A}$	$\mathscr{A}$	$\mathscr{B}$	
03	$\mathscr{C}$	$\mathscr{C}$	$\mathscr{C}$	$\star$	
04	$\mathscr{D}$	$\mathscr{D}$	$\widehat{\mathscr{D}}$	$\star\star$	
05	$\bullet$				

Consider the $(3,4)$ element in Table 10.2, denoted by $\star$ and representing payments regarding policy year 03 in their 4th development year. This is a claim figure for calendar year 06, which is the first future calendar year, and just beyond the edge of the observed figures. Because of the assumed proportionality, the ratio of the elements $\star : \mathscr{C}$ will be about equal to the ratio $\mathscr{B} : \mathscr{A}$. Therefore, a prediction $\widehat{X}_{34}$ of this element $\star$ is

$$\widehat{X}_{34} = \mathscr{C}_{\Sigma} \times \frac{\mathscr{B}_{\Sigma}}{\mathscr{A}_{\Sigma}}. \tag{10.1}$$

Here $\mathscr{B}_{\Sigma}$, for example, denotes the total of the $\mathscr{B}$-elements in Table 10.2, which are observed values. Prediction $\widehat{\mathscr{D}}$ is computed in exactly the same way, multiplying the total of the incremental payments to the left of it by the total above it, and dividing by the total of losses of earlier policy years and development years. The prediction $\star\star$ for $\widehat{X}_{44}$ (policy year 04, calendar year 07, so one year further in the future) can be computed by using the same 'development factor' $\mathscr{B}_{\Sigma}/\mathscr{A}_{\Sigma}$:

$$\widehat{X}_{44} = \mathscr{D}_{\Sigma} \times \frac{\mathscr{B}_{\Sigma}}{\mathscr{A}_{\Sigma}}, \tag{10.2}$$

where the sum $\mathscr{D}_{\Sigma}$ includes $\widehat{\mathscr{D}}$, which is not an actual observation but a prediction constructed as above. By using the fact that $\star = \mathscr{C}_{\Sigma} \times \mathscr{B}_{\Sigma}/\mathscr{A}_{\Sigma}$, it is easy to see that exactly the same prediction is obtained by taking

$$\widehat{X}_{44} = \frac{\mathscr{D}_{\Sigma} \times (\mathscr{B}_{\Sigma} + \star)}{\mathscr{A}_{\Sigma} + \mathscr{C}_{\Sigma}}, \tag{10.3}$$

hence by following the same procedure as for an observation in the next calendar year. In this way, starting with row 2 and proceeding from left to right, the entire lower triangle can be filled with predictions.

Remark 10.2.1 (Mirror property of the chain ladder method)
Note that this procedure produces the same estimates to complete the square if we exchange the roles of development year and year of origin, hence take the mirror image of the triangle around the NW–SE diagonal. ∇

Remark 10.2.2 (Marginal totals property of the chain ladder method)
One way to describe the chain ladder method is as follows: find numbers $\widehat{\alpha}_i, \widehat{\beta}_j$, $i, j = 1, \ldots, t$ such that the products $\widehat{\alpha}_i \widehat{\beta}_j$ (fitted values) for 'observed' combinations

(i, j) with $i + j - 1 \leq t$ have the same column sums and row sums as the actual observations:

$$\sum_{j(i)} \widehat{\alpha}_i \widehat{\beta}_j = \sum_{j(i)} X_{ij} \quad \text{for all } i(j). \tag{10.4}$$

Then predict future values for (i, j) with $i + j - 1 > t$ by $\widehat{\alpha}_i \widehat{\beta}_j$. How to find these numbers $\widehat{\alpha}_i, \widehat{\beta}_j$ will be described later on.

We will illustrate why this procedure leads to the same forecasts as the chain ladder method by looking at Table 10.2. First observe that $\mathscr{B}_\Sigma$ and $\mathscr{C}_\Sigma$ are already column and row sums, but also the sums of claim figures $\mathscr{A}_\Sigma$ needed can be computed from these quantities. For instance in our example, $\mathscr{A}_\Sigma = R_1 + R_2 - (C_5 + C_4)$ when R_i and C_j denote the ith row sum and the jth column sum.

Next, observe that if we replace the past losses X_{ij} by their fitted values $\widehat{\alpha}_i \widehat{\beta}_j$, the row and column sums remain unchanged, and therefore also the quantities like $\mathscr{A}_\Sigma$. When the chain ladder algorithm described above is applied to the new triangle, the numbers $\widehat{\alpha}_i \widehat{\beta}_j$ result as future predictions. $\qquad\qquad\qquad\qquad\qquad \nabla$

The basic principle of the chain ladder method admits many variants. One may wonder if there is indeed proportionality between the columns. Undoubtedly, this is determined by effects that operate along the axis describing the year of origin of the claims. By the chain ladder method, only the run-off pattern can be captured, given that all other factors that have an influence on the proportion of claims settled remain unchanged over time.

The chain ladder method is merely an algorithm, a deterministic method. But there are also stochastic models for the generating process underlying the run-off triangle in which these same calculations lead to an optimal prediction in some sense. See, for example, Section 10.3.1.

10.2.2 Bornhuetter-Ferguson

One of the difficulties with using the chain ladder method is that reserve forecasts can be quite unstable. In Table 10.2, a change of $p\%$ in $\mathscr{C}_\Sigma$ due to sampling variability will generate the same change in all forecasts for this row. So applying this method to a volatile claims experience will produce volatile forecasts. This volatility will show itself by changes in the reserve estimate each year, when a new diagonal of observations is added to the triangle. The Bornhuetter-Ferguson (1972) method provides a procedure for stabilizing such estimates.

Suppose that one has some prior expectation as to the ultimate losses to emerge from each accident period i, specifically, that $E[X_{i1} + \cdots + X_{it}] = M_i$ for some known quantity M_i. This quantity is often referred to as the *schedule* or *budget ultimate losses*. Notably one may have a prior view of the loss ratio M_i/P_i, where P_i is the premium income with accident year i. Combining these prior estimates with the

Table 10.3 Random variables in a run-off triangle

Year of origin	Development year				
	1	$\cdots$	$t-n+1$	$\cdots$	t
1	X_{11}	$\cdots$	$X_{1,t-n+1}$	$\cdots$	X_{1t}
$\vdots$	$\vdots$		$\vdots$		$\vdots$
n	X_{n1}	$\cdots$	$X_{n,t-n+1}$	$\cdots$	X_{nt}
$\vdots$	$\vdots$		$\vdots$		$\vdots$
t	X_{t1}	$\cdots$	$X_{t,t-n+1}$	$\cdots$	X_{tt}

development factors of the chain ladder method, one may form an estimate of the entire schedule of loss development. See Section 10.7 for more details.

It can be shown that the Bornhuetter-Ferguson method can be interpreted as a Bayesian method. The forecasts have the form of a credibility estimator.

10.3 A GLM that encompasses various IBNR methods

Several often used and traditional actuarial methods to complete an IBNR triangle can be described by one Generalized Linear Model. In Table 10.3, the random variables X_{ij} for $i, j = 1, 2, \ldots, t$ denote the claim figure for year of origin i and year of development j, meaning that the claims were paid in calendar year $i + j - 1$. For (i, j) combinations with $i + j - 1 \leq t$, X_{ij} has already been observed, otherwise it is a future observation. As well as claims actually paid, these figures may also be used to denote quantities such as loss ratios. We take a multiplicative model, with a parameter for each row i, each column j and each diagonal $k = i + j - 1$, as follows:

$$X_{ij} \approx \alpha_i \cdot \beta_j \cdot \gamma_k. \tag{10.5}$$

The deviation of the observation on the left hand side from its mean value on the right hand side is attributed to chance. As one sees, if we assume further that the random variables X_{ij} are independent and restrict their distribution to be in the exponential dispersion family, (10.5) is a Generalized Linear Model in the sense of Chapter 9. Year of origin i, year of development j and calendar year $k = i + j - 1$ act as explanatory variables for the observation X_{ij}. The expected value of X_{ij} is the exponent of the linear form $\log \alpha_i + \log \beta_j + \log \gamma_k$, so there is a logarithmic link. Note that the covariates are all dummies representing group membership for rows, columns and diagonals in Table 10.3. We will determine maximum likelihood estimates of the parameters α_i, β_j and γ_k, under various assumptions for the probability distribution of the X_{ij}. It will turn out that in this simple way, we can generate quite a few widely used IBNR techniques.

Having found estimates of the parameters, it is easy to extend the triangle to a square, simply by taking

$$\widehat{X}_{ij} := \widehat{\alpha}_i \cdot \widehat{\beta}_j \cdot \widehat{\gamma}_k. \tag{10.6}$$

A problem is that we have no data on the values of the γ_k for future calendar years k with $k > t$. The problem can be solved, for example, by assuming that the γ_k have a geometric pattern, with $\gamma_k \propto \gamma^k$ for some real number γ.

10.3.1 Chain ladder method as a GLM

The first method that can be derived from model (10.5) is the *chain ladder* method of Section 10.2.1. Suppose that we restrict model (10.5) to:

$$X_{ij} \sim \text{Poisson}(\alpha_i\beta_j) \text{ independent}; \gamma_k \equiv 1. \tag{10.7}$$

If the parameters $\alpha_i > 0$ and $\beta_j > 0$ are to be estimated by maximum likelihood, we have in fact a multiplicative GLM with Poisson errors and a log-link, because the observations X_{ij}, $i, j = 1,\ldots,t$; $i + j \leq t$ are independent Poisson random variables with a logarithmic model for the means; explanatory variables are the factors row number and column number.

By Property 9.3.9 it follows that the marginal totals of the triangle, hence the row sums R_i and the column sums C_j of the observed figures X_{ij}, must be equal to the predictions $\sum_j \widehat{\alpha}_i\widehat{\beta}_j$ and $\sum_i \widehat{\alpha}_i\widehat{\beta}_j$ for these quantities; see (10.4). So it follows from Remark 10.2.2 that the optimal estimates of the parameters α_i and β_j produced by this GLM are equal to the parameter estimates found by the chain ladder method.

One of the parameters is superfluous, since if we replace all α_i and β_j by $\delta\alpha_i$ and β_j/δ we get the same expected values. To resolve this ambiguity, we impose an additional restriction on the parameters. We could use a 'corner restriction', requiring for example $\alpha_1 = 1$, but a more natural restriction to ensure identifiability of the parameters is to require $\beta_1 + \cdots + \beta_t = 1$. This allows the β_j to be interpreted as the fraction of claims settled in development year j, and α_i as the 'volume' of year of origin i: it is the total of the payments made.

Maximizing the likelihood with model (10.7) can be done by an appropriate call of R's function `glm`; see Section 10.5. But by the triangular shape of the data, the system of marginal totals equations admits the following recursive solution method, originally devised by Verbeek (1972) for the case of the arithmetic separation method below.

Algorithm 10.3.1 (Verbeek's algorithm for chain ladder)
Look at Table 10.4. The row and column totals are those of the past X_{ij} in Table 10.3. To solve the marginal totals equations (10.4), we can proceed as follows.

1. From the first row sum equality $\widehat{\alpha}_1(\widehat{\beta}_1 + \cdots + \widehat{\beta}_t) = R_1$ it follows that $\widehat{\alpha}_1 = R_1$. Then from $\widehat{\alpha}_1\widehat{\beta}_t = C_t$ we find $\widehat{\beta}_t = C_t/R_1$.

Table 10.4 The marginal totals equations in a run-off triangle

Year of origin	1	⋯	Development year $t-n+1$	⋯	t	Row total
1	$\alpha_1\beta_1$		$\alpha_1\beta_{t-n+1}$		$\alpha_1\beta_t$	R_1
⋮						⋮
n	$\alpha_n\beta_1$		$\alpha_n\beta_{t-n+1}$			R_n
⋮						⋮
t	$\alpha_t\beta_1$					R_t
Column total	C_1	⋯	C_{t-n+1}	⋯	C_t	

2. Assume that, for a certain $n < t$, we have found estimates $\widehat{\beta}_{t-n+2},\ldots,\widehat{\beta}_t$ and $\widehat{\alpha}_1,\ldots,\widehat{\alpha}_{n-1}$. Then look at the following two marginal totals equations:

$$\begin{aligned}\widehat{\alpha}_n(\widehat{\beta}_1+\cdots+\widehat{\beta}_{t-n+1}) &= R_n;\\ (\widehat{\alpha}_1+\cdots+\widehat{\alpha}_n)\widehat{\beta}_{t-n+1} &= C_{t-n+1}.\end{aligned} \tag{10.8}$$

   By the fact that we take $\widehat{\beta}_1+\cdots+\widehat{\beta}_t = 1$, the first of these equations directly produces a value for $\widehat{\alpha}_n$, and then we can compute $\widehat{\beta}_{t-n+1}$ from the second one.
3. Repeat step 2 for $n = 2,\ldots,t$. ∇

10.3.2 Arithmetic and geometric separation methods

In the separation models, one assumes that in each year of development a fixed percentage is settled, and that there are additional effects that operate in the diagonal direction (from top-left to bottom-right) in the run-off triangle. So this model describes best the situation that there is inflation in the claim figures, or when the risk increases by other causes. This increase is characterized by an index factor for each calendar year, which is a constant for the observations parallel to the diagonal. One supposes that in Table 10.4, the random variables X_{ij} are average loss figures, where the total loss is divided by the number of claims, for year of origin i and development year j.

Arithmetic separation method The arithmetic separation method was described in Verbeek (1972), who applied the model to forecast the number of stop-loss claims reported. As time goes by, due to inflation more claims will exceed the retention, and this effect must be included in the model. In both the arithmetic and the geometric separation method the claim figures X_{ij} are explained by two aspects of time, just as for chain ladder and Bornhuetter-Ferguson. But in this case there is a calendar year effect γ_k, where $k = i+j-1$, and a development year effect β_j. So inflation and run-off pattern are the determinants for the claim figures now. For the *arithmetic separation method* we assume

$$X_{ij} \sim \text{Poisson}(\beta_j \gamma_k) \text{ independent}; \ \alpha_i \equiv 1. \tag{10.9}$$

Again, β_j and γ_k are estimated by maximum likelihood. Since this is again a GLM (Poisson with the canonical log-link), because of Property 9.3.9 the marginal totals property must hold here as well. In model (10.9) these marginal totals are the column sums and the sums over the diagonals, with $i + j - 1 = k$.

The parameter estimates in the arithmetic separation method can be obtained by a variant of Method 10.3.1 (Verbeek) for the chain ladder method computations. We have $E[X_{ij}] = \beta_j \gamma_{i+j-1}$. Again, the parameters β_j, $j = 1, \ldots, t$ describe the proportions settled in development year j. Assuming that the claims are all settled after t development years, we have $\beta_1 + \cdots + \beta_t = 1$. Using the marginal totals equations, see Table 10.4, we can determine directly the optimal factor $\widehat{\gamma}_t$, reflecting base level times inflation, as the sum of the observations on the long diagonal $\sum_i X_{i,t+1-i}$. Since β_t occurs in the final column only, we have $\widehat{\beta}_t = \widehat{X}_{1t} / \widehat{\gamma}_t$. With this, we can compute $\widehat{\gamma}_{t-1}$, and then $\widehat{\beta}_{t-1}$, and so on. Just as with the chain ladder method, the estimates thus constructed satisfy the marginal totals equations, and hence are maximum likelihood estimates because of Property 9.3.9.

To fill out the remaining part of the square, we also need values for the parameters $\gamma_{t+1}, \ldots, \gamma_{2t}$, to be multiplied by the corresponding $\widehat{\beta}_j$ estimate. We find values for these parameters by extending the sequence $\widehat{\gamma}_1, \ldots, \widehat{\gamma}_t$ in some way. This can be done with many techniques, for example loglinear extrapolation.

Geometric separation method The geometric separation method involves maximum likelihood estimation of the parameters in the following statistical model:

$$\log(X_{ij}) \sim \text{N}\left(\log(\beta_j \gamma_k), \sigma^2\right) \text{ independent}; \ \alpha_i \equiv 1. \tag{10.10}$$

Here σ^2 is an unknown variance. We get an ordinary regression model with $E[\log X_{ij}] = \log \beta_j + \log \gamma_{i+j-1}$. Its parameters can be estimated in the usual way, but they can also be estimated recursively in the way described above, starting from $\prod_j \beta_j = 1$.

Note that the values $\beta_j \gamma_{i+j-1}$ in this lognormal model are *not* the expected values of X_{ij}. In fact, they are only the medians; we have

$$\Pr[X_{ij} \le \beta_j \gamma_{i+j-1}] = \frac{1}{2} \quad \text{but} \quad E[X_{ij}] = e^{\sigma^2/2} \beta_j \gamma_{i+j-1}. \tag{10.11}$$

10.3.3 De Vijlder's least squares method

In the least squares method of De Vylder (1978), he assumes that $\gamma_k \equiv 1$ holds, while α_i and β_j are determined by minimizing the sum of squares $\sum_{i,j} (X_{ij} - \alpha_i \beta_j)^2$,

Table 10.5 Run-off data used in De Vylder (1978)

Year of origin	Development year					
	1	2	3	4	5	6
1						4627
2					15140	13343
3				43465	19018	12476
4			116531	42390	23505	14371
5		346807	118035	43784	12750	12284
6	308580	407117	132247	37086	27744	
7	358211	426329	157415	68219		
8	327996	436774	147154			
9	377369	561699				
10	333827					

taken over the set (i, j) for which observations are available. But this is tantamount to determining α_i and β_j by maximum likelihood in the following model:

$$X_{ij} \sim \mathrm{N}(\alpha_i \beta_j, \sigma^2) \text{ independent}; \; \gamma_k \equiv 1. \tag{10.12}$$

Just as with chain ladder, we assume that the mean payments for a particular year of origin/year of development combination result from two effects. First, a parameter characterizing the year of origin, proportional to the size of the portfolio in that year. Second, a parameter determining which proportion of the claims is settled through the period that claims develop. The parameters are estimated by least squares.

In practice it quite often happens that not all data in an IBNR-triangle are actually available. In De Vylder (1978) a 10×10 IBNR-triangle is studied missing all the observations from calendar years $1, \ldots, 5$, as well as those for development years $7, \ldots, 10$. What these numbers represent is not relevant. See Table 10.5.

This paper is the first to mention that in IBNR problems, time operates in three different ways: by policy year i reflecting growth of the portfolio, by development year j reflecting the run-off pattern of the claims, and by calendar year $k = i + j - 1$ reflecting inflation and changes in jurisprudence. De Vijlder proposed to use a multiplicative model $\alpha_i \beta_j \gamma_k$ for the data X_{ij}, and to choose those parameter values α_i, β_j and γ_k that minimize the least squares distance, therefore solving:

$$\min_{\alpha_i, \beta_j, \gamma_k} \sum_{i,j} w_{ij} (X_{ij} - \alpha_i \beta_j \gamma_{i+j-1})^2. \tag{10.13}$$

We multiply by weights $w_{ij} = 1$ if y_{ij} is an actual observation, $w_{ij} = 0$ otherwise, so the sum can be taken over all (i, j) combinations. De Vijlder proceeds by taking the inflation component fixed, hence $\gamma_k = \gamma^k$ for some real γ, and proves that doing this, one might actually have left out inflation of the model altogether, thus taking $\gamma_k \equiv 1$. See Exercise 10.3.4. Next he describes the method of successive substitution to solve the reduced problem, and gives the results for this method. Note that his paper was written in pre-PC times; he used a programmable hand-held calculator.

10.4 Illustration of some IBNR methods

Obviously, introducing parameters for the three time aspects year of origin, year
of development and calendar year sometimes leads to overparameterization. Many
of these parameters could be dropped, that is, taken equal to 1 in a multiplicative
model. Others might be required to be equal, for example by grouping classes having different values for some factor together. Admitting classes to be grouped leads
to many models being considered simultaneously, and it is sometimes hard to construct proper significance tests in these situations. Also, a classification of which
the classes are ordered, such as age class or bonus-malus step, might lead to parameters giving a fixed increase per class, except perhaps at the boundaries or for
some other special class. In a loglinear model, replacing arbitrary parameter values,
associated with factor levels (classes), by a geometric progression in these parameters is easily achieved by replacing the dummified factor by the actual levels again,
or in GLIM parlance, treating this variable as a variate instead of as a factor. Replacing arbitrary values α_i, with $\alpha_1 = 1$, by α^{i-1} for some real α means that we
assume the portfolio to grow, or shrink, by a fixed percentage each year. Doing the
same to the parameters β_j means that the proportion settled decreases by a fixed
fraction with each development year. Quite often, the first development year will
be different from the others, for example because only three quarters are counted
as the first year. In that case, one does best to allow a separate parameter for the
first year, taking parameters $\beta_1, \beta^2, \beta^3, \ldots$ for some real numbers β_1 and β. Instead
of with the original t parameters $\beta_1, \ldots, \beta_t$, one works with only two parameters.
By introducing a new dummy explanatory variable to indicate whether the calendar
year $k = i + j - 1$ with observation X_{ij} is before or after k_0, and letting it contribute
a factor 1 or δ to the mean, respectively, one gets a model for which in one year, the
inflation differs from the standard fixed inflation of the other years. Other functional
forms for the β_j parameters include the Hoerl-curve, where $\beta_j = \exp(\gamma j + \delta \log j)$
for some real numbers γ and δ. These can be used for all rows in common, or for
each row separately (interaction).

In the previous chapter, we introduced the (scaled) deviance as a 'distance' between the data and the estimates. It is determined from the difference of the maximally attainable likelihood and the one of a particular model. Using this, one may
test if it is worthwhile to complicate a model by introducing more parameters. For
a nested model, of which the parameter set can be constructed by imposing linear restrictions on the parameters of the original model, it is possible to judge if
the distance between data and estimates is 'significantly' larger. It proves that this
difference in distance, under the null-hypothesis that the eliminated parameters are
superfluous, is approximately χ^2 distributed, when suitably scaled. In similar fashion, the 'goodness of fit' of non-nested models can be compared by using the Akaike
information criterion, see Remark 9.4.4.

Some regression software leaves it to the user to resolve the problems arising
from introducing parameters with covariates that are linearly dependent of the others, the so-called 'dummy trap' (multicollinearity). The glm function in R is more
user-friendly in this respect. For example if one takes all three effects in (10.5)

Table 10.6 Parameter set, degrees of freedom (= number of observations less number of estimated parameters), and deviance for several models applied to the data of Table 10.1.

Model	Parameters used	Df	Deviance
I	$\mu, \alpha_i, \beta_j, \gamma_k$	15	25.7
II	μ, α_i, β_j	21	38.0
III	μ, β_j, γ_k	21	36.8
IV	$\mu, \beta_j, \gamma^{k-1}$	27	59.9
V	$\mu, \alpha^{i-1}, \beta_j$	27	59.9
VI	$\mu, \alpha_i, \gamma^{k-1}$	27	504.
VII	$\mu, \alpha_i, \beta^{j-1}$	27	504.
VIII	$\mu, \alpha_i, \beta_1, \beta^{j-1}$	26	46.0 ♡
IX	$\mu, \alpha^{i-1}, \beta_1, \beta^{j-1}$	32	67.9
X	$\mu, \alpha^{i-1}, \beta^{j-1}$	33	582.
XI	μ	35	2656

geometric, with as fitted values

$$\widehat{X}_{ij} = \widehat{\mu}\,\widehat{\alpha}^{i-1}\widehat{\beta}^{j-1}\widehat{\gamma}^{i+j-2},\tag{10.14}$$

R does not stop but simply proceeds by taking the last of these three parameters to be equal to 1; see Exercise 10.4.2. Notice that by introducing $\widehat{\mu}$ in (10.14), all three parameter estimates can have the form $\widehat{\alpha}^{i-1}$, $\widehat{\beta}^{j-1}$ and $\widehat{\gamma}^{i+j-2}$. In the same way, we can take $\alpha_1 = \beta_1 = \gamma_1 = 1$ in (10.5). The parameter $\mu = E[X_{11}]$ is the level in the first year of origin and development year 1.

10.4.1 Modeling the claim numbers in Table 10.1

We fitted a number of models to explain the claim figures in Table 10.1. They were actually claim numbers; the averages of the payments are shown in Table 10.8. To judge which model best fits the data, we estimated a few models for (10.5), all assuming the observations to be Poisson($\alpha_i\beta_j\gamma_{i+j-1}$). See Table 10.6. By imposing (loglinear) restrictions like $\beta_j = \beta^{j-1}$ or $\gamma_k \equiv 1$, we reproduce the various models discussed earlier. The reader may verify that in model I, one may choose $\gamma_8 = 1$ without loss of generality. This means that model I has only 6 more parameters to be estimated than model II. Notice that for model I with $E[X_{ij}] = \mu\alpha_i\beta_j\gamma_{i+j-1}$, there are $3(t-1)$ parameters to be estimated from $t(t+1)/2$ observations, hence model I only makes sense if $t \geq 4$.

It can be shown that we get the same estimates using either of the models $E[X_{ij}] = \mu\alpha_i\beta^{j-1}$ and $E[X_{ij}] = \mu\alpha_i\gamma^{i+j-1} = (\mu\gamma)(\alpha_i\gamma^{i-1})(\beta_j\gamma^{j-1})$. Completing the triangle of Table 10.1 into a square by using model VIII produces Table 10.7. The column 'Total' contains the row sums of the estimated future payments, hence exactly the amount to be reserved regarding each year of origin. The figures in the top-left part are estimates of the already observed values, the ones in the bottom-right part are predictions for future payments.

All other models are nested in model I, since its set of parameters contains all other ones as a subset. The estimates for model I best fit the data. About the deviances and the corresponding numbers of degrees of freedom, the following can be said. The chain ladder model II is not rejected statistically against the fullest model I on a 95% level, since it contains six parameters fewer, and the χ^2 critical value is 12.6 while the difference in scaled deviance is only 12.3. The arithmetic separation model III fits the data somewhat better than model II. Model IV with an arbitrary run-off pattern β_j and a constant inflation γ is equivalent to model V, which has a constant rate of growth for the portfolio. In Exercise 10.3.3, the reader is asked to explain why these two models are identical. Model IV, which is nested in III and has six parameters fewer, predicts significantly worse. In the same way, V is worse than II. Models VI and VII again are identical. Their fit is bad. Model VIII, with a geometric development pattern except for the first year, seems to be the winner: with five parameters fewer, its fit is not significantly worse than model II in which it is nested. It fits better than model VII in which the first column is not treated separately. Comparing VIII with IX, we see that a constant rate of growth in the portfolio must be rejected in favor of an arbitrary growth pattern. In model X, there is a constant rate of growth as well as a geometric development pattern. The fit is bad, mainly because the first column is so different.

Note that it cannot be ruled out that there actually is a calendar year effect, if only because model I is already preferred to II if a significance level 6% is chosen instead of 5%. Doing `plot(k,residuals(glm(Nij~i+j, poisson)))` reveals that the residuals for $k = 3$ are all well below zero. Indeed adding one extra parameter γ_3 decreases the deviance substantially. Whether this improvement may be called statistically significant on a 5% level is another matter, because specifically choosing this calendar year $k = 3$ was inspired by looking at the data. So, actually, the deviance difference attained is the maximum of many marginally χ^2 distributed random variables, one for all models implicitly considered. Though probably related, these are not comonotonic. So the probability of rejecting a true hypothesis (Type I error) is no longer 5%.

From model XI, having only a constant term, we see that the 'percentage of explained deviance' of model VIII is more than 98%. But even model IX, which contains only a constant term and three other parameters, already explains 97.4% of the deviation.

The estimated model VIII gives the following predictions:

$$\text{VIII:} \quad \widehat{X}_{ij} = 102.3 \times \alpha_i \times 3.20^{j \neq 1} \times 0.42^{j-1},$$
$$\text{with} \quad \vec{\alpha}' = (1, 0.99, 1.21, 1.47, 1.67, 1.56, 1.81, 1.64). \tag{10.15}$$

Here $j \neq 1$ should be read as a Boolean expression, with value 1 if true, 0 if false (in this case, for the special column with $j = 1$). Model IX leads to:

$$\text{IX:} \quad \widehat{X}_{ij} = 101.1 \times 1.10^{i-1} \times 3.34^{j \neq 1} \times 0.42^{j-1}. \tag{10.16}$$

Table 10.7 The claim figures of Table 10.1 estimated by model VIII. The last column gives the totals for all the future predicted payments.

Year of origin	Development year								Total
	1	2	3	4	5	6	7	8	
2010	102.3	140.1	59.4	25.2	10.7	4.5	1.9	0.8|	0.0
2011	101.6	139.2	59.1	25.0	10.6	4.5	1.9|	0.8	0.8
2012	124.0	169.9	72.1	30.6	13.0	5.5|	2.3	1.0	3.3
2013	150.2	205.8	87.3	37.0	15.7|	6.7	2.8	1.2	10.7
2014	170.7	233.9	99.2	42.1|	17.8	7.6	3.2	1.4	30.0
2015	159.9	219.1	92.9|	39.4	16.7	7.1	3.0	1.3	67.5
2016	185.2	253.8|	107.6	45.7	19.4	8.2	3.5	1.5	185.8
2017	168.0|	230.2	97.6	41.4	17.6	7.4	3.2	1.3	398.7

Table 10.8 Average payments corresponding to the numbers of payments in Table 10.1.

Year of origin	Development year							
	1	2	3	4	5	6	7	8
2010	62	146	117	175	203	212	406	318
2011	133	122	96	379	455	441	429	
2012	148	232	120	481	312	390		
2013	119	185	223	171	162			
2014	93	109	87	190				
2015	33	129	176					
2016	237	179						
2017	191							

10.4.2 Modeling claim sizes

The Poisson distribution with year of origin as well as year of development as explanatory variables, that is, the chain ladder method, is appropriate to model the number of claims. Apart from the numbers of claims given in Table 10.1, we also know the average claim size; it can be found in Table 10.8. For these claim sizes, the portfolio size, characterized by the factors α_i, is irrelevant. The inflation, hence the calendar year, is an important factor, and so is the development year, since only large claims tend to lead to delay in settlement. So for this situation, the separation models are more suitable. We have estimated the average claim sizes under the assumption that they arose from a gamma distribution with a constant coefficient of variation, with a multiplicative model.

The results for the various models are displayed in Table 10.9. As one sees, the nesting structure in the models is $7 \subset 6 \subset 4/5 \subset 3 \subset 2 \subset 1$; models 4 and 5 are both between 6 and 3, but they are not nested in one another. We have scaled the deviances in such a way that the fullest model 1 has a scaled deviance equal to the number of degrees of freedom, hence 15. This way, we can test the significance of the model refinements by comparing the gain in scaled deviance to the critical value of the χ^2 distribution with as a parameter the number of extra parameters estimated.

Table 10.9 Parameters, degrees of freedom and deviance for various models applied to the average claim sizes of Table 10.8.

Model	Parameters used	Df	Deviance	
1	$\mu, \alpha_i, \beta_j, \gamma_k$	15	15	$(\heartsuit)$
2	μ, β_j, γ_k	21	30.2	
3	$\mu, \beta_j, \gamma^{k-1}$	27	36.8	
4	$\mu, \beta^{j-1}, \gamma^{k-1}$	33	39.5	
5	μ, β_j	28	38.7	
6	μ, β^{j-1}	34	41.2	$\heartsuit$
7	μ	35	47.2	

A statistically significant step in both chains is the step from model 7 to 6. Taking the development parameters β_j arbitrary as in model 5, instead of geometric β^{j-1} as in model 6, does not significantly improve the fit. Refining model 6 to model 4 by introducing a parameter for inflation γ^{k-1} also does not lead to a significant improvement. Refining model 4 to model 3, nor model 3 to model 2, improves the fit significantly, but model 1 is significantly better than model 2. Still, we prefer the simple model 6, if only because model 6 is not dominated by model 1. This is because at the cost of 19 extra parameters, the gain in scaled deviance is only 26.2. So the best estimates are obtained from model 6. It gives an initial level of 129 in the first year of development, increasing to $129 \times 1.17^7 = 397$ in the eighth year. Notice that if the fit is not greatly improved by taking the coefficients $\gamma_{i \mid j-1}$ arbitrary instead of geometric or constant, it is better either to ignore inflation or to use a fixed level, possibly with a break in the trend somewhere, just to avoid the problem of having to find extrapolated values of $\gamma_{t+1}, \ldots, \gamma_{2t}$.

By combining estimated average claim sizes by year of origin and year of development with the estimated claim numbers, see Table 10.7, we get the total amounts to be reserved. These are given in the rightmost column of Table 10.10. The corresponding model is found by combining both multiplicative models 6 and IX, see (10.16); it leads to the following estimated total payments:

$$6 \times \text{IX}: \quad \widehat{X}_{ij} = 13041 \times 1.10^{i-1} \times 3.34^{j \neq 1} \times 0.46^{j-1}. \tag{10.17}$$

This model can also be used if, as is usual in practice, one is not content with a square of observed and predicted values, but also wants estimates concerning these years of origin for development years after the one that has last been observed, hence a rectangle of predicted values. The total estimated payments for year of origin i are equal to $\sum_{j=1}^{\infty} \widehat{X}_{ij}$. Obviously, these are finite only if the coefficient for each development year in models 6 and IX combined is less than 1 in (10.17).

Remark 10.4.1 (Variance of the estimated IBNR totals)
To obtain a prediction interval for the estimates in practice, finding an estimate the variance of the IBNR totals is vital. If the model chosen is the correct one and the parameter estimates are unbiased, this variance reflects parameter uncertainty as well as volatility of the process. If we assume that in Table 10.7 the

Table 10.10 Observed and predicted total claims corresponding to the Tables 10.1 and 10.7. Under Total paid are the total payments made so far, under Total est., the estimated remaining payments.

Year of origin	Development year								Total paid	Total est.
	1	2	3	4	5	6	7	8		
2010	6262	22338	6084	2975	2842	636	1624	318		43079	0
2011	13167	14762	7296	12128	4550	1323	429		361	53655	361
2012	16280	42224	9600	9620	6552	780		800	398	85056	1198
2013	19040	36445	18286	6498	3078		1772	881	438	83347	3092
2014	14973	27686	7395	8740		3926	1952	971	483	58794	7331
2015	6105	25929	15136		8696	4324	2150	1069	532	47170	16771
2016	42186	46719		19262	9578	4762	2368	1178	586	88905	37733
2017	32088		42665	21215	10549	5245	2608	1297	645	32088	84224

model is correct and the parameter estimates coincide with the actual values, the estimated row totals are estimates of Poisson random variables. As these random variables have a variance equal to this mean, and the yearly totals are indepen-dent, the total estimated process variance is equal to the total estimated mean, hence $0.8 + \cdots + 398.7 = 696.8 = 26.4^2$. If there is overdispersion present in the model, the variance must be multiplied by the estimated overdispersion factor. The actual vari-ance of course also includes the variation of the estimated mean, but this is harder to come by. Again assuming that all parameters have been correctly estimated and that the model is also correct, including the independence of claim sizes and claim numbers, the figures in Table 10.10 are predictions for compound Poisson random variables with mean $\lambda \mu_2$. The parameters λ of the numbers of claims can be ob-tained from Table 10.7, the second moments μ_2 of the gamma distributed payments can be derived from the estimated means in (10.15) together with the estimated dis-persion parameter. In Section 10.6, we describe a bootstrap method to estimate the predictive distribution. Also we derive a delta method based approximation for the prediction error. ∇

Remark 10.4.2 ('The' stochastic model behind chain ladder)
We have shown that the chain ladder method is just one algorithm to estimate the parameters of a simple GLM with two factors (year of origin and development year), a log-link and a mean-variance relationship of Poisson type ($\sigma^2 \propto \mu$). Mack (1993) describes as 'the' stochastic model behind chain ladder a different set of distribu-tional assumptions under which doing these calculations makes sense. Aiming for a distribution-free model, he cannot specify a likelihood to be maximized, so he sets out to find minimum MSE unbiased linear estimators instead. His model does not require independence, but only makes some assumptions about conditional means and variances, given the past development for each year of origin. They are such that the unconditional means and variances of the incremental observations are the same as in the GLM. ∇

10.5 Solving IBNR problems by R

Since we have shown that the chain ladder and many other methods are actually
GLMs, R's built-in function `glm` can do the necessary calculations. The dependent
variable consists of a vector containing the elements of the triangle of the observed
past losses, for example the aggregate payments in the past. These losses are broken
down by year of origin of the policy and year of development of the claim filing
process, which act as explanatory variables. In other applications, we need the calendar year. In this section we show by a simple example how to get the triangular
data into a usable vector form, as well as how to construct the proper row and column numbers conveniently. Then we show which `glm`-call can be used to produce
the chain ladder estimates, and also how to implement Verbeek's method 10.3.1 to
produce these same estimates.

First, we fill a one-dimensional array `Xij` with, stored row-wise, the 15 incremental observations from the triangle of Exercise 10.3.3. We also store the corresponding row and column numbers in vectors `i` and `j`.

```
Xij <- c(232,106,35,16,2, 258,115,56,27, 221,82,4, 359,71, 349)
  i <- c(  1,  1, 1, 1,1,   2,  2, 2, 2,   3, 3,3,   4, 4,   5)
  j <- c(  1,  2, 3, 4,5,   1,  2, 3, 4,   1, 2,3,   1, 2,   1)
```

In general, if we denote the width of the triangle by `TT`, the length of the vector
`Xij` is `TT*(TT+1)/2`. The row numbers constitute a vector of ones repeated `TT`
times, then twos repeated `TT-1` times, and so on until just the single number `TT`.
The column numbers are the sequence `1:TT`, concatenated with `1:(TT-1)`, then
`1:(TT-2)` and so on until finally just `1`. So from any vector `Xij` containing a
runoff triangle, we can find `TT`, `i` and `j` as follows.

```
TT <- trunc(sqrt(2*length(Xij)))
i <- rep(1:TT,TT:1); j <- sequence(TT:1)
```

Now to apply the chain ladder method to this triangle, and to extract the parameter
estimates for the α_i and β_j in (10.7), we simply call:

```
CL <- glm(Xij~as.factor(i)+as.factor(j), family=poisson)
coefs <- exp(coef(CL)) ##exponents of parameter estimates
alpha.glm <- coefs[1] * c(1, coefs[2:TT])
beta.glm <- c(1, coefs[(TT+1):(2*TT-1)])
```

The resulting values of the coefficients α_i and β_j in the vector `coefs` are:

```
> coefs
  (Intercept) as.factor(i)2 as.factor(i)3 as.factor(i)4 as.factor(i)5
     250.1441        1.1722        0.8315        1.2738        1.3952
as.factor(j)2 as.factor(j)3 as.factor(j)4 as.factor(j)5
       0.3495        0.1264        0.0791        0.0080
```

To apply Verbeek's algorithm 10.3.1 to find these same parameter estimates $\widehat{\alpha}_i$ and
$\widehat{\beta}_j$, we need the row and column sums of the triangle. These sums over all observations sharing a common value of `i` and `j`, respectively, can be found by using
the function `tapply` as below. First, `alpha` and `beta` are initialized to vectors of

length TT. In the loop, we compute alpha[n] using (10.8) and add it to the auxiliary variable aa storing the sum of the α's computed so far. The last line produces the matrix of predicted values $\widehat{\alpha}_i\widehat{\beta}_j$ for all i, j, as the outer matrix product $\widehat{\alpha}\widehat{\beta}'$. So Verbeek's method 10.3.1 can be implemented as:

```
Ri <- tapply(Xij, i, sum); Cj <- tapply(Xij, j, sum)
alpha <- beta <- numeric(TT)
aa <- alpha[1] <- Ri[1]
bb <- beta[TT] <-  Cj[TT] / Ri[1]
for (n in 2:TT) {
    aa <- aa + (alpha[n] <- Ri[n]/(1-bb))
    bb <- bb + (beta[TT-n+1] <- Cj[TT-n+1] / aa)}
pred <- alpha %*% t(beta)
```

Using Verbeek's algorithm 10.3.1 instead of a call of glm to compute parameter estimates is quicker because no iterative process is needed. This is definitely an issue when many bootstrap simulations are done such as in Section 10.6. Also, it is slightly more general since it can also be applied when some of the observations are negative. To get non-negative parameter estimates, all row and column sums must be non-negative, as well as all sums over rectangles such as $\mathscr{A}_\Sigma$ in (10.1). Note that $\mathscr{A}_\Sigma \geq 0$ is not implied by non-negative marginals alone; to see this, consider a 2×2 triangle with $C_1 = R_1 = 1$, $C_2 = R_2 = 2$. Negative numbers in an IBNR-triangle occur in case recuperations, or corrections to case estimates that proved to be too pessimistic, are processed as if they were negative payments in a future development year.

To find the cumulated loss figures, it is convenient to store the IBNR data as a matrix, not as a long array. One straightforward way to construct a TT by TT square matrix containing the IBNR losses at the proper places, and next to construct the row-wise cumulated loss figures from it, is by doing

```
Xij.mat.cum <- Xij.mat <- matrix(0, nrow=TT, ncol=TT)
for (k in 1:length(Xij)) Xij.mat[i[k],j[k]] <- Xij[k]
for (k in 1:TT) Xij.mat.cum[k,] <- cumsum(Xij.mat[k,])
```

For a matrix Xij.mat, the row and column numbers can be found as the matrices row(Xij.mat) and col(Xij.mat) respectively. From these, it is easy to find the calendar years in which a loss occurred. It occurs in the future if its calendar year is past TT. To reconstruct the original long vector containing the past observations from the matrix representation, we have to take the transpose of Xij.mat before extracting the past elements from it, because R stores the elements of arrays in the so-called *column major order*, that is, not by rows but by columns.

```
i.mat <- row(Xij.mat); j.mat <- col(Xij.mat);
future <- i.mat + j.mat - 1 > TT
t(Xij.mat)[!t(future)] ## equals the vector Xij
```

10.6 Variability of the IBNR estimate

The aim of IBNR analysis is to make a prediction for how much remains to be paid on claims from the past. Earlier, we showed how to compute a point estimate of the outstanding claims, using the GLM that underlies the well-known chain ladder method. Point estimates are useful, especially if they have nice asymptotic properties such as the ones resulting from generalized linear models. But often we want to know prediction intervals for the outstanding claims, or for example the 95% quantile. Not only is there a process variance, since future claims constitute a multiple of a Poisson random variable in the chain ladder model, but additionally, there is parameter uncertainty. We can give standard deviations for all estimated coefficients, but from these, we cannot easily compute the variance around the estimated mean of the total outstanding claims. In two papers England and Verrall (1999) and England (2002) describe a method to obtain estimates of the prediction error. Based on the delta method, see Example 9.1.1, they give an approximation that can be derived using quantities produced by a glm call. It involves variances and covariances of the linear predictors and fitted values. They also give a bootstrapping method to estimate the reserve standard errors (the estimation error component of the prediction error). In a subsequent paper, England (2002) proposes not just using the bootstrap estimates to compute a standard deviation, but to actually generate a pseudo-sample of outcomes of the whole future process, in this way obtaining a complete approximate predictive distribution. From this, characteristics such as mean, variance, skewness and medians, as well as other quantiles, are easily derived.

As an example, we use the triangle of Taylor & Ashe (1983). This dataset with 55 incremental losses is used in many texts on IBNR problems.

```
Xij <- scan(n=55)
357848   766940   610542   482940 527326 574398 146342 139950 227229 67948
352118   884021   933894 1183289 445745 320996 527804 266172 425046
290507 1001799   926219 1016654 750816 146923 495992 280405
310608 1108250   776189 1562400 272482 352053 206286
443160   693190   991983   769488 504851 470639
396132   937085   847498   805037 705960
440832   847631 1131398 1063269
359480 1061648 1443370
376686   986608
344014
```

Based on these original data, we compute estimates $\widehat{\alpha}_i, \widehat{\beta}_j$ in a chain ladder model. For that, we invoke the glm-function with Poisson errors and log-link, and as covariates row and column numbers i and j (treated as factors). See also the preceding section. Actually, we take a quasi-Poisson error structure, as if the observations X_{ij} were ϕ times independent Poisson(μ_{ij}/ϕ) random variables, $i, j = 1, \ldots, t$. Here $\mu_{ij} = \alpha_i \beta_j$ for some positive parameters $\alpha_1, \ldots, \alpha_t$ and $\beta_1, \ldots, \beta_t$ with $\beta_1 = 1$. Construct fitted values $\widehat{\alpha}_i \widehat{\beta}_j$, $i, j = 1, \ldots, t$ and compute the sum of the future fitted values, as follows:

```
n <- length(Xij); TT <- trunc(sqrt(2*n))
i <- rep(1:TT, TT:1); i <- as.factor(i)  ## row nrs
j <- sequence(TT:1); j <- as.factor(j)   ## col nrs
```

```
Orig.CL <- glm(Xij~i+j, quasipoisson)
coefs <- exp(as.numeric(coef(Orig.CL)))
alpha <- c(1, coefs[2:TT]) * coefs[1]
beta <- c(1, coefs[(TT+1):(2*TT-1)])
Orig.fits <- alpha %*% t(beta)
future <- row(Orig.fits) + col(Orig.fits) - 1 > TT
Orig.reserve <- sum(Orig.fits[future]) ## 18680856
```

10.6.1 Bootstrapping

England & Verrall (1999) describes a method to create bootstrap estimates. They
are obtained by sampling (with replacement) from the observed residuals in the
past observations to obtain a large set of pseudo-data, and computing an IBNR-
forecast from it. The standard deviation of the set of reserve estimates obtained this
way provides a bootstrap estimate of the estimation error. We will give each step
of the method of England and Verrall, both the theoretical considerations and the
R-implementation.

The Pearson X^2 statistic is the sum of the squared Pearson residuals. In the same
way, the deviance can be viewed as the sum of squared deviance residuals, so the
deviance residual is the square root of the contribution of an observation to the
deviance, with the appropriate sign. See Section 11.4. Though the deviance residual
is the natural choice in GLM contexts, in this case we will use the Pearson residual,
since it is easy to invert:

$$r_{\mathrm{P}} = \frac{x - \mu}{\sqrt{\mu}}, \qquad \text{therefore} \qquad x = r_{\mathrm{P}}\sqrt{\mu} + \mu. \tag{10.18}$$

To calculate the outcomes of the Pearson residuals $(X_{ij} - \widehat{\mu}_{ij})/\sqrt{\widehat{\mu}_{ij}}$, do

```
Prs.resid <- (Xij-fitted(Orig.CL))/sqrt(fitted(Orig.CL))
```

The Pearson residual in (10.18) is unscaled in the sense that it does not include the
scale parameter ϕ. This is not needed for the bootstrap calculations but only when
computing the process error. To estimate ϕ, England and Verrall use the Pearson
scale parameter. It uses a denominator $n - p$ instead of n to reduce bias:

$$\phi_{\mathrm{P}} = \frac{\sum r_{\mathrm{P}}^2}{n - p}, \tag{10.19}$$

where the summation is over all $n = t(t+1)/2$ past observations, and $p = 2t - 1$ is
the number of parameters estimated. In R, do

```
p <- 2*TT-1; phi.P <- sum(Prs.resid^2)/(n-p)
```

We adjust the residuals for bias in the same way as the scale parameter:

$$r'_P = \sqrt{\frac{n}{n-p}}\, r_P. \tag{10.20}$$

This is achieved as follows:

```
Adj.Prs.resid <- Prs.resid * sqrt(n/(n-p))
```

To be able to reproduce our results, we initialize the random number generator so as to get a fixed stream of random numbers:

```
set.seed(6345789)
```

Now run the bootstrap loop many times, for example, 1000 times.

```
nBoot <- 1000; payments <- reserves <- numeric(nBoot)
for (boots in 1:nBoot){ ## Start of bootstrap-loop
```

1. Resample from the adjusted residuals, with replacement:

```
Ps.Xij <- sample(Adj.Prs.resid, n, replace=TRUE)
```

2. Using this set of residuals and the estimated values of $\hat{\mu}_{ij}$, create a new suitable pseudo-history:

```
Ps.Xij <- Ps.Xij * sqrt(fitted(Orig.CL)) + fitted(Orig.CL)
Ps.Xij <- pmax(Ps.Xij, 0) ## Set 'observations' < 0 to 0
```

For convenience, we set negative observations to zero. For the Taylor & Ashe example, about 0.16 negative pseudo-observations were generated in each bootstrap simulation, and setting them to zero obviously induces a slight bias in the results; in other triangles, this effect might be more serious. Note that to obtain feasible estimates $\hat{\alpha}_i, \hat{\beta}_j$, it is not necessary that all entries in the run-off triangle are non-negative, see Verbeek's algorithm 10.3.1 as well as Section 10.5. But this is required in the glm-routine for the poisson and quasipoisson families.

3. From this history, obtain estimates $\hat{\alpha}_i, \hat{\beta}_j$ using a chain ladder model:

```
Ps.CL <- glm(Ps.Xij~i+j, quasipoisson)
coefs <- exp(as.numeric(coef(Ps.CL)))
Ps.alpha <- c(1, coefs[2:TT]) * coefs[1]
Ps.beta <- c(1, coefs[(TT+1):(2*TT-1)])
```

4. Compute the fitted values, and use the sum of the future part as an estimate of the reserve to be held.

```
Ps.fits <- Ps.alpha %*% t(Ps.beta)
Ps.reserve <- sum(Ps.fits[future])
```

5. Then, sample from the estimated process distribution. In this case, this can be done by generating a single Poisson$\left(\sum \hat{\alpha}_i \hat{\beta}_j / \hat{\phi}\right)$ random variable, with the sum taken over the future, and multiplying it by $\hat{\phi}$:

```
Ps.totpayments <- phi.P * rpois(1, Ps.reserve/phi.P)
```

6. At the end of the loop, store the simulated total payments and the estimated reserve to be held.

```
reserves[boots] <- Ps.reserve
payments[boots] <- Ps.totpayments
}  ## Curly bracket indicates end of bootstrap-loop
```

The bootstrap reserve prediction error is computed as

$$\text{PE}_{bs}(R) = \sqrt{\phi_P R + (\text{SE}_{bs}(R))^2},$$ (10.21)

where R is a total reserve estimate (may also be for one origin year only), and $\text{SE}_{bs}(R)$ the bootstrap standard error of the reserve estimate, based on residuals that are adjusted for degrees of freedom as in (10.20). The process variance $\phi_P R$ is added to the estimation variance.

```
PEbs <- sqrt(phi.P*Orig.reserve + sd(reserves)^2) ## 2882413
sd(reserves)^2 / (phi.P * Orig.reserve)           ## 7.455098
```

It proves that the estimation variance in this case is about 7.5 times the process variance. From the simulated values, one may compute various useful statistics. Differences with those given in England (2002) arose because we set negative observations to zero, but are largely due to randomness.

```
payments <- payments/1e6 ## expressed in millions
quantile(payments, c(0.5,0.75,0.9,0.95,0.99))
##       50%      75%      90%      95%      99%
##  18.56828 20.67234 22.35558 23.61801 26.19600
mean(payments)            ## 18.75786
sd(payments)              ## 2.873488
100 * sd(payments) / mean(payments)  ## 15.31885 = c.v. in %
pp <- (payments-mean(payments))/sd(payments)
sum(pp^3)/(nBoot-1)      ## 0.2468513 estimates the skewness
sum(pp^4)/(nBoot-1) - 3  ## 0.2701999 estimates the kurtosis
```

Our results are illustrated in the histogram in Figure 10.1. To the bars, representing fractions rather than frequencies, we added density estimates (the dashed one is a kernel density estimate, the dotted one just a fitted normal density), like this:

```
hist(payments,breaks=21,prob=TRUE)
lines(density(payments), lty="dashed")
curve(dnorm(x, mean = mean(payments), sd = sd(payments)),
    lty="dotted", add=TRUE)
```

Remark 10.6.1 (Caveats using IBNR methods)
If one thing can be learned from this whole exercise and the histogram in Figure 10.1, it is that for the future payments, one should not just give a point estimate prediction like 18.680856 million (the outcome of the chain ladder reserve estimate based on the original data). In a thousand pseudo-replications of the process, payments (in millions) ranged from 7.4 to 29.4, with quartiles 16.7 and 20.7, and one in ten fell outside the bounds (14.3, 23.6). The 'best estimate' is so inaccurate that to

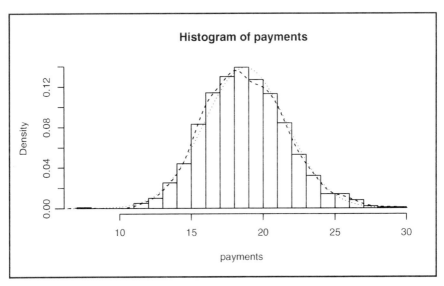

Fig. 10.1 Histogram of payments in millions, with estimated densities

write it in any other way than 19 ± 3 million grossly overstates the precision of the prediction; all digits after the first two are insignificant.

Although some IBNR-methods have ardent supporters, it is very hard to substantiate a claim that any method is superior to the one described here; a lifetime is probably not long enough to prove this in practice. Other things should be considered, such as easy access to software, the possibility to generate convincing and revealing plots easily, control over what one is doing, and adaptability to new rules and situations. All this is not present when one uses a poorly documented black box method. As we have demonstrated here, R is a very suitable tool for the actuary, because it is powerful and closely follows the formulas of the theory. This makes checking, adapting and extending the code very easy.   ∇

10.6.2 Analytical estimate of the prediction error

England and Verrall (1999) also provide a method to approximate the prediction error of an IBNR forecast. It does not need a bootstrap simulation run and uses only by-products stored in a glm object.

The mean squared error of the (unbiased) prediction $\widehat{X}_{ij}$ for each future observation X_{ij}, with $i + j - 1 > t$, can approximately be decomposed into one part process variance and another part estimation variance:

$$
\begin{aligned}
\mathrm{E}[(X_{ij} - \widehat{X}_{ij})^2] &= \left(\mathrm{E}[X_{ij}] - \mathrm{E}[\widehat{X}_{ij}]\right)^2 + \mathrm{Var}[X_{ij} - \widehat{X}_{ij}] \\
&\approx 0 + \mathrm{Var}[X_{ij} - \widehat{X}_{ij}] = \mathrm{Var}[X_{ij}] + \mathrm{Var}[\widehat{X}_{ij}].
\end{aligned}
\tag{10.22}
$$

The squared bias is small and can be neglected as long as the estimators $\widehat{X}_{ij}$, even if not unbiased, are asymptotically unbiased predictors of X_{ij}. The future loss X_{ij} and its forecast $\widehat{X}_{ij}$ computed from past losses are independent random variables, so the variance of their difference is just the sum of their variances.

The process variance in the Poisson case is given by $\mathrm{Var}[X_{ij}] = \phi \mu_{ij}$. For the estimation variance, note that $\mathrm{E}[X_{ij}] = \mu_{ij} = \exp(\eta_{ij})$. Using the delta method, see Example 9.1.1, we see

$$\mathrm{Var}[\widehat{X}_{ij}] \approx \left| \frac{\partial \mu_{ij}}{\partial \eta_{ij}} \right|^2 \mathrm{Var}[\widehat{\eta}_{ij}]. \tag{10.23}$$

Since $\partial \mu / \partial \eta = \mu$ in case of log-link, the last two relations lead to the following approximation for the mean squared error of the prediction of future payment X_{ij}:

$$\mathrm{E}[(X_{ij} - \widehat{X}_{ij})^2] \approx \widehat{\phi} \widehat{\mu}_{ij} + \widehat{\mu}_{ij}^2 \widehat{\mathrm{Var}}[\widehat{\eta}_{ij}], \quad i + j - 1 > t. \tag{10.24}$$

In similar fashion we can show that if $\widehat{X}_{ij}$ and $\widehat{X}_{kl}$ are different estimated future payments, then

$$\mathrm{Cov}[\widehat{X}_{ij}, \widehat{X}_{kl}] \approx \widehat{\mu}_{ij} \widehat{\mu}_{kl} \widehat{\mathrm{Cov}}[\widehat{\eta}_{ij}, \widehat{\eta}_{kl}]. \tag{10.25}$$

Computing the MSE of the prediction $\widehat{R} = \sum \widehat{X}_{ij}$ for future totals $R = \sum X_{ij}$ leads to an expression involving estimated (co-)variances of the various linear predictors. Taking sums over 'future' (i,j) and $(i,j) \neq (k,l)$, we get:

$$
\begin{aligned}
\mathrm{E}[(R - \widehat{R})^2] \\
= \sum \mathrm{E}[(X_{ij} - \widehat{X}_{ij})^2] + \sum \mathrm{E}[(X_{ij} - \widehat{X}_{ij})(X_{kl} - \widehat{X}_{kl})] \\
\approx \sum \mathrm{E}[(X_{ij} - \widehat{X}_{ij})^2] + \sum \mathrm{Cov}[X_{ij} - \widehat{X}_{ij}, X_{kl} - \widehat{X}_{kl}] \\
= \sum \mathrm{E}[(X_{ij} - \widehat{X}_{ij})^2] + \sum \mathrm{Cov}[\widehat{X}_{ij}, \widehat{X}_{kl}] \\
\approx \sum \widehat{\phi} \widehat{\mu}_{ij} + \sum \widehat{\mu}_{ij}^2 \widehat{\mathrm{Var}}[\widehat{\eta}_{ij}] + \sum \widehat{\mu}_{ij} \widehat{\mu}_{kl} \widehat{\mathrm{Cov}}[\widehat{\eta}_{ij}, \widehat{\eta}_{kl}] \\
= \sum \widehat{\phi} \widehat{\mu}_{ij} + \widehat{\mu}' \widehat{\mathrm{Var}}[\widehat{\eta}] \widehat{\mu}.
\end{aligned}
\tag{10.26}
$$

Here $\widehat{\mu}$ and $\widehat{\eta}$ are vectors of length t^2 containing all $\widehat{\mu}_{ij}$ and $\widehat{\eta}_{ij}$. This is one way to implement this using R:

```
Xij.1 <- xtabs(Xij~i+j) ## full square matrix
ii <- row(Xij.1); jj <- col(Xij.1); Xij.1 <- as.vector(Xij.1)
future <- as.numeric(ii+jj-1 > TT)
ii <- as.factor(ii); jj <- as.factor(jj)   ## are now vectors
Full.CL <- glm(Xij.1~ii+jj, fam=quasipoisson, wei=1-future)
Sig <- vcov(Full.CL); X <- model.matrix(Full.CL)
Cov.eta <- X%*%Sig%*%t(X)
mu.hat <- fitted(Full.CL)*future
pe2 <- phi.P * sum(mu.hat) + t(mu.hat) %*% Cov.eta %*% mu.hat
cat("Total reserve =", sum(mu.hat), "p.e. =", sqrt(pe2), "\n")
## Total reserve = 18680856 p.e. = 2945659
```

The use of a cross-tabulation of $\mathtt{Xij}$ by rows $\mathtt{i}$ and columns $\mathtt{j}$ is a quicker way to get from a triangle to a square than the one presented in Section 10.5. This square is stored as a vector of length $\mathtt{TT*TT}$. An estimate of the covariance matrix of the $\vec{\eta} = \mathbf{X}\vec{\beta}$ is $\mathbf{X}\widehat{\Sigma}\mathbf{X}'$, where $\widehat{\Sigma}$ is the estimated variance matrix of the parameters, and $\mathbf{X}$ is the regression matrix (design matrix, model matrix). In $\mathtt{fitted(Full.CL)}$, predictions for future values are automatically included. The approximate prediction error is $2\,945\,659$, the bootstrap prediction error is $2\,882\,413$. To get predictions for just row $\mathtt{r}$, simply set the entries of $\mathtt{mu.hat}$ for elements outside row $\mathtt{r}$ equal to zero. See Exercise 10.6.4.

10.7 An IBNR-problem with known exposures

In this section, we use R to tackle an IBNR problem with given *exposures*, that is, the number of policies n_i for each year of origin is assumed known. The dataset consists of a run-off triangle for a period of eight years; the total number of claims is $\mathtt{Xij}$. To give an estimate of the claims that have yet to be reported, we read the data and compute the row and column numbers just as in Section 10.5.

```
Xij <- scan(n=36)
156   37    6    5    3    2    1    0
154   42    8    5    6    3    0
178   63   14    5    3    1
198   56   13   11    2
206   49    9    5
250   85   28
252   44
221
TT <- trunc(sqrt(2*length(Xij)))
i <- rep(1:TT, TT:1); j <- sequence(TT:1)
ni <- c(28950,29754,31141,32443,34700,36268,37032,36637)
```

Looking at the data, one sees that in the last year of origin, only 221 claims emerged in the first development year, which is appreciably fewer than the losses of the previous years, while the exposure is about the same. This number has no influence on the estimates for the development factors in a chain ladder method (the $\widehat{\beta}$-values), but it is proportional to the estimates of future losses in the final year of origin, so it has considerable influence on the resulting total reserve. See Exercise 10.7.2. This is why it might be better to forecast the future losses using the Bornhuetter-Ferguson method, where it is assumed that the forecasts are not, as with chain ladder, proportional to the row sums, but with other quantities deemed appropriate by the actuary. A good first guess would be that the loss ratios remain about the same, meaning in this case that the number of claims in total for each year of origin is proportional to the exposure. This leads to $M_i = n_i \widehat{\alpha}_1 / n_1$ as a prior mean for the row total of losses corresponding to year i. In the Bornhuetter-Ferguson method, the column parameters used are those of the chain ladder method. So implementing this method, using Verbeek's algorithm 10.3.1 to find estimated parameters $\widehat{\alpha}_i$ and $\widehat{\beta}_j$, goes as follows.

```
Ri <- tapply(Xij, i, sum); Cj <- tapply(Xij, j, sum)
alpha <- beta <- numeric(TT)
aa <- alpha[1] <- Ri[1]
bb <- beta[TT] <- Cj[TT] / Ri[1]
for (n in 2:TT) {
aa <- aa + (alpha[n] <- Ri[n]/(1-bb))
bb <- bb + (beta[TT-n+1] <- Cj[TT-n+1] / aa)}
Mi <- ni * alpha[1] / ni[1]
BF <- Mi %*% t(beta); CL <- alpha %*% t(beta)
future <- row(BF) + col(BF) - 1 > TT
rowSums(BF * future) ## 0.0   0.0   0.5   2.6   6.4 13.2 26.3 76.9
rowSums(CL * future) ## 0.0   0.0   0.6   3.1   7.0 19.2 32.1 90.0
```

The row sums of the future part of the square matrix $\vec{M}\widehat{\beta}'$ represent the estimated total numbers of losses by year of origin. So the reserve estimates for the chain ladder method turn out to be somewhat bigger than those for Bornhuetter-Ferguson. Contrary to what was expected earlier, this also holds for the final year.

One way to describe the chain ladder method is to construct vectors $\widehat{\alpha}$ and $\widehat{\beta}$ (with $\beta_\Sigma = 1$) in such a way that the deviance (9.29) between the data and the 'past' part (upper left triangle) of matrix $\widehat{\alpha}\widehat{\beta}'$ is minimized, and to use the 'future' part of this matrix to determine reserves. For Bornhuetter-Ferguson, use the future part of $\vec{M}\widehat{\beta}'$ instead. It can be shown, see Verrall (2004), that both these methods arise as extreme cases in a Bayesian framework with generalized linear models, with a loose prior for chain ladder and a tight one for Bornhuetter-Ferguson.

In the Bornhuetter-Ferguson method, it is assumed that the effect of the year of origin on the losses in the chain ladder method is captured by some external quantity resembling the portfolio growth, represented in our case by the externally given exposure vector $\vec{n}$. So, instead of $\widehat{\alpha}_i\widehat{\beta}_j$, with $\beta_\Sigma = 1$, the mean for cell (i, j) is estimated as $n_i\widehat{\beta}_j$ with $\widehat{\beta}$ found by the chain ladder method. Evidently, we may get a closer fit to the past data by choosing a different $\widehat{\beta}$. Still assuming the losses to have a (quasi-)Poisson distribution, we get a generalized linear model with about half the number of parameters of the chain ladder method. The fit to the observed data will be worse than with the chain ladder method (which uses optimal $\widehat{\alpha}$ as well as $\widehat{\beta}$), but we can easily judge by a deviance analysis if the fit is significantly worse.

To estimate the β-values using glm, note that the linear predictor $\eta_{ij} = \log n_i + \log \beta_j$ has a 'fixed' component $\log n_i$, which has to be included with a coefficient 1 for each observation in row i. This can be achieved by using the offset mechanism, see Section 9.5. The following R program achieves the fitting:

```
Expo <- ni[i] ## Exposures with each element of Xij
CLi <- glm(Xij~as.factor(i)+as.factor(j), poisson)
CLoff <- glm(Xij~offset(log(Expo))+as.factor(j), poisson)
```

The residual deviance with chain ladder is 34.2 on 21 degrees of freedom, as opposed to 62.0 on 28 with the offset model. Note that the data having an ordinary Poisson structure instead of an overdispersed one (with $\phi > 1$) is not far-fetched; the unscaled deviance equals the $\chi^2(21)$ critical value at level 96.5%. The $\chi^2(7)$ critical value at level 95% being 14.1, we conclude that the data exhibit more change with year of origin than just growth in proportion to the number of contracts. The expo-sure model is of course a restriction of the chain ladder model, with $\alpha_i = n_i/n_1$ fixed. The reader might try other models, by including a calendar year effect, geometric or general, or a geometric effect of year of origin on top of the exposure.

10.8 Exercises

Section 10.1

1. In how many ways can the data in Table 10.1 be organized in a table, by year of origin, year of development and calendar year, vertically or horizontally, in increasing or decreasing order?

Section 10.2

1. Apply the chain ladder method to the given IBNR triangle with cumulated figures. What could be the reason why run-off triangles to be processed through the chain ladder method are usually given in a cumulated form?

Year of	Development year				
origin	1	2	3	4	5
1	232	338	373	389	391
2	258	373	429	456	
3	221	303	307		
4	359	430			
5	349				

Section 10.3

1. Prove (10.11). What is the mode of the random variables X_{ij} in model (10.10)?
2. Apply the arithmetic separation method to the data of Section 10.7. Determine the missing γ values by linear or by loglinear interpolation, whichever seems more appropriate.
3. Which distance between data and predicted values is minimized by the chain ladder method? Which by the separation methods?
4. Why is it not an improvement of the model to use a model $\alpha_i \beta_j \gamma^k$ rather than only $\alpha_i \beta_j$? Making use of R and (10.12), and also by the method of successive substitution, see Section 9.3, verify if the results in De Vylder (1978) have been computed and printed correctly (note that he did only a few iterations). Using the former method, is it a problem to estimate the model including the calendar year parameters? Hint: there is a complication because R asks for starting values if you use glm(..., family=gaussian(link=log),...). Such starting values are supplied, for example, by specifying mustart=mu.st as a parameter, where mu.st contains the fitted values of a Poisson fit with log-link, or of a Gaussian fit with standard (identity) link.

Section 10.4

1. Verify that the same predictions are obtained from the models $E[X_{ij}] = \alpha_i \beta^{j-1}$ and $E[X_{ij}] = \delta_i \gamma^{i+j-2}$. That is, prove that for every positive $\alpha_1, \ldots, \alpha_t$ and β, numbers $\delta_1, \ldots, \delta_t, \gamma$ can be found such that for every (i,j) we have $\alpha_i \beta^{j-1} = \delta_i \gamma^{i+j-2}$, and vice versa.

2. Argue why in model I, where for $i, j = 1, \ldots, t$, we have $E[X_{ij}] = \mu \alpha_i \beta_j \gamma_{i+j-1}$, the parameter γ_t can be taken equal to 1 without loss of generality, meaning that for $t = 8$, model I has only six more parameters to be estimated than model II. Verify that with model I there are $3(t-1)$ parameters to be estimated from $t(t+1)/2$ observations, so model I makes sense only if $t > 3$.

3. Explain why models IV and V are equivalent.

4. For $i = j = 1, 3, 5, 7$, compute the values predicted by models (10.15) and (10.16), and compare these to the actual observations.

5. Verify (10.17). Use it to determine $\sum_{j=1}^{\infty} \widehat{X}_{ij}$.

6. Reproduce Table 10.10. Compare with a direct (quasi-)Poisson model instead of a two-stage model.

7. In the Chain Ladder model, the estimated parameters allow a natural interpretation. Indeed if $\sum \beta_j = 1$ is assumed, the β_j parameters can be identified as the fraction of claims handled in the jth development year, and the α_i are the total claims arising from year of origin i. For the *Extended Chain Ladder* Model I in Table 10.7, such identification is not possible. This is because, see for example Exercise 10.4.2, there are many equivalent parameterizations leading to the same means. Assume that the parameters $\mu, \alpha_i, \beta_j, \gamma_{i+j-1}$ and $\nu, \delta_i, \varepsilon_j, \zeta_{i+j-1}$ produce the same means $\mu_{ij} = \mu \alpha_i \beta_j \gamma_{i+j-1} = \nu \delta_i \varepsilon_j \zeta_{i+j-1}$ for all $i, j = 1, \ldots, t$. Take boundary values $\alpha_1 = \beta_1 = \gamma_1 = 1$ as well as $\delta_1 = \varepsilon_1 = \zeta_1 = 1$. Then show that a real number τ exists such that the parameterizations are related as follows:

$$\nu = \mu, \quad \delta_i = \alpha_i \tau^{i-1}, \quad \varepsilon_j = \beta_j \tau^{j-1}, \quad \zeta_{i+j-1} = \gamma_{i+j-1}/\tau^{i+j-2} \quad \forall i, j.$$

Section 10.5

1. Using for example `help("%*%")` to get inspiration, compare the results of the following calls that are candidates to produce the square of fitted values of a `glm`-call for the chain ladder method, and comment:

```
fitted(CL)
alpha[i]*beta[j]
alpha*beta
alpha%o%beta
alpha%*%beta
outer(alpha,beta)
alpha%*%t(beta)
```

Find parameters `mu`, `alpha[1:TT]` and `beta[1:TT]`, with `alpha[1]` = `beta[1]` = 1, but such that they still lead to the same predictions `mu*alpha[i]*beta[j]`. Also, find equivalent parameter vectors such that the sum of the `beta` elements equals 1.

2. The calendar year corresponding to an observation can be computed simply as `k <- i+j-1`. Using an appropriate call of `glm`, apply the Arithmetic Separation method (10.9) to the data of Exercise 10.2.1. To generate fitted values for the lower triangle of the IBNR-data, plot the coefficients corresponding to the calendar years stored in an array `gamma.sep` by using the call `plot(log(gamma.sep))`, and extrapolate the γ_k-values geometrically. Then generate fitted values for the full IBNR-square.

3. Compute the parameter estimates for the Arithmetic Separation method through a method analogous to Verbeek's algorithm 10.3.1, involving sums over diagonals. Compare the resulting fits with those of the previous exercise.

4. The Hoerl-curve gives a functional form for the β_j parameters, having $\beta_j = \exp(\gamma j + \delta \log j)$ for some real numbers γ and δ. These can be used for all rows in common, or for each row separately (interaction). Apply this to the example in Table 10.1, and test for significance.

Section 10.6

1. Using the method described in this section, construct a predictive distribution of the IBNR-reserve to be held using a gamma error distribution instead of Poisson. Compare the resulting histograms, as well as the first three estimated cumulants of the resulting distributions. [Apply `glm` to the Gamma family with a log-link. Also, check England and Verrall (1999) for when $V(\mu) = \mu^2$ instead of $V(\mu) = \mu$ should be taken into account. Special care should be taken with the random drawing from the future claims. A gamma random variable has to be generated for each cell, rather than for all cells combined such as was possible in the quasi-Poisson case. The final results of the procedure should not deviate too much from the Poisson results, see England and Verrall (1999), Tables 1 and 2.]

2. Repeat the Poisson bootstrapping but now using Verbeek's algorithm 10.3.1 instead of calling `glm`. By calling `Sys.time`, verify that about 20 bootstrap runs can be processed per second using `glm`, and indicate how much faster Verbeek's algorithm 10.3.1 makes the procedure. Also investigate to how much bias the setting zero of negative observations leads, when compared with rejecting the full triangle in case any row or column sum proves to be negative.

3. England and Verrall (1999, formula (3.4)) also compute the contribution to the total prediction error of each separate year of origin i. Essentially, (3.4) equals (3.5) with the summations restricted to the corresponding row of the IBNR predicted triangle. This can be easily achieved by using the same code used for implementing formula (3.5), but now with the $\hat{\mu}_{kj}$ replaced by zero for predictions of origin years with $k \neq i$.
    Reproduce the third column of Tables 1 and 2 in England and Verrall (1999), p. 288.

Section 10.7

1. What happens if both `offset(log(Expo))` and `as.factor(i)` are included as a model term in the example of this section?

2. Find out what happens with the reserves according to chain ladder and Bornhuetter-Ferguson if the number 221 in the last row is replaced by 180 or by 250.

3. Compute the past and future fitted totals in the chain ladder model for the data of this section; you should get the values 2121 and 152.0312. Why does the total of the fitted values for past observations equal `sum(Xij)`? Is this also the case when the factor year of origin in the `glm` call is replaced by the offset term?

4. The cause of the fit of the chain ladder model being rejected at the 5% level (the deviance is 34.2 on 21 df, the critical value `qchisq(0.95,df=21)` is 32.67057) might be that there is an effect in the data that is due to the calendar year. Fit a model that incorporates all three time effects (origin year, development year, calendar year), see also Section 10.3. Do an analysis-of-deviance on the two models: compare the difference in the deviances with an appropriate χ^2-critical value to judge if adding the extra set of covariates (calendar year as a factor) is worthwhile, in the sense that the fit is improved significantly.

A problem with incorporating a calendar year effect is that for the lower right triangle, the calendar year effect cannot be estimated, since no data about that calendar year are available. One way to deal with this problem is to extrapolate the sequence of past calendar year parameters to the future, for example using linear extrapolation, by fitting a straight line through the points (i, γ_i), or geometrically. Do this.

5. In policy year 6, the terms of the policies were a little more consumer friendly, while in calendar year 7, the judges were somewhat more lenient. Inspecting the α_i and γ_k (for the calendar year) estimates confirms this; note that the notion that these years are different from the rest should *not* have been inspired by a peek at the estimation results. For that reason, treat these two years separately from the rest. So try a model that is a restricted version of the chain ladder model in the sense that the effect of year of origin is captured by the known portfolio sizes except for year 6. Also try a model in which the calendar year parameters are restricted to γ_7 arbitrary, $\gamma_k \equiv 1$ otherwise. To estimate the former model, all that has to be done is to replace as.factor(i) in the chain ladder model by (i==6) as a model term. Note the double equality sign in a logical expression; the brackets are also necessary.

Chapter 11
More on GLMs

I do not fear computers. I fear the lack of them —
Isaac Asimov (1920 - 1992)

11.1 Introduction

In this chapter, we first recall the statistical theory of ordinary linear models. Then we define Generalized Linear Models in their full generality, with as a random component any distribution in the exponential dispersion family of densities. This class has a *natural parameter* θ, determining the mean, as well as a dispersion parameter, and contains all the examples we introduced in Chapter 9 as special cases. Starting from the general form of the density, we derive properties of this family, including the mean, variance and mgf. Inference in GLMs is not done by a least squares criterion, but by analysis of deviance; the corresponding residuals are deviance residuals. We discuss some alternatives. We also study the canonical link function. Then we derive the algorithm of Nelder and Wedderburn to determine maximum likelihood estimates for this family, and show how to implement it in R. The algorithm can be applied with any link function. A subfamily of the exponential family of densities not studied in Chapter 9 consists of the compound Poisson–gamma distributions with fixed shape parameter α for the claim sizes. They have a variance equal to $\psi\mu^p$, with μ the mean, ψ the dispersion parameter and $p = \frac{2+\alpha}{1+\alpha} \in (1,2)$. Such mean-variance structures were first studied by Tweedie.

11.2 Linear Models and Generalized Linear Models

In statistics, regression analysis is a technique to model the relationship between certain variables. Output is a random response variable Y being measured, also called dependent variable, response variable, responding variable, explained variable, predicted variable, or regressand. Acting as input are the predictors $x_1,\ldots,x_p$, also called independent variables, controlled variables, manipulated variables, explanatory variables, predictor or predicator variables, control variables, collateral data, covariates or regressors. We assume the latter to be non-random and measurable without error.

We first briefly recall *(weighted) multiple linear regression* models. Assume that $Y_1, \ldots, Y_n$ are independent normal random variables with

$$\mathrm{E}[\vec{Y}] = \vec{\mu} = \mathbf{X}\vec{\beta}; \qquad \mathrm{Var}[\vec{Y}] = \sigma^2 \mathbf{W}^{-1}, \tag{11.1}$$

where the matrix $\mathbf{W}$ is diagonal with diagonal elements $w_{ii} = w_i > 0$ called weights, σ^2 is the dispersion parameter and $\vec{\beta} = (\beta_1, \ldots, \beta_p)'$ is a parameter vector. Just as in the Bühlmann-Straub setting of Chapter 8, the ith observation is regarded as the mean of a sample of w_i iid observations. So our weights are natural weights, though other interpretations for the weights are possible. In GLM contexts, the matrix of regressors $\mathbf{X}$ is usually called the *design matrix*, though statistical problems do not always result from a designed experiment. Its elements x_{ij}, $i = 1, \ldots, n$, $j = 1, \ldots, p$ contain for example the age of object i, or represent a membership indicator (dummy variable) of a subgroup of the observations to which a parameter is attached. The case $w_i \equiv 1$, so $\mathbf{W} = \mathbf{I}$, is called the classical regression model. It is easy to check that if $\vec{Y}$ satisfies a weighted model with design matrix $\mathbf{X}$, then $\mathrm{Var}[Y_i\sqrt{w_i}] = \sigma^2$ for all i, therefore $\mathbf{W}^{1/2}\vec{Y}$ follows a classical model with regressors $\mathbf{W}^{1/2}\mathbf{X}$, and vice versa (where $\mathbf{W}^{1/2}$ is diagonal).

Gauss-Markov theory, to be found in any textbook on econometrics or mathematical statistics, gives the following results for linear models.

1. By setting zero the partial derivatives of the loglikelihood and solving the resulting system of linear normal equations, one may show that the maximum likelihood equations are of linear form $\mathbf{X}'\mathbf{W}\mathbf{X}\vec{\beta} = \mathbf{X}'\mathbf{W}\vec{Y}$, therefore we can determine the ML-estimator of $\vec{\beta}$ explicitly as

$$\widehat{\beta} = (\mathbf{X}'\mathbf{W}\mathbf{X})^{-1}\mathbf{X}'\mathbf{W}\vec{Y}; \tag{11.2}$$

2. Similarly, the weighted mean squared residual

$$\widehat{\sigma}^2 = \frac{1}{n}(\vec{Y} - \mathbf{X}\widehat{\beta})'\mathbf{W}(\vec{Y} - \mathbf{X}\widehat{\beta}) \tag{11.3}$$

can be shown to be the maximum likelihood estimator of σ^2;
3. $\widehat{\beta}$ has a multinormal distribution with mean vector $\vec{\beta}$ and as variance matrix the (Fisher) *information matrix* or *dispersion matrix* $\sigma^2(\mathbf{X}'\mathbf{W}\mathbf{X})^{-1}$;
4. $n\widehat{\sigma}^2/\sigma^2 \sim \chi^2(n-p)$, where p is the length of $\widehat{\beta}$, so $\widetilde{\sigma}^2 := \frac{n}{n-p}\widehat{\sigma}^2$ is an unbiased estimator of σ^2;
5. $\widehat{\beta}$ and $\widehat{\sigma}^2$ are independent;
6. $(\widehat{\beta}, \widehat{\sigma}^2)$ are jointly complete and sufficient statistics for the parameters $(\vec{\beta}, \sigma^2)$.

If the regressand is not normal but from another family in the exponential class, we can use *generalized linear models* (GLM), introduced by Nelder and Wedderburn in 1972. In probability and statistics, the exponential class is an important class of distributions. This is for mathematical convenience, because of their nice algebraic properties. Also, they are often very natural distributions to consider. The normal, gamma, chi-square, beta, Bernoulli, binomial, Poisson, negative binomial and geometric distributions are all subfamilies of the exponential class. So the distribution of the error term may be non-normal and heteroskedastic, having a variance that depends on its mean.

Resorting to a GLM is also a remedy for the problem that the response might be linear in the covariates on some other scale than the identity. The logarithmic scale is an important special case, because linearity on this scale is equivalent to having a multiplicative model rather than an additive one. Very frequently this is essential in insurance applications.

Generalized Linear Models have three characteristics, see Section 9.2:

1. The *stochastic component* states that the observations are *independent* random variables Y_i, $i = 1, \ldots, n$ with a density in the *exponential dispersion family*, see (11.4) below.
2. The *systematic component* of the model connects to every observation a *linear predictor* $\eta_i = \sum_j x_{ij}\beta_j$, linear in the parameters $\beta_1, \ldots, \beta_p$. So $\vec{\eta} = \mathbf{X}\vec{\beta}$.
3. The *link function* connects $\mu_i = E[Y_i]$ uniquely to the linear predictor η_i by $\eta_i = g(\mu_i)$, so $\vec{\eta} = g(\vec{\mu})$.

In the (weighted) linear model, $\vec{\eta}$ and $\vec{\mu}$ coincide, see (11.1); the link function is the identity. Note that in a multiplicative (loglinear) model with $\log \mu_i = \sum \alpha_j \log x_{ij}$, the coefficients α_j are the partial derivatives of $\log \mu_i$ with respect to $\log x_{ij}$. So we may write $\alpha_j = \frac{\partial \mu_i}{\mu_i} / \frac{\partial x_{ij}}{x_{ij}}$, which means that the parameter α_j can be interpreted as the *elasticity* of μ_i with respect to x_{ij}.

We finish this section by giving an example using R.

Example 11.2.1 (Linear and generalized linear models using R)
Assume we have data on the number of trips abroad by groups of w_i, $i = 1, \ldots, 16$, people, of which gender and income are known, as follows:

```
w <- c(1,2,2,1,1,1,1,4,2,4,2,3,2,1,1,2)
Y <- c(0,1,0,8,0,0,0,30,0,1,1,38,0,0,0,26) / w
gender <- c(0,0,0,0,0,0,0,0,1,1,1,1,1,1,1,1)
income <- c(1,2,5,20,1,2,5,20,1,2,5,20,1,2,5,20)
```

We fit an ordinary linear model and a generalized linear model with Poisson distribution and logarithmic link:

```
lm(Y ~ gender+income, weights=w)
glm(Y ~ gender+income, weights=w, family=poisson)
```

The linear model says $Y \approx -2.45 + 1.96 \times \text{gender} + 0.58 \times \text{income}$, the generalized loglinear model is $Y \approx 0.053 \times 1.68^{\text{gender}} \times 1.28^{\text{income}}$. The first model gives negative fitted values in case of low income for gender 0.   ∇

11.3 The Exponential Dispersion Family

In Section 9.2, we listed a number of important distributions to describe randomness in Generalized Linear Models. Below, we give a general definition of the family of possible densities to be used for GLMs. It can be shown that normal, Poisson, gamma and inverse Gaussian random variables are all members of the following family, just as Poisson multiples and binomial proportions.

Definition 11.3.1 (The exponential dispersion family)
The *exponential dispersion family of densities* consists of the densities of the following type:

$$f_Y(y; \theta, \psi) = \exp\left(\frac{y\theta - b(\theta)}{\psi} + c(y; \psi) \right), \quad y \in D_\psi. \tag{11.4}$$

Here ψ and θ are real parameters, $b(\cdot)$ and $c(\cdot; \cdot)$ are real functions. The support of the density is $D_\psi \subset \mathbb{R}$.   ▽

These densities can be discrete or continuous; in some cases, differentials of mixed type are needed. The status of the parameter θ is not the same as that of ψ, because ψ does not affect the mean, in which we are primarily interested. The linear models we described earlier only aimed to explain this mean. Though except in special cases, the value of ψ is fixed and unknown, too, in GLM-literature the above family is referred to as the one-parameter exponential family. The function $b(\cdot)$ is called the cumulant function, see Corollary 11.3.5. The support D_ψ does not depend on θ. The same goes for the function $c(\cdot; \cdot)$ that acts as a normalizing function, enforcing that the density sums or integrates to 1. For the continuous distributions the support is $\mathbb{R}$ for the normal distribution, and $(0, \infty)$ for the gamma and inverse Gaussian distributions. It may also be a countable set, in case of a discrete density. For the Poisson multiples for example, D_ψ is the set $\{0, \psi, 2\psi, \ldots\}$. In the following, we list some examples of members of the exponential dispersion family. For the specific form of the function $b(\cdot)$ as well as the support D_ψ, we refer to Table D. In the exercises, the reader is asked to verify the entries in that table.

Example 11.3.2 (Some members of the exponential dispersion family)
The following parametric families are the most important members of the exponential dispersion family (11.4):

1. The $N(\mu, \sigma^2)$ distributions, with as parameters $\theta(\mu, \sigma^2) = \mu$ and $\psi(\mu, \sigma^2) = \sigma^2$. Note that since the parameter μ denotes the mean here, θ should not depend on σ^2.
2. The Poisson(μ) distributions, with natural parameter $\theta = \log \mu$ and $\psi = 1$.
3. For all natural m, assumed fixed and known, the binomial(m, p) distributions, with as natural parameter the 'log-odds' $\theta = \log \frac{p}{1-p}$ and $\psi = 1$.
4. For all positive r, assumed fixed and known, the negative binomial(r, p) distributions, for $\theta = \log(1 - p)$ and $\psi = 1$.

5. The gamma(α, β) distributions, after the reparameterizations $\theta(\alpha, \beta) = -\beta/\alpha$ and $\psi(\alpha, \beta) = 1/\alpha$. Note that $\theta < 0$ must hold in this case.
6. The inverse Gaussian(α, β) distributions, with parameters $\theta(\alpha, \beta) = -\frac{1}{2}\beta^2/\alpha^2$ and $\psi(\alpha, \beta) = \beta/\alpha^2$. Again, $\theta < 0$ must hold. ∇

Note that there are three different parameterizations involved: the 'standard' parameters used throughout this book, the parameterization by mean μ and dispersion parameter ψ that proved convenient in Section 9.2, and the parameterization with θ and ψ as used in this section. This last parameterization is known as the *natural* or *canonical* parameterization, since the factor in the density (11.4) involving both the argument y and the parameter θ that determines the mean has the specific form $y\theta$ instead of $yh(\theta)$ for some function $h(\cdot)$.

Example 11.3.3 (Gamma distribution and exponential dispersion family)
As an example, we will show how the gamma distributions fit in the exponential dispersion family. The customary parameterization, used in the rest of this text, is by a shape parameter α and a scale parameter β. To find the natural parameter θ, as well as ψ, we equate the logarithms of the gamma(α, β) density and (11.4):

$$-\log \Gamma(\alpha) + \alpha \log \beta + (\alpha - 1) \log y - \beta y = \frac{y\theta - b(\theta)}{\psi} + c(y; \psi). \qquad (11.5)$$

The parameters must be chosen in such a way that θ, ψ and y appear together in the log-density only in a term of the form $\theta y/\psi$. This is achieved by taking $\psi = \frac{1}{\alpha}$ and $\theta = -\frac{\beta}{\alpha} < 0$. To make the left and right hand side coincide, we further take $b(\theta) = -\log(-\theta)$, which leaves $c(y; \psi) = \alpha \log \alpha + (\alpha - 1) \log y - \log \Gamma(\alpha), y > 0$, for the terms not involving θ. In the μ, ψ parameterization, μ is simply the mean, so $\mu = \alpha/\beta$. We see that in the θ, ψ parameterization, the mean of these random variables does not depend on ψ, since it equals

$$\mathrm{E}[Y; \theta] =: \mu(\theta) = \frac{\alpha}{\beta} = \mu = -\frac{1}{\theta}. \qquad (11.6)$$

The variance is
$$\mathrm{Var}[Y; \theta, \psi] = \frac{\alpha}{\beta^2} = \mu^2 \psi = \frac{\psi}{\theta^2}. \qquad (11.7)$$

So the variance is $V(\mu)\psi$, where $V(\cdot)$ is the variance function $V(\mu) = \mu^2$. ∇

The density (11.4) in its general form permits one to derive the mgf of Y. From this, we can derive some useful properties of the exponential dispersion family.

Lemma 11.3.4 (Mgf of the exponential dispersion family)
For each real number t such that replacing θ by $\theta + t\psi$ in (11.4) also produces a density, the moment generating function at argument t of the density (11.4) equals

$$m_Y(t) = \exp \frac{b(\theta + t\psi) - b(\theta)}{\psi}. \qquad (11.8)$$

Proof. We give a proof for the continuous case only; for the proof of the discrete case, it suffices to replace the integrations over the support D_ψ in this proof by summations over $y \in D_\psi$. We can rewrite the mgf as follows:

$$m_Y(t) = \int_{D_\psi} e^{ty} \exp\left[\frac{y\theta - b(\theta)}{\psi} + c(y;\psi)\right] dy \qquad (11.9)$$

$$= \int_{D_\psi} \exp\left[\frac{y\{\theta + t\psi\} - b(\theta + t\psi)}{\psi} + c(y;\psi)\right] dy \times \exp\frac{b(\theta + t\psi) - b(\theta)}{\psi}.$$

So (11.8) follows immediately because we assumed that the second integrand in (11.9) is a density. ▽

Corollary 11.3.5 (Cgf, cumulants, mean and variance)
If Y has density (11.4), its cumulant generating function equals

$$\kappa_Y(t) = \frac{b(\theta + t\psi) - b(\theta)}{\psi}. \qquad (11.10)$$

As a consequence, for the cumulants κ_j, $j = 1, 2, \ldots$ we have

$$\kappa_j = \kappa_Y^{(j)}(0) = b^{(j)}(\theta)\psi^{j-1}. \qquad (11.11)$$

Because of this, the function $b(\cdot)$ is called the *cumulant function*. From (11.11) with $j = 1, 2$, we see that the mean and variance of Y are given by:

$$\mu(\theta) = \mathrm{E}[Y;\theta] = \kappa_1 = b'(\theta),$$
$$\sigma^2(\theta, \psi) = \mathrm{Var}[Y;\theta, \psi] = \kappa_2 = \psi b''(\theta). \qquad (11.12)$$

Note that the mean depends only on θ, but the variance equals the dispersion parameter multiplied by $b''(\theta)$. The *variance function* $V(\mu)$ equals $b''(\theta(\mu))$. ▽

We proved in the previous chapter that for some specific examples, the ML-estimates for $\mathrm{E}[Y_i]$ in the full model all have residual 0, that is, $\widehat{\mu}_i = y_i$, $i = 1, \ldots, n$. This does not always hold; for instance if $Y_i \sim \mathrm{uniform}(0, 2\mu_i)$, the joint density is $\prod \frac{1}{2\mu_i} I_{(0,2\mu_i)}(y_i)$, therefore the ML-estimators for $\mu_i = \mathrm{E}[Y_i]$ are $Y_i/2$. But for all densities in the exponential dispersion family, in the full model the ML-estimates of $\mathrm{E}[Y_i]$ can be proved to have zero residual, or what is the same, the ML-estimator for the mean in a sample of size one from the exponential dispersion family equals the observation.

Property 11.3.6 (ML-estimates of the means in the full model)
The parameter value $\widetilde{\theta}$ maximizing the density $f_Y(y;\theta, \psi)$ of (11.4) is such that $y = \mathrm{E}[Y;\theta, \psi]$, so $\widetilde{\theta} = (b')^{-1}(y)$. ▽

Proof. The result follows easily from (11.4), since $\partial \log f_Y(y;\theta, \psi)/\partial\theta = 0$ must hold for the maximum, so $y = b'(\theta) = \mathrm{E}[Y;\theta, \psi]$ by (11.12). Note that $b''(\theta)$ is the variance function, therefore positive, so $\mathrm{E}[Y;\theta, \psi]$ increases with θ. ▽

Subfamilies of the exponential family (11.4) with different μ and ϕ but the same norming c function and cumulant function b (and hence the same mean-variance relationship, see Section 11.4) are related by two operations: averaging and Esscher transformation. In fact by averaging we can transform a density in the exponential dispersion family into another in the same subfamily with a different dispersion parameter ψ but with the same natural parameter θ, while taking an Esscher transform gives us a density with the same ψ but different θ. To prove this, we use the cgf (11.10).

Property 11.3.7 (Taking sample means)
Let $Y_1, \ldots, Y_m$ be a sample of m independent copies of the random variable Y, and let $\overline{Y} = (Y_1 + \cdots + Y_m)/m$ be the sample mean. If Y is a member of an exponential dispersion subfamily with fixed functions $b(\cdot)$ and $c(\cdot; \cdot)$ and with parameters θ and ψ, then $\overline{Y}$ is in the same exponential dispersion subfamily with parameters θ and ψ/m, if this pair of parameters is allowed.

Proof. By (11.8), we have for the mgf of $\overline{Y}$:

$$m_{\overline{Y}}(t) = \left\{ m_Y \left(\frac{t}{m} \right) \right\}^m = \exp \frac{b(\theta + t\psi/m) - b(\theta)}{\psi/m}. \qquad (11.13)$$

This is exactly the mgf of a member of the exponential dispersion family with parameters θ and ψ/m. ∇

So taking the average of a sample of size m brings us from the parameter pair (θ, ψ) to $(\theta, \psi/m)$. By the same token, we can get from $(\theta, n\psi')$ to (θ, ψ') by taking the average of a sample of size n. Then combining the two operations, we can get from (θ, ψ) to $(\theta, n\psi/m)$ and back. Taking limits, we can get from (θ, ψ) to any other dispersion parameter $(\theta, \alpha\psi)$, $\alpha > 0$. This all provided that these ψ-values lead to a density in the subfamily.

Example 11.3.8 (Poisson multiples and sample means)
For Poisson, only $\phi = 1$ is allowed, therefore we embed these distributions in a two-parameter family, as follows. By Property 11.3.7, the sample means of m Poisson(μ) random variables have a density in the exponential dispersion family (11.4), with $b(\cdot)$, $c(\cdot; \cdot)$ and θ the same as for the Poisson density, but $\psi = 1/m$ instead of $\psi = 1$, and support $D_\psi = \{0, \frac{1}{m}, \frac{2}{m}, \ldots\}$ Such a sample mean is a Poisson($m\mu$) random variable, multiplied by $1/m$. Extending this idea, let $\psi > 0$ be arbitrary, not specifically equal to $1/m$ for some integer m, and look at

$$Y = \psi M, \quad \text{where} \quad M \sim \text{Poisson}(\mu/\psi). \qquad (11.14)$$

It can be shown that Y has density (11.4) with $\theta = \log \mu$ and $b(\theta) = e^\theta$, just as with ordinary Poisson distributions, but with arbitrary $\psi > 0$. The support of Y is $\{0, \psi, 2\psi, \ldots\}$.

As we saw, for $\psi = 1/m$ we get the average of m Poisson(μ) random variables. When $\psi = n$, the resulting random variable has the property that taking the average of a sample of size n of it, we get a Poisson(μ) distribution. So it is natural to call such random variables Poisson sample means. If $\psi = n/m$, (11.14) is the sample average of m random variables of the type with $\psi = n$. So for these values, too, it is rational to call the random variable Y a Poisson average. But in view of (11.14), we also speak of such random variables as Poisson multiples. Note that for $\psi > 1$, we get a random variable in the exponential family with a variance larger than the mean. These variables can be used as an approximation (for which a GLM can be estimated) for an 'overdispersed Poisson' random variable. $\triangledown$

Remark 11.3.9 (Binomial and negative binomial distributions)
For the continuous error distributions, any positive value of ψ leads to a density in the subfamily, but for the (negative) binomial distributions, just as for the Poisson distributions proper, only $\psi = 1$ is allowed. For fixed r or n, just as for Poisson random variables, the device of taking multiples ψ times random variables also leads to an extended class of multiples of (negative) binomial random variables with parameters (μ, ψ). For the binomial case, the interpretation as averages in case of rational multiples ($\psi = n/m$) is not valid in general. $\triangledown$

In Property 11.3.7, we found how to generate exponential family members with the same θ but different ψ. We can also obtain other members of the exponential dispersion family with the same ψ but with different θ. This is done by using the Esscher transformation that we encountered before, for example in Chapter 5.

Corollary 11.3.10 (Exponential dispersion family and Esscher transform)
Recall the Esscher transform with parameter h of a differential $dF(y)$, which is

$$dF_h(y) = \frac{e^{hy}dF(y)}{\int e^{hz}dF(z)}, \tag{11.15}$$

provided the denominator is finite, that is, the mgf with $F(\cdot)$ exists at h. The mgf with the transformed density is easily seen to be equal to $m_h(t) = m(t+h)/m(h)$. For a differential dF in the exponential dispersion family, the cgf of dF_h has the form

$$\begin{aligned}
\kappa_h(t) &= \frac{b(\theta + (t+h)\psi) - b(\theta)}{\psi} - \frac{b(\theta + h\psi) - b(\theta)}{\psi} \\
&= \frac{b(\theta + h\psi + t\psi) - b(\theta + h\psi)}{\psi}.
\end{aligned} \tag{11.16}$$

This is again a cgf of a member of the same exponential dispersion subfamily with the same scale parameter ψ, and location parameter $\theta_h = \theta + h\psi$.

Transforming a density like in (11.15) is also known as *exponential tilting*. $\triangledown$

Corollary 11.3.11 (Generating the exponential dispersion family)
It can be shown that the Esscher transform with parameter $h \in \mathbb{R}$ transforms

1. $N(0,1)$ into $N(\mu, 1)$ when $h = \mu$;
2. Poisson(1) into Poisson(μ) when $h = \log \mu$;
3. binomial($m, \frac{1}{2}$) into binomial(m, p) when $p = \frac{1}{1+e^{-h}}$, so $h = \log \frac{p}{1-p}$;
4. negative binomial($r, \frac{1}{2}$) into negative binomial(r, p) when $p = 1 - \frac{1}{2}e^h$, so $h = \log(2(1-p))$;

5. gamma$(1,1)$ into gamma$(1,\beta)$ when $h = 1 - \beta$;
6. inverse Gaussian$(1,1)$ into inverse Gaussian(α, α^2) when $\alpha = \sqrt{1 - 2h}$, so $h = (1 - \alpha^2)/2$.

So all examples of distributions in the exponential dispersion family that we have given can be generated by starting with prototypical elements of each type, and next taking Esscher transforms and multiples of type (11.14), if allowed. $\qquad\qquad\qquad$ ∇

11.4 Fitting criteria

To judge the quality of a fit, we can look at the residuals and at various variants of the loglikelihood ratio reflecting the distance between fits and observations.

11.4.1 Residuals

Three types of residuals for observation y and (fitted) mean μ are commonly used:

Pearson residuals $\quad r^P = \frac{y-\mu}{\sigma}$, where σ is the standard deviation, dependent on μ through $\sigma^2 = \phi V(\mu)$, with $V(\cdot)$ the variance function; these residuals are very simple, but often remarkably skewed.

Deviance residuals $\quad r^D = \text{sign}(y - \mu)\sqrt{d}$, where d is the contribution of the observation to the deviance $D = -2\phi \log(\widehat{L}/\widetilde{L})$ (see Section 9.4). In other words, the deviance is considered as a sum of squared residuals with the proper sign:

$$D = \sum d_i = \sum (r_i^D)^2 = 2 \sum_i w_i \{ y_i(\widetilde{\theta}_i - \widehat{\theta}_i) - [b(\widetilde{\theta}_i) - b(\widehat{\theta}_i)] \}. \qquad (11.17)$$

Here $\widetilde{\theta}_i$ is such that $\text{E}[Y; \widetilde{\theta}_i, \phi] = y_i$ (the unrestricted maximum, see Property 11.3.6), and $\widehat{\theta}_i$ is the maximum likelihood estimate of θ_i under the current model. See Exercise 11.3.6. Special cases are (9.29) for Poisson and (9.32) for gamma; see also Exercise 11.3.7. For example, the deviance residual for Poisson, with $w = 1$, $\theta = \log \mu$ and $b(\theta) = e^\theta$, is

$$r^D = \text{sign}(y - \mu)\sqrt{2(y\log(y/\mu) - y + \mu)}. \qquad (11.18)$$

Anscombe residuals $\quad$ are based on transformed observations $a(y)$, where the function $a(\cdot)$ is chosen to make $a(Y)$ 'as normal as possible'. It proves that a good choice for $a(\cdot)$ in a GLM with variance function $V(\cdot)$ has

$$a'(y) = V(y)^{-1/3}. \qquad (11.19)$$

By the delta method of Example 9.1.1, a first order approximation for $E[a(Y)]$ is $a(\mu)$, while $\mathrm{Var}[a(Y)] \approx (a'(\mu))^2 \mathrm{Var}[Y]$. This leads to the Anscombe residual

$$r^A = \frac{a(y) - a(\mu)}{a'(\mu)\sqrt{V(\mu)}}. \tag{11.20}$$

Some examples of Anscombe residuals are

$$\text{Poisson:} \quad V(\mu) = \mu^1, \text{ so } r^A = \frac{\frac{3}{2}(y^{2/3} - \mu^{2/3})}{\mu^{1/6}}; \tag{11.21}$$

$$\text{gamma:} \quad V(\mu) = \mu^2, \text{ so } r^A = \frac{3(y^{1/3} - \mu^{1/3})}{\mu^{1/3}}; \tag{11.22}$$

$$\text{inverse Gaussian:} \quad V(\mu) = \mu^3, \text{ so } r^A = \frac{\log y - \log \mu}{\mu^{1/2}}. \tag{11.23}$$

It turns out that Anscombe residuals and deviance residuals are in fact often very similar. As an illustration of this, in Exercise 11.3.9 the reader is asked to compare Taylor expansions for the squares of the three residuals at $y = \mu$ for the case of Poisson errors ($\phi = 1$). The first non-zero term is $(r^P)^2 = (y - \mu)^2/\mu$ for all three, the second is $-(y - \mu)^3/(3\mu^2)$ for both $(r^A)^2$ and $(r^D)^2$.

In Example 9.1.1 it is shown that for the Poisson case, $a(Y) = Y^{2/3}$ has skewness near zero. In the gamma case, the cube-root transformation $a(Y) = Y^{1/3}$ is the well-known Wilson-Hilferty (1931) transformation to normalize χ^2 random variables. In Exercise 2.5.14 the square root of gamma variables was used, which has near-constant variance and third and fourth moments close to those of the normal distribution.

11.4.2 Quasi-likelihood and quasi-deviance

The fit criterion used in GLM theory for statistical inference is the loglikelihood ratio. In case of normality, this is equal to the least squares distance to a 'full' model with a parameter for every observation i. In case of Poisson, gamma and other distributions, it is equal to other suitable distances between observations and fitted values. Every test (Student, F) that can be used in ordinary linear models is valid asymptotically in GLMs.

Assume we have observations y_i, $i = 1, \ldots, n$, from independent Y_i having density (11.4), with natural parameters θ_i and dispersion parameters $\psi_i = \phi/w_i$ with a fixed ϕ and natural weights w_i. For the corresponding means, write $\mu_i = E[Y_i] = b'(\theta_i)$. The means depend on unknown parameters $\vec{\beta} = (\beta_1, \beta_2, \ldots)'$ by $\vec{\mu} = g^{-1}(\mathbf{X}\vec{\beta})$.

Consider the ratio of the sample density with parameters $\tilde{\theta}_i, \psi_i$ and the maximized one in the full model with parameters $\hat{\theta}_i, \psi_i$ such that $y_i = b'(\hat{\theta}_i)$; see Property 11.3.6. The logarithm of this ratio equals

$$\sum_{i=1}^{n} \log f_Y(y_i; \theta_i, \psi_i) - \sum_{i=1}^{n} \log f_Y(y_i; \widetilde{\theta}_i, \psi_i). \tag{11.24}$$

Using (11.4), this can be rewritten as

$$\sum_{i=1}^{n} \left(\frac{y_i \theta_i - b(\theta_i)}{\phi/w_i} - \frac{y_i \widetilde{\theta}_i - b(\widetilde{\theta}_i)}{\phi/w_i} \right) = \sum_{i=1}^{n} \int_{\widetilde{\theta}_i}^{\theta_i} \frac{y_i - b'(\theta)}{\phi/w_i} d\theta. \tag{11.25}$$

Substitute $\mu = b'(\theta)$, so $\frac{d\mu}{d\theta} = b''(\theta) = V(\mu) > 0$. Then (11.25) equals

$$q(\mu_1, \ldots, \mu_n) := \sum_{i=1}^{n} \int_{y_i}^{\mu_i} \frac{y_i - \mu}{V(\mu)\phi/w_i} d\mu. \tag{11.26}$$

Note that of the density (11.4), only the variance function $V(\cdot)$ has remained in (11.26). The function q is called the *quasi-likelihood (QL)*. The QL is the logarithm of the likelihood ratio of the current model and the 'full' model, so it is actually a log-likelihood ratio, not a likelihood.

We also might have obtained relation (11.26) by using

$$\frac{\partial \ell}{\partial \mu} = \frac{\partial \ell}{\partial \theta} \frac{\partial \theta}{\partial \mu} = \frac{y - b'(\theta)}{\phi/w} \frac{1}{b''(\theta)} = \frac{y - \mu}{\phi V(\mu)/w}, \quad y \in R_{\phi/w}. \tag{11.27}$$

Inserting $V(\mu) = \mu^k$ for $k = 0, 1, 2, 3$ in (11.26), we get in (11.24) the likelihood ratios corresponding to a normal, Poisson, gamma and inverse Gaussian GLM.

Remark 11.4.1 (Two-dimensional cross-tables)
In a model with $\mu_{ij} = \alpha_i \beta_j$ as in Section 9.3, we can write down the normal equations to maximize (11.26). We get systems like (9.22), (9.18) and (9.23) by taking $V(\mu) = \mu^k$ for $k = 0, 1, 2$, respectively. But replacing $V(\mu)$ by μ_i in the QL (11.26) leads to the Bailey-Simon method. $\triangledown$

The QL allows other variance functions to be studied, leading to an extended class of possible models. An example is $V(\mu) = \mu^2(1-\mu)^2$, $0 < \mu < 1$, for which the QL can be computed, but for which there is not an actual distribution having this log-likelihood ratio. Also, there is no support restriction in the QL, so the y_i need for example not be integers (or integer multiples of ϕ) in the Poisson case.

The quasi-likelihood can be shown to have enough in common with ordinary log-likelihood ratios to allow many asymptotic results to still remain valid.

The likelihood ratio Λ is the ratio of the maximized likelihood under a model resulting in means $\widehat{\mu}_1, \ldots, \widehat{\mu}_n$ (depending on parameters $\widehat{\beta}_1, \widehat{\beta}_2, \ldots$), divided by the one maximized without imposing any restrictions on the means, i.e., under the 'full' model, and therefore

$$\log \Lambda = q(\widehat{\mu}_1, \ldots, \widehat{\mu}_n). \tag{11.28}$$

The *scaled deviance*, see Section 9.4, is just $-2\log \Lambda$, while $D = -2\phi \log \Lambda$ is the *deviance*. In the same way, the *quasi-deviance* is defined as $-2\phi q$.

Performing the integration in (11.26) for a fixed power variance function $V(\mu) = \mu^p$ and dispersion parameter ϕ, we find the corresponding quasi-deviance. For all $p \notin \{1,2\}$, we get the following general expression:

$$D_p = 2 \sum_{i=1}^{n} w_i \left(\frac{y_i^{2-p} - (2-p)y_i\mu_i^{1-p} + (1-p)\mu_i^{2-p}}{(1-p)(2-p)} \right). \tag{11.29}$$

For $p = 0$, we get the least squares distance D_0. For $p \in (0,1)$, we get a quasi-deviance that is not the deviance with an exponential family density. In the limit for $p \to 1$ we get the Poisson deviance (9.29). For $p \in (1,2)$, D_p is the deviance of the Tweedie distributions of Section 11.7. In the limit for $p \to 2$, the gamma deviance (9.32) arises. The deviance D_3 is the one of the inverse Gaussian distributions.

Maximizing the likelihood with respect to $\vec{\beta}$ for the distributions corresponding to $V(\mu) = \mu^p$ means minimizing the expressions D_p. When estimation is done maximizing a quasi-deviance instead of the likelihood itself, one speaks of quasi-likelihood (QL) estimation. The scaled deviance can be regarded as a distance in $\mathbb{R}^n$ between the vector of fitted values μ_i and the vector of observed values y_i. It is a sum of contributions for each observation taking into account its precision. This contribution gets reduced if the observation is less precise, that is, large.

11.4.3 Extended quasi-likelihood

The deviance is measured in terms of the dispersion parameter ϕ. The variance function also determines the units of measurement for the deviance, so simply differencing these discrepancy measures across variance functions is not feasible. The problem is that the quasi-deviance does not represent a likelihood but a loglikelihood ratio, comparing with the full model. This is fine when only estimation of the means is desired. But to compare different variance functions such as $V(\mu) = \mu^p$ it is necessary to widen the definition of quasi-likelihood. To get a likelihood instead of a likelihood ratio with respect to the saturated model, Nelder and Pregibon (1987) propose to look at the *extended quasi-likelihood*. We confine ourselves to the Tweedie family with a variance function $V(\mu) = \mu^p$.

Definition 11.4.2 (Extended quasi-likelihood)
The *extended quasi-likelihood (EQL)* for the exponential densities having variance functions $V(\mu) = \mu^p$ is defined as

$$Q_p^+(\vec{\mu}, \phi; \vec{y}) = -\frac{1}{2} \sum_{i=1}^{n} \log\left(2\pi\phi y_i^p\right) - \frac{1}{2\phi} D_p(y_1, \dots, y_n; \mu_1, \dots, \mu_n), \tag{11.30}$$

where D_p is as in (11.29), and ϕ is the dispersion parameter. ∇

Note that the estimates of the parameters β obtained by maximizing the EQL Q_p^+ coincide with the ML-estimates. The estimate of ϕ obtained by setting zero the partial derivative of Q_p^+ with respect to ϕ is the mean deviance, without correction for parameters estimated.

When the likelihood itself can be computed, there is not much point in optimizing the EQL. But the EQL can already be computed when only the coefficient p in the variance function is known, thus allowing to compare quasi-likelihood models.

Property 11.4.3 (EQL approximates loglikelihood ratio)
The extended quasi-likelihood (11.30) has values approximating the loglikelihood ratio for current model and saturated model.

Proof. The first term in (11.30) (about) equals the maximized loglikelihood under the saturated model, as can be inferred, for $p = 0, 1, 2, 3$, from:

$$\log f_Y(y; \mu = y, \phi) \begin{cases} = -\frac{1}{2}\log(2\pi\phi) & \text{for } N(\mu, \phi), \\ \approx -\frac{1}{2}\log(2\pi\phi y) & \text{for } Poisson(\mu, \phi), \\ \approx -\frac{1}{2}\log(2\pi\phi y^2) & \text{for } gamma(\frac{1}{\phi}, \frac{1}{\mu\phi}), \\ = -\frac{1}{2}\log(2\pi\phi y^3) & \text{for } IG(\frac{1}{\mu\phi}, \frac{1}{\mu^2\phi}). \end{cases} \quad (11.31)$$

This can easily be verified for the cases $p = 0$ and $p = 3$. For other values of p, the proof is omitted, but for $p = 1, 2$, the approximation is based on Stirling's formula for $y! = y\Gamma(y)$ and $\Gamma(\alpha)$:

$$y! \approx \sqrt{2\pi y}\,(y/e)^y \quad \text{and} \quad \Gamma(\alpha) \approx \sqrt{2\pi/\alpha}\,(\alpha/e)^\alpha. \quad (11.32)$$

For the Poisson case with $\mu = y$ and $\phi = 1$, this gives

$$\log\left(e^{-y}y^y/y!\right) \approx -y + y\log y - \log\left(\sqrt{2\pi y}\left(\frac{y}{e}\right)^y\right) = -\frac{1}{2}\log(2\pi y). \quad (11.33)$$

For the gamma case with $p = 2$, from (11.32) we get

$$\log\left(f_Y(y; \mu = y, \phi)\right) = \log\left(\frac{1}{\Gamma(\alpha)}\beta^\alpha y^{\alpha-1}e^{-\beta y}\right)\Big|_{\alpha = \frac{1}{\phi}, \beta = \frac{1}{y\phi}}$$

$$\approx -\log\left(\sqrt{2\pi\phi}\,(\phi e)^{-1/\phi}\right) + \frac{1}{\phi}\log\frac{1}{y\phi} + \left(\frac{1}{\phi} - 1\right)\log y - \frac{1}{\phi} \quad (11.34)$$

$$= -\frac{1}{2}\log\left(2\pi\phi y^2\right).$$

Adding up the values (11.31) for all i gives the first term of the EQL. ∇

Remark 11.4.4 (Improved EQL for Poisson)
An improved Stirling approximation for $y!$ is obtained by adding an extra term:

$$y! \approx \sqrt{2\pi(y + 1/6)}\,(y/e)^y. \quad (11.35)$$

For the quasi-Poisson random variables with $p = 1$ and arbitrary ϕ, the correction term in the EQL (11.30) is then equal to $-\frac{1}{2}\sum\log\left(2\pi\phi(y_i + 1/6)\right)$. ∇

11.5 The canonical link

In the definition of the exponential dispersion family we gave, the parameterization used leads to a term of the form $y\theta$ in the loglikelihood. Because of this property, θ is called the *natural* or *canonical* parameter. There is also a natural choice for the link function.

Definition 11.5.1 (Canonical link function)
The *standard link* or *canonical link* is defined as the link function $\eta = g(\mu)$ with the property that the natural parameter θ coincides with the linear predictor η. ∇

Note that $\eta(\theta) = g(\mu(\theta))$, so $\eta \equiv \theta$ holds if the link function $g(\mu)$ is the inverse function of $\mu(\theta) = b'(\theta)$. This canonical link has several interesting properties. Property 9.3.9 shows that in a Poisson GLM with log-link, the marginal fitted and observed totals coincide. This result can be extended.

Property 11.5.2 (Canonical link and marginal totals)
If in any GLM with covariates x_{ij} and canonical link $g(\cdot)$, the fitted value for observation $i = 1,\ldots,n$ under a maximum likelihood estimation is $\widehat{\mu}_i = g^{-1}(\widehat{\eta}_i) = g^{-1}\left(\sum_{j=1}^{p} x_{ij}\widehat{\beta}_j\right)$, it can be proved that the following equalities hold:

$$\sum_i w_i y_i x_{ij} = \sum_i w_i \widehat{\mu}_i x_{ij}, \quad j = 1,\ldots,p. \tag{11.36}$$

If the x_{ij} are dummies characterizing membership of a certain group like a row or a column of a table, and the y_i are averages of w_i iid observations, on the left hand side we see the observed total, and on the right the fitted total.

Proof. To prove the equalities (11.36), recall that the $\widehat{\beta}_j$ that maximize the loglikelihood must satisfy the normal equations. The loglikelihood of the parameters when y is observed equals

$$\ell(\beta_1,\ldots,\beta_p;y) = \log f_Y(y;\beta_1,\ldots,\beta_p). \tag{11.37}$$

An extremum of the total loglikelihood based on the entire set of observations $Y_1 = y_1,\ldots,Y_n = y_n$ satisfies the conditions:

$$\sum_i \frac{\partial}{\partial \beta_j}\ell(\beta_1,\ldots,\beta_p;y_i) = 0, \quad j = 1,\ldots,p. \tag{11.38}$$

For the partial derivative of ℓ with respect to β_j we have by the chain rule and by the fact that $\theta \equiv \eta$ for the canonical link:

$$\frac{\partial \ell}{\partial \beta_j} = \frac{\partial \ell}{\partial \theta}\frac{\partial \theta}{\partial \beta_j} = \frac{\partial \ell}{\partial \theta}\frac{\partial \eta}{\partial \beta_j}, \quad j = 1,\ldots,p. \tag{11.39}$$

With dispersion parameter ϕ and known a priori weights w_i, using (11.4) and $\mu(\theta) = b'(\theta)$, see (11.12), we get for observation $i = 1,\ldots,n$:

$$\frac{\partial \ell}{\partial \beta_j} = \frac{w_i(y_i - \mu_i)x_{ij}}{\phi}, \quad j = 1,\ldots,p. \tag{11.40}$$

The loglikelihood of the whole sample $y_1,\ldots,y_n$ is obtained by summing over all observations $i = 1,\ldots,n$. Setting the normal equations equal to zero then directly leads to maximum likelihood equations of the form (11.36). $\quad\triangledown$

A related property of the standard link is the following.

Property 11.5.3 (Sufficient statistics and canonical link)
In a GLM, if the canonical link $\theta_i \equiv \eta_i = \sum_j x_{ij}\beta_j$ is used, the quantities $S_j = \sum_i w_i Y_i x_{ij}$, $j = 1,\ldots,p$, are a set of sufficient statistics for $\beta_1,\ldots,\beta_p$.

Proof. We will prove this by using the factorization criterion, hence by showing that the joint density of $Y_1,\ldots,Y_n$ can be factorized as

$$f_{Y_1,\ldots,Y_n}(y_1,\ldots,y_n;\beta_1,\ldots,\beta_p) = g(s_1,\ldots,s_p;\beta_1,\ldots,\beta_p)h(y_1,\ldots,y_n), \tag{11.41}$$

for $s_j = \sum_i w_i y_i x_{ij}$, $j = 1,\ldots,p$ and suitable functions $g(\cdot)$ and $h(\cdot)$. Write

$$f_{Y_1,\ldots,Y_n}(y_1,\ldots,y_n;\beta_1,\ldots,\beta_p) = \prod_{i=1}^{n} \exp\left(\frac{y_i\theta_i - b(\theta_i)}{\phi/w_i} + c(y_i;\phi/w_i)\right)$$

$$= \exp\sum_i \frac{y_i\sum_j x_{ij}\beta_j - b\left(\sum_j x_{ij}\beta_j\right)}{\phi/w_i} \exp\sum_i c(y_i;\phi/w_i) \tag{11.42}$$

$$= \exp\frac{1}{\phi}\left[\sum_j \beta_j \sum_i w_i y_i x_{ij} - \sum_i w_i b\left(\sum_j x_{ij}\beta_j\right)\right] \times \exp\sum_i c\left(y_i;\frac{\phi}{w_i}\right).$$

From this representation, the required functions $g(\cdot)$ and $h(\cdot)$ in (11.41) can be derived immediately. The fact that the support of Y does not depend on θ or on the β_j parameters is essential in this derivation. $\quad\triangledown$

Remark 11.5.4 (Advantages of canonical links)
Sometimes it happens in actuarial practice that the observations have been aggregated into a table, of which, for reasons of confidentiality, only the marginals (row and column sums) are available. Also, it often happens that policies have been grouped into cells, to save time and space. If one uses a standard link, the marginal totals, as well as the cell totals, apparently are sufficient statistics, hence knowing only their outcomes, the maximum likelihood parameter estimates can still be determined. The standard link also has advantages when the optimization algorithm of Nelder and Wedderburn is used. It leads to somewhat fewer iteration steps being necessary, and also divergence is much more exceptional. $\quad\triangledown$

Example 11.5.5 (Canonical links for various error distributions)
As stated above, the canonical link $\theta(\mu)$ makes the natural parameter θ linear in the β parameters. Because $\mu(\theta) = b'(\theta)$, the canonical link is nothing but $g(\mu) = (b')^{-1}(\mu)$. The canonical links are listed in Table D. For the normal distributions

with $b(\theta) = \frac{1}{2}\theta^2$, the canonical link is the identity function. For the Poisson and the Poisson multiples, we have $\mu(\theta) = e^{\theta}$ and hence the log-link is the standard link. For the gamma, the canonical link is the reciprocal. For the binomial it is the logit link $\theta = \log\frac{p}{1-p}$. This is also known as the log-odds, being the logarithm of the so-called *odds-ratio* $p/(1-p)$.

If $\theta \equiv \eta$ and moreover $\eta \equiv \mu$, then apparently $b'(\theta) = \theta$ holds, and the sequence of cumulants (11.13) implied by this belongs to the normal distribution. ∇

Example 11.5.6 (Threshold models: logit and probit analysis)
Assume that the observations Y_i denote the fractions of successes in n_i independent trials, $i = 1,\dots,n$, each with probability of success p_i. Further assume that a trial results in a success if the 'dose' d_i administered in trial i exceeds the tolerance X_{ik}, $k = 1,\dots,n_i$, which is a random variable having an $N(\mu_i, \sigma^2)$ distribution. Here μ_i is a linear form in the ancillary variables. Apparently

$$p_i = \Phi(d_i; \mu_i, \sigma^2) = \Phi\left(\frac{d_i - \mu_i}{\sigma}\right). \qquad (11.43)$$

Therefore, we have a GLM with a binomial distribution for the random component and with $\eta = \Phi^{-1}(p)$ as a link function. For the binomial distribution we have the following canonical link function:

$$\theta = \eta = \log\frac{p}{1-p}. \qquad (11.44)$$

Solving this for p leads to $p = e^{\eta}/(e^{\eta}+1)$. Now if we replace the distribution of the tolerances X_{ik} by a logistic(μ_i, σ) distribution with cdf $F_{X_{ik}}(d) = e^{d^*}/(e^{d^*}+1)$ for $d^* = (d - \mu_i)/\sigma$, we get a binomial GLM with standard link.

In case the thresholds X_i have a normal distribution, we speak of probit analysis, in case of a logistic distribution, of logit analysis. The second technique is nothing but a binomial GLM involving a multiplicative model not for the probability of success p itself, but rather for the odds-ratio $p/(1-p)$, therefore a canonical link. Probit analysis can be applied in the same situations as logit analysis, and produces similar results.

Logit and probit models can be applied with credit insurance. Based on certain characteristics of the insured, the probability of default is estimated. Another application is the problem to determine probabilities of disability. In econometrics, analyses such as these are used for example to estimate the probability that some household owns a car, given the number of persons in this household, their total income, and so on. ∇

11.6 The IRLS algorithm of Nelder and Wedderburn

The GLMs introduced in Nelder and Wedderburn (1972), though encompassing a whole range of useful models, have a formulation sufficiently tight to allow for one general solution method to exist to solve all of them. This method is called the *iteratively reweighted least squares* (IRLS) algorithm.

11.6.1 Theoretical description

To maximize the loglikelihood (11.4) for $\beta_1, \ldots, \beta_p$, we must solve the normal equations for the maximum likelihood parameter estimates $\widehat{\beta}_j, j = 1, \ldots, p$:

$$\sum_{i=1}^{n} \frac{\partial}{\partial \beta_j} \ell(\beta_1, \ldots, \beta_p; y_i) = 0, \qquad j = 1, \ldots, p. \tag{11.45}$$

One way to solve such a set of equations is to use Newton-Raphson iteration. In this technique, to approximate the root of $f(x) = 0$ we solve $h(x) = 0$ instead, with $h(x) = f(x_t) + f'(x_t)(x - x_t)$ a linear approximation to $f(x)$ at the current best guess x_t for the root. In a one-dimensional setting, we then find a better approximation x_{t+1} as follows:

$$x_{t+1} = x_t - (f'(x_t))^{-1} f(x_t). \tag{11.46}$$

For an n-dimensional optimization, this technique leads to the same formula, but with the points x replaced by vectors. Also, in our case where the $f(x_t)$ are first derivatives, the reciprocal is the inverse of the matrix of second partial derivatives of l, that is, of the *Hessian matrix*. The algorithm of Nelder and Wedderburn replaces the Hessian by its expected value, so it uses the *information matrix*. The technique that arises in this way is called *Fisher's scoring method*. We will show that the system of equations to be solved in the iteration step in this case equals the one of a particular weighted regression problem.

To this end, temporarily consider one sample element and drop its index i. Recall that for mean, variance function and linear predictors, with g denoting the link function, we have, see (11.12):

$$\mu(\theta) = b'(\theta); \quad V(\mu) = \partial \mu / \partial \theta = b''(\theta); \quad \eta = \sum_j x_j \beta_j = g(\mu). \tag{11.47}$$

Applying the chain rule to the loglikelihood ℓ corresponding to the density in (11.4) then leads to

$$\frac{\partial \ell}{\partial \beta_j} = \frac{\partial \ell}{\partial \theta} \frac{\partial \theta}{\partial \mu} \frac{\partial \mu}{\partial \eta} \frac{\partial \eta}{\partial \beta_j} = \frac{y - b'(\theta)}{\phi/w} \frac{1}{b''(\theta)} \frac{\partial \mu}{\partial \eta} x_j = \frac{y - \mu}{\phi/w} \frac{1}{V(\mu)} \frac{\partial \mu}{\partial \eta} x_j. \tag{11.48}$$

Just as in (11.26), of the likelihood, only the variance function has remained. The link function $\eta = g(\mu)$ determines $\partial \mu / \partial \eta = \partial g^{-1}(\eta)/\partial \eta$. For the second partial derivatives we have:

$$\frac{\partial^2 \ell}{\partial \beta_j \partial \beta_k} = \frac{\partial^2 \ell}{\partial \eta^2} x_j x_k. \tag{11.49}$$

Using the chain rule and the product rule for differentiation we get

$$
\frac{\partial^2 \ell}{\partial \eta^2} = \frac{\partial}{\partial \eta}\left(\frac{\partial \ell}{\partial \theta}\frac{\partial \theta}{\partial \eta}\right) = \frac{\partial \theta}{\partial \eta}\left(\frac{\partial^2 \ell}{\partial \theta \partial \eta}\right) + \frac{\partial \ell}{\partial \theta}\frac{\partial^2 \theta}{\partial \eta^2}
$$
$$
= \frac{\partial^2 \ell}{\partial \theta^2}\left(\frac{\partial \theta}{\partial \eta}\right)^2 + \frac{\partial \ell}{\partial \theta}\frac{\partial^2 \theta}{\partial \eta^2}.
$$
(11.50)

By the equality $\partial \ell / \partial \theta = w(y - \mu)/\phi$, its derivative $\partial^2 l / \partial \theta^2 = -w/\phi \; \partial \mu / \partial \theta$ and $\partial \mu / \partial \theta = b''(\theta) = V(\mu)$, we get

$$
\frac{\partial^2 \ell}{\partial \eta^2} = \frac{w}{\phi}\left[-V(\mu)\left(\frac{\partial \theta}{\partial \mu}\right)^2\left(\frac{\partial \mu}{\partial \eta}\right)^2 + (y - \mu)\frac{\partial^2 \theta}{\partial \eta^2}\right]
$$
$$
= \frac{w}{\phi}\left[-\frac{1}{V(\mu)}\left(\frac{\partial \mu}{\partial \eta}\right)^2 + (y - \mu)\frac{\partial^2 \theta}{\partial \eta^2}\right].
$$
(11.51)

In case of a canonical link, hence $\theta \equiv \eta$ and therefore $\partial^2 \theta / \partial \eta^2 = 0$, only the first term remains. In Fisher's scoring method, the actual Hessian in the Newton-Raphson iteration is replaced by its expected value, that is, the negative of the Fisher information matrix $\mathscr{I}$. In this case, too, the second term of (11.51) vanishes. To get the expectation over the entire sample, we take the sum over all observations, and the following expression for the (j, k) element of $\mathscr{I}$ remains:

$$
\mathscr{I}_{jk} = \mathrm{E}\left[-\frac{\partial^2 \ell}{\partial \beta_j \partial \beta_k}\right] = \sum_i \frac{w_i}{\phi}\frac{1}{V(\mu_i)}\left(\frac{\partial \mu_i}{\partial \eta_i}\right)^2 x_{ij} x_{ik}
$$
$$
= \sum_i \frac{W_{ii} x_{ij} x_{ik}}{\phi} = \frac{1}{\phi}\left(\mathbf{X}'\mathbf{W}\mathbf{X}\right)_{jk}.
$$
(11.52)

Here $\mathbf{W}$ is a diagonal weight matrix with

$$
W_{ii} = \frac{w_i}{V(\mu_i)}\left(\frac{\partial \mu_i}{\partial \eta_i}\right)^2,
$$
(11.53)

depending on the μ_i. Since $\eta = g(\mu)$, we have $\partial \eta_i / \partial \mu_i = g'(\mu_i)$.

In view of (11.48), using these same weights and taking $u_i = (y_i - \mu_i)g'(\mu_i)$, the lhs of the normal equations $\partial \ell / \partial \beta_j = 0$, $j = 1, \ldots, p$, can be written as:

$$
\frac{\partial \ell}{\partial \beta_j} = \sum_i \frac{W_{ii} x_{ij} (y_i - \mu_i)}{\phi} g'(\mu_i), \qquad \text{or} \qquad \frac{\partial \ell}{\partial \beta} = \frac{1}{\phi}\mathbf{X}'\mathbf{W}\vec{u}.
$$
(11.54)

Using Fisher's scoring method amounts to finding an improved estimate $\vec{b}^*$ for $\vec{\beta}$ from the old one $\vec{b}$ as follows:

$$
\vec{b}^* = \vec{b} + \mathscr{I}^{-1}\frac{\partial \ell}{\partial \beta}, \qquad \text{or equivalently} \qquad \mathscr{I}\left(\vec{b}^* - \vec{b}\right) = \frac{\partial \ell}{\partial \beta}.
$$
(11.55)

Let $\widehat{\eta}$ and $\widehat{\mu}$ be the vectors of linear predictors and fitted values when the parameter vector equals $\vec{b}$, so

$$\widehat{\eta} = \mathbf{X}\vec{b} \quad \text{and} \quad \widehat{\mu} = g^{-1}(\widehat{\eta}) \tag{11.56}$$

then by (11.52)

$$\mathscr{I}\vec{b} = \frac{1}{\phi}\mathbf{X}'\mathbf{W}\mathbf{X}\vec{b} = \frac{1}{\phi}\mathbf{X}'\mathbf{W}\widehat{\eta}, \tag{11.57}$$

so we can rewrite the Fisher scoring iteration equation (11.55) as follows:

$$\mathscr{I}\vec{b}^* = \frac{1}{\phi}\mathbf{X}'\mathbf{W}\vec{z} \quad \text{where} \quad z_i = \widehat{\eta}_i + (y_i - \widehat{\mu}_i)g'(\widehat{\mu}_i). \tag{11.58}$$

The elements z_i are called the modified dependent variable. Note that $z_i = g^*(y_i; \mu_i)$ with $g^*(\cdot; \mu_i)$ the linear approximation to $g(\cdot)$ at μ_i.

So, noting that the factor $1/\phi$ cancels out, a maximum likelihood estimate of β is found by the following iterative process:

> Repeat
> $$\vec{b}^* := (\mathbf{X}'\mathbf{W}\mathbf{X})^{-1}\mathbf{X}'\mathbf{W}\vec{z};$$
> using $\vec{b}^*$, update the 'working weights' $\mathbf{W}$, $\qquad$ (11.59)
> as well as the 'working dependent variable' $\vec{z}$
> until convergence.

In the special case of an ordinary linear model, we have $\eta = g(\mu) = \mu$ and $V(\mu) \equiv 1$, so the normal equations $\partial \ell/\partial \beta_j = 0$ are a linear system, and have as a solution the above expression with $\vec{y}$ replacing $\vec{z}$ and weights $W_{ii} = w_i$; no iteration steps are necessary. Note that Fisher's scoring method summarized in (11.59) indeed is an iteratively reweighted least squares algorithm.

For a good initial solution, we do not need a guess for $\vec{b}$ directly, but we can also proceed as follows. First compute good guesses for μ_i and η_i by taking $\mu_i = y_i$, $i = 1, \ldots, n$ (take care when at the boundary in a binomial or Poisson model) and constructing η_i by applying the link-function to μ_i. From these, compute initial values for W_{ii} and take $z_i = \eta_i$. Now use regression to find a first guess for the b_j.

11.6.2 Step-by-step implementation

As an example, assume that we want to estimate a Poisson GLM on 10 observations, with an intercept and a trend as covariates, and no weights. The design matrix $\mathbf{X}$ then has a vector of ones as its first column; the second is $(1, 2, \ldots, 10)'$.

- Initialize:
  $\mathbf{X} \leftarrow$ design matrix $\quad$ [column of ones, column $1, \ldots, 10$]
  $\vec{y} \leftarrow$ dependent variable

$\vec{z} \leftarrow g(\vec{y})$ (link function)

$\mathbf{W} \leftarrow$ diagonal matrix containing prior weights   [all equal to 1]

$\vec{b} \leftarrow (\mathbf{X'WX})^{-1} \mathbf{X'W}\vec{z}$ (find starting values by weighted regression)

All this is accomplished by the R commands:

```
X <- cbind(rep(1,10),1:10)
y <- c(14,0,8,8,16,16,32,18,28,22)
z <- log(y+(y==0))
W <- diag(10)
b <- solve(t(X) %*% W %*% X) %*% t(X) %*% W %*% z
cat("Start:", b[1], b[2], "\n")
```

In R, `solve(A)` produces the inverse of matrix A. Note that we cannot just take `z <- log(y)` in case `y==0`, see Exercise 11.6.1.

- Repeat the following steps until convergence:

$\vec{\eta} \leftarrow \mathbf{X}\vec{b}; \quad \vec{\mu} \leftarrow g^{-1}(\vec{\eta})$ [see (11.56)]

$W_{ii} \leftarrow \frac{1}{\mathrm{Var}[Y_i]} \left(\frac{\partial \mu_i}{\partial \eta_i}\right)^2$ [see (11.53)]

$\vec{z} \leftarrow \mathbf{X}\vec{b} + g'(\vec{\mu}) \cdot (\vec{y} - \vec{\mu})$ [entrywise product; see (11.58)]

$\mathbf{S} \leftarrow (\mathbf{X'WX})^{-1}; \quad \vec{b} \leftarrow \mathbf{SX'W}\vec{z}$ [see (11.59)]

Implementation of this iterative process requires:

```
for (it in 1:5){
  eta <- as.vector(X %*% b)
  mu <- exp(eta)                    ## eta = g(mu) = log(mu)
  W <- diag(mu)              ## (g'(mu))^(-2)/V(mu) = mu^2/mu
  z <- X %*% b + (y-mu)/mu   ## d eta/d mu = g'(mu) = 1/mu
  S <- solve(t(X) %*% W %*% X)
  b <- S %*% t(X) %*% W %*% z
  cat("it =", it, b[1], b[2], "\n")}
```

Of course when to stop the iteration is not a trivial problem, but simply doing five iteration steps suffices in this case; see below.

- In the end, the following results are produced:

$\vec{b} =$ estimate of $\vec{\beta}$

$\mathbf{S} =$ estimate of $\mathrm{Var}[\vec{b}]$: the Fisher information matrix.

This is because, asymptotically, $\vec{b} - \vec{\beta} \sim N(\vec{0}, \mathscr{I}^{-1})$.

Note how closely the R-commands to run this example, with naively implemented loop and some output, follow the description of the IRLS algorithm given.

The values of $\vec{b}$ produced by this R code are:

```
Start: 1.325353 0.2157990
it = 1 1.949202 0.1414127
it = 2 1.861798 0.1511522
it = 3 1.859193 0.1514486
it = 4 1.859191 0.1514489
it = 5 1.859191 0.1514489
```

To make R do the computations, simply do

```
coef(glm(y~I(1:10),poisson))
(Intercept)       I(1:10)
  1.8591909     0.1514489
```

11.7 Tweedie's Compound Poisson–gamma distributions

In actuarial applications, the relevant random variables generally are claim totals. Such claim totals are zero with positive probability, but, in many models, continuously distributed otherwise. Usually, a compound Poisson model is suitable to model such totals. By cleverly choosing the claims distribution, we can get exponential families of claim distributions with a variance function $V(\mu) = \mu^p$ for any $p \in (1,2)$. This enables us to tackle the problem of estimating and testing the parameters using GLM-methods. These distributions are in the so-called Tweedie class. We already saw the cases μ^p with $p = 0,1,2,3$.

For any exponent $p \in (1,2)$, by a suitable choice of the parameters of the counting distribution and the claim size distribution we can achieve that the compound Poisson–gamma distributions form a subfamily of the exponential dispersion family with mean μ and variance function of the form $V(\mu) = \mu^p$. First we look at a parameterization with parameters $\mu > 0$ and $\psi > 0$. Assume that $Y \sim$ compound Poisson(λ) with gamma(α, β) claim sizes, and parameters satisfying

$$\lambda = \frac{\mu^{2-p}}{\psi(2-p)}; \qquad \alpha = \frac{2-p}{p-1}; \qquad \frac{1}{\beta} = \psi(p-1)\mu^{p-1}. \qquad (11.60)$$

Using (3.60), it is easy to verify that in this way we get a family of distributions with mean $\lambda\alpha/\beta = \mu$ and variance $\lambda(\alpha/\beta^2 + (\alpha/\beta)^2) = \lambda\alpha(\alpha+1)/\beta^2 = \psi\mu^p$. Note that all claim sizes have common α, hence the same shape, dictated by the value of the exponent p in the variance function; the expected claim numbers and the scale vary to generate all possible (μ, ψ) combinations.

We want to show that for this particular choice of parameters (11.60), the 'density' of Y can be written as in Definition 11.3.1, both for $y = 0$ and for $y > 0$. Note that the cdf of Y is a mixed distribution in the sense of Section 2.2. We have

$$f_Y(y) = e^{-\beta y} e^{-\lambda} \sum_{n=1}^{\infty} \frac{\beta^{n\alpha}}{\Gamma(n\alpha)} y^{n\alpha-1} \frac{\lambda^n}{n!}, \qquad y > 0. \qquad (11.61)$$

Now because of the choice in (11.60), it is not difficult to see that $\lambda\beta^\alpha$ does not depend on μ, only on the parameter ψ and the constant p. Therefore the sum in (11.61) depends on ψ and y, but not on μ. To check that (11.61), together with $dF_Y(0) = \Pr[Y = 0] = e^{-\lambda}$, is of the form (11.4), we define $c(y, \psi)$, $y > 0$, as the logarithm of the sum in (11.61), and $c(0, \psi) = 0$. Next, we equate $-\beta = \theta/\psi$ as well as $\lambda = b(\theta)/\psi$. This gives

$$\theta = -\beta\psi = \frac{1}{(1-p)\mu^{p-1}}, \tag{11.62}$$

so

$$\mu(\theta) = (\theta(1-p))^{-1/(p-1)}, \tag{11.63}$$

and

$$b(\theta) = \lambda\psi = \frac{\mu^{2-p}}{2-p}. \tag{11.64}$$

The reader may check that, as it should be, $V(\mu(\theta)) = \mu'(\theta)$ holds. Note that the cases of a Poisson multiple ($p = 1$) and a gamma variate ($p = 2$) can be obtained as limits of this class. This fact may be verified by taking limits of the mgfs, or understood as follows. For $p \downarrow 1$, in the limit we get Poisson(μ/ψ) many claims that are degenerate on ψ. When $p \uparrow 2$, at the same time $\lambda \to \infty$ and $\alpha \downarrow 0$, in such a way that $\lambda\alpha \to \mu/\psi$. The resulting limit distribution is the gamma($\mu/\psi, 1/\psi$) distribution.

Actual distributions in the exponential dispersion family with a variance function $V(\mu) = \mu^p$ exist for all $p \notin (0,1)$. For $p \in (0,1)$, still the quasi-likelihood (11.26) can be maximized to obtain parameter estimates.

Remark 11.7.1 (Negative binomial random variables)
For any fixed r, the negative binomial(r, p) distribution also has a variance function 'between' those of Poisson and gamma, with $V(\mu) = \mu + \mu^2/r$. Being integer-valued, it might be a better choice to model overdispersed counts, see Examples 3.3.1 and 3.3.2, but it is somewhat more awkward to use than the quasi-Poisson family. ∇

11.7.1 Application to an IBNR problem

A package `tweedie` extending R exists that enables one to easily estimate a GLM from Tweedie's family of distributions. It was contributed by Peter Dunn and Gordon Smyth, see, for example, Dunn and Smyth (2005), and offers density (`dtweedie`), tail probability (`ptweedie`), quantile (`qtweedie`) and random sample (`rtweedie`) for any given Tweedie distribution with parameters μ, ϕ and p, where $\mathrm{Var}[Y] = \phi\mu^p$. Also, it contains R code to produce a GLM family object with any power variance function and any power link $g(\mu) = \mu^q$, where $q = 0$ denotes the log-link. It includes the Gaussian, Poisson, gamma and inverse Gaussian families as special cases with $p = 0$, $p = 1$, $p = 2$ and $p = 3$, and provides access to a range of generalized linear model response distributions that are not otherwise provided by R. It is also useful for accessing distribution-link combinations that are not supported by the `glm` function.

To illustrate the actuarial use of Tweedie's class of distributions, we generated an IBNR triangle containing drawings from Y_{ij}, $i, j = 1, \ldots, 10$, $i + j \leq 11$, having Tweedie distributions with mean $\mu_{ij} = \mu r_i c_j \gamma^{i-1}\delta^{j-1}$ and variance $V(\mu_{ij}) = \psi\mu_{ij}^p$.

Table 11.1 Estimation results for some values of p in $[1,2]$, selected to find the ML-estimate

p	$\widehat{\phi}$	Reserve	loglikelihood
1	86.85	17287	$-\infty$
1.25	12.83	17300	-376.844
1.5	2.08	17369	-368.463
1.75	0.40	17605	-367.764
2	0.10	18233	-374.902
1.384	4.75	17325	-370.883
1.616	0.94	17447	-367.312
1.759	0.38	17619	-367.867
1.654	0.73	17484	-367.229

The parameter values chosen were $p = 1.5$, $\psi = 2$, $\mu = 10^4$, $\gamma = 1.03$, $\delta = 0.9$. The r_i were known relative exposures for each row, the c_j given development factors for each column in the IBNR-triangle. Expressed as percentages, their values are

$$r = (100, 110, 115, 120, 130, 135, 130, 140, 130, 120);$$
$$c = (30, 30, 10, 20, 5, 3, 1, 0.5, 0.3, 0.2).$$
(11.65)

To estimate the parameters γ, δ and μ we used a log-link and a power variance function $V(\mu) = \mu^p$, generating a deviance D_p as in (11.29). In Table 11.1, we find the estimation results with some values of p, the last few chosen by R's function optimize. We list the estimated dispersion parameter (taken to be the mean deviance). Using the function dtweedie, we can actually compute the likelihood. An alternative is to use a quick-and-dirty implementation of formula (11.61). Unless $p = 1$ or $p = 2$, the following function only works for scalar arguments, and it is not optimized for speed or guaranteed to be numerically stable:

```
dTweedie <- function (y, power, mu, phi)
{ if (power==2) s <- dgamma(y, 1/phi, 1/(phi*mu)) else
  if (power==1) s <- dpois(y/phi, mu/phi) else
  { lambda <- mu^(2-power)/phi/(2-power)
    if (y==0) s <- exp(-lambda) else
    { alpha <- (2-power)/(power-1)
      beta <- 1 / (phi * (power-1) * mu^(power-1))
      k <- max(10, ceiling(lambda + 7*sqrt(lambda)))
      s <- sum(dpois(1:k,lambda) * dgamma(y,alpha*(1:k),beta))
  } }
  return(s) }
```

Next to the resulting loglikelihood, we list in Table 11.1 the estimate of the IBNR-reserve to be held, which is equal to the sum over all future predicted values $\widehat{\mu}_{ij} = \widehat{\mu} r_i c_j \widehat{\gamma}^{i-1} \widehat{\delta}^{j-1}$, for those $i, j = 1, \ldots, 10$ with $i + j > 11$. Note that the value of $\widehat{\phi}$ varies very strongly with p, being about right ($\phi = 2$) only for p close to the actual value 1.5. The loglikelihood is maximal at $p = 1.654$, but the actual value $p = 1.5$ leads to an acceptable value as well. The ML-estimate of the reserve equals 17484. Observe that the required reserve increases with the exponent p, but that it is not very sensitive to the value of p, just as, it turns out, are the parameter estimates $\widehat{\gamma}, \widehat{\delta}$ and $\widehat{\mu}$. To fit a Tweedie GLM, you can use the `tweedie` family object as found in the `statmod` package contributed by Gordon Smyth. A family object specifies the name of the family, the variance function μ^p, the link function μ^q or $\log \mu$, as well as its name and inverse function, the deviance residuals function, the AIC function, the function $\mu(\eta)$, the initialization calls needed, and which μ and η values are valid. The R code to produce these results is given below.

```
require(tweedie)###If "FALSE" results, download it from CRAN first
TT <- 10; i <- rep(1:TT, each=TT); j <- rep(1:TT, TT)
past <- i + j - 1 <= TT; n <- sum(past)
Expo <- c(100, 110, 115, 120, 130, 135, 130, 140, 130, 120)
Runoff <- c(30, 30, 10, 20, 5, 3, 1, 0.5, 0.3, 0.2)
Off <- rep(Expo, each=TT) * rep(Runoff, TT); lOff <- log(Off)
##note that future values are input as 0.01; they get weight 0 anyway
Xij <- scan(n=100)
4289.93 3093.71 1145.72 1387.58 293.92 189.17   42.36 11.41 4.31 12.39
3053.09 2788.81  682.44 1475.69 253.31 100.58   79.35 15.48 8.06  0.01
4388.93 2708.67  688.42 2049.57 353.20 266.43 109.42 47.90 0.01  0.01
4144.15 2045.63 1642.27 1310.97 548.97 159.87   69.86  0.01 0.01  0.01
2912.73 4078.56 1652.28 2500.94 394.99 220.89    0.01  0.01 0.01  0.01
5757.18 5200.83 1177.65 2486.30 580.29   0.01    0.01  0.01 0.01  0.01
4594.18 3928.15 1236.01 2729.68   0.01   0.01    0.01  0.01 0.01  0.01
3695.03 3688.23 1300.97   0.01   0.01   0.01    0.01  0.01 0.01  0.01
3967.13 4240.97    0.01   0.01   0.01   0.01    0.01  0.01 0.01  0.01
4933.06    0.01   0.01   0.01   0.01   0.01    0.01  0.01 0.01  0.01
round(xtabs(Xij~i+j)) ## produces a table of the input values
y <- Xij[past]
Tweedie.logL <- function(pow)
{ gg <- glm(Xij~i+j+offset(lOff), tweedie(pow,0), wei=as.numeric(past))
  reserve <- sum(fitted.values(gg)[!past])
  dev <- deviance(gg); phi.hat <- dev/n
  mu <- fitted.values(gg)[past]; hat.logL <- 0
  for (ii in 1:length(y))
  { hat.logL <- hat.logL + log(dTweedie(y[ii], pow, mu[ii], phi.hat)) }
  cat("Power =", round(pow,3), "\tphi =", round(phi.hat,2),
      "\tRes. =", round(reserve), "\tlogL =", round(hat.logL,3), "\n")
  hat.logL   }
for (pow in c(1,1.25,1.5,1.75,2)) Tweedie.logL(pow)
oo <- optimize(Tweedie.logL, c(1.01,1.99), maximum=T, tol=1e-4)
```

If `pow=1`, the density of the observed values is zero, since they are not multiples of Poisson outcomes. R warns that the observations are not integer.

11.8 Exercises

Section 11.2

1. Describe the design matrices $\mathbf{X}$ used in both fits in the example at the end of this section. Note that an intercept is implied, so the first column of $\mathbf{X}$ is a column of ones. Check with `model.matrix(lm.)`.
   To check if `vcov(lm.)` is the ML-estimated Fisher information matrix $\widehat{\sigma}^2 (\mathbf{X}'\mathbf{WX})^{-1}$, do

   ```
   w <- c(1,2,2,1,1,1,1,4,2,4,2,3,2,1,1,2)
   Y <- c(0,1,0,8,0,0,0,30,0,1,1,38,0,0,0,26) / w
   gender <- c(0,0,0,0,0,0,0,0,1,1,1,1,1,1,1,1)
   ```

```
income <- c(1,2,5,20,1,2,5,20,1,2,5,20,1,2,5,20)
lm. <- lm(Y ~ gender+income, weights=w)
mean(lm.$residuals^2*w) *
  solve(t(model.matrix(lm.))%*%diag(w)%*%model.matrix(lm.)) /
    vcov(lm.)
```

Here `solve(A)` computes the inverse of matrix A, `t(A)` gives its transpose, `diag(w)` is a diagonal matrix with vector w as its diagonal, `%*%` does a matrix multiplication and `/` does an elementwise division.

Instead of $\hat{\sigma}^2$, which quantity is used by R to compute the variance/covariance matrix? In the output of `summary(glm.)`, find where the quantities `sqrt(diag(vcov(glm.)))` occur. In this case, `diag(A)` produces a vector of the diagonal elements of matrix A; yet another use of this function is `diag(n)` for scalar n, producing an $n \times n$ identity matrix.

Section 11.3

1. Prove the relations $E\left[\frac{\partial \ell(\theta,Y)}{\partial}\right] = 0$ as well as $E\left[\frac{\partial^2 \ell(\theta,Y)}{\partial\theta^2}\right] + E\left[\left(\frac{\partial \ell(\theta,Y)}{\partial\theta}\right)^2\right] = 0$, where $\ell(\theta,y) = \log f_Y(y;\theta,\phi)$ for f_Y as in (11.4). From these relations, find $E[Y;\theta]$ and $Var[Y;\theta]$.

2. Check the validity of the entries in Table D for all distributions listed. Verify the reparameterizations, the canonical link, the cumulant function, the mean as a function of θ and the variance function. Also determine the function $c(y;\phi)$.

3. The marginal totals equations are fulfilled, by (11.36), for the Poisson distribution in case of a log-link. Prove that the same holds for all power link functions $g(\mu) = \mu^\alpha$, $\alpha > 0$, by adding up the ML-equations, weighted by β_j. What is the consequence for the deviance of Poisson observations with this link function?

4. The same as the previous exercise, but now for gamma observations.

5. Show that in general, the scaled deviance satisfies (11.17).

6. From the expression in the previous exercise, derive expressions for the scaled deviances for the normal, Poisson, binomial, gamma and inverse Gaussian distributions.

7. Prove the statements about Esscher transforms in Corollary 11.3.11.

8. Show that the Anscombe residuals for Poisson, gamma and inverse Gaussian are as given in (11.21)–(11.23).
    Show that (11.17) reduces to (11.18) when the deviance residual for Poisson is computed.
    Compare Taylor expansions at $y = \mu$ for the case of Poisson errors ($\phi = 1$) for the squares of the three residuals. The first term should be $(r^P)^2 = \frac{(y-\mu)^2}{\mu}$ for all three, the second is $-\frac{(y-\mu)^3}{3\mu^2}$ for both $(r^A)^2$ and $(r^D)^2$.
    For $\mu = 1$ and $y = 0, .2, .4, .6, 1, 1.5, 2, 2.5, 3, 4, 5, 10$, compare the values of r^P, r^A and r^D in case of a Poisson distribution.
    Draw a sample $Y_1, \ldots, Y_{1000}$ from a gamma distribution with shape $\alpha = 5$ and scale $\beta = 1$. For $p = 1/3$ and $p = 1/2$, do a visual test for normality of Y^p by inspecting normal Q-Q plots, see Figure A.3.
    Now let the gamma shape parameter be $\alpha = 0.1$. Draw a histogram of the values $a(Y) = \sqrt[3]{Y}$. Are the resulting residuals $3(y^{1/3} - \mu^{1/3})/\mu^{1/3}$ indeed more or less symmetric?

9. Investigate the standard normality of the Anscombe residual (11.22) of $X \sim \text{gamma}(\mu, 1)$, with $\mu = 1, 2, 5, 10, 20, 50$. Take a sample of size 1000 and do a Student test for the mean, an F-test for the variance, and the test known as the Jarque-Bera test, see Appendix (A.1), for the third and fourth moment. Inspect Q-Q plots.

10. In the same way as the previous exercise, analyze standard normality of the Anscombe residual (11.23) for the case $X \sim$ Inverse Gaussian$(\mu, 1)$.

Section 11.4

1. Plot the variance as a function of the mean for some values of p in a compound Poisson–gamma GLM with parameters (11.62). Compare with the same for a negative binomial(r, p) GLM.
2. Check that $V(\mu(\theta)) = \mu'(\theta)$ holds when $\mu(\theta)$ is as in (11.62).
3. Verify that the mgf of the compound Poisson–gamma distribution is of the form (11.8).
4. Find the quasi-likelihood for the case $V(\mu) = \mu^2(1 - \mu)^2$.
5. When $\mu_{ij} = \alpha_i \beta_j$ as in (9.7), write down the normal equations corresponding to (11.26). Also, verify the statements in Remark 11.4.1.
6. Verify the formula for D_p. Verify D_1 and D_2 by direct integration, and compare with the corresponding relations (9.29) and (9.32). Verify that they also follow as limits from D_p. Verify D_0.
7. Verify that the maximum extended QL estimators of the β parameters as well as for ϕ are as described. Also verify if the ML estimator of ϕ coincides with the maximum extended QL estimator.
8. Compare extended QL and ordinary loglikelihood ratio for the (inverse) Gaussian case ($p = 0$ and $p = 3$).

Section 11.5

1. What does Property 11.5.2 imply about the sum of the residuals if there is a constant term in the model? And if there is a factor in the model? (See also Exercise 11.3.3.)

Section 11.6

1. Why the exception for the case $\texttt{y==0}$ in the initialization phase of the example at the end of this section?
2. In a Poisson GLM with standard link, what are the values of the working dependent variable and weights? If $\vec{\beta}$ is *known*, what are mean and variance of the working dependent variables Z_i?
3. Compare $\texttt{vcov(g)}$ and S. The same for $\texttt{model.matrix(g)}$ and X. Compare the standard errors of the parameter estimates in the output of $\texttt{summary(g)}$ with $\texttt{sqrt(diag(S))}$.
4. In our example in this section, iteration is stopped after 5 iterations, which turns out to be adequate. How should the code be changed to allow the iteration to be stopped after a fixed maximum number of iterations, or earlier when the state 'convergence' has been reached, taken to mean that both coefficients b have exhibited a relative change of 10^{-6} or less? Hint: use $\texttt{break}$ to jump out of a for-loop, as in Section 3.5. Define the state 'convergence' as $\texttt{abs(dev-devold)}$ / $\texttt{(abs(dev)+0.1)}$ < $\texttt{tol}$ for some suitable scalar quantity dev and some small value of tol.

5. How should the code be changed for a binomial distribution with a standard (logit) link? Take the same y-vector and fix the number of experiments at 50.

6. If we take an ordinary linear model with identity link for y instead of a Poisson GLM, what changes should be made in the R code above, and what happens to the output?

7. Using the fact that ML-estimates of the parameters satisfy the marginal totals equations in a Poisson GLM with standard link, determine the optimal parameter values by successive substitution. Compare the likelihoods after each iteration step with those of IRLS.

8. A certain credit insurer has classified a number of insureds into groups of size about 60 on the basis of certain characteristics, has given each group a score based on these characteristics and counted how many defaulted, as follows:

```
Size   <- c(59,60,62,56,63,59,62,60)
Score  <- c(69,72,76,78,81,84,86,88)
Def    <- c( 6,13,18,28,52,53,61,60)
```

   Adapt the IRLS algorithm above to estimate the default probability as a function of the score, using a logit link $\eta = \log \frac{\mu}{1-\mu}$. Check your algorithm with the glm output.
   Hint: do not just set z <- log(Def/(Size-Def)) to initialize but add 0.5 to both numerator and denominator.

9. Show that for a normal GLM with identity link, no iterations of Fisher's scoring method are necessary. Does the same hold for a normal GLM with some other link?

10. The canonical link $\theta(\mu)$ expresses the natural parameter θ as a function of the mean μ in a GLM. Show that $d\theta(\mu)/d\mu = 1/V(\mu)$ as well as $db(\theta(\mu))/d\mu = \mu/V(\mu)$. Verify this for the cases $V(\mu) = \mu^j$, $j = 0,1,2,3$.

Section 11.7

1. Determine the maximized likelihood in the full model, see (11.31), for a Tweedie distribution with $V(\mu) = \mu^p$, $1 < p < 2$. Incorporate the refined Stirling formula (11.35). What does the EQL look like?

2. Run the following R code:

```
hist(rtweedie(1000, power=1.001, mu=1, phi=1), breaks=41)
sum(dtweedie((1:1999-.5)/100, 1.5, 1, 1)) / 100 +
  dtweedie(0, 1.5, 1, 1)
```

   Explain the results.

3. Generate a 6×6 IBNR triangle of Tweedie distributed claim totals, with variance function $V(\mu) = \mu^p$ for $p = 1.2$. Take the mean of observation X_{ij} to be proportional to α^{i-1} for $\alpha = 1.03$ and to β_j, with $\beta = (1,.9,.4,.2,.1,.05)$, and take $E[X_{11}] = 10$. Do an analysis similar to the one in this section, and report your results.

Appendix A
The 'R' in Modern ART

Of fundamental importance is that S is a language. This makes S much more useful than if it were merely a "package" for statistics or graphics or mathematics. Imagine if English were not a language, but merely a collection of words that could only be used individually – a package. Then what is expressed in this sentence is far more complex than any meaning that could be expressed with the English package —
Patrick J. Burns (S Poetry, 1998)

This appendix provides an informal introduction to the concepts behind the software R and the programming language S it implements. The aim is to get a feeling of how R operates and how its output should be interpreted when applied to problems from actuarial risk theory. Many texts about using R are available, see the references. The first section of this appendix was inspired by Burns' (2005) guide for the unwilling S user, the second shows how to do some exploratory data analysis on stock prices. In the third section, we illustrate the use of R by generating a portfolio of automobile insurance risks, to be used for testing purposes. For the analysis of such a portfolio using Generalized Linear Models, see Section 9.5.

A.1 A short introduction to R

R is a programming environment that is well suited for solving statistical and econometric problems. As regards risk theory, it is important that it offers many useful mathematical functions like the Fast Fourier Transform (FFT). Another essential capability of R is solving Generalized Linear Models (GLM) in the sense of Nelder and Wedderburn (1972). See Chapters 9–11. Both R and S-Plus implement the S language to fit and analyze statistical models, and currently they differ only on minor details. R is also known as GNU-S, where the acronym GNU is short for "GNU's Not Unix". Apart from being stable, fast, always up-to-date and very versatile, the chief advantage of R is that it is available to everyone free of charge. And because it is open-source software, its sourcetexts are available, so it is not a black box system where one has to guess how programmers have interpreted and implemented the theory. It has extensive and powerful graphics abilities, and is developing rapidly, being the statistical tool of choice in many academic environments.

Our basic environment is Windows, but R is available on many more platforms. To start up R, download a copy of the program on www.r-project.org (the CRAN website) and install it. Put a shortcut to the program on your desktop, and click on it to enter R's graphical user interface. To quit, press for example Alt-F4.

S is an interactive language. User programs are not compiled, but interpreted and executed. The program R prompts the user for input, displaying its prompt >. After a command is entered, R responds. Just typing the name of an object will cause R to print it, so a "hello world" program involves simply typing in the string

```
"Hello world!"
```

Statements to be executed can also be stored in a script file. To execute them, use cut-and-paste, or open the file in R's own script editor by choosing File → Open script. Then use one of the Run commands that have become available under Edit, or click an appropriate icon with the same effect. Or just press Ctrl-R to run the current line or selection.

Objects Almost everything in R is an "object". Object oriented programming (OOP) is a programming paradigm that uses objects and their interactions to design applications and computer programs. Many modern programming languages support OOP. R objects may hold a single value, a vector of values, a matrix or a more general record. The most common object is a *vector* containing real, integer or complex numbers, character strings, or Boolean variables (TRUE or FALSE). A *matrix* also has a single type of entry, but has rows and columns. A vector is not always the same as a matrix with just one column. A *data frame* is a matrix that may have different types in different columns. There is only one type within each column of a data frame. The components of a *list* can be any sort of object including another list. The result of a regression, for example, is an object containing coefficients, residuals, fitted values, model terms and more.

Vectors and lists can have *names*, and each element or component gets a name. The rectangular objects have *dimnames*: names for each row and each column. Data frames must have dimnames, matrices may or may not.

A *factor* is a division into categories. For example, drivers might be of type "Rural", "City" or "Other", called the levels of this factor. Also, a quantitative variable like a policy year or a bonus-malus class might be treated as a factor.

Objects that must be saved need to get a name. One name identifies a whole object; "names" and "dimnames" just specify pieces of objects. Valid names are most combinations of letters, digits, the period (".") and the underscore ("_"). For example .a, ._ and just . are acceptable names, though _a and _ are not. For obvious reasons, a name .1 is also not legal, but ..1 is.

Many people will experience trouble with the fact that names are *case-sensitive*. So T and t are two different names; T and TRUE are Boolean values 'true', t(B) is the transpose of a matrix B. The family of gamma distributions in a generalized linear model is to be referred to as Gamma, to avoid confusion with gamma(y) denoting the gamma function $\Gamma(y)$ (factorials). But the Poisson family is just poisson.

Some names are *reserved*, like return, break, if, TRUE, FALSE. The short versions T and F are not reserved in R, though redefining them might lead to problems. The same holds for c, t and q for example.

The way to assign names to objects is to give the name, then the two characters <- (the '*gets arrow*', pronounced 'gets' and best surrounded by spaces), then the command that creates the object:

```
const <- 1/sqrt(2*pi)
```

The 'gets' operator also has a rightward form: `a <- 1` and `1 -> a` have the same effect. Another way to do an assignment is simply `a = 1`, but this is not allowed under all circumstances.

Reading from a file To read a large object produced for example by Excel, first write it to a text file. Having its roots in UNIX, R treats the backslash \ as an escape character, so type filename and path, for example, as

```
fn <- "C:\\R-files\\somefile.text"  ## or equivalently:
fn <- "C:/R-files/somefile.text"
```

Simply using a single forward slash / instead works fine in Windows systems. Note that the remainder of a line after a # symbol is treated as *comment*.

To put the numbers written to a file into a vector, use the `scan` function. It expects numbers separated by white space, and reads on until the end of the file. It can also be used for keyboard input, or input from a script; then it reads on until a blank line. The command could be:

```
> vec <- scan("C:/Documents and Settings/Rob Kaas/
+ Desktop/anyfile.txt")
```

Here, we list not just the command to be given to R, but also R's prompt >. Commands can be split up over several lines. Since the first command line above ended in the middle of a character string, R assumes it is not finished, so it gives its continuation prompt + and waits for the remainder of the command. But quite often, this continuation prompt appearing indicates a typing error, notably when a closing bracket is omitted somewhere.

Later on, you can inform R that you actually want to store the numbers in `vec` in a matrix, for example by

```
mat <- matrix(vec, ncol=5, byrow=TRUE)
```

This instructs R to stack the numbers in `vec` in a matrix with 5 columns, but not in *column major order* such as is standard in R and some other languages, but by row.

To read from a file saved by a spreadsheet program as a comma separated file, do one of these:

```
pp <- scan("thefile.csv", sep=",")  ## or:
pp <- scan("thefile.csv", dec=",", sep=";")
```

The second form applies in systems where the decimal point is replaced by a comma. Then the separation character in a `.csv` file is a semicolon. In this text, we will generally work with small enough datasets to include the numbers in the script file, or type them in at the keyboard, using `Xij <- scan(n=15)` to store 15 numbers in `Xij`.

Combining numbers into a vector To *combine* elements into a vector, use c ( ) :

```
> numbers <- c(0, 3:5, 20, 0)
> numbers
[1]  0  3  4  5 20 0
```

In this listing, there is input echoed by R, which is preceded by a screen prompt >, as well as output. The [1] starting the response says that the line begins with the first element. By 3 : 5 we denote the sequence of numbers 3, 4 and 5.

The c function may also concatenate other things than numbers:

```
words <- c("Testing", "testing", "one", "two", "three")
```

One way of creating matrices is to bind vectors together. The rbind function treats the vectors as rows:

```
> a.matrix <- rbind(numbers, 1:6)
> a.matrix
          [,1] [,2] [,3] [,4] [,5] [,6]
numbers     0    3    4    5   20    0
            1    2    3    4    5    6
```

While other languages involve heavy use of for-loops, they are best avoided in R, as it is an interpreted language rather than a compiled one. This is often possible when doing the same operation on all elements of a vector, for instance taking the sum or the mean of a vector of values. The first way given below to store the first 10^7 squares in a takes about half a second, the second one more than a minute:

```
n <- 1e7; a <- (1:n)^2
for (i in 1:n) a[i] <- i^2
```

Matrix multiplication in a programming language like Pascal is not a trivial matter. In R, we simply do

```
> a.matrix %*% 2:7
          [,1]
numbers   170
          112
```

From the output one sees that the result is a 2×1 matrix, with two rows and one column, which is not quite the same as a 2-vector. Note that %*% is used for matrix multiplication; a.matrix * numbers does elementwise multiplication when applied to matrices or vectors. Operators with % signs around them are variants of the ones without. Compare the following ways to multiply vector $(1, 2, 3)$ with itself:

```
>      1:3 * 1:3         ## elementwise product; equals (1:3)^2
[1] 1 4 9
>      t(1:3)*1:3        ## elementwise product (as a matrix)
       [,1] [,2] [,3]
[1,]     1    4    9
>                        ## inner product (as a matrix);
>      1:3 %*% 1:3       ## equals t(1:3)%*%1:3 and crossprod(1:3)
       [,1]
```

```
[1,]    14
>                           ## outer product;
>     1:3 %*% t(1:3)    ## equals 1:3%o%1:3 and crossprod(t(1:3))
      [,1] [,2] [,3]
[1,]    1    2    3
[2,]    2    4    6
[3,]    3    6    9
```

As a matrix multiplication of a 1×3 matrix (row vector) by a 3×1 matrix (column vector), the inner product is a scalar (1×1 matrix): $(1,2,3)(1,2,3)' = 1^2 + 2^2 + 3^2 = 14$; the outer product $(1,2,3)'(1,2,3)$ is a 3×3 matrix with the cross products ij for $i, j = 1, 2, 3$.

Extracting parts of objects Extracting pieces of objects is done by subscripting, using square brackets. There are four common ways of indexing.

1. Positive numbers select the index numbers that you want:

   ```
   > words[c(3:5, 1)]
   [1] "one"          "two"          "three"      "Testing"
   ```

2. In R, vectors are always numbered with 1 as their first element. This makes it possible to use negative numbers to give the indices to be left out:

   ```
   > numbers[-4]
   [1]   0   3   4 20   0
   ```

3. If there are names, you can select the names that you want.
4. Logicals: select the locations that you want.

   ```
   > numbers[numbers < 10]
   [1] 0 3 4 5 0
   ```

Testing equality with real numbers needs to be done with care, since the inexactness of computed numbers can cause equality not to hold exactly. Our computer gives:

```
> sqrt(2)^2 == 2
[1] FALSE
```

From the [1] we see that R is actually printing a vector (of length 1). Note that a double symbol '==' means equality, so a==1 results in FALSE or TRUE, but a=1 is an assignment, or a named argument of a function. Similarly, a < -1 is Boolean, but a<-1, without a space between the < and the -, denotes an assignment.
    It is possible to replace the values of part of an object:

```
> numbers2 <- numbers
> numbers2[4:5] <- c(17,19)
> numbers2
[1]   0   3   4 17 19   0
```

In matrices and data frames the rows and columns are subscripted separately:

```
> a.matrix[2:1, numbers>4]
        [,1] [,2]
          4   5
numbers   5   20
```

The result is a submatrix of a.matrix with the second and the first row (in that order), and all columns where the numbers vector has an element larger than 4.

Leave a blank in a dimension to indicate that you want *all* the rows or columns:

```
> a.matrix[, c(1,3,5)]
        [,1] [,2] [,3]
numbers   0   4   20
          1   3   5
```

Lists are created with list, almost always with the tag=object form. The $ operator is used to extract a component out of a list.

```
> list1 <- list(num=numbers, char=words)
> list1$char
[1] "Testing" "testing" "one"      "two"      "three"
```

Doing arithmetic In arithmetic, the usual order of operations applies. Use parentheses to modify the default order of computation; spaces are generally ignored:

```
>   9*3 ^-2
[1] 1
> (9*3)^-2
[1] 0.001371742
> -2^-.5
[1] -0.7071068
> (-2)^-.5
[1] NaN
```

The vectors need not have the same length. Consider the command:

```
> c(1,0)+(-1:4)
[1] 0 0 2 2 4 4
```

We can visualize this as below. The first two columns of this table show the original problem. The next two columns show the expanded form of the problem; the shorter vector is copied down its column until it is as long as the other vector. If the length of the longer vector is not a multiple of the one of the shorter vector, R gives a warning. Once the expansion is done, the resulting vectors can be added. The answer is shown in the final column of the table.

original		expanded		answer
c(1,0)	-1:4	c(1,0)	-1:4	c(1,0)+(-1:4)
1	−1	1	−1	0
0	0	0	0	0
	1	1	1	2
	2	0	2	2
	3	1	3	4
	4	0	4	4

Missing values are denoted by NA and propagate through calculations. Use is.na to test for missing values; the first command below compares elements to the value NA, and the result of these comparisons is not true or false, but simply NA.

```
> c(1, 7, NA, 3) == NA
[1] NA NA NA NA
> is.na(c(1, 7, NA, 3))
[1] FALSE FALSE  TRUE FALSE
```

Parameters for functions We already met a few R-functions: c, scan and sqrt for example. Usually not all of a function's arguments need to be given, as many of them will have default values. The call scan("thefile.csv", sep=",") used earlier was invoked with two arguments. The first is not named so it must refer to the first argument of the function. The second is named by sep. Since it is named, it need not be the second argument of the function.

The name used to specify the sep argument could have been shortened to se, but not to just s because there are other arguments to scan that begin with s; use args(scan) to see this function's arguments and default values, and just scan to see the full source text of this function. Special rules apply to functions like c that have an arbitrary number of arguments.

Generic functions For functions like print, plot and summary, what happens depends on the parameters given. A data frame is printed much like a matrix. When print sees a data frame, it invokes a function designed to print data frames. What generic functions do depends on the class of the object on which they operate.

Making plots When you just plot an object, the result often makes sense. For example plot(numbers) plots the values against the index numbers. To add a line $a + bx$ with intercept $a = 10$ and slope $b = 20$ to the plot, call abline(10, 20). Another useful graphics function for exploratory data analysis is hist to plot a histogram. Right-click on a plot to save it, to the clipboard or as a PostScript file.

Using R's help facility To get help with specific R functions, type a question mark followed by the name of the function. For example, ?objects, or, equivalently, help(objects). More information is found by help.search("objects").

At the end of help pages, there are illustrative examples that can be pasted into the R-console and run, or reproduced, for example, by example(glm). For the interested reader, there are many texts on the internet about using R. Or type `help.start()` in R and click away.

A.2 Analyzing a stock portfolio using R

To illustrate how R can be used in risk management, assume that we are interested in buying a portfolio consisting of stocks of one company manufacturing tyres and another producing liquor. Our historical data consist of the weekly stock prices in the latest three calendar years, and on the basis of that, we want to predict the future behavior of a portfolio consisting of a mix of these two stocks. R has a lot of functionality to handle time series, but we will use standard functions only here.

The prices were as follows:

```
Tyres <- scan(n=157)
307 315 316 314 324 310 311 295 278 295 318 343 342 323 328 303
309 307 315 296 313 316 317 306 307 326 330 330 333 341 337 353
356 359 349 351 359 360 363 342 337 334 352 357 360 368 363 366
366 365 381 401 401 421 422 425 417 427 436 440 432 406 401 420
420 424 416 403 400 392 391 390 406 415 429 420 415 420 417 445
447 449 447 450 460 470 495 507 518 516 522 524 484 497 490 500
464 458 446 450 471 485 486 501 506 502 494 497 465 478 490 496
517 506 497 483 474 495 499 483 477 481 474 479 431 438 431 436
434 453 442 445 461 463 481 490 470 480 497 507 503 508 485 492
490 519 506 539 542 553 558 562 532 510 512 504 474
Liquor <- scan(n=157)
 781  784  757  741  728  726  743  746  768  752  758  754  779
 777  780  815  791  779  802  797  800  860  873  854  855  846
 824  833  838  851  827  847  859  853  926  935  952  962  958
 943  938  949 1000 1003 1018 1022 1026 1019 1037 1026  999 1011
 994 1036 1030 1028 1005 1006 1005  970  983  984  980  996  975
 976  987 1008 1057 1054 1040 1045 1057 1101 1096 1094 1108 1102
1104 1105 1092 1098 1113 1076 1060 1054 1054 1057 1078 1077 1091
1093 1079 1086 1064 1134 1167 1217 1155 1171 1174 1213 1218 1250
1329 1350 1334 1318 1315 1310 1368 1370 1389 1420 1377 1366 1424
1455 1475 1478 1458 1430 1409 1407 1414 1409 1402 1386 1392 1391
1445 1448 1462 1474 1482 1495 1525 1547 1538 1436 1465 1454 1460
1476 1521 1588 1581 1587 1539 1574 1537 1518 1530 1500 1554 1532
1498
```

To show their behavior in time, we look at some standard plots, see Figure A.1.

```
par(mfrow=c(1,2))
plot(Tyres, xlab="Week", type="l")
plot(Liquor, xlab="Week", type="l")
```

The first line serves to produce two plots next to each other; see `?par` for more possibilities to change the way plots look.

If S_t, $t = 1, 2, \ldots$, are the stock prices, one may look at the simple, net or arithmetic returns $(S_{t+1} - S_t)/S(t)$, but here we look at the logarithms of the returns (geometric returns). In the celebrated Black-Scholes setting based on geometric Brownian motion, the log-returns are independent and normally distributed. Using the fact that `Tyres[-1]` is the `Tyres` vector without its first element, we can calculate the log-returns without using a for-loop:

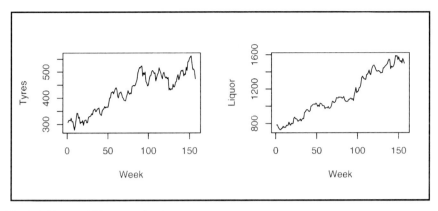

Fig. A.1 Tyres and Liquor stock prices

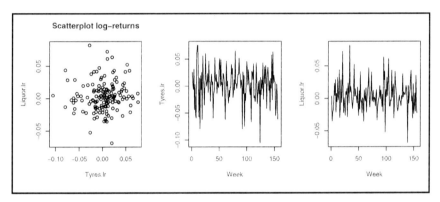

Fig. A.2 Scatterplot and Time series plot of logreturns

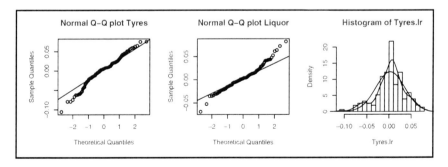

Fig. A.3 Normal Q-Q plots and a histogram of logreturns

```
Tyres.lr <- log(Tyres[-1]/Tyres[-length(Tyres)])
Liquor.lr <- log(Liquor[-1]/Liquor[-length(Liquor)])
```

To produce plots of the log-returns, for Tyres against Liquor and for both against time, see Figure A.2, do:

```
par(mfrow=c(1,3))
plot(Tyres.lr, Liquor.lr, main="Scatterplot log-returns")
plot(Tyres.lr, xlab="Week", type="l")
plot(Liquor.lr, xlab="Week", type="l")
```

To judge if the marginal distributions are indeed normal, we can do a visual test for normality by inspecting normal Q-Q plots, see Figure A.3. These plots are produced by the calls

```
par(mfrow=c(1,3))
qqnorm(Tyres.lr, main="Normal Q-Q plot Tyres")
qqline(Tyres.lr)
qqnorm(Liquor.lr, main="Normal Q-Q plot Liquor")
qqline(Liquor.lr)
```

In Q-Q plots, the sample quantiles are plotted against the theoretical quantiles. If the normal Q-Q plots are close to a straight line, the marginals might well be normal. In our cases, the fit is not very good; the diagonal line through the first and third quartile (produced by qqline) is not followed closely. Since the sample quantiles in the tails are larger than in case of normality, the tails of the distributions are too 'thick' for normal distributions.

Other graphical means to judge normality are plotting a histogram and comparing it to a fitted normal density. The third graph in Figure A.3 can be produced as follows:

```
hist(Tyres.lr, prob=T, breaks=21)
curve(dnorm(x, mean=mean(Tyres.lr), sd=sd(Tyres.lr)), add=T)
lines(density(Tyres.lr))
```

In the final line, we add a so called *kernel density estimate*, see ?density. A kernel density estimator of a density is easy to visualize conceptually. As a kernel function we take for example a normal density with some suitably chosen standard deviation and divided by n. We place it centered at each of the n data points, and then simply add all of them together to form a kernel density estimate. In fact if $\{y_1, \ldots, y_n\}$ are the realizations of a random sample $Y_1, \ldots, Y_n$ from a continuous random variable Y, and X is a discrete r.v. having probabilities $\Pr[X = y_i] = \frac{1}{n}, i = 1, \ldots, n$, then the cdf of X is the empirical cdf of $Y_1, \ldots, Y_n$. If additionally, U is a standard normal random variable, independent of X, then the cdf of $X + \sigma U$ is a kernel density estimate for the cdf of Y. If the estimated kernel density is very similar to the fitted normal density, we can conclude that the log-returns might well be normal. For log-returns of weekly stock prices, the normality assumption is of course more plausible than for daily prices, because of the Central Limit Theorem.

A statistical test for normality of a sample $X_1, \ldots, X_n$ is the test known as the Jarque-Bera test. Its test statistic *JB* is defined as

$$JB = \frac{n}{6}\left(S^2 + \frac{K^2}{4}\right). \tag{A.1}$$

Here n is the number of observations (or degrees of freedom in general), S is the sample skewness and K is the sample kurtosis. They are computed as

$$S = \frac{\widehat{\mu}_3}{\widehat{\sigma}^3} = \frac{\frac{1}{n}\sum_{i=1}^{n}\left(X_i - \overline{X}\right)^3}{\left(\frac{1}{n}\sum_{i=1}^{n}\left(X_i - \overline{X}\right)^2\right)^{3/2}};$$

$$K = \frac{\widehat{\mu}_4}{\widehat{\sigma}^4} - 3 = \frac{\frac{1}{n}\sum_{i=1}^{n}\left(X_i - \overline{X}\right)^4}{\left(\frac{1}{n}\sum_{i=1}^{n}\left(X_i - \overline{X}\right)^2\right)^2} - 3, \tag{A.2}$$

where $\overline{X} = \frac{1}{n}\sum X_i$ is the sample mean and $\widehat{\sigma}^2 = \frac{1}{n}\sum(X_i - \overline{X})^2$. The statistic JB approximately has a $\chi^2(2)$ distribution; asymptotically, under the null-hypothesis both its terms can be shown to be independent squares of standard normal random variables. If either skewness or kurtosis deviates from its 'normal' value 0, JB will be large. The critical value at 95% level is 6.0. To test if our Tyres logreturns possibly arose from a normal distribution, do

```
x <- Tyres.lr - mean(Tyres.lr)
m2 <- mean(x^2); m3 <- mean(x^3); m4 <- mean(x^4)
S2 <- m3^2/m2^3; K <- m4/m2^2 - 3
JB <- length(x)/6 * (S2 + K^2/4)
p.value <- 1-pchisq(JB, df=2)
##JB = 9.305275; p.value = 0.009536414
```

The test statistic $JB = 9.3$ is larger than 6.0, so normality of the logreturns is rejected. The Liquor logreturns produce an even larger $JB = 18.9$.

In the sequel, however, we will proceed as if the log-returns are not just marginally normal, but in fact bivariate normal. The plots in Figure A.3 say nothing about the stocks having a joint normal distribution. But to see if this is a valid assumption, we can look at scatterplots of the log-returns, see Figure A.2.

The parameters $\mu_T, \sigma_T, \mu_L, \sigma_L$ and $\rho_{T,L}$ of the bivariate normal distribution can easily be estimated, for example by:

```
Tyres.lr.mean <- mean(Tyres.lr)
Tyres.lr.sd <- sd(Tyres.lr)
Liquor.lr.mean <- mean(Liquor.lr)
Liquor.lr.sd <- sd(Liquor.lr)
Tyres.Liquor.lr.corr <- cor(Tyres.lr, Liquor.lr)
## Results in 0.00278 0.03226 0.00418 0.02235 0.06142
```

Suppose we want to purchase a portfolio of equal parts of Tyres stocks at the current price 474 and Liquor stocks at price 1498. Then we are interested, for example, in its future performance over a horizon of two calendar years, or 104 weeks. Based on our data, we assume that the weekly log-returns (X_i, Y_i), $i = 158, \ldots, 261$, have a bivariate normal distribution with parameters as computed above, and are independent for different i. Writing

$$X = X_{158} + \cdots + X_{261}; \quad Y = Y_{158} + \cdots + Y_{261}, \tag{A.3}$$

the random variable to be predicted is

$$S = 474e^X + 1498e^Y. \tag{A.4}$$

By the fact that $\mathrm{Cov}(\sum X_i, \sum Y_i) = \sum \mathrm{Cov}(X_i, Y_i)$ if X_i, Y_j are independent when $i \neq j$, estimates of the parameters of (X, Y) can be computed by:

```
Periods <- 104
mean.X <- Periods * Tyres.lr.mean
mean.Y <- Periods * Liquor.lr.mean
sd.X <- sqrt(Periods * Tyres.lr.sd^2)
sd.Y <- sqrt(Periods * Liquor.lr.sd^2)
cov.XY <- Periods * Tyres.Liquor.lr.corr *
          Tyres.lr.sd * Liquor.lr.sd
r.XY <- cov.XY / sd.X / sd.Y
```

Since X and Y are bivariate normal, S in (A.4) is a sum of dependent lognormal random variables. To compute the cdf and quantiles of S is a tough problem, see also Section 7.7. One way to proceed is by just simulating a lot of outcomes of S and looking at sample quantiles instead of theoretical quantiles. For that, we need a method to generate drawings from (X, Y)

To generate a sample of multivariate random normal n-tuples with arbitrary mean and variance matrix, we can use the R function mvrnorm. It is to be found in the library MASS consisting of objects associated with Venables and Ripley's (2002) book 'Modern Applied Statistics with S'. Below, as an illustration, we will explain the Cholesky decomposition method in a bivariate setting with $n = 2$.

Let U and V be independent for all i standard normal. Then for all real α, we have

$$r(U, U + \alpha V) = \frac{\mathrm{Cov}[U, U + \alpha V]}{\sigma_U \sigma_{U+\alpha V}} = \frac{1}{\sqrt{1 + \alpha^2}}, \tag{A.5}$$

meaning that for $r > 0$, the correlation of U and $U + \alpha V$ equals r if we take

$$\alpha = \sqrt{1/r^2 - 1}. \tag{A.6}$$

Then $W = r \cdot (U + \alpha V)$ is standard normal, too. It has correlation r with U also in case $r < 0$; for $r = 0$, take $W = V$. Finally let

$$X' = \mathrm{E}[X] + U \sqrt{\mathrm{Var}[X]}; \quad Y' = \mathrm{E}[Y] + W \sqrt{\mathrm{Var}[Y]}, \tag{A.7}$$

then $S' := 474e^{X'} + 1498e^{Y'} \sim S$ with S as in (A.4).

So the problem of drawing pseudo-random outcomes from the distribution of S has been reduced to generating streams of univariate independent standard normal random variables U and V. This can be achieved in R by simply calling the function rnorm. In its standard mode, it applies the inverse standard normal cdf $\Phi^{-1}(\cdot)$ to

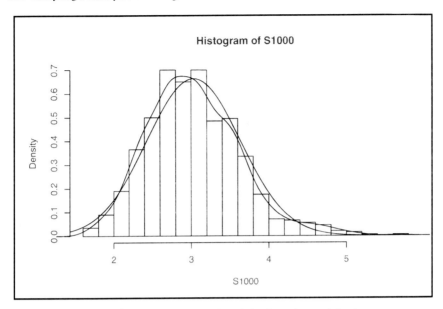

Fig. A.4 Histogram of $S/1000$, with estimated kernel density and normal density

iid uniform$(0,1)$ arguments $U_1, U_2, \ldots$; the latter can be obtained by invoking the function `runif`.

As a result of all this, the following code generates a sample of one thousand outcomes of S, in which (X, Y) has a bivariate normal distribution with parameters as computed above:

```
set.seed(2525) ## initialize the random number generator
U <- rnorm(1000); V  <- rnorm(1000)
alpha <- sqrt(1/r.XY^2-1)  * sign(r.XY)
X <- mean.X + sd.X * U
Y <- mean.Y + sd.Y * (U + alpha * V) * r.XY
S <- Tyres[length(Tyres)]*exp(X) +
     Liquor[length(Liquor)]*exp(Y)
```

See Figure A.4 for the histogram with estimated kernel density and fitted normal density, produced by

```
par(mfrow=c(1,1)); S1000 <- S/1000
hist(S1000, breaks=21, prob=T)
lines(density(S1000))
curve(dnorm(x, mean=mean(S1000), sd=sd(S1000)), add=T)
```

Sample quantiles of S, as a percentage of the purchase price, are found as follows:

```
> pr <- c(2.5, 5, 10, 25, 50, 75, 90, 95, 97.5, 99, 99.5)/100
> S <- S/(Tyres[length(Tyres)] + Liquor[length(Liquor)])
> round(100*quantile(S, pr))
 2.5%    5%   10%   25%   50%   75%   90%   95% 97.5%   99% 99.5%
  102   109   117   133   152   173   191   207   227   243   252
```

So judging from the 2.5%, 50% and the 97.5% quantiles, the bi-annual returns are roughly 50% ± 25%. And a portfolio of one stock Tyres and one stock Liquor, with current price 474 + 1498 = 1972, will be worth at least 152% of that in two years with probability 50%, between 109% and 207% with probability 90%, and it will decrease in value in about one in forty cases. But in one out of 200 cases, it will sell for more than 2.5 times the original value. This is all under the stipulation that future stock prices will follow our estimated model for the past prices. Apart from the fact that (joint) normality of the log-returns was firmly rejected, a quick look at Figure A.1 shows that for the Tyres stock, for example, it is entirely plausible that the increasing trend in the price stopped around week 100.

A.3 Generating a pseudo-random insurance portfolio

Assume we want to generate pseudo-random data for an automobile insurance portfolio of which the policies have the following risk factors:

- sx two gender levels, in proportions 6 : 4;
- jb job class, three levels in proportions 3 : 2 : 1;
- re region of residence, with three levels, in proportions 3 : 3 : 4;
- tp type of car, with three levels, in proportions 7 : 5 : 8.

Most factors are independent. For example, the probability of having both jb=1 and re=3 is $\frac{3}{3+2+1} \cdot \frac{4}{3+3+4} = \frac{1}{5}$. The last two factors, however, are mutually *correlated*, which means that the probabilities of levels for re depend on the level of tp, and vice versa. The combined factor levels have the following joint probabilities:

	tp	1	2	3
re	1	0.10	0.05	0.15
	2	0.15	0.10	0.05
	3	0.10	0.10	0.20

Drawing the first two risk factors is easy to do using R's sample function.

```
n.obs <- 10000; set.seed(4)
sx <- as.factor(sample(1:2, n.obs, repl=T, prob=c(6,4)))
jb <- as.factor(sample(1:3, n.obs, repl=T, prob=c(3,2,1)))
```

The first argument of sample gives the possible values, the second the number of drawings to be generated. Drawing without replacement is the default, so we must add repl=T. The prob=c(6,4) parameter indicates the relative probabilities of the values, so the probability of gender level 1 above equals 60%. All risk factors are to be treated as *factors*, that is, their values 1, 2, ... are merely class labels.

To draw the values of the correlated factors tp and re with the given probabilities, we first draw a combination of these two factors (re.tp, with 9 levels), and recode levels 1,4,7 to `tp=1`, levels 2,5,8 to `tp=2` and levels 3,6,9 to `tp=3`, and analogously for `re`. To show that what we did is what we wanted, we make a table of the numbers of policies, broken down by these two factors. As we no longer need it, we delete the auxiliary vector `re.tp` at the end.

```
re.tp <- sample(1:9, n.obs, repl=T, prob=c(2,1,3,3,2,1,2,2,4))
tp <- as.factor(c(1,2,3,1,2,3,1,2,3)[re.tp])
re <- as.factor(c(1,1,1,2,2,2,3,3,3)[re.tp])
table(list(region=re, type=tp)); rm(re.tp)
```

Many policies are terminated before the end of the policy year. In that case, the exposure-to-risk time is some fraction of a year. In fact, we want, whatever the value of the other factors, 80% of the policies to have an exposure time `mo` of 12 months (four quarters), 10% 6 months, and the remaining 10% 3 months.

```
mo <- 3 * sample(1:4, n.obs, repl=T, prob=c(1,1,0,8))
```

For each policy, let μ_i denote the average number of claims per policy year. It is a function of the risk factors given only; each factor operates multiplicatively and independently, so there is no *interaction* in the sense that certain combinations of risk factors lead to a value different from that brought on by the main effects.

We assume the claim numbers of each policy to follow a Poisson process, see Chapter 4. To get the actual Poisson parameter with policy i, each μ_i must be multiplied by the exposure in years, see (3.73). As one sees below, in our fictitious portfolio, an increase by one in any factor level leads to 20% more claims on average, except that `jb` does not have any influence. The base level is 0.05.

```
mu <- 0.05 * c(1,1.2)[sx] *
             c(1,1,1)[jb] *
             c(1,1.2,1.44)[re] *
             c(1,1.2,1.44)[tp]
y <- rpois(n.obs, mu * mo/12)
```

In our example, the sample size was 10000. In practice, much larger portfolio sizes occur, even millions. R holds objects it is using in memory. So then the amount of random access memory (RAM) in the computer starts playing a role. Also for example fitting a generalized linear model (GLM) takes minutes instead of milliseconds. Therefore it would be advantageous if we could, without too much loss of information, look at group totals instead of the individual policies. We split up our portfolio into cells of policies having the same levels for all four risk factors, therefore also the same average annual number of claims. The number of such cells is $2 \times 3 \times 3 \times 3 = 54$. If the weights w_k denote the total number of exposure years in cell k, and μ_k the Poisson parameter for each policy in force one full year in that cell, it is easy to show that the total number of claims has a Poisson$(w_k\mu_k)$ distribution.

In this situation, the individual claim numbers do not give more information about the value of the Poisson parameter in a cell than their total does; it can be shown that, given the total number of claims in a cell, the distribution of the individual claim numbers, or any statistic derived from them, does not depend on this parameter. This last property characterizes a *sufficient statistic* with respect to a parameter. So if we assume the model (Poisson process) to be correct and only the parameters to be unknown, to estimate these parameters we may work with an aggregated version of the portfolio just as well as with the full portfolio.

All this only makes sense if the risk factors have only few levels such as here; the original price of the vehicle would, for example, have to be rounded to multiples of 1000, and weight classes containing intervals of actual car weights would have to be introduced. To condense the data, we use R's aggregate function:

```
aggr <- aggregate(list(Expo=mo/12,nCl=y,nPol=1),
                  list(Jb=jb,Tp=tp,Re=re,Sx=sx),  sum)
aggr[sort(sample(1:54,20)),]
```

For each combination of the risk factors in the second parameter of aggregate, a cell is created and stored as a row of the data frame aggr. In addition the sum function is applied to the list of quantities specified in the first parameter. The last line prints all columns of the data frame for a sorted sample of 20 of the 54 cells. There is a row for each cell, and the following columns:

- Jb, Tp, Re, Sx are the factor levels for the cell. Jb has levels $1,2,3,1,2,3,\ldots$; the first 27 cells have level 1 for Sx;
- Expo is the accumulated exposure time (in years) of all policies in the cell;
- nCl represents the total number of claims made in a particular year;
- nPol counts the number of policies with these levels of the risk factors.

Note that, for example, jb is a vector of length n.obs, while the column aggr$Jb of the data frame has length only 54. In Section 9.5, a quite similar portfolio is studied using the techniques of Chapter 9.

Appendix B
Hints for the exercises

*I was gratified to be able to answer promptly. I said I don't
know — Mark Twain (1835 - 1910)
It's so much easier to suggest solutions when you don't know
too much about the problem — Malcolm Forbes (1919 - 1990)*

CHAPTER 1

Section 1.2

1. Use the following characterization of convexity: a function $v(x)$ is convex if, and only if, for every x_0 a line $l_0(x) = a_0 x + b_0$ exists, such that $l_0(x_0) = v(x_0)$ and moreover $l_0(x) \le v(x)$ for all x [usually, $l_0(\cdot)$ is a tangent line of $v(\cdot)$]. Take $x_0 = \mathrm{E}[X]$. If $v(x) = x^2$, we get $\mathrm{Var}[X] \ge 0$.

2. Consider especially the random variables X with $\Pr[X = x_0 \pm h] = 0.5$.

3. Apply the previous exercise to prove that both functions $+v$ and $-v$ are convex.

4. Examine the inequalities $\mathrm{E}[u(X)] > \mathrm{E}[u(Y)]$ and $\mathrm{E}[u(X)] > u(w)$. X is preferred over w for $w < 625$.

5. Apply Jensen's inequality to (1.11).

6. $P^+ = 20$; $P^+ \approx 19.98$.

7. $W = 161.5$.

8. Taking $w = 0$, $u(0) = 0$ and $u(1) = 1$ gives $u(2) > 2$, $u(4) = 2u(2)$ and $u(8) < 2u(4)$. There are x with $u''(x) > 0$ and with $u''(x) < 0$.

9. $P^+[X] = 3/4$, $P^+[2X] = 4/3$.

Section 1.3

4. Use $\lim_{\alpha \downarrow 0} \frac{\log(m_X(\alpha)) - \log(m_X(0))}{\alpha} = \frac{d \log(m_X(\alpha))}{d\alpha}\big|_{\alpha=0}$ or use a Taylor series argument.

5. See Table A at the end for the mgf of X.

6. $P_X = 412.5 < P_Y$; $\alpha > 0.008$.

7. Logarithmic. $P = 4$ solves $\mathrm{E}[u(w - P + 2^N)] = u(w)$.

8. $\alpha \ge 0.05$. Dimension of α is money^{-1}.

9. What is the mgf of X?

10. Logarithmic utility. Use l'Hôpital's rule, or write $w^c = e^{c \log w}$ and use a Taylor expansion.

11. Power utility functions. What is $\frac{d \log u'(w)}{dw}$?

12. Assume $\text{Var}[X] > 0$ and $\alpha > 0$, so $\text{Var}[e^{\alpha X}] > 0$, and use Jensen's inequality.

13. $P^+ = 50.98$, $P^- = 50.64$.

Section 1.4

1. Linear on $(-\infty, 2]$ with $\pi(0) = 2.5$, $\pi(2) = 0.5$; $\pi(x) = (4-x)^2/8$ on $[2,4]$, and $\pi(x) = 0$ on $[4, \infty)$. In your sketch it should be visible that $\pi'(2+0) \neq \pi'(2-0)$.

2. Use (1.38) to find $f_S(d)$.

3. Use partial integration.

4. Use that when the variance is fixed, stop-loss is optimal; next apply the previous exercise.

5. Use (1.39).

6. $E[I(X)] = E[(X-d)_+] + d\Pr[X \geq d]$.

7. Based on a sample of $1\,000\,000$ standard normal elements, the R-program computes estimates of $E[X\,|\,X>1]$, $E[XI_{X>1}]$ and $E[(X-1)_+]$. The theoretical values can be derived using the previous exercise.

CHAPTER 2

Section 2.2

1. a) $E[X] = 1/2$; $\text{Var}[X] = 9/4$; b) $E[X] = 1/2$; $\text{Var}[X] = 37/12$.

2. $E[Y] = 7/4$, $\text{Var}[Y] = 77/48$.

3. $P^+ = 5.996 \neq 1.1 \times 100\log(19/18)$, so not quite perfectly.

4. $E[X] = 60$; $m_X(t) = 0.9e^0 + 0.02e^{1000t} + \int_0^{1000} 0.00008e^{tx}\mathrm{d}x = \cdots$

5. Condition on $I = 1$ and $I = 0$.

6. $IX + (1-I)Y$ for $I \sim$ Bernoulli(0.5), $X \equiv 2$ and $Y \sim$ uniform$(2,4)$, independent.

7. $c = 1/3$, $\mathrm{d}G(1) = \mathrm{d}G(2) = 1/2$, $\mathrm{d}H(x) = \mathrm{d}x/2$ on $(0,1) \cup (2,3)$.

8. $E[T] = E[Z]$, $E[T^2] \neq E[Z^2]$.

9. $N(0, q^2 + (1-q)^2)$ and $N(0,1)$.

10. Use that $W = \frac{2}{3}X_1 + \frac{1}{3}W'$ with $W' \sim W$ independent of X_1, see (2.25). For the mgf, use $m_{W-1/2}(t) = \prod E\left[\exp(t(2X_i - 1)/3^i)\right] = \ldots$.

Section 2.3

1. Cf. Table 2.1.

2. Total number of multiplications is quadratic: $6n^2 - 15n + 12$, $n \geq 3$.

3. Write (2.30) as $\varphi(s; \mu_1 + \mu_2, \sigma_1^2 + \sigma_2^2) \times \int \varphi(x; \mu_3, \sigma_3^2)\mathrm{d}x$.

4. For the second part, use induction, the convolution formula, and the relation $\binom{n}{h} + \binom{n}{h-1} = \binom{n+1}{h}$ for all n and h.

Section 2.4

1. $f_S(s) = 2(e^{-s} - e^{-2s})$; $m_S(t) = \frac{2}{1-t} - \frac{2}{2-t}$.

3. $\kappa_4 = E[(X - \mu_1)^4] - 3\sigma^4$.

4. See Table A at the end of the book for the mgfs.

5. $\kappa_3 = 0$, $\kappa_4 = -0.1$.

6. Use (2.49) and Table A.

8. X is symmetric around μ if $X - \mu \sim \mu - X$. There are two solutions: $p = 0.04087593$ and $p = 0.4264056$.

9. $(1 - 2q)/\sqrt{q(1-q)}$ (see Table A). If X is symmetric, then the third central moment equals 0, therefore $q \in \{0, 0.5, 1\}$ must hold. Symmetry holds for all three of these q-values.

10. Use (2.50) and Table A.

11. The cumulants are the coefficients of $t^j/j!$ in the cgf.

12. Their pgfs are polynomials of degree n, and they are identical only if all their coefficients are the same.

13. Show that X/δ and Y/δ have the same pgf.

14. Where is the mgf defined, where the characteristic function? Sometimes this function can be extended to all complex numbers, like $(1 - t)^{-1}$ for the exponential distribution. $E[e^{itX}] = E[e^{-itX}]$ implies that the imaginary part of the functions must be equal to zero.

15. Use Exercise 11. For symmetry, $\Pr[Z = 0] = \Pr[Z = 10]$ is necessary. Prove that Z is symmetric whenever this is the case.

16. $\delta = \sqrt[3]{2}$.

17. Show that $g_X^{(n)}(1) = E[X(X - 1)\cdots(X - n + 1)]$, and argue that the raw moments can be computed from these so-called factorial moments. See (2.50).

Section 2.5

1. You should get the following results (fill in the question marks yourself):

Argument:	3 − 0	3 + 0	3.5	4 − 0	4 + 0
Exact:	.080	?	.019	?	.004
NP:	.045	?	.023	?	.011
Gamma:	.042	?	.021	?	.010
CLT:	.023	?	.006	?	.001

2. Solve $x = s + \gamma(s^2 - 1)/6$ for s. Verify if this inversion is allowed!

4. Use l'Hôpital's rule to prove that $\lim_{\gamma \downarrow 0} \left[\sqrt{9/\gamma^2 + 6x/\gamma + 1} - 3/\gamma\right] = x$. Take $X^* = (X - \mu)/\sigma$, then approximate $\Pr[X^* \le z]$ by $\Pr[Z^* \le z]$, where $Z^* = (Z - \alpha/\beta)\beta/\sqrt{\alpha}$, and $Z \sim$ gamma(α, β) with skewness γ, therefore $\alpha = 1/\gamma^2$. Then for $\alpha \to \infty$, we have $\Pr[Z^* \le z] \to \Phi(z)$ because of the CLT.

5. Using (2.59), we see that the critical value at $1 - \varepsilon$ is $18 + 6(y + (y^2 - 1)/9)$ if $\Phi(y) = 1 - \varepsilon$.

6. The χ^2 distribution is a special case of the (translated) gamma approximation.

7. If α is integer, Poisson-probabilities can be used to find gamma-cdfs.

8. For example for $\varepsilon = 0.95$: a table gives 28.9, (2.26) gives 28.59. The NP approximation from Exercise 5 gives 28.63.

9. Loading = 21.14%.

10. Loading = 21.60%.

11. For $x = -1$ we find $(3/\gamma - 1)^2 \geq 0$ under the square-root sign.

12. Using Table A one finds $\gamma = 0, 2, 4, 6, 14, \infty$ for the skewnesses in cases (i)–(vi).

13. Let X_1 be a claim of type 1, then $\Pr[X_1 = 0] = 1 - q_1$, $\Pr[X_1 = j] = q_1 p_1(j)$, $j = 1, 3$. $E[S_1] = 20$, $\mathrm{Var}[S_1] = 49.6$, capital is $E[S] + 1.645\sqrt{\mathrm{Var}[S]} = 99.999$.

14. Use that $E[U'] = \int u' f_U(u) \, du = \dots$.

16. You should get 0.0103, 0.0120, 0.0113 in cases (i)–(iii) for translated gamma, and 0.0120, 0.0138, 0.0113 in cases (iii)–(v) for NP.

17. Avoid taking square roots of negative numbers.

21. E.g. $E[Y^2] = E[U^2] + \frac{\gamma}{3}E[U(U^2 - 1)] + \frac{\gamma^2}{36}E[(U^2 - 1)^2] = \dots$

Section 2.6

1. $1 - \Phi(g(d))$ with $g^2(d) - \frac{\{B - 380 + 10d\}^2}{297 + 49.5d^2}$, $d \in [2, 3]$.

2. Maximize $g^2(d)$.

3. $1 - \Phi(1.984) = .0235$.

CHAPTER 3

Section 3.2

1. For Poisson(λ): $E[S] = \lambda\mu_1$, $\mathrm{Var}[S] = \lambda\mu_2$, $m_S(t) = e^{\lambda(m_X(t)-1)}$.

2. Use (2.50).

3. Let N' denote the number of females, then we have $N' = B_1 + \cdots + B_N$, if N is the number of eggs and $B_i = 1$ if from the ith egg, a female hatches. Now use (3.5) to prove that $N' \sim$ Poisson(λp).

4. $\Pr[S = 0, 1, 2, 3, 4] = e^{-2}, 0.2e^{-2}, 0.42e^{-2}, 0.681333e^{-2}, 1.0080667e^{-2}$.

5. $f(4, 5, 6) = 0.2232, 0.1728, 0.0864$; $E[S] = 3.2$, $\mathrm{Var}[S] = 3.04$.

6. $E[S] = 3.2$ (see Exercise 5), $\mathrm{Var}[S] = 5.6$.

7. Use mathematical induction. Or: prove that lhs=rhs by inspecting the derivatives as well as one value, for example at $x = 0$.

Section 3.3

1. Examine the mgfs; fill in $p = 1 - \lambda/r$ in the negative binomial mgf and let $r \to \infty$. Use that $\lim_{r \to \infty}(1 - \lambda/r)^r = e^{-\lambda}$.

2. In 3.3.1 if Λ is degenerate; in 3.3.2 if $c \downarrow 0$.

3. Compare the claim numbers and the claim sizes.

Section 3.4

1. See Example 3.4.3.

2. If $x_1 = x_2$, the frequency of this claim amount is $N_1 + N_2$.

3. $p(0) = p(1) = p(2)/2 = p(3) = p(4)$.

4. Show that $\Pr[N' = n, N - N' = m] = \Pr[N' = n] \Pr[N - N' = m]$ for all n and m. Or: apply Theorem 3.4.2 with $x_1 = 0$ and $x_2 = 1$.

5. Show that $\Pr[N_1 = 1] \Pr[N_2 = 1] = 0.3968 \times 0.3792 \neq \Pr[N_1 = 1, N_2 = 1] = 0.144$. (Note that Theorem 3.4.2 was proved for the Poisson-case only.)

6. $S = x_1 N_1 + x_2 N_2 + \cdots$ and $S_0 = 0 N'_0 + x_1 N'_1 + x_2 N'_2 + \cdots$ with $N_j, N'_j \sim \cdots$

7. Adding the second column takes $n^2/2$ operations, the third $n^2/3$, in total . . .

Section 3.5

1. $f(s) = \frac{1}{s}[0.2f(s-1) + 0.8f(s-2) + 1.8f(s-3) + 3.2f(s-4)]; f(0) = e^{-2}$

2. Verify $s = 0$ separately; for $s > 0$, use (3.15) and induction.

3. Check if every point in the (a,b) plane has been dealt with. Make a sketch.

4. There are $2t$ multiplications $\lambda h p(h)$ for $h = 1, \ldots, t$. For $m > t$: $t(t+1)/2$, for $m \le t$: $m(m + 1)/2 + m(t - m)$. Asymptotically the number of operations increases linearly with t if the maximal claim size is finite, and quadratically otherwise.

5. $E[N] = \sum_n n q_n = \sum_{n \ge 1} n(a + b/n)q_{n-1} = a \sum_{n \ge 1} (n-1)q_{n-1} + a + b$, and so on.

6. Interpolate between $\pi(2)$ and $\pi(3)$. $d = 2.548$. [The stop-loss premiums are linear because the cdf is constant.]

7. Use Panjer and interpolation.

8. $S_2 \sim N_1 + 3N_3$ with $N_1 \sim$ Poisson(2) and $N_3 \sim$ Poisson(1). Should you interpolate to determine the cdf?

9. $\lambda p(1) = \alpha$, $2\lambda p(2) = 2\alpha$ and $p(1) + p(2) = 1 - p(0)$.

10. $\pi(2.5) = 1.4014$.

11. Subtract (3.34) from $E[(S - 0)_+]$.

12. Start with $E[(S - (d-1))_+^2] - E[(S - d)_+^2] = \cdots$. If $p(d)$ is the expression to be computed, then $p(d-1) - p(d) = 2\pi(d-1) - 1 + F(d-1)$.

13. Implement (3.27) instead of (3.31).

14. Logarithmic$(1 - p)$, see (3.16); use $p^r = e^{r \log p} = 1 + r \log p + O(r^2)$.

16. Use relation (3.34). The function rev reverses a vector, while cumsum computes its cumulative sums.

Section 3.6

1. Think dice. After the first two function calls, y[1:64] contains the probabilities of one die showing $\{0,\dots,63\}$ pips.

Section 3.7

1. CLT: 0.977, gamma: 0.968, NP: 0.968.
2. $\alpha = 15$, $\beta = 0.25$, $x_0 = -20$. NP: $F_S(67.76) \approx 0.95$, and $F_S(E[S]+3\sqrt{\text{Var}[S]}) \approx 0.994$ (Note: $\Phi(3) = 0.999$).

Section 3.8

1. If S^* is the collective model approximation with $\lambda_j = -\log(1-q_j)$, prove that $\lambda_j > q_j$, hence $E[S^*] = \sum_j \lambda_j b_j > E[S]$; analogously for the variance.
2. $\tilde{S}$: E $= 2.25$, Var $= 3.6875$, $\gamma = 6.41655\sigma^{-3} = 0.906$. $S \sim$ compound Poisson with $\lambda = 1.5$, $p(1) = p(2) = 0.5$, therefore E $= 2.25$, Var $= 3.75$, $\gamma = 6.75\sigma^{-3} = 0.930$. $\tilde{S}$: $\alpha = 4.871$, $\beta = 1.149$, $x_0 = -1.988$. S: $\alpha = 4.630$, $\beta = 1.111$ and $x_0 = -1.917$.
4. The second. The ratio of the resulting variances is approximately 80%.
5. Use the fact that the first factor of the terms in the sum decreases with x.
6. Max$[S] = 3000$, Max$[T] = 4000$; $E[S] = E[T] = 30$; Var$[S] = 49.5$, Var$[T] = 49.55$; the claim number distribution is binomial$(2000, 0.01)$ for both; $S \sim$ weighted sum of binomial random variables, $T \sim$ compound binomial. If $B_i \sim$ Poisson, then $S \equiv T \sim$ compound Poisson.
7. Compound Poisson$(10\Pr[X > \beta])$, with claims $\sim$ uniform$(0, 2000 - \beta)$. Equivalently, it is compound Poisson(10) with claims $\sim (X - \beta)_+$.
8. P1: $z_1^2 n_1 q_1(1 - q_1) + \cdots$; P2: larger. 'The' collective model: equal. The 'open' collective model: different.
9. Replacing the claims on a contract of type 1 by a compound Poisson(1) number of such claims leads to a random variable $\sim N_1 + 3N_3$ with $N_k \sim$ Poisson$(q_1 p_1(k))$, $k = 1, 3$. So $T \sim M_1 + 2M_2 + 3M_3$ with $M_1 \sim$ Poisson(25), $M_2 \sim$ Poisson(20), $M_3 \sim$ Poisson(5). Panjer: $f(s) = \frac{1}{s}\sum_h h\lambda p(h)f(s-h) = \frac{1}{s}[25f(s-1) + 40f(s-2) + 15f(s-3)]$. Apply NP or gamma.
10. Binomial(n,q), Poisson(nq), Poisson$(-n\log(1-q))$, no.

Section 3.9

1. Additionally use that $E[Y^j] = m_{\log Y}(j)$, where the mgfs can be found in Table A.
2. $\beta X \sim$ gamma$(\alpha, 1)$ if $X \sim$ gamma(α, β); $X/x_0 \sim$ Pareto$(\alpha, 1)$ if $X \sim$ Pareto(α, x_0); $e^{-\mu}X \sim$ Lognormal$(0, \sigma^2)$ if $X \sim$ Lognormal(μ, σ^2); $\beta X \sim$ IG$(\alpha, 1)$ if $X \sim$ IG(α, β).
4. A vital step is that $p(x) \geq p(0)e^{-\beta x}$ for all $x \geq 0$.
5. $\Pr[Z - z > y | Z > z] = q(z)e^{-\alpha y} + (1 - q(z))e^{-\beta y}$ if $q(z) = \frac{qe^{-\alpha z}}{qe^{-\alpha z}+(1-q)e^{-\beta z}}$. $q(\cdot)$ is monotonous with $q(0) = q$, $q(\infty) = 1$.

6. The median of the lognormal distribution is e^μ. Mode: $f'(x) = 0$ holds for $x = e^{\mu - \sigma^2}$.

7. Mode of IG is $(-3 + \sqrt{9 + 4\alpha^2})/2\alpha$.

9. You should find 6.86647, 6.336969, 5.37502, 4.919928 for these quantiles.

21. Think numbers of successes/failures. Recall Example 2.5.2.

22. Use the factorization theorem and (3.86).

23. $X U^{1/\alpha} \sim$ gamma$(\alpha, 1)$.

24. We get Bernoulli(p), binomial(n, p), N$(0, 1)$, N(μ, σ^2), lognormal(μ, σ^2), exponential(1), exponential(β), gamma(n, β), Pareto$(1, 1)$, Pareto(β, x_0), Weibull(α, β) and Gompertz(b, c) r.v.'s (in this order).

25. By Example 3.3.1, the N_i r.v.'s are negative binomial with parameters $\alpha, (\beta/w_i)/(\beta/w_i + 1)$, with $\alpha = 1/40$, $\beta = 1/2$.

28. $X \sim$ logistic(μ, σ) if $F_X(d) = e^t/(e^t + 1)$, where $t = (d - \mu)/\sigma$, so $F_X(d) = p \iff d = \ldots$ Then use $Y = F_X^{-1}(U) = \ldots$ with $U \sim$ uniform$(0, 1)$.

Section 3.10

1. 1000×0.0004; $1000 \times .0070$ (NP) or $1000 \times .0068$ (Translated gamma).

2. Subtract from $E[(S - 0)_+] = E[S]$.

3. Work with $(X - \mu)/\sigma$ rather than with X. $E[(X - \mu)_+] = \sigma\varphi(0) = \cdots$

4. To compute $E[(S - (S - d)_+)^2]$, approximate $E[(S - d)^2]$ just as in (3.29).

5. $\lambda x_0^\alpha d^{1-\alpha}/(\alpha - 1)$.

6. Write $X = e^Y$, so $Y \sim$ N(μ, σ^2), and find $E[(e^Y - d)_+]$. Or verify the derivative.

7. $\pi(d)$ is convex.

8. Determine the left and right hand derivatives of $E[(N - d)_+]$ at $d = 1$ from difference ratios. $\Pr[N = 1] = 0.2408$.

9. Use the fact that U is symmetric.

10. Use Exercises 3.2.1 and 3.9.9.

11. Use (3.105) for the gamma distribution, (3.104) for the normal distribution, Exercise 3.10.6 for the lognormal.

Section 3.11

1. Use partial integration and $\int_0^\infty (\mu - t)_+ dt = 0.5\mu^2$. The function $(\mu - t)_+$ consists of two tangent lines to the stop-loss transform.

4. Use and prove that $(x - t)_+ + (y - d)_+ \geq (x + y - (t + d))_+$, and apply induction. Further, use the given rule of thumb to show that the premiums are about equal.

5. Var$[T]/$Var$[S] = 1.081$; $E[(T - d)_+]/E[(S - d)_+] = 1.022, 1.177, 1.221$ for $d = 0.5, 1, 1.5$. Note that $d = 0.5 \approx \mu + \sigma/6$, $d = 1 \approx \mu + \sigma$, $d = 1.5 \approx \mu + 2\sigma$.

6. Take $f(x) = E[(U - x)_+] - E[(W - x)_+]$ and $\delta = 1$; we have $f(x) = 0$, $x \leq 0$.

CHAPTER 4

Section 4.2

1. $\frac{f(t-s)dt}{1-F(t-s)}$.

2. $p_n(t+dt) = p_n(t) + p'_n(t)dt = (1 - \lambda dt)p_n(t) + \lambda dt\, p_{n-1}(t)$. Both sides denote the probability of n claims in $(0, t+dt)$.

Section 4.3

1. See further the remarks preceding (4.11).

2. Use (4.10).

3. $m_X(t) < \infty$ for $t < 3$, and $R = 1$.

4. $\theta = 2.03$.

5. $c = \log m_S(R)/R$ with $R = |\log \varepsilon|/u$.

6. $R = 0.316$.

7. $R - 1; \theta - 1.52 > 0.4$ (or use $dR/d\theta > 0$).

8. Solve θ and R from $1 + (1+\theta)\mu_1 R = m_X(R) = (1-R)^{-2}$ for $0 < R < 1$; this produces $R = [3 + 4\theta - \sqrt{9 + 8\theta}]/[4(1+\theta)]$. No: $R < 1$ must hold.

9. $m_Y(r)$ is finite for $r \le 0$ and $r \le \beta/2$, respectively, and infinite otherwise.

10. Consider $\frac{dR}{dc}$. Then use $\frac{dc}{dR} \ge 0$.

Section 4.4

1. Compare the surpluses for $\theta < 0$ and $\theta = 0$, using the same sizes and times of occurrence of claims.

2. See (4.32): $(1+\theta)^{-1} = 0.5$, therefore $\theta = 1$, $\theta/\{(1+\theta)\mu_1\} = 1$ gives $\mu_1 = 0.5$, hence $X \sim$ exponential(β) with $\beta = 2$; λ is arbitrary. Or: claims $\sim IX$ with $I \sim$ Bernoulli(q).

3. Because of Corollary 4.4.2 we have $R = 1$; no; $(\alpha + \beta e^{-u})^{-1}$.

4. $1 - \psi(0) > \Pr\left[\text{no claim before } \varepsilon/c \text{ \& no ruin starting from } \varepsilon\right] > 0$. Or: $\psi(\varepsilon) < 1$, therefore $R > 0$, therefore $\psi(0) < 1$ by (4.26).

5. $R = 6$ is ruled out since $m_X(6) = \infty$; $R = 0$ is also not feasible. Then, look at $\psi(0)$ and the previous exercise, and at $\psi(u)$ for large u.

6. $R = 0.5$; $c = \frac{2}{5}$.

Section 4.5

1. $U(\tilde{T}) = -1$; $\tilde{\psi}(u) = e^{-\tilde{R}(u+1)}$ with $\tilde{R} = \log(p/q)$.

2. Processes with adjustment coefficient $\tilde{R}$ apparently are only profitable (as regards expected utility) for decision makers that are not too risk averse.

3. It is conceivable that ruin occurs in the continuous model, but not in the discrete model; the reverse is impossible; $\Pr[T \leq \tilde{T}] = 1$ implies that $\tilde{\psi}(u) \leq \psi(u)$ for all u.

4. Use (4.32). $\tilde{\psi}(u) \leq e^{-\tilde{R}u} = e^{-Ru}$ with $R = 1$. But a better bound is $\tilde{\psi}(u) \leq \psi(u)$.

Section 4.6

1. $R_h = \{\theta - \alpha\xi\}/\{(1-\alpha)(1+\theta-\alpha(1+\xi))\}$; relative safety loading after reinsurance: $\{\theta - \alpha\xi\}/\{1-\alpha\}$. α must satisfy $0 \leq \alpha \leq 1$ and $\alpha < \theta/\xi$.

2. Safety loading after reinsurance: $\{\theta - \xi e^{-\beta}\}/\{1 - e^{-\beta}\}$.

3. $\tilde{R} = \{5 - 8\alpha\}/\{2(1-\alpha)^2\}$ is maximal for $\alpha = 0.25$.

4. $R = (1-2\alpha)/\{(1-\alpha)(3-4\alpha)\}$, so $0 \leq \alpha < 0.5$.

Section 4.7

1. $L_1 \sim$ uniform$(0,b)$; $L_1 \sim$ exponential with the same parameter as the claims.

2. $L = 0$ means that one never gets below the initial level.

3. $\psi(0) = \ldots$ as well as $\ldots$, and hence $\ldots$

6. See also Exercise 4.3.1.

Section 4.8

1. $\gamma = 1 + \theta$.

3. $R = 2$, $\theta = -1 + 1/(0.5 + \alpha)$. $\theta > 0 \Longrightarrow \cdots$. Use that $\psi(u)$ decreases if $\psi'(0) < 0$.

4. $\psi(u) = \frac{4}{9}e^{-2u} + \frac{1}{9}e^{-4u}$.

5. e^{-2u}; $I \sim$ Bernoulli$(\frac{1}{9})$.

6. $\psi(u) = \frac{2}{5}e^{-0.3u} - \frac{1}{15}e^{-0.8u}$.

7. $\psi(u) = \frac{5}{8}e^{-u} - \frac{1}{24}e^{-5u}$.

8. One gets a non-increasing step function, see (4.60). A density like this is the one of a mixture of uniform$(0,x)$ distributions; it is unimodal with mode 0.

10. $p(x) = 2(e^{-x} - e^{-2x})$; take care when $R = 2.5$.

11. $c = c_1 + c_2$, $\lambda = \lambda_1 + \lambda_2 = 9$, $p(x) = \frac{1}{9}p_1(x) + \frac{8}{9}p_2(x) = \cdots$, and so on.

Section 4.9

1. $\beta = E[L]/\{E[L^2] - (1+\theta)E^2[L]\}$; $\alpha = (1+\theta)\beta E[L]$.

CHAPTER 5

Section 5.2

1. Take the derivative of (5.6) and set zero.
2. Portfolio premium = 49.17; optimal $u = 104.21$; optimal $R = 0.0287$; premiums for A and B are 5.72 and 1.0287 (variance premium) and 5.90 and 1.0299 (exponential premium).

Section 5.3

1. (a) 1   (b)–(d) $1 + \alpha$ (e) $-\log(1-\alpha)/\alpha$ (h) $-\log\varepsilon$ (i) ∞ (j) $(1-h)^{-1}$.
2. Show that $(\alpha^2 \pi'[X;\alpha])' = \alpha \text{Var}[X_\alpha]$, with X_α the Esscher transform of X with parameter $\alpha > 0$.
3. If $N \sim \text{Poisson}(\lambda)$ and $X \sim \text{gamma}(\alpha,\beta)$, then the premium is $1.1\lambda\alpha/\beta$.
4. $\frac{\lambda}{\gamma}[m_X(\gamma) - 1]$.
5. $\frac{\lambda\alpha}{\beta}[1 + \frac{\gamma}{\beta}(1 + \alpha)]$.
6. Use $\kappa_X(h) = \log \text{E}[e^{hX}]$.
7. Members of the same family with different parameters result.
8. Show: derivative of the Esscher premium = variance of the Esscher transform.
9. $\lambda \text{E}[Xe^{hX}]$.
10. Use Exercise 5.3.6 and a Taylor expansion.
11. Use $\text{E}[e^{\alpha X}] \geq e^{\alpha(b-\varepsilon)} \Pr[X \geq b - \varepsilon]$.
15. Such a mixture is additive.
16. Such a condition is: 'X and Y not positively correlated'.
    $\pi[X+Y] - \pi[X] - \pi[Y] = \cdots \geq 0 \iff |\rho_{XY}| \leq 1$.

Section 5.5

2. Cauchy-Schwarz; check this in any text on mathematical statistics.

Section 5.6

1. TVaR = 9.022945; ES = 0.1651426
2. Use Figures 5.1 and 5.2 and the definitions (5.35) and (5.41).
5. $\text{CTE}[X+Y;0.9] = 0.95$, while $\text{CTE}[X;0.9] = 0.95$.
6. Let $U = (S - \alpha)/\beta$, and use $\text{VaR}[U;p] = p$, $\text{TVaR}[U;p] = 1 + \frac{p}{2}$, $\text{ES}[U;p] = \frac{1}{2}(1-p)^2$.
7. $\text{TVaR}[U;p] = \frac{\varphi(\Phi^{-1}(p))}{1-p}$ if $U = \frac{S-\mu}{\sigma}$. Substitute $u = \Phi^{-1}(t)$ in (5.41) to prove this.
8. Use the result in Exercise 3.9.6.

10. Use partial fractions to find a, b such that $\frac{1}{z-t}\frac{1}{t^2} = \frac{a}{z-t} + \frac{b+ct}{t^2}$.

11. $\pi[2S; h] = \cdots = 2\pi[S; 2h] > 2\pi[S; h]$ if $h > 0$.

CHAPTER 6

Section 6.2

1. 45%; 760% vs. 900%

Section 6.3

1. See the text before (6.9).
2. All rows of $\mathbf{P}^2$ are (p, pq, q^2).
3. $l(\infty) = (p, q)$, $e(\lambda) = \lambda e^{-\lambda}(c - a)/[c(1 - e^{-\lambda}) + a e^{-\lambda}]$.
4. Use $b(\lambda) > (c - b)(1 - e^{-2\lambda})$, and $u e^{-u}/(1 - e^{-u}) = u/(e^u - 1) \leq 1$ for $u \geq 0$.
    $e(\lambda) \approx 1$ for $b \ll c$ and λ small.
7. $\alpha = 0.3$.
8. $e(0.050) \approx 0.50$
9. $s/(1 - t) = p/(1 - p)$.

CHAPTER 7

Section 7.2

1. Use Theorem 7.2.3.
2. Use the previous exercise, or the additivity/multiplicativity properties of gamma random variables.
3. Compare B_i to $A_i B_i$ for suitable Bernoulli variables.
6. Verify that $F_N(i) \geq F_M(i)$ for $i = 0, 1, 2, 3$. Why is that also sufficient for $N \leq_{\text{st}} M$?
7. Take $Y = X + I$ with $I = I(X) = 0$ if $X \in \{0, 2\}$, and $I = 1$ otherwise.
    Alternatively, fill a table with probabilities $\Pr[X = i, Y = j]$ such that the marginals are correct and $\Pr[X = i, Y = j] = 0$ for $i > j$.

Section 7.3

1. Look at the ratio of the densities. To avoid convergence problems, write the stop-loss premiums as finite sums: $\Sigma_d^{\infty} = \Sigma_0^{\infty} - \Sigma_0^{d-1}$.
2. Use the previous exercise and Exercise 7.2.3.

3. Use that $(x+y-d)_+ \le (x-\frac{1}{2}d)_+ + (y-\frac{1}{2}d)_+$ for all non-negative x, y and d. From this, it follows that $E[(X_1 + X_2 - d)_+] \le 2E[(X_1 - 0.5d)_+] = E[(2X_1 - d)_+]$; independence is not necessary.

4. $E[(X-a_1)_+] = E[(Y-a_1)_+]$; $E[(X-a_3)_+] = E[(Y-a_3)_+]$; so $E[(X-d)_+]$ and $E[(Y-d)_+]$ cannot cross. Or: the cdfs cross once, the densities twice. For such a counterexample, see for example Section 7.4.2.

5. Let $G(x) = p$ on (a,b) and let $F(x_0+0) \ge p$, $F(x_0-0) < p$. Then $F \le G$ on $(-\infty, x_0)$, $F \ge G$ on (x_0, ∞). Note that unless $F \equiv G$, $F \ge G$ everywhere nor $F \le G$ everywhere can hold, otherwise unequal means result.

7. If H is the uniform$(0,3)$ cdf, consider G with $G = F$ on $(-\infty, 1.5)$, $G = H$ on $[1.5, \infty)$.

8. See Exercise 4.

9. a) Consider the ratio of the densities; b) use Example 7.3.4 for Poisson$(E[M])$.

10. $X \le_e Y$.

14. Consider a series expansion for $m_Y(t) - m_X(t)$.

15. No, no, no. [Why is it sufficient to prove only the last case?]

16. If $d \ge 0.5$ then $V \le_{SL} W$. If $d < 0.5$, we never have $V \le_{SL} W$. If $E[W] \le E[V]$, hence $d \le 1 - \sqrt{1/2}$, then $W \le_{SL} V$, since the cdfs cross once.

17. See Theorem 7.6.2.

18. $X_1 \le_{SL} X_2 \le_{SL} X_3 \le_{SL} X_4$ because of earlier exercises. $X_2 \le_{SL} X_5$ by dispersion. $X_3 \le_{SL} X_5$ nor $X_5 \le_{SL} X_3$ since Var$[X_5] >$ Var$[X_3]$ but Pr$[X_3 > 15] > 0$, and the same for X_4. To show that exponential order does not hold, consider $e^{5(e^t-1)}/\{2/3 + 1/3e^{5t}\}$ as $t \to \infty$, or use a similar argument as above.

19. No: $E[X_p] \equiv 1$. The mgf of X_p is $\{1 - t + t^2 p(1-p)\}^{-1}$. Use $(p(Y-d) + (1-p)(Z-d))_+ \le p(Y-d)_+ + (1-p)(Z-d)_+$, as well as $E[X_p | X_{1/2}] \equiv X_{1/2}$ and Corollary 7.3.16.

20. G and V are cdfs of compound distributions with claim size $\sim F(\cdot)$. So determine q_i such that $(q_0, q_1, q_2) \le_{SL} (1/4, 1/4, 1/4, 1/4)$.

21. By counting frequencies of claim amounts, write $S_2 = 2N_2 + \cdots + 5N_5$ and $S_1 = N_1 + S_2$. Or: $S_3 \sim S_2$ if $\lambda_3 = 5$, $p_3(x) = 1/5$ for $x = 0,2,3,4,5$. Note: only compare compound Poisson distributions with the same λ.

22. $E[X] = E[Y]$ rules out stochastic order. $E[(Y-d)_+] = \frac{1}{2}\{E[(\frac{3}{2}X - d)_+] + E[(\frac{1}{2}X - d)_+]\} = \frac{3}{4}E[(X - \frac{2}{3}d)_+] + \frac{1}{4}E[(X - 2d)_+] \ge E[(X - d)_+]$ because of convexity.

23. $\lambda = 0$; λ such that $e^{-\lambda} \le 1/4$ and $(1+\lambda)e^{-\lambda} \le 3/4$, hence $\cdots$; $\lambda \ge 1$.

25. If $p_j < \bar{p} < p_k$, replace A_j and A_k by $B_j \sim$ Bernoulli$(\bar{p})$ and $B_k \sim$ Bernoulli$(p_j + p_k - \bar{p})$, and use Exercise 7.3.8. Proceed by induction.

26. Examine when the densities cross once, when twice. There is stochastic order when $p^5 \ge \frac{1}{6}$ or $(1-p)^5 \ge \frac{1}{6}$, stop-loss order when $p^5 \ge \frac{1}{6}$, hence $p \ge 0.699$, and stop-loss order the other way when $p \le 0.5$. Verify that for $p \in (\frac{1}{2}, \sqrt[5]{1/6})$, neither $X \le_{SL} Y$ nor $Y \le_{SL} X$ holds.

Section 7.4

1. The cdf is monotonous in p as well as μ. The stop-loss premiums are $p\mu e^{-d/\mu}$. In case of equal means $p\mu$, there is stop-loss monotony in μ.

2. Use earlier results found on order between binomial random variables.

3. $m_M(\log m_X(t)) \le m_M(\log m_Y(t)) \le m_N(\log m_Y(t))$, $t \ge 0$.

4. $X \leq_e Y$.

5. One recognizes the stop-loss premiums at d of the retained claims after reinsurance of type stop-loss, excess of loss, and proportional, all with equal expected value.

6. The reasoning that larger skewness implies fatter tails implies larger stop-loss premiums breaks down because of (3.116).

7. $E[S] = E[T]$ precludes stochastic order. To show that $S \leq_{SL} T$, find the sign changes in $f_S(x) - f_T(x)$, $x = 0,1,2,3,4$, under the assumption that instead of 1000 policies, there is only one policy in class $i = 1, 2$, and use Theorem 7.3.13.

8. Compare the means, that is, the stop-loss premiums at $d = 0$, and also look at the ratio of the stop-loss premiums for large d, using l'Hôpital's rule.

9. See the final sentence of this section.

10. $m_X(t) - m_Y(t) = \cdots$

Section 7.5

1. If $E[(X-d)_+] > E[(Y-d)_+]$, then $E[(X-t)_+] > E[(Y-t)_+]$ for all t because of the form of the stop-loss transform of X. This is impossible in view of (3.115).

3. If (7.36) applies, it is the maximum, otherwise it is the best of $\{0, \bar{0}\}$ and $\{b, \bar{b}\}$.

5. $\{b, \bar{b}\}$ resp. $\{0, \bar{0}\}$. Express the third raw moment in $t = (d - \mu)/\sigma$.

6. Consider $E[(X-d)(X-\bar{d})^2]$ for $d = 0$ and $d = b$.

7. Use concentration and dispersion. Variances: $\lambda E^2[Y]$, $\lambda E[Y^2]$, $\lambda b E[Y]$.

Section 7.6

3. $X \sim$ Bernoulli(0.5), $Y \equiv 1$.

4. By the Rule of thumb 3.10.6, the ratio of the stop-loss premiums is about 5 to 3.

6. For point (x, y) with $x > y$ to be in the support of the comonotonic joint cdf $H(x, y) = \min\{F_X(x), F_Y(y)\}$, we must have $H(x, y) > H(y, y)$. This is impossible because of $F_X(y) \geq F_Y(y)$.

7. What does a table of the joint probabilities look like?

9. For TVaR, use the corresponding properties of VaR. To prove c), take $d' = \text{VaR}[Y; p]$ in characterization (5.43) of TVaR.

10. Use that $\text{Var}[X^U + Y^U]$ is maximal. What does it mean that $\Pr[X \leq x]\Pr[Y \leq y] = \min\{\Pr[X \leq x], \Pr[Y \leq y]\}$ for all x, y?

11. Recall that $X + Y$ as well as $X^U + Y^U$ are normal r.v.'s.

Section 7.7

2. The conditional distribution of X, given $Z = z$ is again normal, with as parameters
$E[X|Z=z] = E[X] + \rho \frac{\sigma_X}{\sigma_Z}(z - E[Z])$ and $\text{Var}[X|Z=z] = \sigma_X^2(1 - \rho^2)$ for $\rho = r(X, Z)$.

4. $\Pr[g_i(U, Z) \leq x] = \int \Pr[F_{X_i|Z=z}^{-1}(U) \leq x] dF_Z(z) = \cdots$.

5. Conditionally on $X = x$, the first term of S'_u equals x with probability one, the second has the conditional distribution of Y, given $X = x$.

Section 7.8

1. $\Pr[X \leq x|Y \leq y] = \Pr[X \leq x, Y \leq y]/\Pr[Y \leq y] \geq \cdots$, hence $\cdots$
2. Use $\Pr[X \leq x_1, X \leq x_2] \geq \Pr[X \leq x_1]\Pr[X \leq x_2]$, and condition on $Y = y$ and $Z = z$.
3. Prove that $C(1,1) = 1$ and $c(u,v) \geq 0$ if $|\alpha| \leq 1$, and that the marginal cdfs are uniform$(0,1)$. To determine the Spearman rank correlation of X and Y, compute $\iint uv\,c(u,v)\,du\,dv$ to show that this correlation is $\alpha/3$. Kendall's τ equals $\tau = 2\alpha/9$.
5. $r(X,Y) = \{e^\sigma - 1\}/\sqrt{(e^{\sigma^2} - 1)(e - 1)}$. Since $\sigma \to \infty$ implies $r(X,Y) \downarrow 0$, there exist perfectly dependent random variables with correlation arbitrarily close to zero. But $\rho = \tau = 1$ for any value of σ, hence Kendall's and Spearman's association measures are more well-behaved than Pearson's.
6. $r(X^2,(X+b)^2) = (1+2b^2)^{-\frac{1}{2}}$.

CHAPTER 8

Section 8.2

1. The basis for all these covariance relations is that $\mathrm{Cov}[\Xi_i + \Xi_{it}, \Xi_j + \Xi_{ju}] = 0$ if $i \neq j$; $= a$ if $i = j, t \neq u$; $= a + s^2$ if $i = j, t = u$.
4. a) Minimize $\mathrm{Var}[\{X_{j,T+1} - m - \Xi_j\} + \{m + \Xi_j - z\overline{X}_j - (1-z)\overline{X}\}]$. b) $z(\overline{X}_j - \overline{X})$.
5. $az + a(1 - z^2)/(Jz)$; $s^2 + a(1 - z)\{1 + (1 - z)/(Jz)\}$; $a(1 - z)\{1 + (1 - z)/(Jz)\}$.
6. $z\overline{X}_j + (1 - z)\overline{X}/[1 + a/(Jzm^2)] \leq (8.9)$, hence biased downwards.
7. Sum of premiums paid is $J\overline{X}$.
8. Use $\mathrm{E}[(\overline{X}_j - \overline{X})^2] = (a + s^2/T)(1 - 1/J)$ and $\mathrm{E}[(X_{jt} - \overline{X}_j)^2] = s^2(1 - 1/T)$.
9. Set $\frac{d}{dp}\mathrm{E}[(Y - p)^2] = 0$, or start from $\mathrm{E}[\{(Y - \mu) + (\mu - p)\}^2] = \cdots$
10. Block-diagonal with blocks $aJ + s^2I$, with I the identity matrix and J a matrix of ones.

Section 8.3

1. Take expectations in $\mathrm{Cov}[X,Y|Z] = \mathrm{E}[XY|Z] - \mathrm{E}[X|Z]\mathrm{E}[Y|Z]$.

Section 8.4

1. The Lagrangian for this constrained minimization problem is $\sum_t \alpha_t^2 s^2/w_t - 2\lambda(\alpha_\Sigma - 1)$. Setting the derivatives with respect to α_i equal to zero gives $\alpha_i/w_i = \lambda/s^2$ for all i.
   Or: $\sum_t \alpha_t^2/w_t = \sum_t(\{\alpha_t - w_t/w_\Sigma\} + w_t/w_\Sigma)^2/w_t = \cdots$

2. See the remarks at the end of this section.

3. Follow the proof of Theorem 8.4.1, starting from the MSE of a linear predictor of m instead of $m + \Xi_j$.

4. Analogous to Theorem 8.2.4; apply Exercise 8.2.9.

9. See Remark 8.4.3.

10. $\tilde{s}^2 = 8$, $\tilde{a} = 11/3 \Longrightarrow \tilde{z} = \cdots \Longrightarrow \widetilde{m + \Xi_j} = 12.14, 13.88, 10.98$

Section 8.5

2. Use Bayes' rule.

3. Use Exercise 8.3.1 to determine $\text{Cov}[X_1, X_2]$.

4. Take the derivative of the density and set zero.

5. $\tilde{\alpha} = 1.60493$; $\tilde{\tau} = 15.87777$; $\chi^2 = 0.22$.

6. Use that $\Lambda \equiv E[N|\Lambda]$.

CHAPTER 9

Section 9.2

1. For example the mean of a gamma$(\alpha = 1/\psi_i, \beta = 1/(\psi_i\mu_i))$ r.v. equals $\alpha/\beta = \mu_i$. For the variances: ψ_i; μ_i; $\psi_i\mu_i$; $\psi_i\mu_i(1 - \mu_i)$; $\psi_i\mu_i^2$; $\psi_i\mu_i^2$. See Table D.

2. Coefficient of variation: s.d./mean $= \cdots = \sqrt{\phi}$; skewness $= 2\sqrt{\phi}$.

Section 9.3

4. Constant coefficient of variation.

6. The same values result for $x_i y_j$, but 0.1 times the χ^2 value.

8. Negative with BS, 0 with marginal totals method.

Section 9.4

1. $\hat{L}$ and $\tilde{L}$ can be found by filling in $\mu_i = \hat{\mu}_i$ and $\mu_i = y_i$ in (9.28).

3. Take the sum in (9.17) over both i and j.

CHAPTER 10

Section 10.1

1. 24.

Section 10.2

2. The mode of the lognormal(μ, σ^2) distribution is $e^{\mu - \sigma^2}$, see Exercise 3.8.6.
5. See the previous chapter.

Section 10.3

1. $X_{ij} \sim$ lognormal; see Table A and Exercise 3.8.6.

Section 10.4

1. Replace the α_i in the first model by $\alpha_i \gamma^{i-1}$.
2. $3(t-1) < t(t+1)/2$ if $t \geq 4$.
4. (10.14) implies $\hat{X}_{11} = 102.3 \times 1.00 \times 1 \times 0.42^0 = 102.3$; (10.16) implies $\hat{X}_{11} = 101.1$.
5. Use that $\sum_{j=1}^{\infty} \beta^{j-1} = (1 - \beta)^{-1}$.

CHAPTER 11

Section 11.3

1. Start from $e^{\ell(y)} \frac{\partial \ell}{\partial \theta} = \frac{\partial e^{\ell(y)}}{\partial \theta}$, and exchange the order of integration and differentiation.
2. See also Example 11.3.3.
3. Use $\frac{\partial \ell}{\partial \beta_j} = \frac{\partial \ell}{\partial \theta} \frac{d\theta}{d\mu} \frac{d\mu}{d\eta} \frac{\partial \eta}{\partial \beta_j} = \cdots$, see (11.39). Add up the ML-equations weighted by β_j, and use $\eta_i = \sum_j x_{ij} \beta_j$.
5. Fill in $\tilde{\theta}_i$ and $\hat{\theta}_i$ in $\log f_Y(y; \theta, \psi)$, see (11.4), and (9.27), (9.29) and (9.32).
6. Derive $b(\theta(\mu))$ from $b(\theta)$ and $\theta(\mu)$.
7. Compute the densities, or look at the mgfs.

Appendix C
Notes and references

When a thing has been said and well, have no scruple. Take it and copy it — Anatole France (1844 - 1924)

In spite of the motto of this chapter (another famous saying by the same author is *"If a million people say a foolish thing, it is still a foolish thing"*), we will try to give credit where credit is due. Additional material on many subjects treated in this book, apart from the references indicated by chapter, can be looked up in your library in two recent encyclopedias: Melnick & Everitt (Eds., 2008) as well as Teugels & Sundt (Eds., 2004). Also, search engines and online encyclopedias often provide access to useful material. Required for studying the book is a course on mathematical statistics on the level of Bain & Engelhardt (1992) or a similar text.

CHAPTER 1

Basic material in the actuarial field on utility theory and insurance is in Borch (1968, 1974). The utility concept dates back to Von Neumann & Morgenstern (1944). The Allais paradox is described in Allais (1953). For a description of Yaari's (1987) dual theory of risk, see Wang & Young (1998) and Denuit *et al.* (1999), but also the book Denuit *et al.* (2005). Both utility theory and Yaari's dual theory can be used to construct risk measures that are important in the framework of solvency, both in finance and in insurance, see for example Wason *et al.* (2004).

CHAPTER 2

Since the seminal article of Panjer (1981) on recursions for the collective model, also many recursion relations for (approximate and exact) calculation of the distribution in the individual model were given. All these methods assume that the portfolio consists of rather large groups of identical policies. We refer to De Pril (1986), Dhaene & De Pril (1994), Dhaene *et al.* (2006), Sundt (2002) as well as Sundt & Vernic (2009) for an overview.

CHAPTER 3

For collective models, see also the textbook Bowers *et al.* (1986, 1997). Many other books cover this topic, for example Seal (1969), Bühlmann (1970), Gerber (1979), Goovaerts *et al.* (1990), Heilmann (1988) and Sundt (1999), as well as the work by Rolski *et al.* (1998). Beard *et al.* (1977, 1984) contains a lot of material about the NP approximation. While collective risk models assume independence of the claim severities, a new trend is to study the sum of dependent risks, see for example Denuit *et al.* (2005). A text on statistical aspects of loss distributions is Hogg & Klugman (1984), or more recently Klugman *et al.* (1997). Some references propagating the actuarial use of the inverse Gaussian distributions are Ter Berg (1980a, 1980b, 1994). The rejection method was an idea of Von Neumann (1951); for a description of this and many other numerical methods, see also Press *et al.* (2007). Panjer's recursion was introduced in Panjer (1981), extended by Sundt & Jewell (1981). The function `Panjer.Poisson` in Remark 3.5.7 was inspired by the package `actuar` by Vincent Goulet. Bühlmann (1984) compares Panjer's recursion and Fast Fourier Transform based methods by counting the number of multiplications needed.

CHAPTER 4

Ruin theory started with F. Lundberg (1909), Cramér (1930, 1955), as well as O. Lundberg (1940). An interesting approach based on martingales can be found in Gerber (1979). The ruin probability as a stability criterion is described in Bühlmann (1970). The book by Beekman (1964) gives an early connection of Poisson processes and Wiener processes. A more recent book is Embrechts *et al.* (1997). Many papers have been published concerning the numerical calculation of ruin probabilities, starting with Goovaerts & De Vylder (1984). Note that F.E.C. De Vijlder published under the *nom de plume* De Vylder. Gerber (1989) derives the algorithm (4.79) to compute ruin probabilities for discrete distributions. Babier & Chan (1992) describe how to approximate ruin probabilities using the first three moments of the claims; see also Kaas & Goovaerts (1985). Seal (1978) calls survival probabilities 'The Goal of Risk Theory'.

CHAPTER 5

The section connecting premium principles to the discrete ruin model is based on Bühlmann (1985); insurance risk reduction by pooling was described in Gerber (1979). In the 1970s premium principles were a hot topic in actuarial research. The basics were introduced in Bühlmann (1970). See also Gerber (1979, 1983) and Goovaerts *et al.* (1984) for characterizations of premium principles. Goovaerts &

Dhaene (1998) give a characterization of Wang's (1996) class of premium principles. The notion of coherent risk measures is advocated in Artzner *et al.* (1997,
1999), but see also Huber (1981). The optimality of VaR in the sense of Remark
5.6.2 and the fact that the optimal cost then is a TVaR was demonstrated in Dhaene
et al. (2008). Note that in the literature, there is confusion about the terms TVaR,
CVaR, CTE, and ES. See Remark 5.6.9.

CHAPTER 6

Pioneering work in the theoretical and practical aspects of bonus-malus systems can
be found in Bichsel (1964), as well as in Loimaranta (1972). Denuit *et al.* (2007)
gives a comprehensive description of the insurance aspects of bonus-malus systems.
The study that led to the Dutch bonus-malus system described in this chapter was
described in De Wit *et al.* (1982). Bonus-malus systems with non-symmetric loss
functions are considered in Denuit & Dhaene (2001).

CHAPTER 7

The notion of stop-loss order entered into the actuarial literature through the paper by Bühlmann *et al.* (1977). In the statistical literature many results generalizing
stop-loss order are available in the context of convex order. See for example Karlin
& Studden (1966). A standard work for stochastic orders is Shaked & Shanthikumar (1994). Applications of ordering principles in operations research and reliability
can be found in Müller & Stoyan (2002). Recently, the concept of convex order has
been applied in the financial approach to insurance where the insurance risk and
the financial risk are integrated. Comonotonic risks play an important role in these
dependency models. Review papers about this topic are Dhaene *et al.* (2002a,b).
Forerunners to Chapter 7 are the monograph by Kaas *et al.* (1994), based on the
Ph.D. thesis by Van Heerwaarden (1991). See also the corresponding chapters of
Goovaerts *et al.* (1990). A comprehensive treatment of the actuarial theory of dependent risks and stochastic orders can be found in Denuit *et al.* (2005). The choice
of the conditioning variable V in Section 7.7 is discussed in Vanduffel *et al.* (2008).

CHAPTER 8

The general idea of credibility theory can be traced back to the papers by Mowbray
(1914) and Whitney (1918). A sound theoretical foundation was given by Bühlmann
(1967, 1969). See also Hickman & Heacox (1999), which is the source of the motto
with this chapter. Several approaches can be taken to introduce credibility theory.

One is the Bayesian approach using a least squares error criterion and a risk parameter Θ that is a random variable characterizing some hidden risk quality. Another approach applies projections in Hilbert spaces, as in De Vylder (1996). In this text we use a variance components model such as often encountered in econometrics. The advantage of this approach, apart from its simplicity and elegance, is the explicit relationship with ANOVA, in case of normality. A textbook on variance components models is Searle *et al.* (1992). We have limited ourselves to the basic credibility models of Bühlmann, because with these, all the relevant ideas of credibility theory can be illustrated, including the types of heterogeneity as well as the parameter estimation. For a more complete treatment of credibility, the reader is referred to Dannenburg *et al.* (1996), which was the basis for our Chapter 8, or to the Ph.D. thesis of Dannenburg (1996). The interpretation of a bonus-malus system by means of credibility theory was initiated by Norberg (1976); for the negative binomial model, we refer to Lemaire (1985).

CHAPTER 9

The paper by Nelder & Wedderburn (1972) introduces the generalized linear model (GLM). The textbook McCullagh & Nelder (1989) contains some applications in insurance rate making. A textbook on Generalized Linear Models geared towards an actuarial audience is De Jong & Heller (2008). The heuristic rate making techniques are treated more fully in Van Eeghen *et al.* (1983). Antonio & Beirlant (2007) describes credibility theory as well as GLMs as special cases of generalized linear mixed models. The cube-root transformation in Example 9.1.1 is the well-known Wilson-Hilferty (1931) transformation to normalize χ^2 random variables.

CHAPTER 10

The first statistical approach to the IBNR problem goes back to Verbeek (1972). The three dimensions of the problem were introduced in De Vylder (1978). An encyclopedic treatment of the various methods is given in Taylor (1986); see also Taylor (2000). The triangle of Taylor & Ashe (1983) is used in many texts on IBNR problems. The relation with generalized additive and multiplicative linear models is explored in Verrall (1996, 2000). *The* model behind the chain ladder method is defended in Mack (1993, 1994). Doray (1996) gives UMVUEs of the mean and variance of IBNR claims for a model with lognormal claim figures, explained by row and column factors. Research currently goes in the direction of determining the economic value of run-off claims with discounting. The statistical framework gives the extrapolated claim figures as a cash flow. The calendar year is of a different nature than the development year and the year of origin, because it includes inflation and discounting. A reference is Goovaerts & Redant (1999). Both the analytical estimate of the prediction error of the chain ladder method in Section 10.6 and the bootstrap method were proposed by England & Verrall (1999) and England (2002). The Bornhuetter-Ferguson method of IBNR reserving was introduced in Bornhuetter & Ferguson (1972). Verrall (2004) shows that this method as well as the chain ladder method can be described as generalized linear models in a Bayesian framework. The data in Section 10.7 are from a contest organized by ASTIN NL.

CHAPTER 11

The Gauss-Markov theory can be found in Bain & Engelhardt (1992, Ch. 15). See also McCullagh & Nelder (1989) for the theory behind generalized linear models. The IRLS algorithm was developed in Nelder & Wedderburn (1972). The name Tweedie has been associated with this family by Jørgensen in honor of M.C.K. Tweedie (1984). The extended quasi-likelihood was introduced in Nelder & Pregibon (1987). Dunn & Smyth (2005) describe the numerical aspects of the functions d/p/q/rtweedie they contributed to R. The use of Tweedie's class of exponential dispersion models for claims reserving is described in Wüthrich (2003).

APPENDIX A

Next to the informal introduction by Burns (2005) that inspired Appendix A.1, there is a wealth of other useful material on R on http://cran.r-project.org, the official CRAN website. For instance Paradis (2005) is a good text for beginners. A recent text describing the use of R in statistics is Crawley (2007). The MASS library is based on Venables & Ripley (2002). The Jarque-Bera test is described in Bera & Jarque (1980). Racine & Hyndman (2002) describe how to use R to teach econometrics.

REFERENCES

Allais M. (1953). "Le comportement de l'homme rationnel devant le risque: critique des postulats et axiomes de l'Ecole Americaine", *Econometrica*, 21, 503–546.

Antonio K. & Beirlant J. (2007). "Actuarial statistics with generalized linear mixed models", *Insurance: Mathematics and Economics*, 40, 58–76.

Artzner P., Delbaen F., Eber J.M. & Heat, D. (1997). "Thinking coherently". *Risk*, 10(11), 68-71.

Artzner P., Delbaen F., Eber J.M. & Heat, D. (1999). "Coherent Measures of Risk", *Mathematical Finance* 9.3, 203–228.

Babier J. & Chan B. (1992). "Approximations of ruin probability by di-atomic or di-exponential claims", *ASTIN Bulletin* 22-2, 235–246.

Bain L.J. & Engelhardt M. (1992). "Introduction to probability and mathematical statistics", Duxbury, Belmont.

Beard R.E., Pentikäinen T. & Pesonen E. (1977, 1984). "Risk theory", Chapman & Hall, London.

Beekman J.A. (1964). "Two stochastic processes", Halsted Press, New York.

Bera A.K. & Jarque C.M. (1980). "Efficient tests for normality, homoscedasticity and serial independence of regression residuals", *Economics Letters* 6, 255-259.

Bichsel F. (1964). "Erfahrungs-Tarifierung in der Motorfahrzeughaftplicht-Versicherung", *Mitteilungen der Vereinigung Schweizerischer Versicherungsmathematiker*, 64, 119–130.

Borch K. (1968). "The economics of uncertainty", Princeton University Press, Princeton.

Borch K. (1974). "The mathematical theory of insurance", Lexington Books, Toronto.

Bornhuetter R.L. & Ferguson R.E. (1972). "The actuary and IBNR", *Proceedings of the Casualty Actuarial Society*, LIX, 181–195.

Bowers N.L., Gerber H.U., Hickman J.C., Jones D.A. & Nesbitt C.J. (1986, 1997). "Actuarial mathematics", Society of Actuaries, Itasca, Illinois.

Bühlmann H. (1967). "Experience rating and credibility I", *ASTIN Bulletin*, 4, 199–207.

Bühlmann H. (1969). "Experience rating and credibility II", *ASTIN Bulletin*, 5, 157–165.

Bühlmann H. (1970). "Mathematical methods in risk theory", Springer Verlag, Berlin.

Bühlmann H., Gagliardi B., Gerber H.U. & Straub E. (1977). "Some inequalities for stop-loss premiums", *ASTIN Bulletin*, 9, 169–177.

Bühlmann H. (1984). "Numerical evaluation of the compound Poisson distribution: Recursion or Fast Fourier Transform?", Scandinavian Actuarial Journal, 116–126.

Bühlmann H. (1985). "Premium calculation from top down", *ASTIN Bulletin*, 15, 89–101.

Burns P. (2005). "A Guide for the Unwilling S User",
http://www.burns-stat.com/pages/Tutor/unwilling_S.pdf

Cramér H. (1930). "On the mathematical theory of risk", Skandia Jubilee Volume, Stockholm.

Cramér H. (1955). "Collective risk theory, a survey of the theory from the point of view of stochastic processes", Skandia Jubilee Volume, Stockholm.

Crawley M.J. (2007). "The R book", Wiley, Chichester.

Dannenburg D.R., Kaas R. & Goovaerts M.J. (1996). "Practical actuarial credibility models", Institute of Actuarial Science, Amsterdam.

Dannenburg D.R. (1996). "Basic actuarial credibility models — Evaluations and extensions", Ph.D. Thesis, Thesis/Tinbergen Institute, Amsterdam.

Denuit M., Dhaene J. & Van Wouwe M. (1999). "The economics of insurance: a review and some recent developments", *Mitteilungen der Vereinigung Schweizerischer Versicherungsmathematiker*, 99, 137–175.

Denuit M. & Dhaene J. (2001). "Bonus-malus scales using exponential loss functions", *Blätter der Deutsche Gesellschaft für Versicherungs–mathematik*, 25, 13–27.

Denuit M., Dhaene J., Goovaerts M.J. & Kaas R. (2005). "Actuarial Theory for Dependent Risks: Measures, Orders and Models", Wiley, New York.

Denuit M., Maréchal X., Pitrebois S. & Walhin J.-F. (2007). "Actuarial Modelling of Claim Counts: Risk Classification, Credibility and Bonus-Malus Systems", Wiley, New York.

De Jong P. & Heller G.Z. (2008). "Generalized linear models for insurance data", Cambridge University Press, Cambridge.

De Pril N. (1986). "On the exact computation of the aggregate claims distribution in the individual life model", *ASTIN Bulletin*, 16, 109–112.

De Vylder, F. (1978). "Estimation of IBNR claims by least squares", Mitteilungen der Vereinigung Schweizerischer Versicherungsmathematiker, 78, 249-254.

De Vylder F. (1996). "Advanced risk theory, a self-contained introduction", Editions de l'Université de Bruxelles, Brussels.

De Wit G.W. *et al.* (1982). "New motor rating structure in the Netherlands", ASTIN-groep Nederland.

Dhaene J. & De Pril N. (1994). "On a class of approximative computation methods in the individual risk model", *Insurance : Mathematics and Economics*, 14, 181–196.

Dhaene J., Denuit M., Goovaerts M.J., Kaas R. & Vyncke D. (2002a). "The concept of comonotonicity in actuarial science and finance: Theory", *Insurance: Mathematics & Economics*, 31, 3-33.

Dhaene J., Denuit M., Goovaerts M.J., Kaas R. & Vyncke D. (2002b). "The concept of comonotonicity in actuarial science and finance: Applications", *Insurance: Mathematics & Economics*, 31, 133–161.

Dhaene J., Ribas C. & Vernic (2006). "Recursions for the individual model", *Acta Mathematica Applicatae Sinica*, English Series, 22, 631–652.

Dhaene J., Laeven R.J.A., Vanduffel S., Darkiewicz G. & Goovaerts M.J. (2008). "Can a coherent risk measure be too subadditive?," *Journal of Risk and Insurance* 75, 365–386.

Doray L.G. (1996). "UMVUE of the IBNR Reserve in a lognormal linear regression model", *Insurance: Mathematics & Economics*, 18, 43–58.

Dunn P.K. & Smyth, G.K. (2005). "Series evaluation of Tweedie exponential dispersion model densities", *Journal Statistics and Computing*, 15, 267–280.

Embrechts P., Klüppelberg C. & Mikosch T. (1997). "Modelling extremal events for insurance and finance", Springer-Verlag, Berlin.

England P. & Verrall R. (1999). "Analytic and bootstrap estimates of prediction errors in claims reserving", *Insurance: Mathematics and Economics*, 25, 281-293.

England P. (2002). "Addendum to Analytic and bootstrap estimates of prediction errors in claims reserving", *Insurance: Mathematics and Economics*, 31, 461-466.

Gerber H.U. (1979). "An introduction to mathematical risk theory", Huebner Foundation Monograph 8, distributed by Richard D. Irwin, Homewood Illinois.

Gerber H.U. (1985). "On additive principles of zero utility", *Insurance: Mathematics & Economics*, 4, 249–252.

Gerber H.U. (1989). "From the convolution of uniform distributions to the probability of ruin", *Mitteilungen der Vereinigung Schweizerischer Versicherungsmathematiker*, 89, 249–252.

Goovaerts M.J. & De Vylder F. (1984). "A stable recursive algorithm for evaluation of ultimate ruin probabilities", *ASTIN Bulletin*, 14, 53–60.

Goovaerts M.J., De Vylder F. & Haezendonck J. (1984). "Insurance premiums", North-Holland, Amsterdam.

Goovaerts M.J., Kaas R., Van Heerwaarden A.E. & Bauwelinckx T. (1990). "Effective actuarial methods", North-Holland, Amsterdam.

Goovaerts M.J. & Dhaene J. (1998). "On the characterization of Wang's class of premium principles", *Transactions of the 26th International Congress of Actuaries*, 4, 121–134.

Goovaerts M.J. & Redant R. (1999). "On the distribution of IBNR reserves", *Insurance: Mathematics & Economics*, 25, 1–9.

Heilmann W.-R. (1988). "Fundamentals of risk theory", Verlag Versicherungswirtschaft e.V., Karlsruhe.

Hickman J.C. & Heacox L. (1999). "Credibility theory: The cornerstone of actuarial science", North American Actuarial Journal, 3.2, 1–8.

Hogg R.V. & Klugman S.A. (1984). "Loss distributions", Wiley, New York.

Huber P. (1981). "Robust Statistics", Wiley, New York.

Jørgensen B. (1987). "Exponential dispersion models", *J. R. Statist. Soc. B*, 49, 127–162.

Jørgensen B. (1997). "Theory of Dispersion Models", Chapman & Hall, London.

Kaas R. & Goovaerts M.J. (1985). "Bounds on distribution functions under integral constraints", *Bulletin de l'Association Royale des Actuaires Belges*, 79, 45–60.

Kaas R., Van Heerwaarden A.E. & Goovaerts M.J. (1994). "Ordering of actuarial risks", Caire Education Series, Amsterdam.

Kaas, R. & Gerber, H.U. (1995) "Some alternatives for the individual model", *Insurance: Mathematics & Economics*, 15, 127-132.

Karlin S. & Studden W.J. (1966). "Tchebycheff systems with applications in analysis and statistics", Interscience Publishers, Wiley, New York.

Klugman, S.A., Panjer H.H. & Willmot, G.E. (1998). "Loss models — From data to decisions", Wiley, New York.

Lemaire J. (1985). "Automobile insurance: actuarial models", Kluwer, Dordrecht.

Loimaranta K. (1972). "Some asymptotic properties of bonus systems", *ASTIN Bulletin*, 6, 233–245.

Lundberg F. (1909). "Über die Theorie der Rückversicherung", *Transactions of the first International Congress of Actuaries*, 2, 877–955.

Lundberg O. (1940). "On random processes and their applications to sickness and accidents statistics", Inaugural Dissertation, Uppsala.

Mack T. (1993). "Distribution-free calculation of the standard error of chain ladder reserve estimates", *ASTIN Bulletin*, 23, 213–225.

Mack T. (1994). "Which stochastic model is underlying the chain ladder model?", *Insurance: Mathematics and Economics*, 15, 133-138.

McCullagh P. & Nelder J.A. (1989). "Generalized Linear Models", Chapman & Hall, London.

Melnick E. & Everitt B. (Eds.) (2008). "Encyclopedia of Quantitative Risk Analysis and Assessment", Wiley, New York.

Mowbray A.H. (1914). "How extensive a payroll exposure is necessary to give a dependable pure premium", *Proceedings of the Casualty Actuarial Society*, 1, 24–30.

MÜLLER A. & STOYAN D. (2002). *Comparison Methods for Stochastic Models and Risks.* Wiley, New York.

Nelder J.A. & Wedderburn, R.W.M. (1972). "Generalized Linear Models", Journal of the Royal Statistical Society, A, 135, 370–384.

Nelder J.A. & Pregibon D. (1987). "An extended quasi-likelihood function", *Biometrika*, 74, 221–232.

Norberg R. (1976). "A credibility theory for automobile bonus systems", *Scandinavian Actuarial Journal*, 92–107.

Panjer H.H. (1981). "Recursive evaluation of a family of compound distributions", *ASTIN Bulletin*, 12, 22–26.

Paradis E. (2005). "R for beginners",
`http://cran.r-project.org/doc/contrib/Paradis-rdebuts_en.pdf`

Press W.H., Teukolsky S.A., Vetterling W.T. & Flannery B.P. (2007). "Numerical Recipes 3rd Edition: The Art of Scientific Computing", Cambridge University Press.

Racine J. & Hyndman R. (2002). "Using R to teach econometrics", *Journal of Applied Econometrics*, 17, 175-189 (2002)

Rolski T., Schmidli H., Schmidt V. & Teugels J. (1998). "Stochastic Processes for Insurance and Finance", Wiley, Chichester.

Searle S.R., Casella G. & McCulloch C.E. (1992). "Variance components", Wiley, New York.

Seal H.L. (1969). "Stochastic theory of a risk business", Wiley, New York.

Seal H.L. (1978). "Survival Probabilities: The Goal of Risk Theory", Wiley, New York.

Shaked M. & Shanthikumar J.G. (1994). "Stochastic orders and their applications", Academic Press, New York.

Stoyan D. (1983). "Comparison methods for queues and other stochastic models", Wiley, New York.

Sundt B. & Jewell W.S. (1981). "Further results of recursive evaluation of compound distributions", *ASTIN Bulletin* 12, 27–39.

Sundt, B. (1992). "On some extensions of Panjer's class of counting distributions", *ASTIN Bulletin* 22, 61–80.

Sundt B. (1999). "An introduction to non-life insurance mathematics", 4th edition, Verlag Versicherungswirtschaft GmbH, Karlsruhe.

Sundt B. (2002). "Recursive evaluation of aggregate claims distributions". *Insurance: Mathematics and Economics* 30, 297–322.

Sundt B. & Vernic R. (2009). "Recursions for convolutions and compound distributions with insurance applications", EAA Lecture Notes, Springer Verlag, Heidelberg.

Taylor, G.C. & Ashe, F.R. (1983). "Second Moments of Estimates of Outstanding Claims", *Journal of Econometrics* 23, 37–61.

Taylor G.C. (1986). "Claims reserving in non-life insurance", North-Holland, Amsterdam.

Taylor G.C. (2000). "Loss Reserving: An Actuarial Perspective", Kluwer Academic Publishers, Dordrecht.

Ter Berg P. (1980a), "On the loglinear Poisson and Gamma model", *ASTIN-Bulletin* 11, 35–40.

Ter Berg P. (1980b), "Two pragmatic approaches to loglinear claim cost analysis", *ASTIN-Bulletin* 11, 77–90.

Ter Berg P. (1994), "Deductibles and the Inverse Gaussian distribution", *ASTIN-Bulletin* 24, 319–323.

Teugels J. & Sundt B. (Eds.) (2004). "Encyclopedia of Actuarial Science I–III", Wiley, New York.

Tweedie M.C.K. (1984). "An index which distinguishes between some important exponential families". In Ghosh J.K & Roy J. (Eds.), *Statistics: Applications and New Directions. Proceedings of the Indian Statistical Institute Golden Jubilee International Conference*, Indian Statistical Institute, Calcutta, 579-604.

Van Eeghen J., Greup E.K. & Nijssen J.A. (1983). "Rate making", Nationale-Nederlanden N.V., Rotterdam.

Van Heerwaarden A.E. (1991). "Ordering of risks — Theory and actuarial applications", Thesis Publishers, Amsterdam.

Vanduffel, S., Chen X., Dhaene J., Goovaerts M., Henrard L. & Kaas R. (2008). "Optimal approximations for risk measures of sums of lognormals based on conditional expectations", *Journal of Computational and Applied Mathematics*, to be published.

Venables W.N. & Ripley B.D. (2002, 4th ed.). "Modern Applied Statistics with S", Springer, New York.

Verbeek H.G. (1972). "An approach to the analysis of claims experience in motor liability excess of loss reinsurance", *ASTIN Bulletin* 6, 195–202.

Verrall R. (1996). "Claims reserving and generalized additive models", *Insurance: Mathematics & Economics* 19, 31–43.

Verrall R. (2000). "An investigation into stochastic claims reserving models and the chain-ladder technique", *Insurance: Mathematics & Economics* 26, 91–99.

Verrall, R.J. (2004). "A Bayesian generalized linear model for the Bornhuetter-Ferguson method of claims reserving", *North American Actuarial Journal* 8(3), 67–89.

Von Neumann J. & Morgenstern O. (1944). "Theory of games and economic behavior", Princeton University Press, Princeton.

Von Neumann J. (1951), "Various techniques used in connection with random digits. Monte Carlo methods", Nat. Bureau Standards, 12, pp. 3638.

Wang S. (1996). "Premium calculation by transforming the layer premium density", *ASTIN Bulletin* 26, 71–92.

Wang S. & Young V. (1998). "Ordering risks: expected utility theory versus Yaari's dual theory of risk", *Insurance: Mathematics & Economics* 22, 145–161.

Wason S. *et al.* (2004). "A Global Framework for Insurer Solvency Assessment", Report of the Insurer Solvency Assessment Working Party, International Actuarial Association.

Wilson, E. B. & Hilferty, M. M. (1931). "The distribution of chi-square", *Proceedings of the National Academy of Sciences of the United States of America* 17, 684–688.

Whitney A.W. (1918). "The theory of experience rating", *Proceedings of the Casualty Actuarial Society* 4, 274–292.

Wüthrich, M.V. (2003). "Claims Reserving Using Tweedie's Compound Poisson Model", *ASTIN Bulletin* 33, 331–346.

Yaari M.E. (1987). "The dual theory of choice under risk", *Econometrica* 55, 95–115.

Appendix D
Tables

Facts are stupid things — Ronald Reagan (1911 - 2004)

Table A The most frequently used discrete and continuous distributions

Distribution	Density & support	Moments & cumulants	Mgf
Binomial(n,p) $(0 < p < 1,\, n \in \mathbb{N})$	$\binom{n}{x} p^x (1-p)^{n-x}$ $x = 0,1,\ldots,n$	$\mathrm{E} = np,\, \mathrm{Var} = np(1-p),$ $\gamma = \dfrac{np(1-p)(1-2p)}{\sigma^3}$	$(1 - p + pe^t)^n$
Bernoulli(p)	$\equiv$ Binomial$(1,p)$		
Poisson(λ) $(\lambda > 0)$	$e^{-\lambda}\dfrac{\lambda^x}{x!},\, x = 0,1,\ldots$	$\mathrm{E} = \mathrm{Var} = \lambda,$ $\gamma = 1/\sqrt{\lambda},$ $\kappa_j = \lambda,\, j = 1,2,\ldots$	$\exp[\lambda(e^t - 1)]$
Negative binomial(r,p) $(r > 0,\, 0 < p < 1)$	$\binom{r+x-1}{x} p^r (1-p)^x$ $x = 0,1,2,\ldots$	$\mathrm{E} = r(1-p)/p$ $\mathrm{Var} = \mathrm{E}/p,$ $\gamma = \frac{(2-p)}{p\sigma}$	$\left(\dfrac{p}{1 - (1-p)e^t}\right)^r$
Geometric(p)	$\equiv$ Negative binomial$(1,p)$		
Uniform(a,b) $(a < b)$	$\dfrac{1}{b-a};\, a < x < b$	$\mathrm{E} = (a+b)/2,$ $\mathrm{Var} = (b-a)^2/12,$ $\gamma = 0$	$\dfrac{e^{bt} - e^{at}}{(b-a)t}$
$N(\mu,\sigma^2)$ $(\sigma > 0)$	$\dfrac{1}{\sigma\sqrt{2\pi}}\exp\dfrac{-(x-\mu)^2}{2\sigma^2}$	$\mathrm{E} = \mu,\, \mathrm{Var} = \sigma^2,\, \gamma = 0$ $(\kappa_j = 0,\, j \geq 3)$	$\exp(\mu t + \tfrac{1}{2}\sigma^2 t^2)$
Gamma(α,β) $(\alpha,\beta > 0)$	$\dfrac{\beta^\alpha}{\Gamma(\alpha)}x^{\alpha-1}e^{-\beta x},\, x > 0$	$\mathrm{E} = \alpha/\beta,\, \mathrm{Var} = \alpha/\beta^2,$ $\gamma = 2/\sqrt{\alpha}$	$\left(\dfrac{\beta}{\beta - t}\right)^\alpha$ $(t < \beta)$
Exponential(β)	$\equiv$ gamma$(1,\beta)$		
$\chi^2(k)$ $(k \in \mathbb{N})$	$\equiv$ gamma$(k/2, 1/2)$		
Inverse Gaussian(α,β) $(\alpha > 0,\, \beta > 0)$	$\dfrac{\alpha x^{-3/2}}{\sqrt{2\pi\beta}}\exp\left(\dfrac{-(\alpha - \beta x)^2}{2\beta x}\right)$ $F(x) = \Phi\left(\dfrac{-\alpha}{\sqrt{\beta x}} + \sqrt{\beta x}\right) + e^{2\alpha}\Phi\left(\dfrac{-\alpha}{\sqrt{\beta x}} - \sqrt{\beta x}\right),\quad x > 0$	$\mathrm{E} = \alpha/\beta,\, \mathrm{Var} = \alpha/\beta^2,$ $\gamma = 3/\sqrt{\alpha}$	$e^{\alpha(1 - \sqrt{1 - 2t/\beta})}$ $(t \leq \beta/2)$
Beta(a,b) $(a > 0, b > 0)$	$\dfrac{x^{a-1}(1-x)^{b-1}}{B(a,b)},\, 0 < x < 1$	$\mathrm{E} = \dfrac{a}{a+b},\, \mathrm{Var} = \dfrac{\mathrm{E}(1-\mathrm{E})}{a+b+1}$	
Lognormal(μ,σ^2) $(\sigma > 0)$	$\dfrac{1}{x\sigma\sqrt{2\pi}}\exp\dfrac{-(\log x - \mu)^2}{2\sigma^2},$ $x > 0$	$\mathrm{E} = e^{\mu + \sigma^2/2},\, \mathrm{Var} = e^{2\mu + 2\sigma^2} - e^{2\mu + \sigma^2},$ $\gamma = c^3 + 3c$ with $c^2 = \mathrm{Var}/\mathrm{E}^2$	
Pareto(α,x_0) $(\alpha,x_0 > 0)$	$\dfrac{\alpha x_0^\alpha}{x^{\alpha+1}},\, x > x_0$	$\mathrm{E} = \dfrac{\alpha x_0}{\alpha - 1},\, \mathrm{Var} = \dfrac{\alpha x_0^2}{(\alpha-1)^2(\alpha-2)}$	
Weibull(α,β) $(\alpha,\beta > 0)$	$\alpha\beta(\beta y)^{\alpha-1}e^{-(\beta y)^\alpha},\, x > 0$	$\mathrm{E} = \Gamma(1 + 1/\alpha)/\beta,$ $\mathrm{Var} = \Gamma(1 + 2/\alpha)/\beta^2 - \mathrm{E}^2,$ $\mathrm{E}[Y^t] = \Gamma(1 + t/\alpha)/\beta^t$	

Table B Standard normal distribution; cdf $\Phi(x)$ and stop-loss premiums $\pi(x)$

	+0.00		+0.05		+0.10		+0.15		+0.20	
x	$\Phi(x)$	$\pi(x)$	$\Phi(x)$	$\pi(x)$	$\Phi(x)$	$\pi(x)$	$\Phi(x)$	$\pi(x)$	$\Phi(x)$	$\pi(x)$
0.00	0.500	0.3989	0.520	0.3744	0.540	0.3509	0.560	0.3284	0.579	0.3069
0.25	0.599	0.2863	0.618	0.2668	0.637	0.2481	0.655	0.2304	0.674	0.2137
0.50	0.691	0.1978	0.709	0.1828	0.726	0.1687	0.742	0.1554	0.758	0.1429
0.75	0.773	0.1312	0.788	0.1202	0.802	0.1100	0.816	0.1004	0.829	0.0916
1.00	0.841	0.0833	0.853	0.0757	0.864	0.0686	0.875	0.0621	0.885	0.0561
1.25	0.894	0.0506	0.903	0.0455	0.911	0.0409	0.919	0.0367	0.926	0.0328
1.50	0.933	0.0293	0.939	0.0261	0.945	0.0232	0.951	0.0206	0.955	0.0183
1.75	0.960	0.0162	0.964	0.0143	0.968	0.0126	0.971	0.0111	0.974	0.0097
2.00	0.977	0.0085	0.980	0.0074	0.982	0.0065	0.984	0.0056	0.986	0.0049
2.25	0.988	0.0042	0.989	0.0037	0.991	0.0032	0.992	0.0027	0.993	0.0023
2.50	0.994	0.0020	0.995	0.0017	0.995	0.0015	0.996	0.0012	0.997	0.0011
2.75	0.997	0.0009	0.997	0.0008	0.998	0.0006	0.998	0.0005	0.998	0.0005
3.00	0.999	0.0004	0.999	0.0003	0.999	0.0003	0.999	0.0002	0.999	0.0002
3.25	0.999	0.0002	1.000	0.0001	1.000	0.0001	1.000	0.0001	1.000	0.0001
3.50	1.000	0.0001	1.000	0.0000	1.000	0.0000	1.000	0.0000	1.000	0.0000

Table C Selected quantiles of the standard normal distribution

x	1.282	1.645	1.960	2.326	2.576	3.090	3.291	3.891	4.417
$\Phi(x)$	0.900	0.950	0.975	0.990	0.995	0.999	0.9995	0.99995	0.999995

Examples of use: $\Phi(1.17) \approx 0.6\Phi(1+0.15) + 0.4\Phi(1+0.20) \approx 0.879$;
$\Phi^{-1}(0.1) = -1.282$; $\Phi(-x) = 1 - \Phi(x)$; $\pi(-x) = x + \pi(x)$.

NP approximation: If S has mean μ, variance σ^2 and skewness γ, then

$$\Pr\left[\frac{S-\mu}{\sigma} \leq x\right] \approx \Phi\left(\sqrt{\frac{9}{\gamma^2} + \frac{6x}{\gamma} + 1} - \frac{3}{\gamma}\right)$$

$$\text{and} \quad \Pr\left[\frac{S-\mu}{\sigma} \leq s + \frac{\gamma}{6}(s^2 - 1)\right] \approx \Phi(s)$$

Translated gamma approximation: If $G(\cdot\,; \alpha, \beta)$ is the gamma cdf, then

$$\Pr[S \leq x] \approx G(x - x_0; \alpha, \beta) \quad \text{with} \quad \alpha = \frac{4}{\gamma^2}; \beta = \frac{2}{\gamma\sigma}; x_0 = \mu - \frac{2\sigma}{\gamma}.$$

Table D The main classes of distributions in the GLM *exponential dispersion family*, with the customary parameters as well as the (μ,ϕ) and (θ,ϕ) reparameterizations, and more properties

Distribution	Density	(μ,ϕ) reparameterization	Cumulant function $b(\theta)$
	Domain	Canonical link $\theta(\mu)$	$E[Y;\theta]=\mu(\theta)=b'(\theta)$
		Variance function $V(\mu)$	
$N(\mu,\sigma^2)$	$\frac{1}{\sigma\sqrt{2\pi}}e^{-\frac{(y-\mu)^2}{2\sigma^2}}$	$\phi=\sigma^2$	$\frac{\theta^2}{2}$
		$\theta(\mu)=\mu$	θ
		$V(\mu)=1$	
$Poisson(\mu)$	$e^{-\mu}\frac{\mu^y}{y!}$	$\phi=1$	e^θ
	$y=0,1,2,\ldots$	$\theta(\mu)=\log\mu$	e^θ
		$V(\mu)=\mu$	
$Poisson(\mu,\phi)$	$e^{-\mu/\phi}\frac{(\mu/\phi)^{(y/\phi)}}{(y/\phi)!}$	$\theta(\mu)=\log\mu$	e^θ
	$y=0,\phi,2\phi,\ldots$	$V(\mu)=\mu$	e^θ
$Binomial(m,p)$	$\binom{m}{y}p^y(1-p)^{m-y}$	$\mu=mp;\ \phi=1$	$m\log(1+e^\theta)$
$(m\in\mathbb{N}$ fixed$)$	$y=0,\ldots,m$	$\theta(\mu)=\log\frac{\mu}{m-\mu}$	$\frac{me^\theta}{1+e^\theta}$
		$V(\mu)=\mu(1-\frac{\mu}{m})$	
$Negbin(r,p)$	$\binom{r+y-1}{y}p^r(1-p)^y$	$\mu=\frac{r(1-p)}{p};\ \phi=1$	$-r\log(1-e^\theta)$
$(r>0$ fixed$)$	$y=0,1,\ldots$	$\theta(\mu)=\log\frac{\mu}{r+\mu}$	$\frac{re^\theta}{1-e^\theta}$
		$V(\mu)=\mu(1+\frac{\mu}{r})$	
$Gamma(\alpha,\beta)$	$\frac{1}{\Gamma(\alpha)}\beta^\alpha y^{\alpha-1}e^{-\beta y}$	$\mu=\frac{\alpha}{\beta};\ \phi=\frac{1}{\alpha}$	$-\log(-\theta)$
	$y>0$	$\theta(\mu)=-\frac{1}{\mu}$	$-\frac{1}{\theta}$
		$V(\mu)=\mu^2$	
$IG(\alpha,\beta)$	$\frac{\alpha y^{-3/2}}{\sqrt{2\pi\beta}}\exp\frac{-(\alpha-\beta y)^2}{2\beta y}$	$\mu=\frac{\alpha}{\beta};\ \phi=\frac{\beta}{\alpha^2}$	$-\sqrt{-2\theta}$
	$y>0$	$\theta(\mu)=-\frac{1}{2\mu^2}$	$\frac{1}{\sqrt{-2\theta}}$
		$V(\mu)=\mu^3$	
$Tweedie(\lambda,\alpha,\beta)$	$\sum\limits_{n=1}^{\infty}\frac{\beta^{n\alpha}y^{n\alpha-1}e^{-\beta y}}{\Gamma(n\alpha)}\frac{\lambda^n e^{-\lambda}}{n!}$	$\mu=\frac{\lambda\alpha}{\beta};\ \phi=\frac{\alpha+1}{\beta}\mu^{1-p}$	$\frac{\{(1-p)\theta\}^{(2-p)/(1-p)}}{2-p}$
$(\alpha$ fixed;	for $y>0$;	$\theta(\mu)=\frac{\mu^{1-p}}{p-1}$	$\{(1-p)\theta\}^{1/(1-p)}$
$p=\frac{\alpha+2}{\alpha+1})$	$e^{-\lambda}$ for $y=0$	$V(\mu)=\mu^p$	

Index